PRIMER ON MECHANICS OF MATERIALS
VOLUME 2

PRIMER ON MECHANICS OF MATERIALS VOLUME 2

N.J. MASSON

TABLE OF CONTENTS

DEDICATION

****

This book is dedicated to the Shell International Petroleum Company Limited, London who, through their local subsidiary the United British Oilfields Limited, Point Fortin, awarded me a scholarship in 1954 to pursue tertiary technical education and training in the United Kingdom.

Whatever my subsequent brief service with Shell Trinidad Limited, and by comparison much longer service with the Government of Trinidad and Tobago might have contributed to national development, is attributed wholly to my benefactor.

N J Masson

This page is intentionally left blank.

PREFACE

True education is self-education, is application. It has been said that most men fail not because of a lack of capacity but rather because of failure to apply themselves; a shortcoming of which the distinguished jurist Edward Abbott Parry inferred was a failure "... to read and learn and digest beneath the lamp of industry."

The lectures, symposia and whatnots of formal education are but mere adjuncts to the self education in which every student aspiring to professional status in engineering no less than in any other field of endeavour, must earnestly and fervently engage.

This book is offered as a supplementary text to students of Mechanics of Materials. It was designed as an aid to self-study of the topics which would normally be included in any undergraduate curriculum. However, it was not written to satisfy the requirements of any syllabus in particular but hopefully in such a style as to make the subject delightful for any reader and thereby to facilitate the process of digestion while in the refulgent rays of Judge Parry's third lamp.

The reader may wonder why in the face of an astounding number of books on the subject at hand, consisting as they do of numerous ideas and manner of expressing them, the author found it necessary to add another. The answer is simply that it was felt one was needed which consisted of his own selection of topics, the order of their arrangement and method of teaching; and in doing so to point out some of the dodges and pitfalls along the way.

Much of the science of Mechanics of Materials is expressed in the language of the Queen and Handmaiden of the Sciences, viz. Mathematics[1], a subject with a distinct hierarchical structure. Consequently the topics dealt with in this text are presented in a logical, 'vertical sequence' akin to the arrangement of the floors of a ziggurat: each one relying on the strength and support provided by preceding constructive effort. Inclusion of computer software for the solution of problems was avoided on the ground that while computer programs "... are capable of quite refined analysis" "... these have not led to an improvement in the reliability of engineering design." I have however drawn attention to the importance of the need for skill in the use of the electronic digital computer and for proficiency in computer programming in applications involving the Finite Element Method. This is so because invariably the solution of engineering problems using this technique literally involves an enormous number of calculations.

---

[1] "One reason", said Albert Einstein, "why Mathematics enjoys special esteem above all other sciences is that its propositions are absolutely certain and indisputable while those of the other sciences are to some extent debatable and in constant danger of being overthrown by newly discovered facts."

In a work such as this, total originality in respect of the treatment of subject matter is well nigh impossible. Having myself ploughed through and quarried hard in an extraordinarily large number of volumes on Strength of Materials, Mechanics, Applied Physics and other books devoted to special topics related to the main subject, it is inevitable that much of what was retained would probably have unconsciously manifested itself here and there on the written page. That notwithstanding, it is hoped that the book survives what Isaac Asimov called the "Road-to-Xanadu Test". Readers with a knowledge of the laws of statical equilibrium, elementary calculus, and of matrix algebra should have no difficulty in following the text which is replete with drawings, illustrated examples and solved problems, some based on questions set at engineering examinations of external authorities. I shall be most grateful for notifications (nojose@tstt.net.tt) of any errors in the text and for constructive criticism of the work that would aid in its improvement. The bibliography in the Appendix lists the authors of the referenced texts upon whose shoulders I stand in sincere acknowledgement of their kind assistance.

As a consequence of the large number of drawings and detailed solutions of many illustrated examples and solved problems, the text had of necessity to be divided into three parts: **Volume 1** containing Chapters 1-5; **Volume 2**, Chapters 6-12 and **Volume 3**, Chapters 13-20.

All units used in the text except those in which some quantities are expressed in so-called permitted units e.g. angular measurement in degrees and pressure in bars, are based on the Systeme Internatinale (SI) Metric System.

I end this Preface in a manner similar to that of its beginning with a quotation attributed this time to the great French aphorist Francois de Rochefoucauld: "Our capacity exceeds our will power and it is often only to excuse ourselves that we hug the belief that things are impossible." I urge, implore, beg, beseech you to let your will power be the equal of your capacity which is limitless.

CHAPTER 6

SHEARING FORCE AND BENDING MOMENT

INTRODUCTION

Mention the word 'beam' and many would most likely conjure up images of a ray of light or a smile or even a scriptural passage or even Scotty of Gene Roddenberry's "Star Trek" fame. But 'beam' in the context of engineering technology has a somewhat unique connotation. It means a structural component, be it made out of wood, or iron or plastic, or concrete and steel, or fashioned from some other material, but nonetheless one capable of supporting forces transverse (shear forces), lateral loads (direct forces) and, the effects of bending. It is generally assumed that such shear forces, loads and bending moments act in a plane containing the beam's longitudinal axis. Crankshafts, camshafts, motor vehicle axles and turbine shafts are all examples of beams, from a mechanical-engineering standpoint. Typically, beams in civil engineering and building-construction practice are joists and girders.

Elementary elastic stress analysis of beams which is considered in the chapter following will be limited to the stresses arising from the internal forces and moments necessary to maintain equilibrium conditions at any transverse section.

It is with reference to beams that I introduce the topics of shear force and bending moment. When plane frameworks were considered in Chapter 2 it was pointed out there, that the *raison d'etre* of the Method of Sections was that having divided (figuratively) a framework which was in equilibrium into, say, two parts by an imaginary plane passing through the entire structure, in order that each part of the framework should remain in a state of equilibrium, (i) the internal forces exposed in the members 'cut' by the imaginary plane, in each part, (ii) the imposed loads acting on each part, and (iii) any external reactions acting on each part, had to be in balance. That is to say, the system of forces comprising these three elements for each part must satisfy the laws of statical equilibrium; viz. $\sum F_{X,Y} = 0$; $\sum M_{X,Y} = 0$.

SHEAR FORCE

Let us now apply the Method of Sections to the beam with the load configuration shown at Fig. 1. The end 'A' of the beam is on a roller.

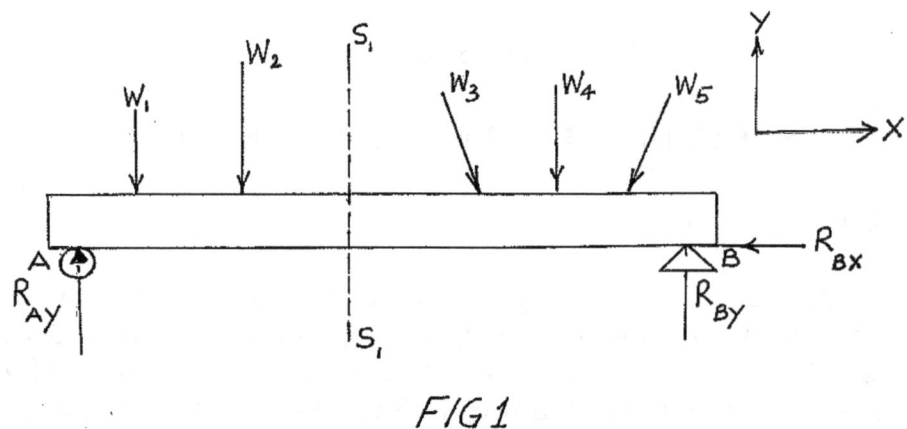

FIG 1

Cut the beam with the imaginary transverse plane, the trace of which is S_1S_1. On the left-hand side, assuming that $R_{AY} > W_1 + W_2$, it is clear that the part of the beam marked Section 1 will move upwards. To restore vertical equilibrium to this part an internal force V must be applied acting downwards as shown in the left-hand diagram of Fig. 2.

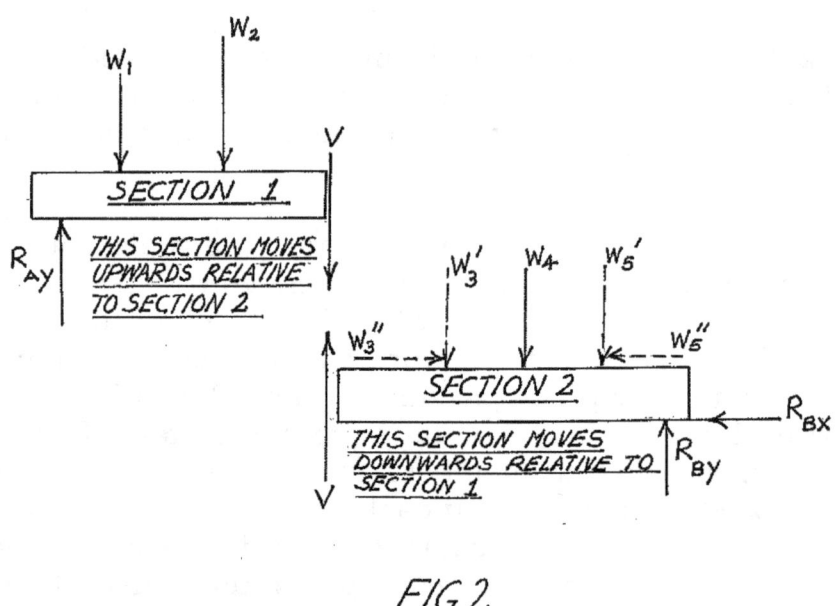

FIG 2

It is evident that to the right of S_1S_1 an internal force V, this time acting upwards must be applied as shown, for when sections 1 and 2 of the beam are brought together, the force V on the left must equalize its counterpart V on the right. Observe that at Section 2 in Fig. 2, I have converted the inclined loads W_3 and W_5 into their vertical and horizontal components. It is the vertical components W_3' and W_5' that contribute to the shear. The internal force V is what is referred to as shearing force or shear force or just plain shear. If it had been assumed initially that $R_{AY} < W_1 + W_2$, then for equilibrium of section 1, force 'V' would have had to act upwards; for section 2, downwards. Thus the situation as described previously would be reversed. The foregoing implies that a convention is necessary to indicate what sign 'V' should take when the section of the beam to the left goes upwards, in which case 'V' is downwards; and when the section to the left goes downwards in which case 'V' is upwards; and vice versa for the section to the right.

The convention to be followed here for shear is summarized thus:

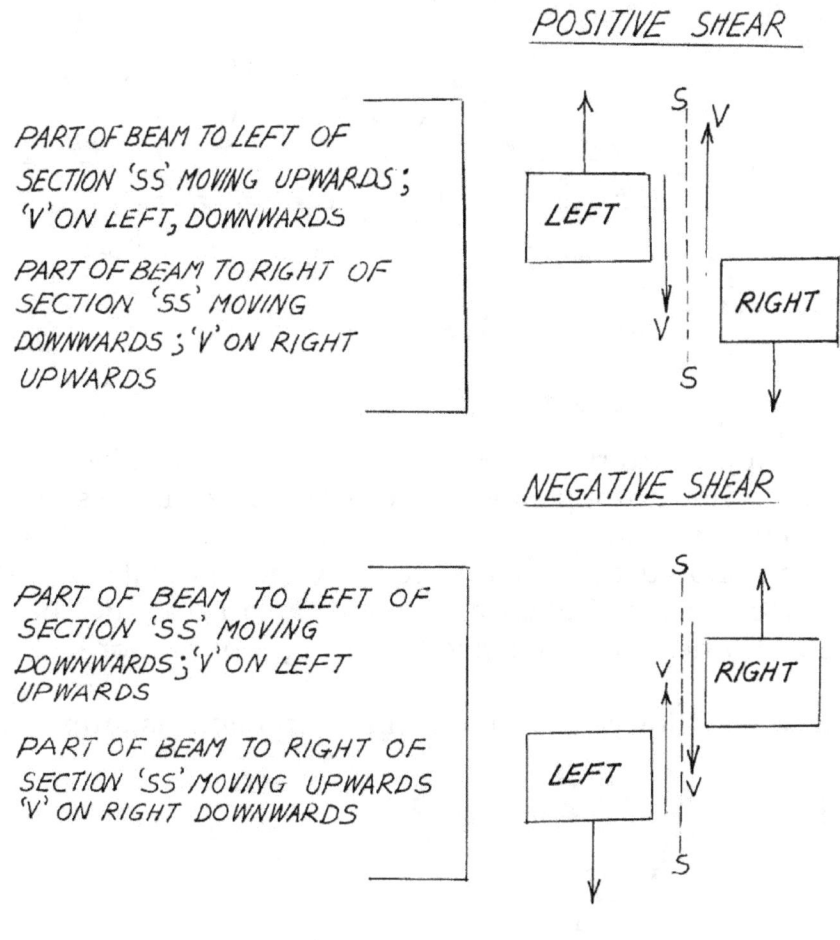

POSITIVE SHEAR

PART OF BEAM TO LEFT OF SECTION 'SS' MOVING UPWARDS; 'V' ON LEFT, DOWNWARDS

PART OF BEAM TO RIGHT OF SECTION 'SS' MOVING DOWNWARDS; 'V' ON RIGHT UPWARDS

NEGATIVE SHEAR

PART OF BEAM TO LEFT OF SECTION 'SS' MOVING DOWNWARDS; 'V' ON LEFT UPWARDS

PART OF BEAM TO RIGHT OF SECTION 'SS' MOVING UPWARDS 'V' ON RIGHT DOWNWARDS

SHEAR FORCE CONVENTION

It is the internal shear force 'V' which causes shear stresses in a beam: one of the elastic stresses referred to earlier.

A formal definition of shear or shearing force at any section of a beam may therefore be stated as follows: the algebraic sum of all imposed loads and external reaction/s acting on either side of the section in a direction vertical to the axis of the beam.

Let us turn our attention to a consideration of moments at section S_1S_1.

BENDING MOMENT

Referring again to Fig. 1, suppose the moment of R_{AY} about a point at that part of the beam where S_1S_1 cuts it, is greater than the combined moment of W_1 and W_2 about the same point. Clearly therefore in order to restore equilibrium to section 1, an anti-clockwise internal resisting moment 'M'.

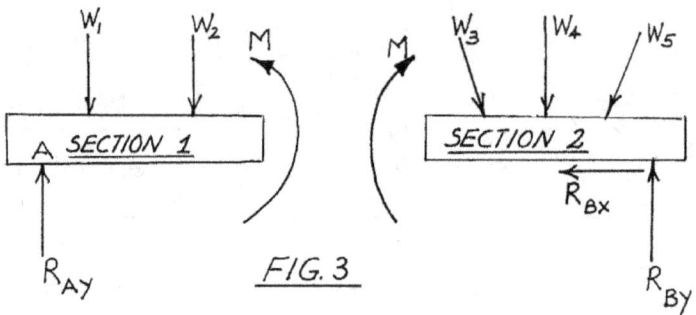

FIG. 3

would have to be applied to this part, at S_1S_1 as shown in Fig. 3. Equilibrium of section 2, requires that an opposite or clockwise internal resisting moment 'M' be applied as shown. As was the case with shear, a convention is required to indicate what sign should be assigned to 'M' which is called bending moment, when it is anti-clockwise to the left and clockwise to the right of section 1; and clockwise to the left and anti-clockwise to the right of section 2; and vice-versa.

The convention to be followed here for bending moment is summarized thus:

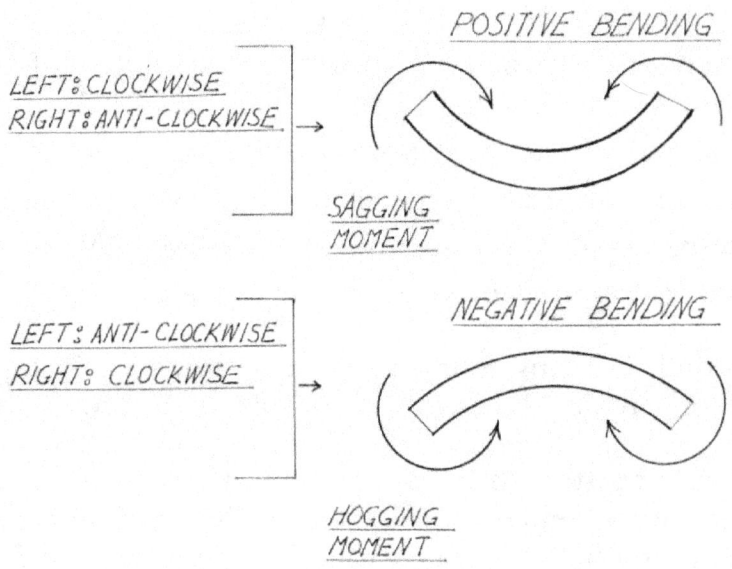

POSITIVE BENDING

LEFT: CLOCKWISE
RIGHT: ANTI-CLOCKWISE →

SAGGING
MOMENT

NEGATIVE BENDING

LEFT: ANTI-CLOCKWISE
RIGHT: CLOCKWISE →

HOGGING
MOMENT

BENDING MOMENT CONVENTION

It is the internal resisting moment within the section of the beam cut by S_1S_1 that causes bending stresses in the beam: another of the elastic stresses which we shall be considering shortly.

A formal definition of bending moment at any section of a beam could be stated in the following terms: the algebraic sum of the moments of all imposed loads and external reactions acting on either side of the section.

Going back again to Fig. 1, resolve imposed or applied load W_3 into a longitudinal or lateral component W_3'' and, a transverse component W_3'; similarly, W_5 into W_5'' and W_5'. Now, cut the beam with a plane perpendicular to its axis, its trace being S_2S_2. Refer to Fig. 4.

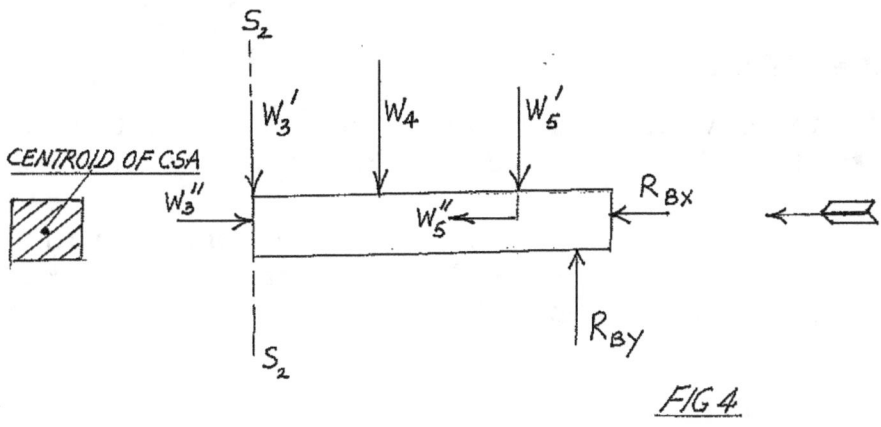

FIG 4

Equilibrium along the axis of the beam demands that W_3'', W_5'' and R_{Bx}, all of which are assumed to be passing through the centroid of the beam's cross sectional area, must be in equilibrium, i.e. $W_3'' - W_5'' = R_{Bx}$. So, for the configuration shown at Fig. 4, that part of the beam between W_3' and W_5' and that part alone would be under compression, if $W_3'' > W_5''$. If the directions of W_3'', W_5'' and R_{Bx} were reversed, then assuming as before that $W_3'' > W_5''$ that part of the beam and that part alone would be under tension.

The convention which is being followed here is that tensile forces and stresses are positive; and compressive forces and compressive stresses negative.

What we have demonstrated thus far is that at any section of a beam, there may be three reactive elements necessary in order to maintain equilibrium viz. (i) shear force; (ii) bending moment; and (iii) thrust or axial force. This is one way of saying that whereas when an imaginary 'cut' is made in a framework the internal forces exposed as it were, are straightaway in equilibrium with the imposed or applied loads and external reactions, in the case of a beam when an imaginary 'cut' is made through it, it becomes necessary to apply: (i) an external shear force, (ii) an external resisting moment, and (iii) an external axial force in order to maintain equilibrium.

SHEAR FORCE -, AXIAL THRUST/COMPRESSION -, AND BENDING MOMENT - DIAGRAMS

It is assumed that most students are familiar with the preparation of diagrams showing, at any transverse section of a beam, the shear force, axial thrust or compression, and bending moment acting on such a section. However, for the purposes of review, one example will be treated here to illustrate the procedures involved: a refresher perhaps for those who know already the ropes, as it were; a useful reference source for the newly initiated. I shall take pains to explain.

Illustrative Example 1

It is required to draw the axial force, shear force and bending moment diagrams for the beam loaded at its centroidal axis as shown in Fig. 5a.

The first order of business is to determine the reactions. For this purpose the inclined loads at B and D are resolved into their vertical and horizontal components as shown in Figs. 5b and 5c.

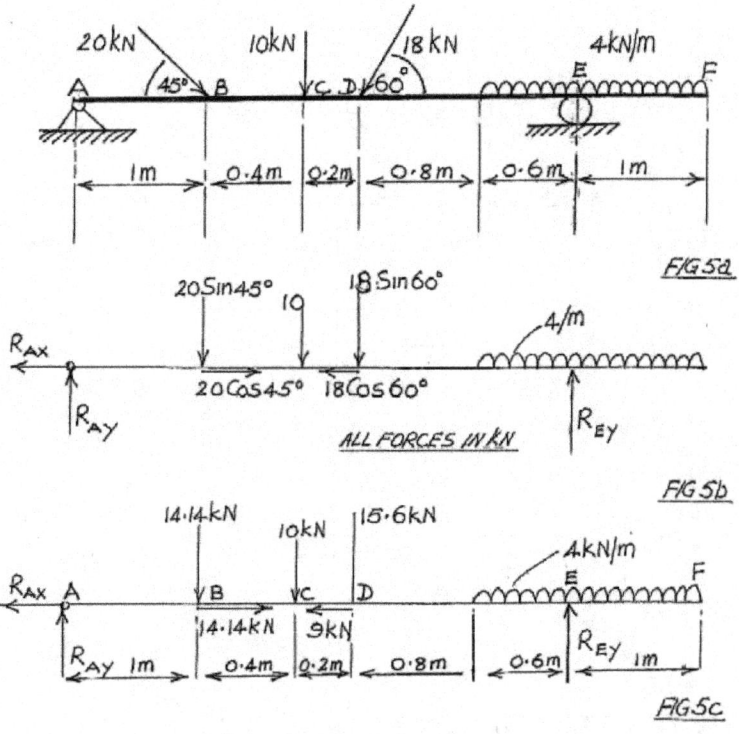

FIG 5a

FIG 5b

FIG 5c

Referring to Fig. 5c, we have for

$$\sum F_X = 0, \ R_{AX} + 9 - 14.14 = 0$$
$$\therefore \quad R_{AX} = 5.14kN$$

With this information, the axial force diagram may be drawn as in Fig. 5d

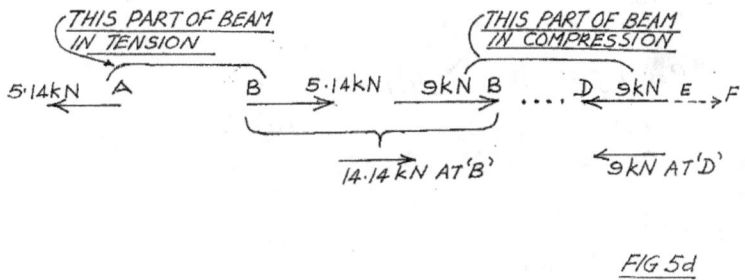

FIG 5d

The portion of the beam *'AB'* is subjected to an axial pull of *5.14kN*; and portion *'BD'* to axial compression of *9kN.* See Fig. 6

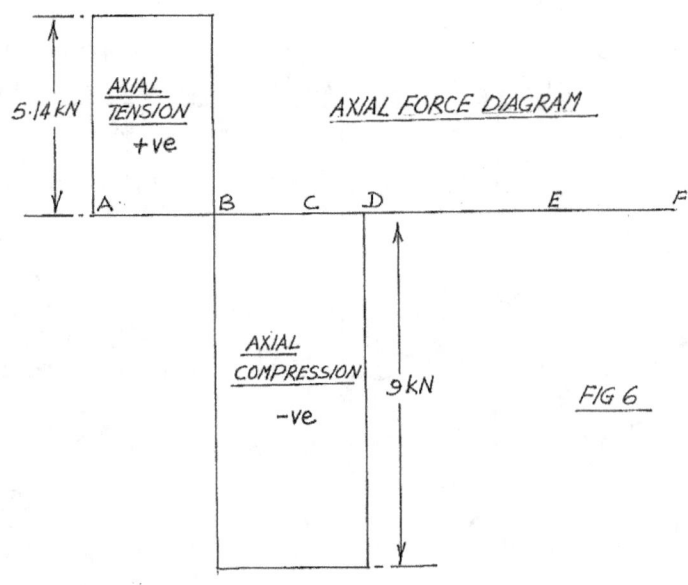

AXIAL FORCE DIAGRAM

5.14 kN

AXIAL TENSION +ve

A   B   C   D   E   F

AXIAL COMPRESSION -ve

9kN

FIG 6

For $\sum F_y = 0$

$$R_{AY} - 14.14 - 10 - 15.6 - 4(1.6) + R_{EY} = 0$$
$$\therefore \qquad R_{AY} + R_{EY} = 46.14 \, kN$$

For $\sum M_A = 0$

$$-14.14(1) - 10(1.4) - 15.6(1.6) - 4(1.6)(3.2) + R_{EY}(3) = 0$$

i.e. $-14.14 - 14 - 24.96 - 20.48 + 3_{REY} = 0$

i.e. $3R_{EY} = 73.58$

or $R_{EY} = 24.53kN$, say

$R_{EY} = 24.5kN$

$\therefore \qquad R_{AY} = 21.61kN$, say,

$R_{AY} = 21.6kN$

Check this value of R_{AY}, by taking $\sum M_E = 0$, i.e.

$$R_{AY}(3) - 14.14(2) - 10(1.6) - 15.6(1.4) - 4(0.6)(0.3) + 4(1)(0.5) = 0;$$

i.e. $3R_{AY} = 64.8kN$ or $R_{AY} = 21.6kN$. Check

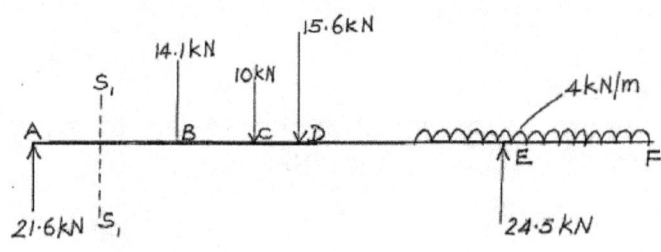

14.1kN   10kN   15.6kN   4kN/m

S_1

A   B   C   D   E   F

21.6kN S_1 24.5 kN

FIG 7

~ 8 ~

Now that all the transverse loads on the beam are known, the shear force diagram may be prepared. Note that the horizontal forces at A, B and D have been excluded from Fig. 7 because these are not needed for the shear force diagram. These were already shown in the Axial Force Diagram, Fig. 6. Let us begin the shear force determinations, starting from end 'A'.

Taking a section at S_1S_1, i.e. making an imaginary transverse cut at S_1S_1, the 21.6kN force at 'A' would push that part of the beam between 'A' and S_1S_1, upwards and as we know, a downwards force of 21.6kN would be necessary at S_1S_1 to establish equilibrium, as shown at Fig. 7a

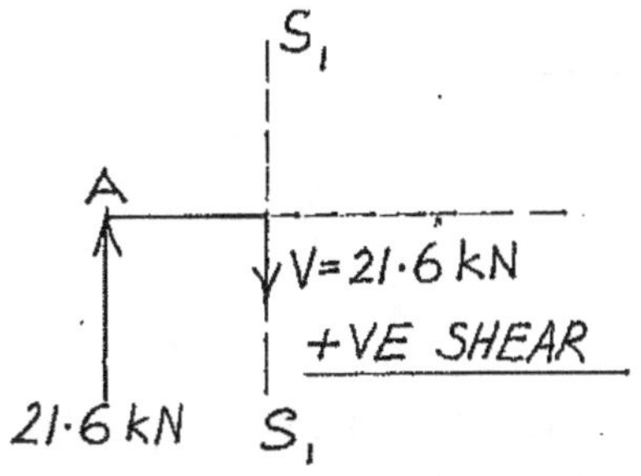

FIG 7a

If S_1S_1 were moved just immediately to the left of B, the situation as represented at Fig. 7a would continue to obtain, until just immediately to the right of B, the value of the positive shear is now only (21.6 – 14.1) kN, i.e. 7.5kN, as shown at Fig. 7b

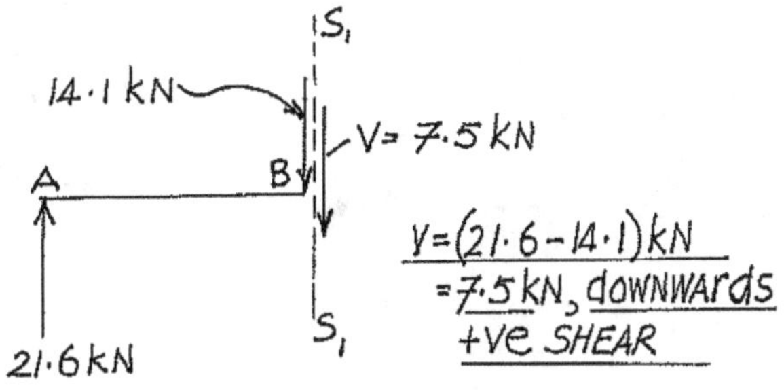

FIG 7b

$V = (21.6 – 14.1)kN$
   $= 7.5kN$, downwards, i.e.
+ve shear

As S_1S_1 moves up to a position immediately to the left of C, the state of positive shear of 7.5kN continues. At position 'C' there is a concentrated downward load of 10kN and just immediately to the right of 'C', 'V' now has to act upwards in order to restore equilibrium; that is to say a state of negative shear now exists. See Fig. 7c.

The point the student must note is that shear force went from a positive value just to the left of the concentrated load at *C*, to a negative value just to the right of the said load at *C*. Evidently, shear force had to pass through a zero value to achieve this: to move from positive to negative.

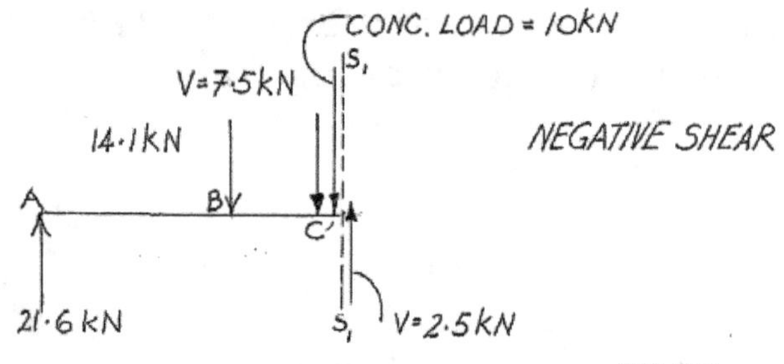

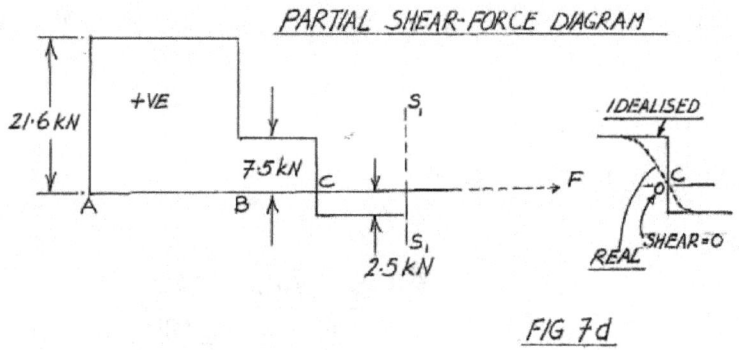

FIG 7C

This is a point which is sometimes missed by students who fail to appreciate that a maximum or minimum bending moment could occur at such a point as '*C*'; but more about that later on. Then again, the beautiful shear force diagrams with their right-angled corners at concentrated loads are but an idealization of a practical state of affairs.

FIG 7d

Because the phrase "load at a point" and one of its many variants, for example, "concentrated load at point so and so" convey an idealized concept which does not exist in nature, loads being in reality spread over some finite area, the shift of shear from a positive to a negative value as at *C*, is reflected in the shape of the shear force diagram by the dotted line as indicated in the sketch to the right of Fig. 7d.

As S_1S_1 moves beyond *C* to *D* the negative shear becomes more negative; -2.5kN just to the right of '*C*' becoming *–18.1 kN* at *D* and remaining at this value up to just immediately left of the start of the uniformly distributed load *(udl)* of *4k/m*. See Fig. 7e.

~ 10 ~

14·1kN
10kN
15·6kN
UDL = 4kN/m
A
B
C
D
E
F
0·6m
R_E
21·6kN
V=18·1kN, UPWARDS;
NEGATIVE SHEAR.

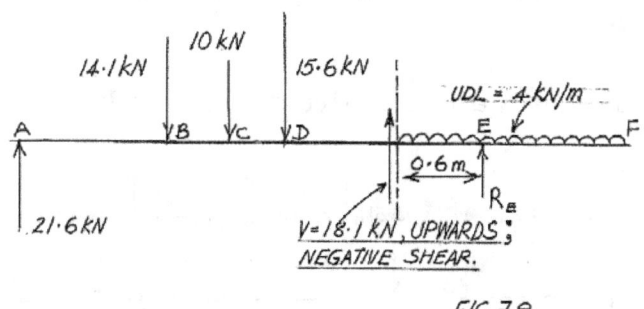

FIG 7e

The shear force diagram at this stage, taking note of the idealized representation of the situation at '*C*', is as shown at Fig. 7f.

0·6m
0·3m
4kN/m
E
A
B
C D
18·1kN
R_E = 24·5kN

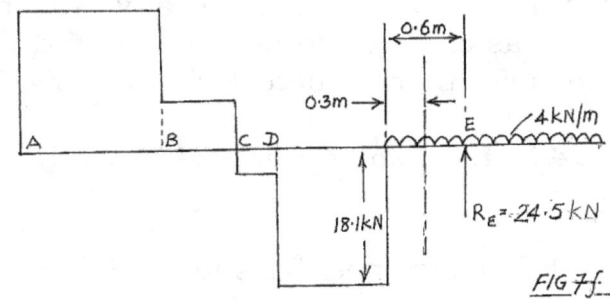

FIG 7f

Consider now a position for S_1S_1 midway between the start of the *udl* and *E,* as at Fig. 7g

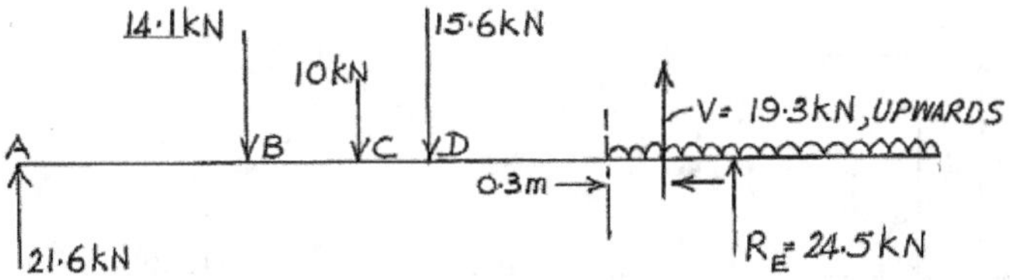

14·1kN
15·6kN
10kN
V= 19·3kN, UPWARDS
A
B
C
D
0·3m
21·6kN
R_E= 24·5kN

FIG 7g

Remember that by definition, *SF* = algebraic sum of <u>all</u> forces to the right or left of a section. We have been working from the left, so we continue:

$$V = 21.6 - 14.1 - 10 - 15.6 - 4(0.3)$$
$$= -19.3kN$$

∴ *V = is upwards and according to our shear force convention, negative*

As S_1S_1 takes up a position just immediately to left of R_E

$$V = 21.6 - 14.1 - 10 - 15.6 - 4(0.6)$$
$$= -20.5kN$$

∴ *V* is upwards as shown in Fig. 7(h). Now just to the right of, *E,*

$$V = 21.6 - 14.1 - 10 - 15.6 - 4(0.6) + 24.5$$
$$\therefore \quad V = +4kN$$

$\therefore$ V downwards $= 4kN$ and positive shear results.

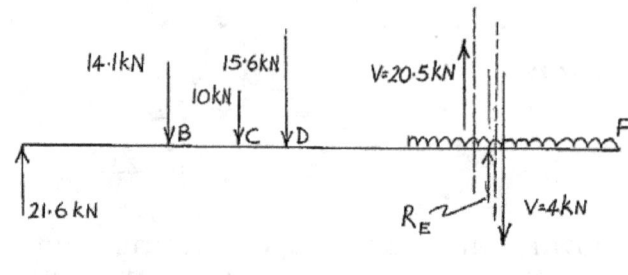

So it is seen that just to the left of E there is negative shear and immediately to the right of 'E' the shear has changed to positive. Could there be a maximum or minimum bending moment there, where V changes sign? Midway between E and F?

$$V = 21.6 - 14.1 - 10 - 15.6 - 4(0.6) + 24.5 - 4(0.5)$$
$$V = +2kN$$

i.e. V downwards to the left; hence positive shear. Since 'EF' is an overhanging portion of the beam i.e. without any reaction at F, the shear force at $F = 0$.

The complete shear force diagram may now be drawn. It is shown at Fig. 8.

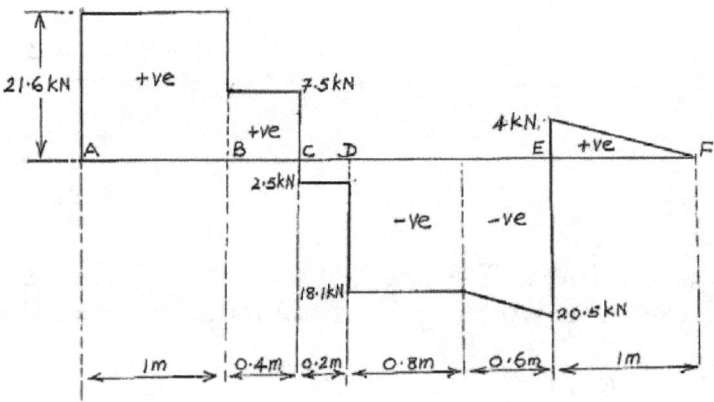

FIG 8 : COMPLETE SHEAR FORCE DIAGRAM
(NOT DRAWN TO SCALE)

We come now to the determination of bending moment and preparation of the bending moment diagram.

~ 12 ~

Taking moments from left:

$$B.M_A = 0$$
$$B.M_B = 21.6 \times 1 = 21.6kNm, \text{ clockwise, } +ve$$
$$B.M_C = 21.6 \times 1.4 - 14.1(0.4) = 24.6kNm, \text{ clockwise, } +ve$$
$$B.M_D = 21.6 \times 1.6 - 14.1(0.6) - 10(0.2) = 24.1kNm, \text{ clockwise } +ve$$

Remember that bending moment is defined as the algebraic sum of the moments of all imposed loads and external reactions acting on either side of a section of the beam; also that in our convention, clockwise moments on the left or anticlockwise moments on the right are positive.

Accordingly, right up to that part of the beam, approaching from the left, where the *udl* starts, the bending moment there is given by:

$$B.M = 21.6(2.4) - 14.1(1.4) - 10(1.0) - 15.6(0.8)$$
$$= 51.84 - 19.74 - 10 - 12.48$$
$$= 9.62kNm, \text{ clockwise, } +ve$$

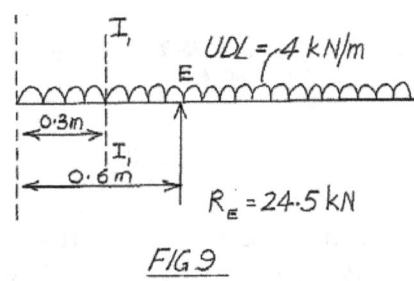

FIG 9

With reference to Fig. 9, bending moment at section I_1I_1, midway between start of the *udl* from the left and $E = BM_{I1I1} = 21.6(2.7) - 14.1(1.7) - 10(1.3) - 15.6(1.1 = 58.32 - 23.97 - 13 - 17.16 = 4.19kNm$.

Bending moment just immediately to left of $E = 21.6(3) - 14.1(2) - 10(1.6) - 15.6(1.4) - 4(0.6)\left(\dfrac{0.6}{2}\right) = 64.8 - 28.2 - 16 - 21.84 - 0.72 = -1.96kNm$

∴ Bending moment just immediately to left of $E = -1.96kNm$, say, -2kNm. See Fig. 9 Consider bending moment at a point midway between E and F. The bending moment at this point is

$$B.M_{E/F} = 21.6(3.5) - 14.1(2.5) - 10(2.1) - 15.6(1.9) -$$
$$- 4(1.1)\dfrac{(1.1)}{2} + 24.5(0.5)$$
$$= 75.6 - 35.25 - 21 - 29.64 - 2.42 + 12.25$$
$$= -0.46kNm, \text{ say } 0.5kNm$$

Bending moment at F, BM_F is given by

$$B.M_F = 21.6(4) - 14.1(3) - 10(2.6) - 15.6(2.4)$$
$$4(1.6)\left(\frac{1.6}{2}\right) + 24.5(1)$$
$$= 86.4 - 42.3 - 26 - 37.44 - 5.12 + 24.5$$
$$\therefore \quad B.M_F = 0.04kNm, \text{ say, } zero$$

The bending moment diagram may now be drawn. This is shown at Fig. 10.

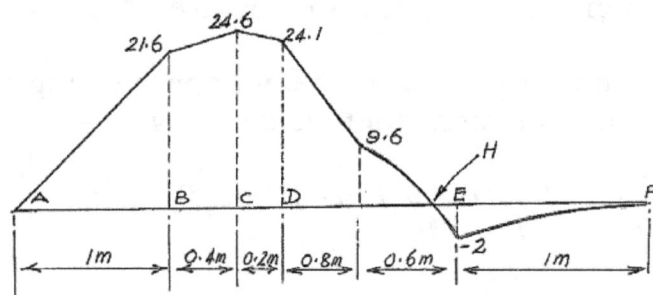

BENDING MOMENT DIAGRAM (NOT DRAWN TO SCALE)
(BENDING MOMENTS IN kN m

FIG 10

The loaded beam at Fig. 5a, axial force diagram at Fig. 6, shear force diagram at Fig. 8, and bending moment diagram at Fig. 10 were drawn approximately to scale. These diagrams, drawn to scale are shown at Figs. 11, 12, 13 and 14 respectively. Before we leave this section, let me demonstrate for the benefit of the recently initiated recruits how it can be shown analytically that the shear force varies linearly with distance in that part of the beam with the uniformly distributed load.

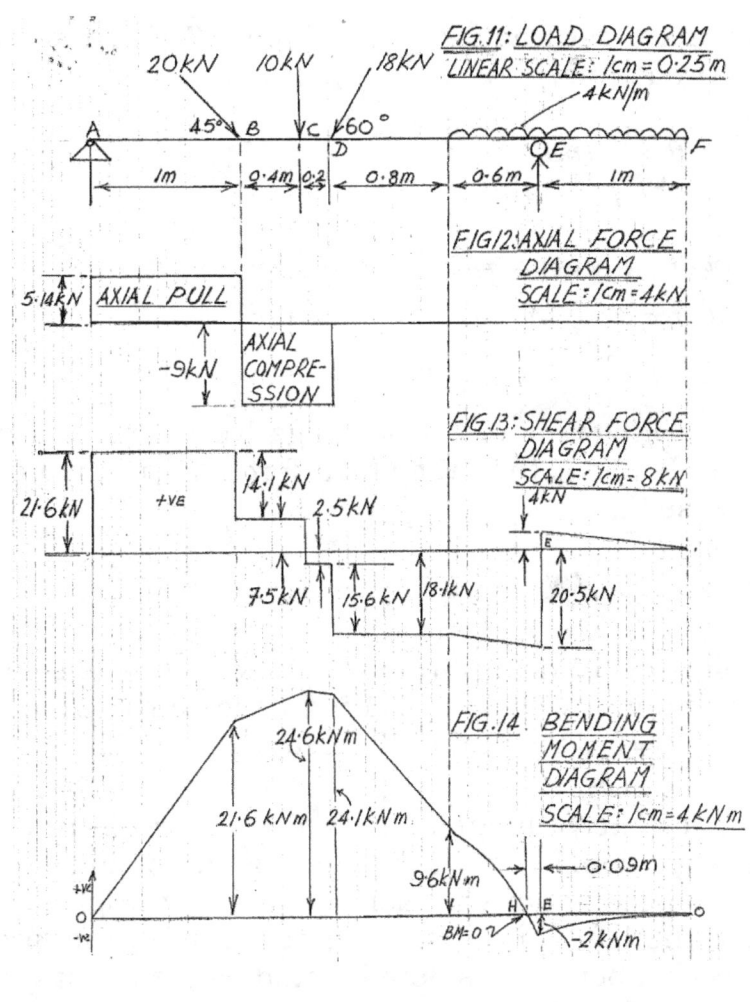

FIG.11: LOAD DIAGRAM
LINEAR SCALE: 1cm = 0.25m

FIG.12: AXIAL FORCE DIAGRAM
SCALE: 1cm = 4kN

FIG.13: SHEAR FORCE DIAGRAM
SCALE: 1cm = 8kN

FIG.14. BENDING MOMENT DIAGRAM
SCALE: 1cm = 4kNm

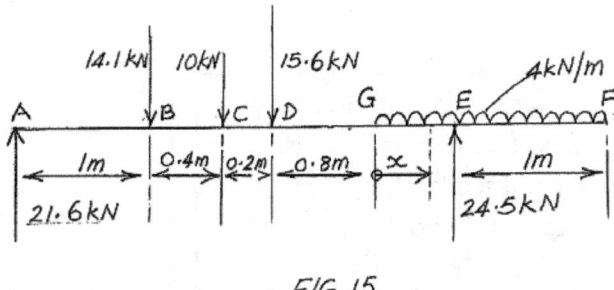

FIG. 15

At a distance 'x' from G as indicated at Fig 15,

$$V_x = 21.6 - 14.1 - 10 - 15.6 - 4x$$

i.e. $V_x = (-4x - 18.1)\ kN$

which demonstrates that the shear force varies linearly with distance from 'G' as we move from left to right. Not only that! The slope of the line reflecting the variation in shear is negative, i.e. the line slopes downwards: $\dfrac{dV}{dx} = -4$.

~ 15 ~

Evidently, just the minutest fraction to the left of E which is distant $0.6m$ from G, shear force is approaching.

$$V_x = -4(0.6) - 18.1$$
$$= -20.5kN$$

At E, where $R_E = 24.5kN$ and upwards
$$\therefore \quad V_E = (-20.5 + 24.5)kN$$
$$= +4kN$$

so that shear force changes sign – from negative to positive at E. If now we transfer the origin of 'x' to "E", then the shear force 'V_x' at any position between E and F may be expressed as:

$$V_x = (+4 - 4x)kN$$

This again demonstrates a linear variation of shear force with distance from E; also the slope is negative as before; heading downwards from a positive shear of $4kN$ at E to zero at $x = 1m$, the end of the beam.

It might be useful for the student to note that in working out the values of shear force, we always considered at every stage a section of the beam to the left. We could just as easily have chosen to work from right, but we arbitrarily chose to work from the left. And when the nett sum of the forces to the left of the point on the section being considered would cause an upward movement of the section, a downwards 'V' is required to restore equilibrium; hence positive shear.

| If nett result of forces acting on this section of beam is to cause it to move upwards, relative to right hand section then 'V; must act downwards to restore equilibrium. | V ↓ Positive shear |

| If nett result of forces acting on this section of the beam is to cause it to move downwards,, relative to the left hand section, the 'V' must act upwards to restore equilibrium | ↑ Negative shear V |

If we wish to develop an analytical expression for bending moment for that part of the beam between 'G' and E in Fig 15, then we may write, considering moments about the point distant 'x' from G

$$BM_x = 21.6(2.4 + x) - 14.1(1.4 + x)$$
$$- 10(1+x) - 15.6(0.8+x)$$
$$= 4x\left(\frac{x}{2}\right)$$

i.e. $$BM_x = 9.62 - 18.1 - 18.1x - 2x^2$$

At E where $x = 0.6m$
$$BM_E = 9.62 - 10.86 - 0.72$$
i.e. $$BM_E = -1.96kNm, \text{ say, } -2kN$$

Let us confirm that *B.M.* achieves a value of zero between G and E.

Writing $0 = 9.62 - 18.1x - 2x^2$

and solving this quadratic equation we find that x has 2 values: $x = +0.51m$; and $x = -9.5m$. Clearly the only feasible solution here is $x = 0.51m$ or a distance 2.91m from A.

Let us very quickly confirm some of the results we obtained for shear force and bending moment by considering analyses starting at the right-hand most part of the beam. For example, taking a section, say, *0.75m* from *F*, the downward load equals *4(0.75)*kN = *3kN*; and downward on the right is positive. Just immediately before 'E', the shear is *4(1)kN = 4kN*, positive; and just immediately after *E* it is *(4- 24.5)kN = -20.5kN*. Thus as we saw previously shear force changes from a value of *+4kN to -20.5kN* at '*E*'.

Proceeding in this manner, the student should confirm the values obtained in the earlier analysis from the left.

Similarly, at a section *0.75m* from *F*, the bending moment = $4(0.75)\left(\dfrac{0.75}{2}\right)$ kNm = *1.125kNm.* This is a clockwise moment on the right; that is a hogging moment if you will, and therefore negative. At '*E*' itself, the minimum bending moment of *-2kNm*, i.e. $-4(1)\left(\dfrac{1}{2}\right)$ kNm is achieved. Further, if '*x*'' were the distance, this time measured from '*F*' and extending to a section between '*E*' and '*G*', then,

$$BM_x = -4(x)\left(\frac{x}{2}\right) - 24.5(x-1)$$
$$= -2x^2 - 24.5x + 24.5 = 0$$

Where *B.M*$_x$ = *0*, $2x^2 - 24.5x + 24.5 = 0$ from which the only possible solution here is $x = 1.1m$, i.e. *0.1m* to left of '*E*' or *0.5m* from '*G*'. This confirms the previous determination from the left. Note that moments clockwise on the right are negative; and anticlockwise on the right, positive.   I also leave for the students' benefit the exercise of confirming the bending moment values previously obtained for *B, C, D and 'G'*, by considering sections from the right.

GENERAL PROCEDURES FOR PREPARATION OF SHEAR FORCE, AXIAL FORCE, AND BENDING MOMENT DIAGRAMS

While the variety of problems of shear-force and bending-moment determination is virtually limitless, it is nevertheless possible to map out a set of general procedures which when applied methodically would lead to the correct solution.   Here we shall concern ourselves only with statically determinate beams.   The steps are:

(i)   prepare a sketch of beam, and mark position and line of action of all loads and external reactions; if there are any inclined loads resolve them into their horizontal and vertical components; if there are any externally applied moments indicate their position and magnitude; if for example a beam is carrying an offset load as in Fig. 16, then the loading configuration could be converted to that shown in Fig. 17: an axial load '*W*' and an anti-clockwise moment or couple '*Wu*'. Purely for amplification of this loading configuration, the shear force and bending moment diagrams are shown at Figs. 18 and 19 respectively

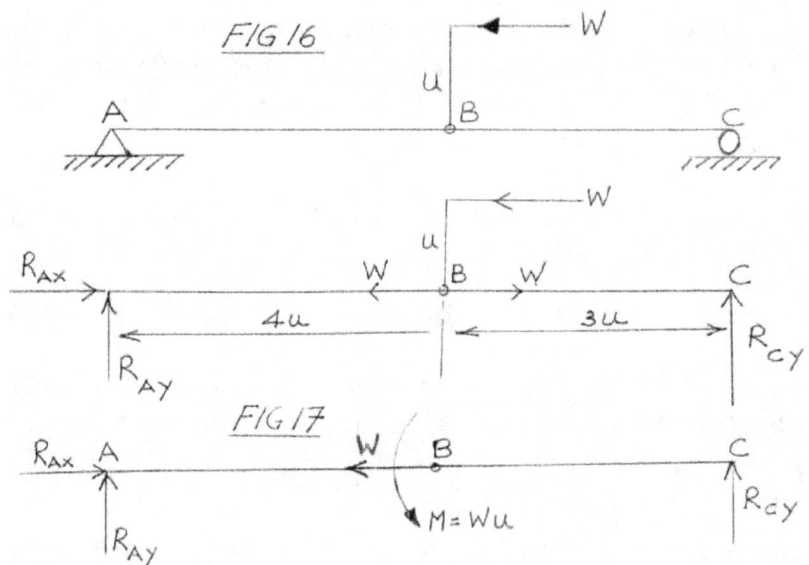

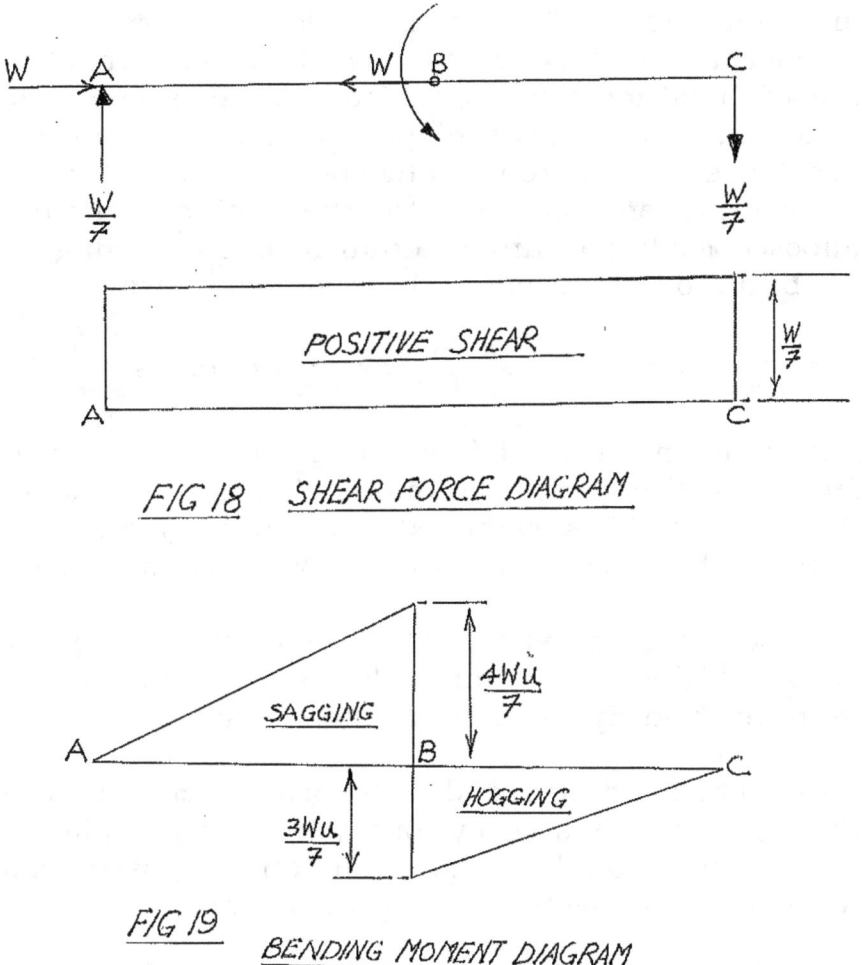

FIG 18   SHEAR FORCE DIAGRAM

FIG 19   BENDING MOMENT DIAGRAM

(ii)   determine reactions using the laws of statical equilibrium: $\sum F_y = 0$; $\sum F_x = 0$; $\sum M = 0$. Invariably one of the reactions is obtained by taking moments about some convenient point, in which case $\sum F_y = 0$, gives the value of the other reaction. However, as a check on the latter calculation it is advisable that the value of the other reaction be determined by considering moments about some other convenient point on the beam;

(iii)   indicate on sketch the magnitude and direction of axial and transverse forces including external reactions and any external moments;

(iv)   determine the value of shear 'V' at various sections along the beam, commencing at either the left end or right end.

    (a)   If the left end is chosen, consider a section just immediately inside it, due attention being given of any imposed load or external reaction on this section. Ask the question: What value and

~ 19 ~

direction must *'V'* have in order to restore the section to equilibrium? If *'V'* is downwards, then that part of the beam from the left-hand-most part of the beam to the section just immediately inside it, is in a state of positive shear; if upwards a state of negative shear. More formally you would recall that shear force at a section taken from the left is the algebraic sum of all forces, i.e. imposed loads and any external reactions, acting on that part of the beam to left of section. Therefore you have

$$V = \sum \text{Upward forces} - \sum \text{Downward forces}$$

Note that any upward force on the beam is +ve; any downward force, negative; hence the minus sign in the equation. Clearly, if the value of *"V"* as evaluated is positive, then shear is positive; if negative then shear is negative, as was stated earlier.

Take several sections along the beam and compute the value of *'V'* at each of them. Remember always to work out *'V'* at sections just to the left and right of concentrated loads.

(b)   If the right-hand-most end of the beam is chosen for the conduct of the analysis, then start by considering a section immediately inside it, due attention being given to any imposed load or external reaction on this section. Compute *'V'* from:

$$V = \sum \text{Downward Forces} - \sum \text{Upward Forces}$$

In this equation, downward forces are positive; upward forces negative. If *V* is positive, shear is positive; if *V* is negative, then shear is negative.

Take several sections along the beam and compute the value of *'V'* for each of them. As was stated in the case of treating shear from the left, always work out *'V'* at sections just to the right and left of concentrated loads.

(v)   based on the computed values of shear force at the sections considered, plot the shear force diagram to a suitable scale and label clearly.

(vi)   if there are horizontal components of forces on the beam, then solving $\sum F_y = 0$ would find any unknown horizontal force.

Having identified all horizontal forces the beam should be inspected to determine which parts are in tension or compression; for any part in tension the axial forces at each end of that part must be equal in magnitude and opposite in direction, tending to pull it apart; whereas for any part in compression the axial forces at each end of that part must be equal in magnitude and opposite in direction, tending to compress the part.

(vii)   determine the value of bending moment at various sections along the beam commencing at either the left end or right end.

   (a)   if the left end is chosen, then at the section of beam considered from that end,

$$B.M = \sum \text{clockwise moments left of the section} - \sum \text{anti-clockwise moments left of the section}$$

   Note that clockwise moments are positive on the left; anti-clockwise moments on the left are negative; hence the minus sign in the equation.

   (b)   If the right end is chosen, then at the section of beam considered from that end

$$B.M = \sum \text{anti-clockwise moments right of section} - \sum \text{clockwise moments right of section.}$$

   Note that anti-clockwise moments are positive on the right; clockwise moments on the right are negative; hence the minus sign in the equation.

(viii)   based on the computed values of bending moment at the sections considered, plot the bending moment diagram to a suitable scale; if there were any imposed external bending moments on the beam these could be taken into account at this stage and the initial bending moment diagram adjusted accordingly.

POINTS OF INFLEXION OR POINTS OF CONTRAFLEXURE

Analysis of the bending moment distribution in the case of the Illustrative Example 1 demonstrated that one part of the beam was subjected to a positive or sagging moment and the other part to a negative or hogging moment in our convention. Evidently at the part of the beam where the bending moment changed sign its value is zero. This section was found to be approximately *2.9m* from *A or 0.1m* to the left of *E* or *0.5m* from *G*. Because it is assumed that all loads and moments were applied to the beam at its centroidal axis, where the plane through the section cuts this axis identifies a point. Such points at which bending moment goes through zero either from positive to negative or vice-versa are called points of inflexion or points of contraflexure.

Referring again to the Illustrative Example 1, the centroidal axis at a point at the immovable support at *'A'* to the point of contraflexure *'H'* would have deflected as shown at Fig. 20a, because the bending moment throughout that distance was a sagging moment.

ESSENTIALLY FIG 14 (REPEATED)

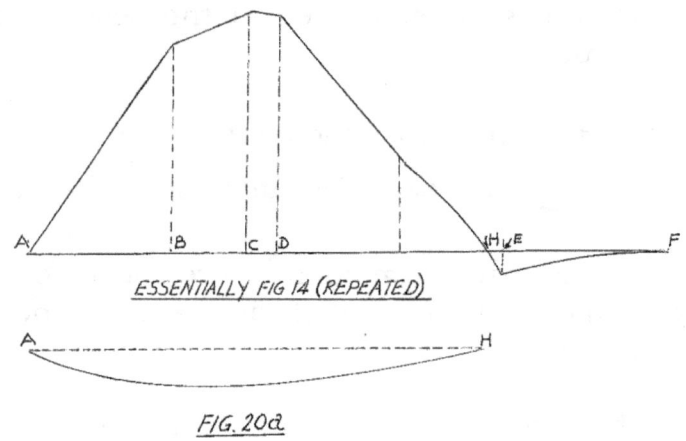

FIG. 20a

There being no deflection at *E*, the centroidal axis between the point of contrareflexure *'H'* and *E* the location of the other immovable support, would have deflected as shown in Fig. 20b, because of the 'hogging' moment between *'H'* and *'E'*

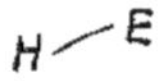

FIG 20b

And between *E* and *F*, the bending moment on this section of the beam being everywhere of negative value or 'hogging', the shape of the centroidal axis for this part would be as shown in Fig. 20c.

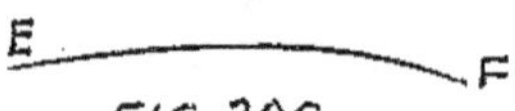

FIG 20C

The deflected centroidal axis of the entire beam is therefore as shown in 20d. This deflection curve is called the elastic curve or elastic line of the beam for the beam loaded as in Fig. 5a.

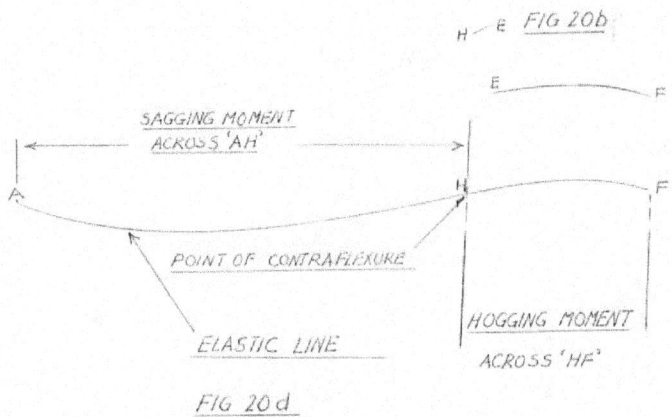

FIG 20d

RELATIONSHIP BETWEEN INTENSITY OF LOADING, SHEARING FORCE AND BENDING MOMENT

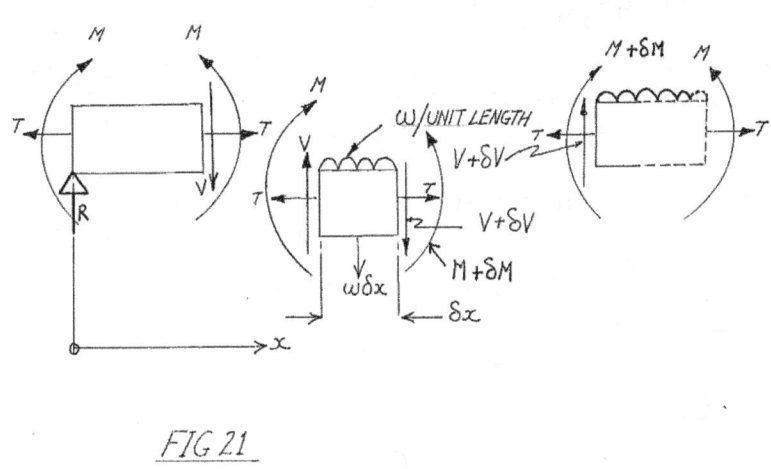

FIG 21

In Fig. 21, a section of beam of length 'δx' carrying an intensity of loading of 'w' newtons per metre is being analysed to determine shear force and bending moment intensity on it.

Considering only shear at this juncture, it should be clear that for vertical equilibrium between the portion of the beam to the left of the section and the left-hand end of the beam of length 'δx' to the right, the shear 'V' must act as shown. At the right-hand end of the portion of beam of length 'δx' it should be evident that because of the vertical load of $w\delta x$ acting through its center of gravity, the downward shear force there must increase by δV.

Considering therefore the vertical equilibrium of the portion of the beam of length 'δx'

$$V - w\delta x - (V + \delta V) = 0$$
$$\therefore \quad V - w\delta x - V - \delta V = 0$$

~ 23 ~

$$\text{i.e.} \qquad \frac{\delta V}{\delta x} = -w$$

In the limit as $\delta x \to 0 \quad \dfrac{\delta V}{\delta x} \to \dfrac{dV}{dx}$

$$\therefore \qquad \frac{dV}{dx} = -w$$

$$\text{or} \qquad \int dV = -\int w\, dx$$

$$\text{i.e.} \qquad V = -\int w\, dx$$

Similarly, considering moments about right-hand end of the portion of the beam of length 'δx', for by the laws of static equilibrium $\sum$ Moments $=0$, we have:

$$M + V\delta x - w\delta x\left(\frac{\delta x}{2}\right) - (M + \delta M) = 0$$

$$\text{i.e.} \qquad M + V\delta x - w\left(\frac{(\delta x)^2}{2}\right) - M - \delta M = 0$$

Neglecting second-order small quantities: $w\left(\dfrac{\delta x}{2}\right)^2 = 0$

$$\therefore \qquad V\delta x - \delta M = 0$$

$$\text{or} \qquad \frac{\delta M}{\delta x} = V$$

As $\delta x \to 0; \quad \dfrac{\delta M}{\delta x} \to \dfrac{dM}{dx}$

$$\therefore \qquad \frac{dM}{dx} = V$$

$$\text{or} \qquad \int dM = \int V\, dx$$

$$\text{i.e.} \qquad M = \int V\, dx$$

Evidently $\dfrac{d}{dx}\left(\dfrac{dM}{dx}\right) = \dfrac{d}{dx}(V)$ or $\dfrac{d^2 M}{dx^2} = \dfrac{dV}{dx} = -w$. So that bending moment 'M' may also be expressed as

$$M = -\iint w\, dx\, dx.$$

How may these equations be applied: $V = -\int w\, dx$ and $M = \int V\, dx$?

Load intensity w is generally in terms of force per unit length, e.g. newton/metre *(N/m)* or kilonewton/metre *(kN/m)* or similar. Thus, shear force 'V' which is upwards at the left end of the section of beam of length 'δx', and therefore positive, is the area under the 'w' vs 'x' characteristic. This is what

$\int w\,dx$ means. Similarly $M = \int V\,dx$ means that bending moment is the area of the 'V' vs 'x' characteristic.

Illustrative Example 2

A joist used at a cantilever has an intensity of loading per metre including its own weight approximated by the distribution reflected in the following tabulation:

X (in metres):	0	0.5	1.0	1.5	2.0	2.5	3.0	3.5	4.0
W(newton/m):	300	265	234	200	168	132	90	48	0

Distance 'x' is measured from the fixed end of the cantilever as shown in Fig. 22.

Determine the maximum shear force and maximum bending moment on the cantilever

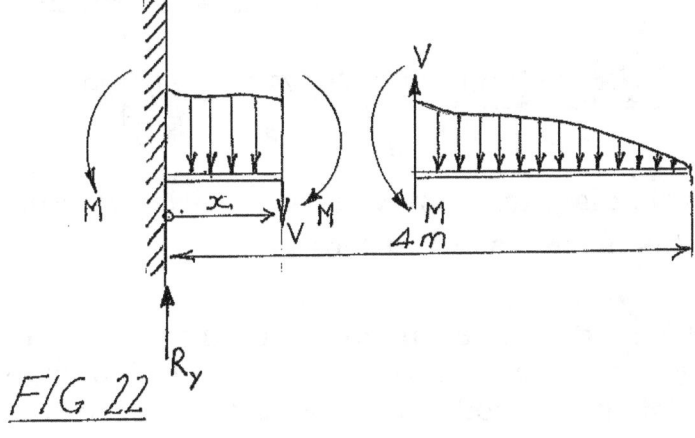

FIG 22

The load intensity diagram is plotted in Fig. 23

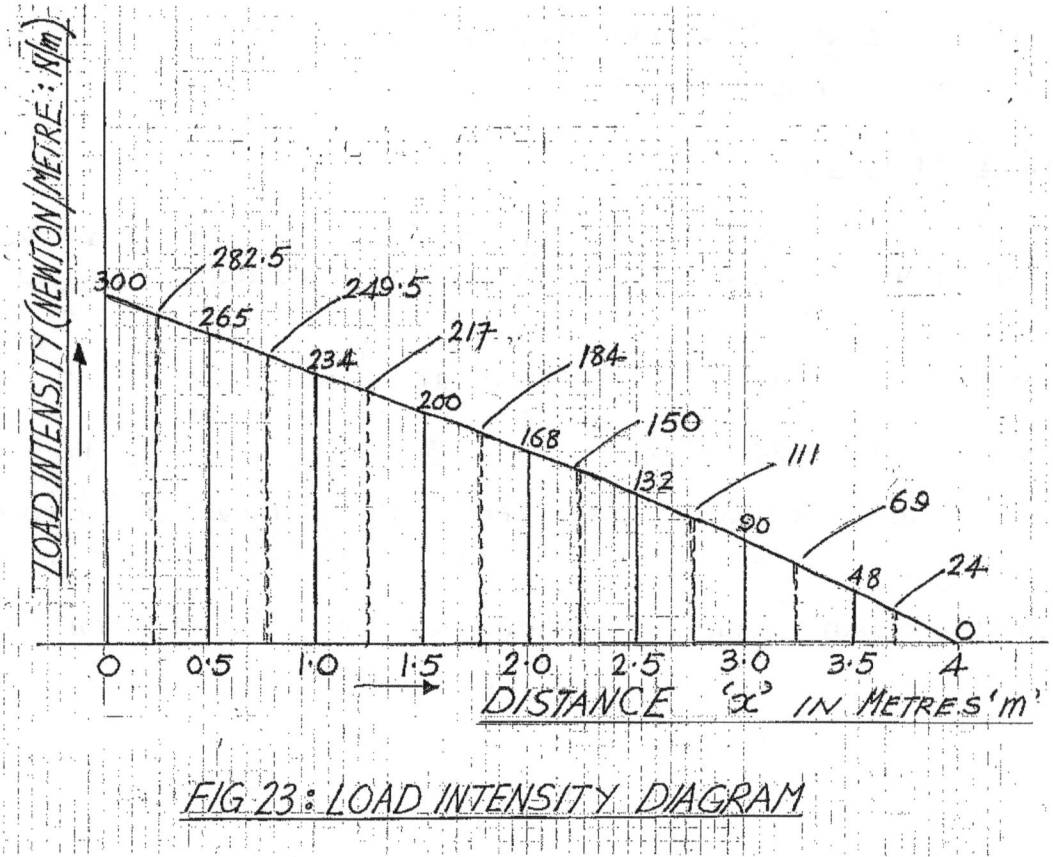

FIG 23 : LOAD INTENSITY DIAGRAM

The area under this diagram from $x = 0$ to $x = 4m$ gives the maximum shear on the cantilever. This is equivalent to $\int w dx$.

This area may be readily determined by using either the trapezoidal rule or mid-ordinate rule or Simpson's rule. Although Simpson's rule is the most accurate of the strip methods for the determination of irregularly curved figures, the mid-ordinate rule will be used because of is inherent simplicity.

By the mid-ordinate rule

Area of Figure = average of mid-ordinates x length of base where mid-ordinate is defined as the average of the sum of the ordinates delimiting a strip. Thus the mid-ordinate of the first strip starting from the fixed end of the cantilever is

$= \left(\dfrac{300 + 265}{2} \right) = 282.5 N/m.$

Accordingly the mid-ordinates of the load-intensity distribution shown in Fig. 23 are:

x (metres)	0.25,	0.75,	1.25,	1.75,	2.25,	2.75
mid-ordinate 'w'(N/m)	282.5,	249.5,	217,	184,	150,	111

x (metres)	3.25	3.75
mid-ordinate 'w'(N/m)	69	24

$$\text{Average mid-ordinate} = \frac{(282.5 + 249.5 + 217 + 184 + 150 + 111 + 69 + 24)}{8}$$

$$= 1287/8$$

$$= 160.875 \text{ N/m, say } 161 \text{ N/m}$$

$$\therefore \quad V = \int_0^4 w dx$$

$$= 161 \text{ N/m x 4m}$$

$$V = 644N$$

So that the maximum shear occurs at the cantilever support and is *644N*. Evidently the maximum bending moment also occurs at the fixed end of the cantilever.

$$M = \int_0^4 V dx$$

$$= 282.5 \times 0.5 \times 0.25 + 249.5 \times 0.5 \times 0.75 +$$
$$217 \times 0.5 \times 1.25 + 184 \times 0.5 \times 1.75 + 150 \times 0.5 \times$$
$$2.25 + 111 \times 0.5 \times 2.75 + 69 \times 0.5 \times 3.25 + 24 \times 0.5 \times$$
$$3.75)\text{Nm}$$

$$= (35.3125 + 93.5625 + 135.625 + 161 + 168.75 +$$
$$152.625 + 112.125 + 45)\text{Nm}$$

$$M_{max} = 904\text{Nm}$$

Referring to Fig. 22, it is seen that in order for the portion of the cantilever to the right of the section to be in equilibrium, shear force V must act upwards. Therefore V on the left must act downwards as shown. Clearly therefore shear is positive. Note also that there is a vertical reaction at the support, a fact often overlooked by many. Similarly, there is a support moment at 'A' to balance the negative bending moment of maximum value *904Nm* on the cantilever causing it to bend thus

We consider next the stresses due to the three possible reactive elements at any section of a loaded beam, namely the stresses due to any axial forces, bending moment and shear.

Axial stresses

When there are any axial forces i.e. pulls or thrusts acting on any part of the beam as was demonstrated in the example of the loaded beam in Fig. 5, 5a – 5c, the relevant stresses are determined simply by dividing the magnitude of the forces by the cross-sectional area of the relevant section of the beam; tensile stresses are regarded as positive and compressive stresses negative in our convention.

If there are other tensile or compressive stresses acting on the beam say, due to bending then these must be taken into account in determining the overall stress distribution in some of the solved problems. Bending stresses and shear stresses are treated in separate chapters.

SOLVED AND OTHER PROBLEMS

Q1. A beam ABCD 5 metres long carries a uniformly distributed load of 1000N per metre run. The beam supports at B and C are symmetrical about the centre of the span, each being located 1 metre from each end. Draw the bending moment diagram for the beam.

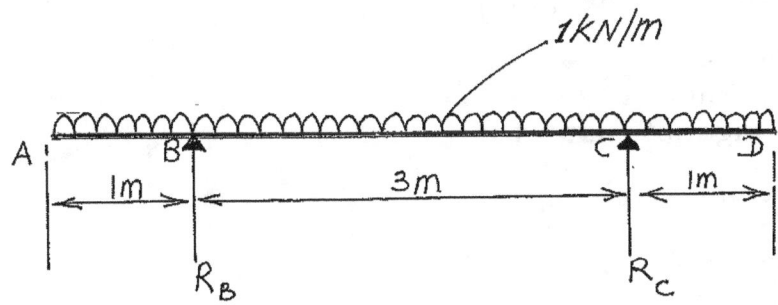

FIG 1

R_B and R_C are the reactions at A and C respectively.

$\therefore$ R_B = R_C = 2500N

Now,

$$BM_A = 0$$

BM half-way between A and B;

$$= 1000\left(\frac{1}{2}\right) \cdot \frac{1}{4}$$
$$= 125Nm$$

$$BM_B = 1000(1) \cdot \frac{1}{2}$$
$$= 500Nm$$

B.M)

BM at 0.5 to right of support 'B'

$$= R_B(0.5) - 1000(1.5)\left(\frac{1.5}{2}\right)$$
$$= 1250 - 1125$$
$$= +125Nm$$

BM at 1m to right of support B

$$= R_B(1) - 1000(2)\left(\frac{2}{2}\right)$$

$$= 2500 - 2000$$

$$= +500Nm$$

BM at 1.5m to right of support B

$$= 2500(1.5) - 1000(2.5\left(\frac{2.5}{2}\right)$$

$$= 3750 - 3125$$

$$= +625Nm$$

BM at 2m to right of support B

$$= 2500(2) - 1000(3)\left(\frac{3}{2}\right)$$

$$= 5000 - 4500$$

$$= +500 \ Nm$$

and so on.

The student is left to work out the BM values at distances 2.5m and 3m to right of B and at 0.5 beyond C.

The bending moment diagram is at Fig. 2.

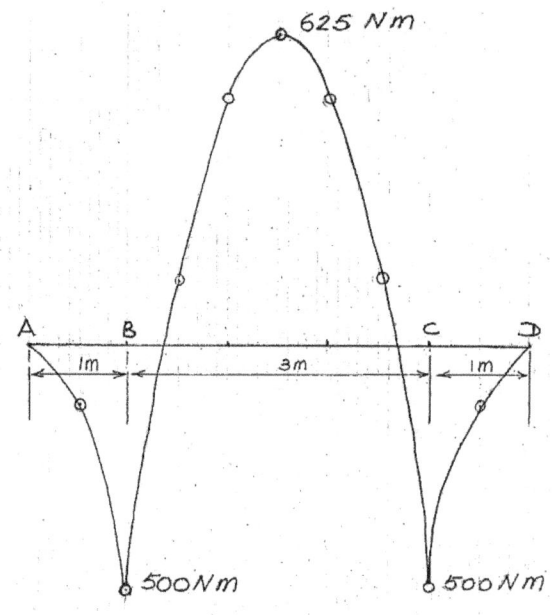

FIG 2: BENDING MOMENT DIAGRAM

SCALE: 1cm = 80 N/m

Q2. The load on a cantilever varies uniformly along its length from a value of 1kN/metre at its fixed end to zero at its free end. The length of the cantilever is 4m.

Draw the shear force and bending moment diagrams and state their maximum values.

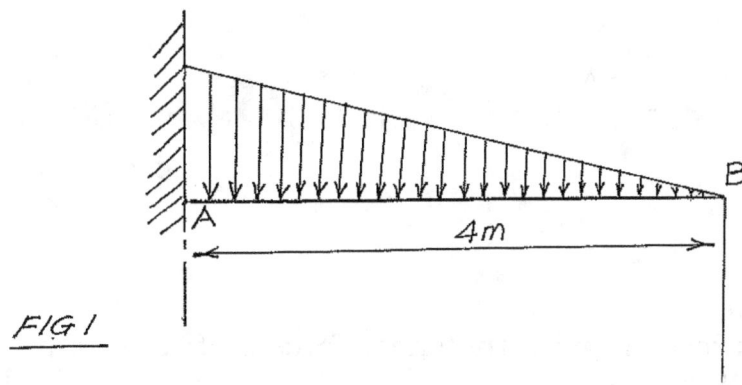

FIG 1

The loaded cantilever is shown in Fig. 1. I have not shown them in Fig. 1, but it is clear there must be a vertical reaction at A and also a fixing moment there to maintain equilibrium. We shall find the values of this reaction and fixing moment shortly.

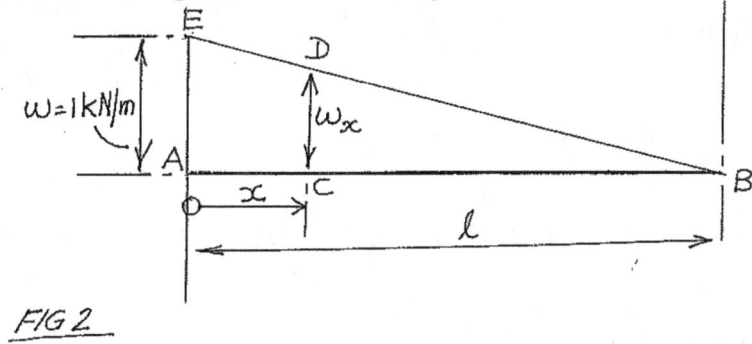

FIG 2

In Fig. 2, the load intensity per metre run at a section distant 'x' metres from 'A' is w_x

Considering the load intensity distribution represented by triangles ABE and CBD it is evident, as was indicated in the foregoing, that

$$w_x = \frac{w}{\ell}(\ell - x)$$

Recall,

$$\frac{dF}{dx} = -w_x$$

Therefore at section 'x' from 'A'

~ 31 ~

$$\frac{dF}{dx} = -\frac{w}{\ell}(\ell - x)$$

i.e. $$F = -\int_0^\ell \frac{w}{\ell}(\ell - x)$$

$$F = -wx + \frac{wx^2}{2\ell} + K$$

At $x = \ell$, $F = 0$

$$\therefore \quad 0 = -w\ell + \frac{w\ell}{2} + K$$

or $$K = \frac{w\ell}{2}$$

$$\therefore \quad F_x = -wx + \frac{wx^2}{2\ell} + \frac{w\ell}{2}$$

The latter expression gives the shear force distribution across the beam. Given $w = 1\text{kN/m}$; $\ell = 4\text{m}$,

$$\therefore \quad F_x = \left(-x + \frac{x^2}{8} + 2\right)kN$$

which expression indicates a parabolic distribution. Using this expression, a table of values of F for 'x' at intervals of 0.5m from A is drawn up as follows:

Distance 'x' in metres	0	0.5	1	1.5	2	2.5	3	3.5	4
Shear Force 'F_x' in kN	2	1.53	1.13	0.78	0.5	0.28	0.13	0.03	0

It is also known that

$$\frac{d^2M}{dx^2} = -w_x$$

$$= -\frac{w}{\ell}(\ell - x)$$

so that

$$M_x = -\frac{wx^2}{2} + \frac{wx^3}{6\ell} + K_1 x + K_2$$

At $x = \ell$, $M = 0$

$$\therefore \quad 0 = -\frac{w\ell^2}{2} + \frac{w\ell}{6\ell} + K_1\ell + K_2$$

$$= -\frac{w\ell^2}{2} + \frac{w\ell^2}{6} + K_1\ell + K_2$$

$$0 = -\frac{w\ell^2}{3} + K_1\ell + K_2$$

Now the shear force $F_x = \dfrac{dM}{dx} = 0$ at $x = \ell$

$$\therefore \quad -wx + \frac{wx^2}{2\ell} + K_1 = 0 \text{ at } x = \ell$$

$$\therefore \quad K_1 = w\ell - \frac{w\ell}{2} = \frac{w\ell}{2}$$

which means that $K_2 = \dfrac{w\ell^2}{3} - \dfrac{w\ell^2}{2} = -\dfrac{w\ell^2}{6}$

$$\therefore \quad M_x = -\frac{wx^2}{2} + \frac{wx^3}{6\ell} + \frac{w\ell x}{2} - \frac{w\ell^2}{6}$$

When $w = 1\text{kN/m}$ and $\ell = 4$

$$M_x = \left(-\frac{x^2}{2} + \frac{x^3}{24} + 2x - \frac{8}{3}\right) Nm$$

Distance 'x' in metres	0	0.5	1	1.5	2.0	2.5	3.0	3.5	4
Bending Moment B.M$_x$ in kNm	-2.7	-1.82	-1.16	-0.69	-0.37	-0.18	-0.08	-0.04	0

Bending moment distribution is given in the above tabulation.

The shear force and bending moment diagrams are plotted in Fig. 3 and 4. The maximum shear = 2kN occurs at support 'A' at which location the minimum bending moment of –2.7kNm occurs.

Observe that the shear force at A = 2kN is in fact the value of the reaction at A of which I spoke at the outset.

Similarly, the fixing bending moment at A = -2.7kNm; the bending moment is a hogging moment and consequently is negative (in our convention).

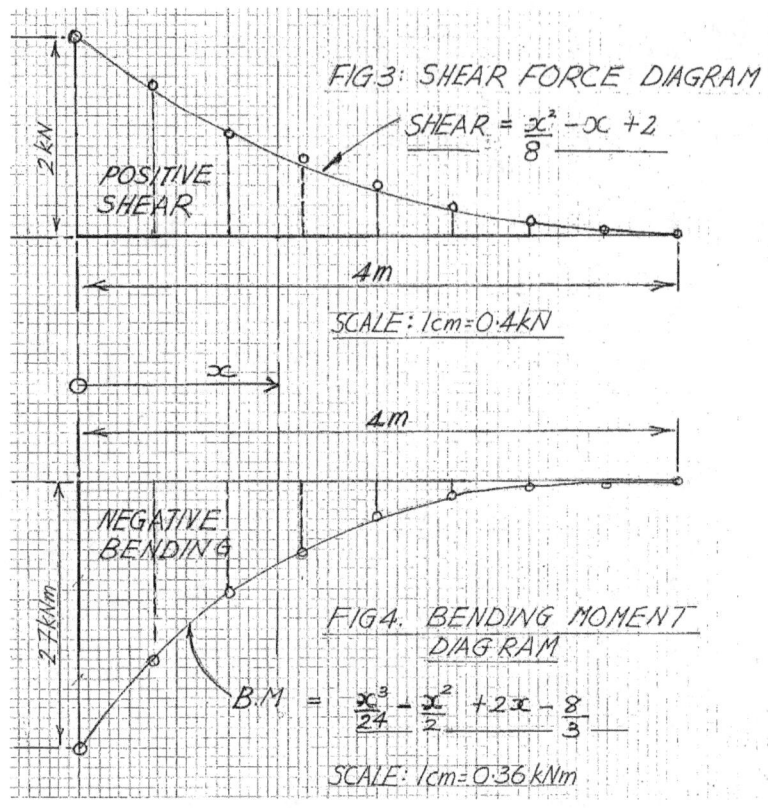

FIG 3: SHEAR FORCE DIAGRAM

SHEAR $= \dfrac{x^2}{8} - x + 2$

POSITIVE SHEAR

4m

SCALE: 1cm = 0.4kN

4m

NEGATIVE BENDING

FIG 4. BENDING MOMENT DIAGRAM

$B.M = \dfrac{x^3}{24} - \dfrac{x^2}{2} + 2x - \dfrac{8}{3}$

SCALE: 1cm = 0.36 kNm

Q3. Draw to a scale of 1cm = 10kNm the bending moment diagram for the beam carrying the concentrated load of 40kN as shown in Fig. 1. The portions of the beam AB and ED are joined to the section BD by means of pin connections.

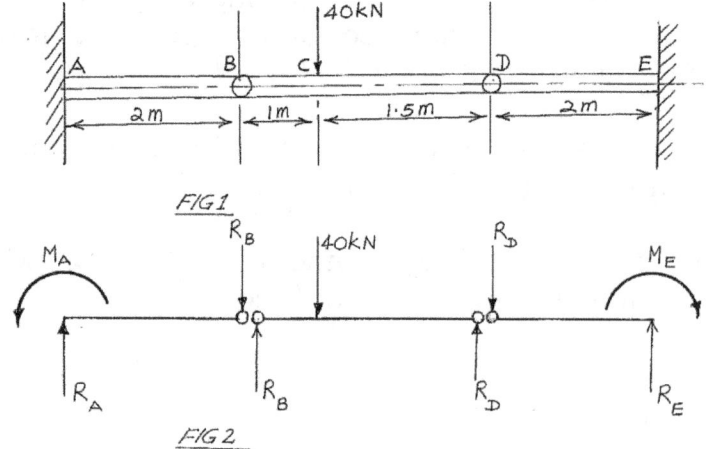

40kN

FIG 1

FIG 2

Fig. 2 is the free-body diagram (FBD). Because there are pins at B and D the only reactions there in the FBD are the forces R_B and R_D. Remember that a pin cannot resist a moment acting about its axis, but it can resist

~ 34 ~

motion normal to it. Hence in the FBD diagram vertical reactions are shown at B and D for the three beam sections; but there are no moments at either of these points; but there are fixing moments at 'A' and 'E' as shown.

Considering beam section AB in Fig. 2, and taking moments about B.

$$-M_A + R_A(2) = 0$$
$$\therefore \quad M_A = 2R_A \qquad \ldots\ldots\ldots \quad (i)$$

For vertical equilibrium

$$R_A = R_B \qquad \ldots\ldots\ldots \quad (ii)$$

Taking now beam section BD and moments about D we have

$$R_B(2.5) - 40(1.5) = 0 \qquad \ldots\ldots\ldots \quad (iii)$$

from which

$$R_B = 24kN$$

and because $R_B + R_D = 40$ $\qquad \ldots\ldots\ldots \quad (iv)$

$$R_D = 16kN$$

Similarly

$$+ M_E - R_E(2) = 0$$
$$\therefore \quad M_E = 2R_E \qquad \ldots\ldots\ldots \quad (v)$$

Considering vertical equilibrium of DE

$$R_E = R_D$$
$$\therefore \quad R_E = 16kN$$

By equation (ii)

$$R_A = 24kN$$

and by (i) $M_A = 48kNm$

Similarly by (v) $M_E = 32kNm$

Collecting our results

$$R_A = 24kN$$
$$R_E = 16kN$$
$$M_A = 48kNm \text{ (anticlockwise)}$$
$$M_E = 32kNm \text{ (clockwise)}$$

The bending moment diagram is shown at Fig. 3.

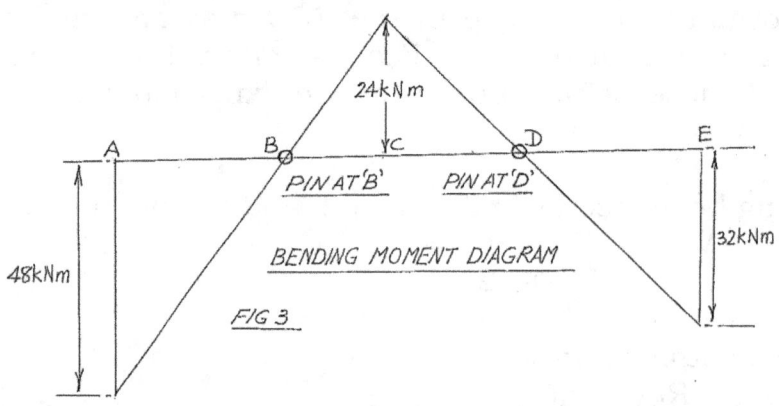

24kNm

PIN AT 'B'   PIN AT 'D'

BENDING MOMENT DIAGRAM

32kNm

48kNm

FIG 3

Q4. A sloping wooden beam AD is 5m long, freely supported at A and pin jointed at D. The beam carries two concentrated loads : 15kN at B and 18kN at C. See Fig. 1. Draw the shearing force, bending moment and beam thrust diagrams.

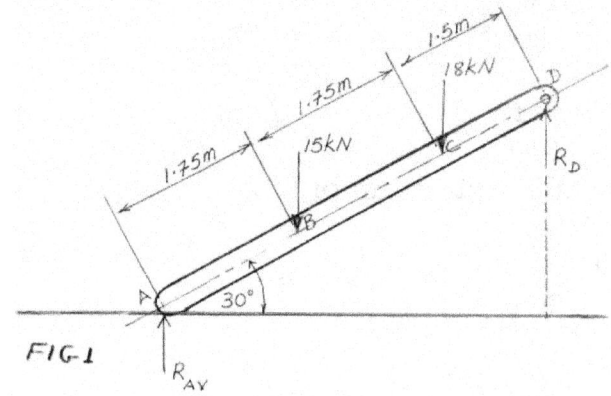

FIG 1

It should be evident that because the external forces at B and C are vertical the reactions at A and D must also be vertical.

Taking moments at D

$$R_{AV}.5Cos30° = 15(3.25Cos30°) + 18(1.5Cos30°)$$

i.e. $5R_{AV} = 15(3.25) + 18(1.5)$

$$= 48.75 + 27$$

$$= 75.75kN$$

∴ $R_{AV} = 15.15$ say 15.2kN

and $R_D = 33 - 15.2$

$$= 17.8kN$$

The student must remember that by definition, "the shearing force at any point along the span of a beam is the algebraic sum of all the <u>perpendicular</u> components of the forces acting on the portion of the beam to the right or left of that point."

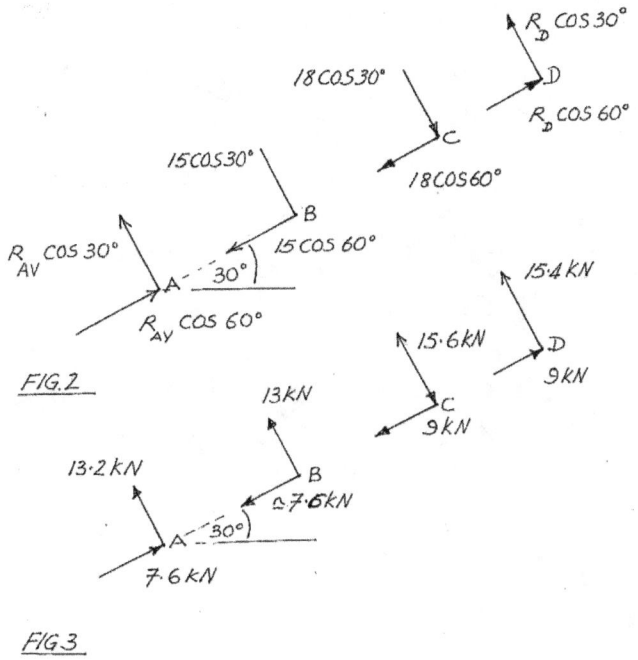

$R_D \cos 30°$

$18\cos30°$

$R_D \cos 60°$

$15\cos30°$

D

C

$18\cos60°$

$R_{AV}\cos30°$

B

$15\cos60°$

A   30°

$R_{AV}\cos60°$

FIG.2

15·4 kN

15·6kN

D

13kN

C

9kN

9kN

13·2 kN

B

7·6kN

A   30°

7·6 kN

FIG.3

In Figs. 2 and 3 the algebraic and numerical values of all the components perpendicular to and in line with the axis of the cantilever are stated.

We now draw the shear force and thrust diagrams:

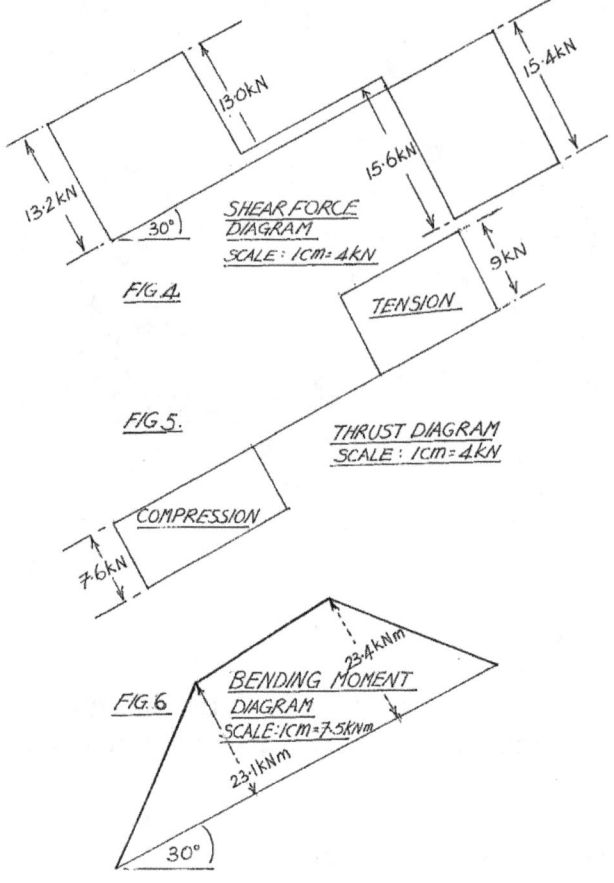

FIG 4.

SHEAR FORCE DIAGRAM
SCALE : 1cm = 4kN

13.2 kN   30°   13.0 kN   15.6kN   15.4kN

TENSION   9kN

FIG 5.

THRUST DIAGRAM
SCALE : 1cm = 4kN

COMPRESSION

7.6kN

FIG 6.   BENDING MOMENT DIAGRAM
SCALE : 1cm = 7.5kNm

23.4kNm

23.1kNm

30°

Referring to Fig. 3
BM_A = 0
BM_B = 13.2(1.75) = 23.1 kNm
BM_C = 1.32(3.5) − 13(1.75) = 46.2-22.75 = 23.45kNm
BM_D = 0; Check : 13.2(5) − 13(3.25) − 15.6(1.5) = 0.35($\approx$0)

The student should also observe that in the case of a sloping beam with vertical reactions the B.M diagram is the same as that for a horizontal beam of length equal to the projected length of the sloping beam. This will now be demonstrated.

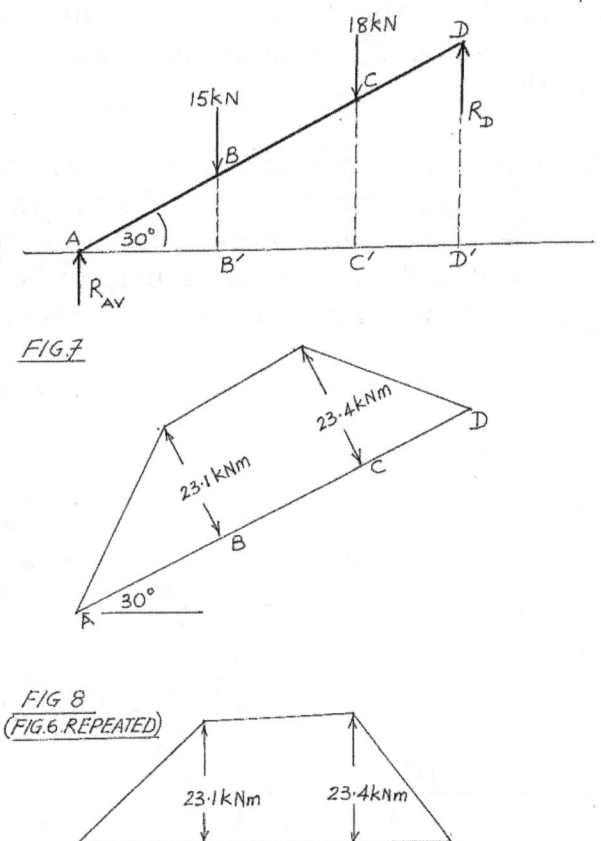

FIG 7

FIG 8
(FIG.6.REPEATED)

FIG 9

The original loading configuration is shown at Fig. 7.
Considering moments on projected length AD[1].

Moment at B' = $R_{AV}(1.75\text{Cos}30)$
                 = $15.2(1.52)$
                 = 23.1kNm
Moment at C' = $15.2(3.5\text{Cos}30) - 15(1.75\text{Cos}30)$
                 = $46.1 - 22.7$
                 = 23.4kNm
Moment at D' = 0 ;

Check : $15.2(5\text{Cos}30) - 15(3.25\text{Cos}30) - 18(1.5\text{Cos}30)$
             = $0.25(\approx 0)$

Considering moments on sloping beam AD
$BM_B = 13.2(1.75) = 23.1\text{kNm}$
$BM_C = 13.2(3.5) - 13(1.75) = 23.45\text{kNm}$
$BM_D = 13.2(5) - 13(3.35) - 15.6(1.5) \approx 0$

Refer to Fig. 8 which is the bending moment diagram drawn on the sloping beam AD. In Fig. 9 the bending moment diagram is drawn on the projected length *AD'*.

Q5. A bent bar ABC, wholly in a horizontal plane, is acted upon by an anti-clockwise couple of magnitude 30kNm at C of Fig. 1, which represents the plan view of the bar. The couple acts in a vertical plane passing through the axis of CB and its axis being perpendicular to AB. Calculate shear-, twisting-, and bending moments along AB and BC

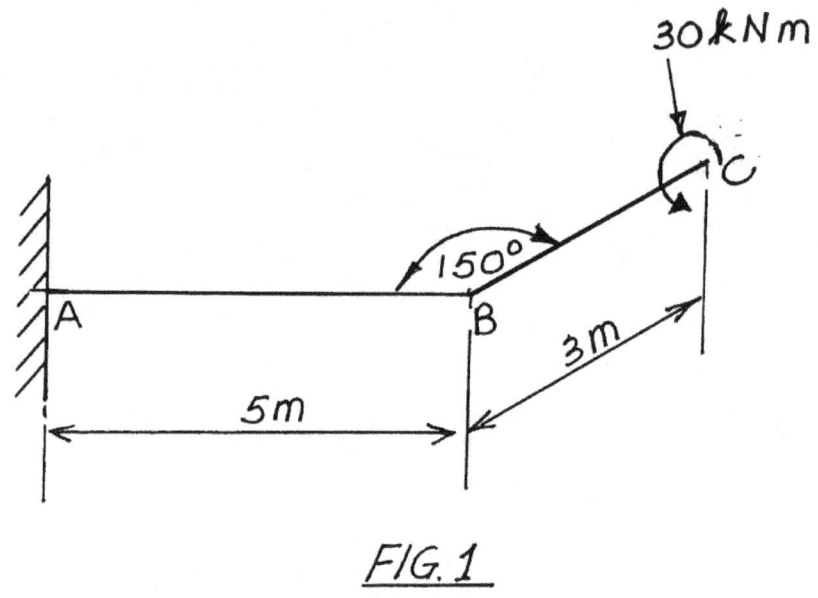

FIG. 1

Let us start by separating BC from the rest of the bar and doing what is necessary to keep it in equilibrium as a separate entity.

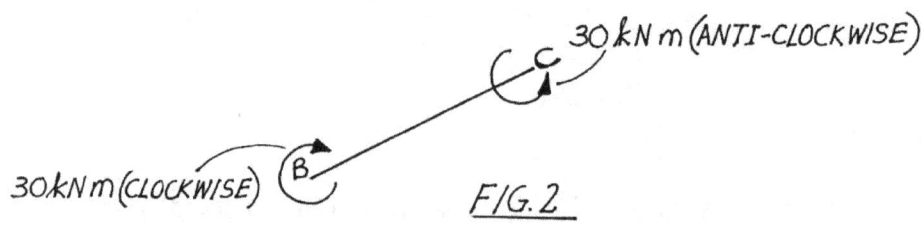

FIG. 2

Apart from the anti-clockwise couple at C nothing else is acting on BC. Therefore in order to keep it in equilibrium, a clockwise couple 30kNm (in the BC-plane, let us called it) must be applied at B as indicated at Fig. 2. Because the axis of couple at C is normal to BC, there is no twisting moment on CB. Let us now consider AB.

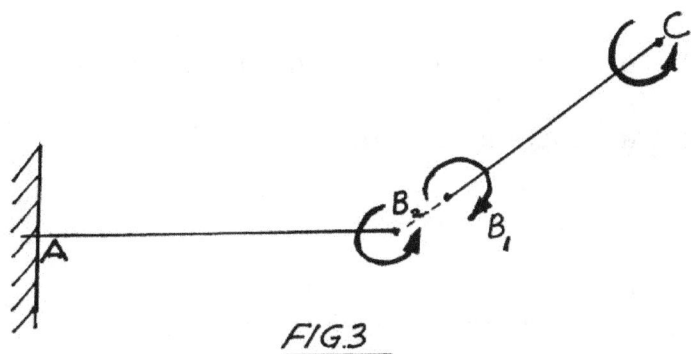

FIG.3

Equilibrium will be achieved at 'joint' B on AB by an anti-clockwise couple of 30kNm. Remember however that this anti-clockwise couple at B is acting in the BC-plane which is at 150° to the plane of AB. Evidently, the component couple perpendicular to the axis of AB is anti-clockwise and of magnitude $30Cos60° kNm = 15kNm$, and the component couple parallel to the axis of $AB = 30Sin60° kNm$, anti-clockwise.

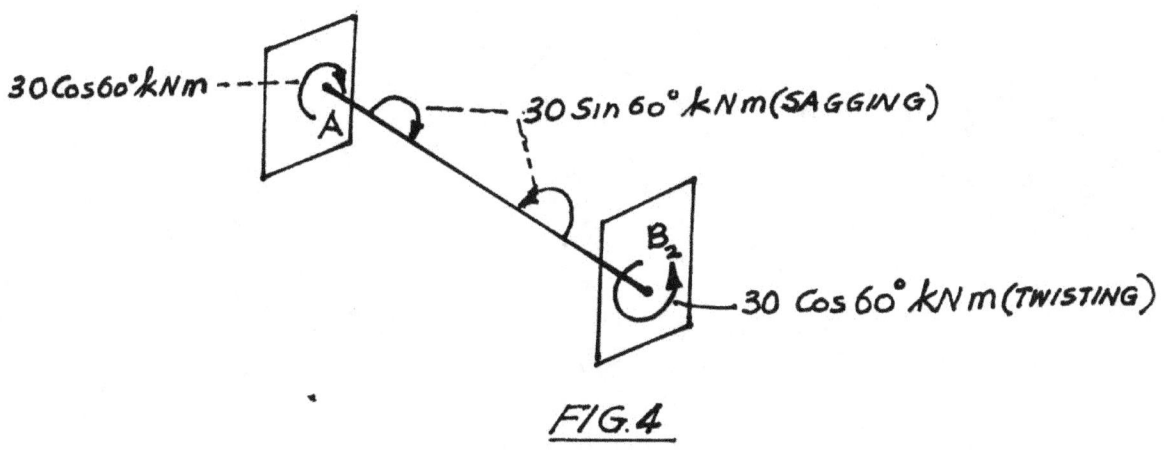

FIG.4

Evidently the couple $30Cos60 = 15kNm$ which is perpendicular to the axis of AB is a twisting moment; whereas the anti-clockwise couple $30Sin60° kNm = 15\sqrt{3}kNm = 25.98kNm$ (say) 26kNm, which is in the vertical plane passing through the axis of AB is a sagging bending moment. The couples at B_2 must be balanced by equal and opposite couples at A, as shown in Fig. 4.

We may therefore sum up as follows:

<u>For AB</u>
Shear Force = 0
Twisting Moment : anti-clockwise at A; clockwise at
B = 15kNm
Bending Moment : sagging i.e. positive = 26kNm

<u>For BC</u>
Shear Force = 0
Twisting Moment = 0
Bending Moment : Sagging i.e. positive = 30kNm

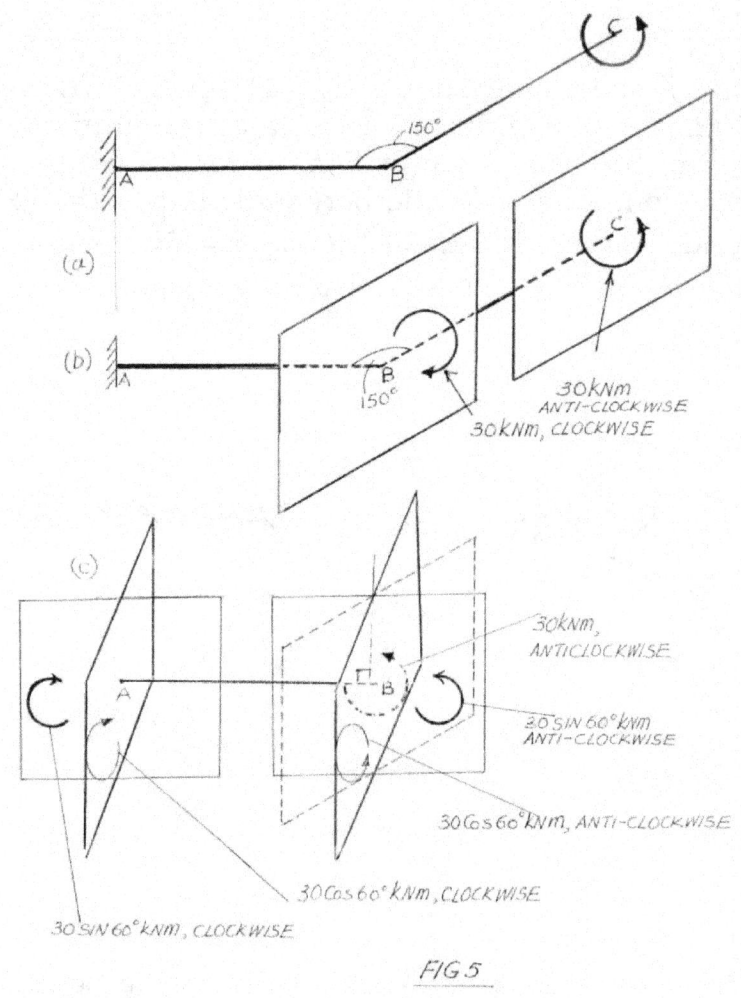

FIG 5

In Fig. 5 I have shown a composite diagram comprising three parts:

(a)   the bent bar with anti-clockwise bending moment 30kNm acting at
        'C' in the plane BC, the axis of this moment being perpendicular to
        the plane;

(b)   BC shown separated from the bent bar. To maintain BC in equilibrium a <u>clockwise</u> bending moment of magnitude 30kN must act at 'B' in the BC plane; and,

(c)   AB shown separated from the bent bar. This separation means that for equilibrium at 'B' there must be an <u>anti-clockwise</u> moment of magnitude 30kNm <u>acting in the BC-plane</u>. Plane BC is at 150° to the AB-plane. This <u>anti-clockwise</u> moment of 30kNm at B must now be resolved into an anti-clockwise moment in planes parallel and perpendicular to the AB-plane : perpendicular to AB-plane of magnitude 30Cos60°kNm i.e. 15kNm and that parallel to AB plane of 30Sin60°kNm or 26kNm.

Therefore for equilibrium of AB there must be at A a <u>clockwise</u> bending moment of 26kNm in the plane of AB and a <u>clockwise</u> bending moment 15kNm in a plane perpendicular to that of plane AB. That is to say overall there is on AB a sagging moment of 26kNm on the bar AB and also a torque of 15kNm : 26kNm clockwise at A and 26kNm anti-clockwise at B; also a clockwise twisting moment of 15kNm at A and a anti-clockwise twisting moment of 15kNm at B.

Q6.   Derive expressions for shear force, thrust and bending moment at point 'P' in the structure shown in Fig. 1, in terms of W, x and 'θ'

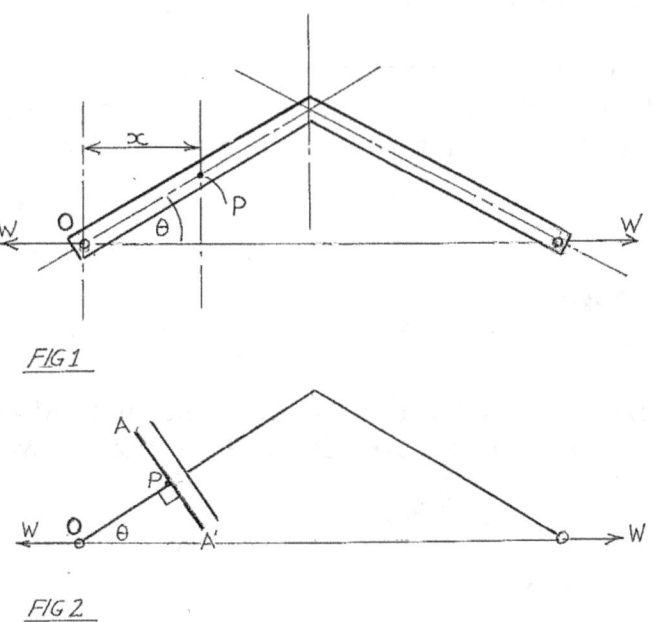

FIG 1

FIG 2

In order to determine the shear force, thrust and bending moment at point 'P' we must first recall that, by definition, the section at P must be normal or perpendicular to the member. Therefore, angle OPA' in Fig. 2 is 90°.

Evidently, shear force must act parallel to AA' and thrust perpendicular to AA. We will put in the directional arrows later.

The next step is to resolve W into components parallel and perpendicular to AA.

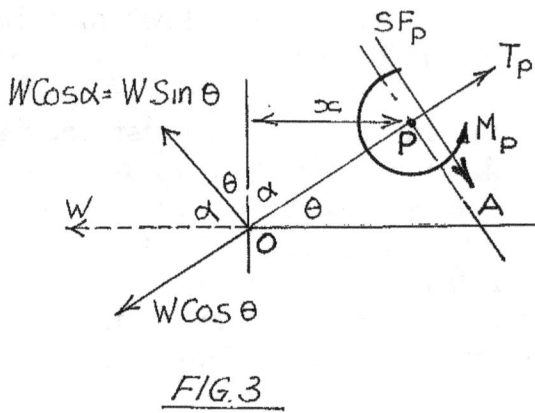

FIG. 3

In Fig. 3, W is resolved into : component $WCos\alpha = WSin\theta$ perpendicular to the member; and axial component $WCos\theta$.

Considering equilibrium of the portion of the structure isolated in Fig. 3, it is clear that SF$_P$ and T$_P$ i.e. the shear force and thrust force respectively at P must act in the directions indicated by the arrowed lines.

Evidently,
$$SF_P = WSin\theta$$
And $$T_P = WCos\theta$$

It only remains to evaluate the bending moment M$_P$ at P. Because W is replaced by $WSin\theta$ and $WCos\theta$, it is clear that if we consider moments about P,
$$MP = WSin\theta \times OP$$

But $$\frac{x}{OP} = Cos\theta$$

i.e. $$OP = \frac{x}{Cos\theta}$$

So that $$M_P = WSin\theta \cdot \frac{x}{Cos\theta}$$

$$\therefore \qquad M_P = Wx\tan\theta$$

Collecting results:

Shear Force$_P$	=	$W Sin\theta$
Thrust$_P$	=	$W Cos\theta$
Bending Moment$_P$	=	$Wx\tan\theta$

Q7. An overhanging beam 20 metres long is simply supported by two rigid props each 4 metres from each end. For a distance of 8 metres from the left-hand-most end of the beam a uniformly distributed load of 20kN/m is carried, and from the point where this load ends another uniformly distributed load of unknown intensity is carried right up to the right-hand support prop. A point load equivalent to 10kN is carried 4m to the right of this same support. Determine the intensity of the unspecified uniformly distributed load which would make the centre of the beam have a shear force of zero, and calculate the position of any points of contraflexure.

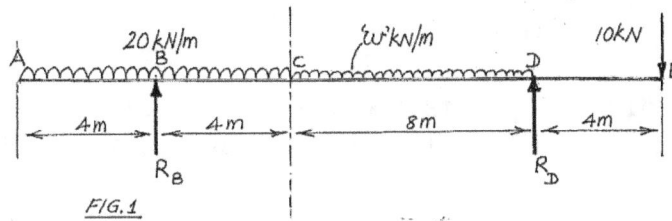

FIG. 1

The beam, its loading and supports are sketched at Fig. 1.

Total load is given by the sum of:

(i)   Uniformly distributed load (udl) of 20kN/m for 8m = 160kN.
(ii)   Taking unknown udl as wkN/m, then load for 8m   = 8 wkN
(iii)   Point load   = 10kN

$$\therefore \qquad \text{Total load} = 160 + 8w + 10$$
$$= (170 + 8w)\text{kN}$$
$$\therefore \qquad R_B + R_D = 170 + 8w \qquad \dots\dots\dots\dots \text{(i)}$$

Taking moments about B
$$10(16) - R_D(12) + 8w(8) = 0$$
$$\text{i.e.} \quad 12R_D = 64w + 160$$

$$\therefore \qquad R_D = \left(\frac{16w + 40}{3}\right) \qquad \dots\dots\dots\dots \text{(ii)}$$

Considering shear force from the right we have, referring to Fig. 2

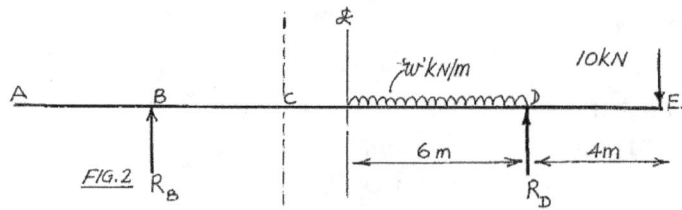

FIG.2

Shear Force at centre of beam

$$= 10 - R_D + 6w$$

$$= 10 - \left(\frac{16w+40}{3}\right) + 6w$$

For this shear force to be equal to zero,

$$30 - 16w - 40 + 18w = 0$$

$$\text{i.e. } 2w = 10$$

$$\text{or } w = 5\text{kN/m}$$

Substituting this result in (i) and (ii) we get

$$R_D = 40\text{kN}$$

and

$$R_B = 170\text{kN}$$

Remembering that for points of contraflexure bending moment is zero, we must now examine B.M. distribution along the beam. Refer to Fig. 3

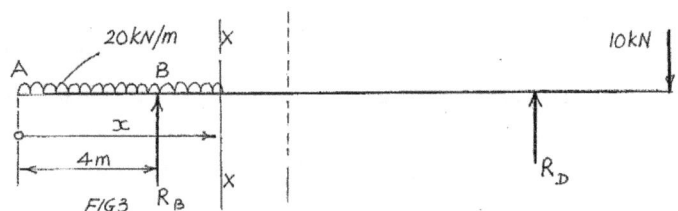

FIG3

Considering B.M at XX distant x from A and within the 20kN/m regime of loading, we have

$$BM_{XX} = 20x\left(\frac{x}{2}\right) - R_B(x-4)$$

$$= 10x^2 - 170(x-4)$$

$$= 10x^2 - 170x + 680$$

For point of contraflexure, $BM_{XX} = 0$

$$\text{i.e. } x^2 - 17x + 68 = 0$$

Solving for x,

$$x = \frac{17 \pm \sqrt{(17)^2 - 4(68)}}{2}$$

$$= \frac{17 \pm 4.12}{2}$$

~ 46 ~

$$x_{1,2} = 10.56m \ or \ 6.44m$$

Evidently, x = 10.56m takes us outside the regime of loading we considered for udl = 20kN/m and therefore must be rejected; not so, x = 6.44m, for this is within the regime.

Let us now extend x into the load regime of 5kN/m. Refer to Fig. 4.

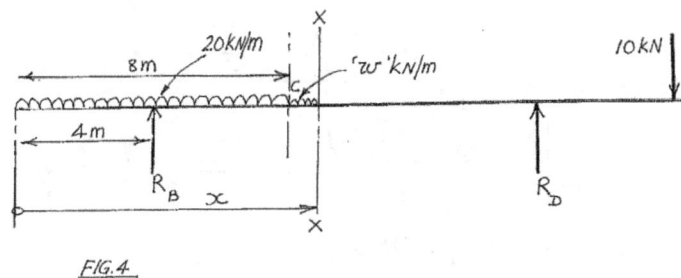

FIG.4.

As before, we consider bending moments from the left and

$$BM_{XX} = 20(8)(x-4) - 170(x-4) + 5(x-8)\left(\frac{x-8}{2}\right)$$

$$= 160x - 640 - 170x + 680 + \frac{5}{2}(x^2 - 16x + 64)$$

$$= 40 - 10x + \frac{5}{2}(x^2 - 16x + 64)$$

For point of contraflexure BM_{XX} = 0

$$\therefore \quad 80 - 20x + 5x^2 - 80x + 320 = 0$$
$$i.e. \quad 5x^2 - 100x + 400 = 0$$
$$or \quad x^2 - 20x + 80 = 0$$
$$\therefore \quad x_{1,2} = \frac{20 \pm \sqrt{(20)^2 - 320)}}{2} \quad or \quad \frac{20 \pm \sqrt{\{(20)^2 - 320\}}}{2}$$
$$= \frac{20 \pm 4\sqrt{5}}{2}$$

Only possible solution is:

$$x = \frac{20 + 8.94}{2}$$
$$= 14.47mm$$

There is thus another point of contraflexure, *10.47m* to right of prop R_B.

Q8. A mechanical engineer is in the process of designing a small hand-powered crane and in order to determine the size of *I-beam* required sets out to calculate the maximum bending moment M_{max} on the beam and its position. The arrangement being considered by the designer is as shown at Fig. 1. What is M_{max} and its position on the beam given that the maximum load on each wheel is *8kN*, span of *I-beam* is *5 metres*, and the distance between the centre-line of the axles = *1m?*

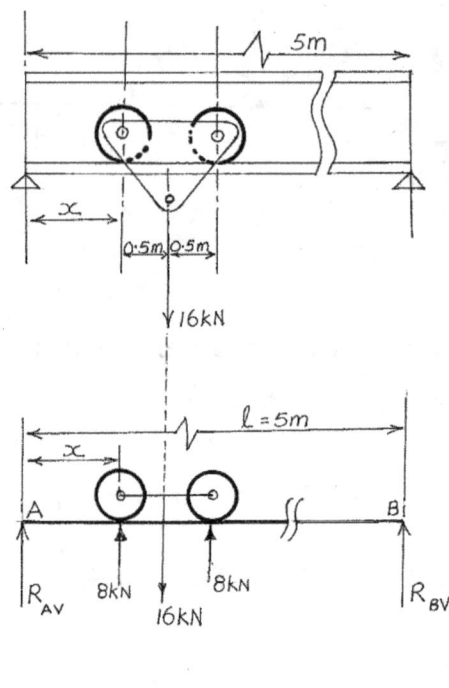

FIG 1

The load on the beam caused by the loads on the wheels is equivalent to 16kN. This resultant force is 0.5m from each wheel.

Let x be the distance between A and the wheel closer to this point where the maximum bending moment occurs.

Considering moments about
$$R_{AV}(5) = 16(5 - x - 0.5)$$
$$= 16(4.5 - x)$$
$$\therefore \quad 5R_{AV} = 72 - 16x$$
$$or \quad R_{AV} = 14.4 - 3.2x$$

so that bending moment M_x, x metres from A is
$$+R_{AV}(x)$$
positive, because in our convention for bending moment, clockwise moments on the left of section i.e. a sagging moment, are positive.

Symbolically

$$M_x = +R_{AV}(x)$$
$$= (14.4 - 3.2x)x$$
$$M_x = 14.4x - 3.2x^2 \qquad \ldots\ldots\ldots\ldots \text{(ii)}$$

For maximum B.M $\dfrac{dMx}{dx} = 0$

$\therefore \qquad \dfrac{dM}{dx} = 14.4 - 6.4x$

for $\qquad \dfrac{dM}{dx} = 0$

$\qquad 6.4x = 14.4$

i.e. $\qquad x = 2.25m$

Accordingly, when we substitute this value of x in (ii)

$$B.M_{max} = 14.4(2.25) - 3.2(2.25^2)$$
$$= 31.5 - 16.2$$
$\therefore \qquad \underline{BM_{max} = 15.3kNm}$

Q9. A spreader bar AB 1 metre long is used as part of a device for lifting heavy loads. The length of chain AC is 600 cm and of chain CB 800m. The hoisting cable is attached to the middle of the bar at D. Neglecting the mass of the bar and chains, draw the shear force, thrust, and bending moment diagrams for the bar when the downward force due to a load being lifted is 100kN. Fig. 1 shows the load resting on the floor.

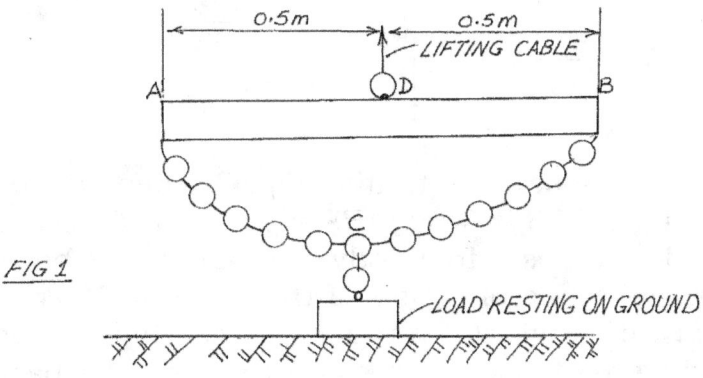

FIG 1

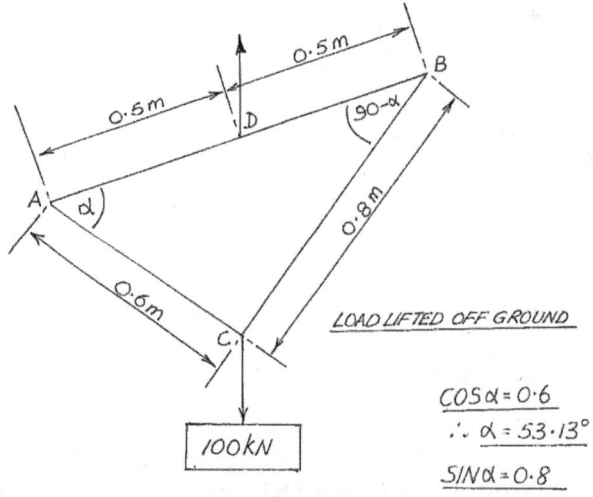

LOAD LIFTED OFF GROUND

$Cos\alpha = 0.6$
$\therefore \alpha = 53.13°$
$Sin\alpha = 0.8$

FIG. 2

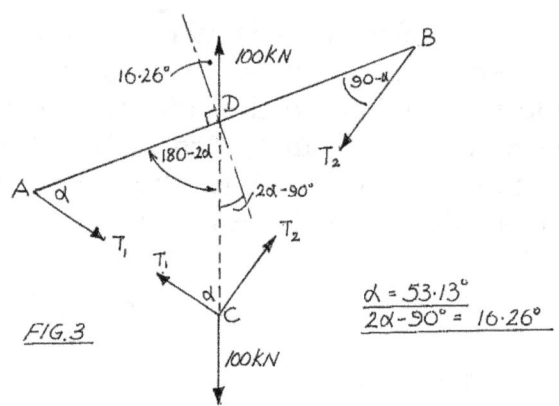

$\alpha = 53.13°$
$2\alpha - 90° = 16.26°$

FIG.3

The first thing to observe is that in Fig. 2 when the load is lifted off the ground, the angle ACB is $90°$: $(AB)^2 = (AC)^2 + (BC)^2$, i.e. $1^2 = (0.6)^2 + (0.8)^2$. Also, in Fig. 3 which is a free-body diagram of the bar and the chains, it is evident that the line of action of the force of *100kN* downwards being the equilibrant of tensions T_1 and T_2 in the chains must be the same as that of the upwards force of *100kN* on the spreader bar. Again this latter force is the equilibrant of T_1 and T_2 on the bar.

Referring to Fig. 2 and employing the cosine formula:
$$CD^2 = (0.6)^2 + (0.5)^2 - 2.(0.6)(0.5)Cos\alpha$$
$$= 0.36 + 0.25 - 2(0.6)(0.5)(0.6)$$
$$= 0.36 + 0.25 - 0.36$$
$$= 0.25$$
$$\therefore \quad CD = 0.5m$$
Accordingly, *angle ACD* $= \alpha$

Referring to Fig. 3

$$T_1 Cos\alpha + T_2 Sin\alpha = 100$$

i.e. $$0.6T_1 + 0.8T_2 = 100$$ . . . . (i)

Also

$$T_1 Sin\alpha = T_2 Cos\alpha$$

i.e. $$0.8T_1 = 0.6T2$$ . . . . (ii)

∴ $$T_1 = 0.75T_2$$

Substituting this result in (i)

$$0.6(0.75)T_2 + 0.8T_2 = 100$$

or $$1.25T_2 = 100$$

∴ $T_2 = 80kN$

so that

$$T_1 = 60kN$$

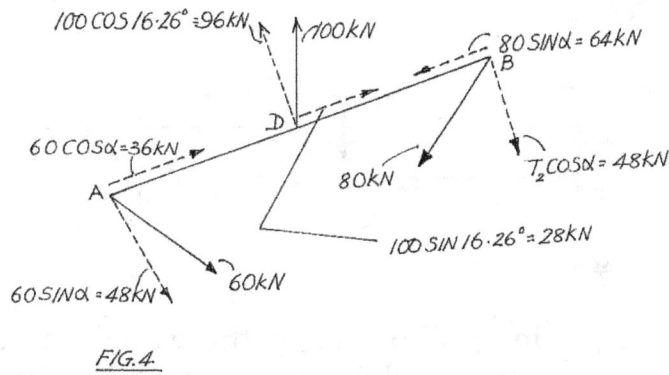

FIG.4.

In Fig. 4, the forces perpendicular and parallel to the bar are indicated. Evidently the thrust in the spreader bar is zero.

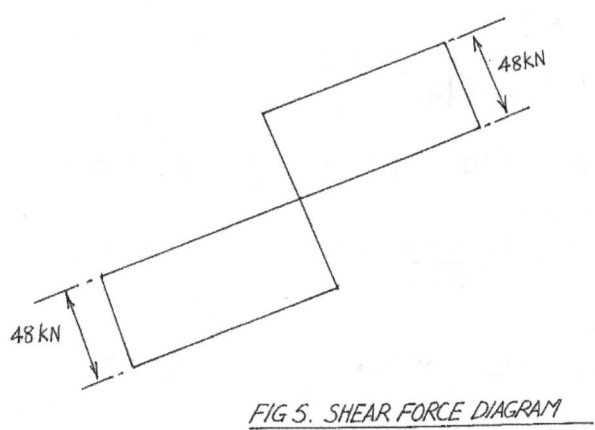

FIG 5. SHEAR FORCE DIAGRAM

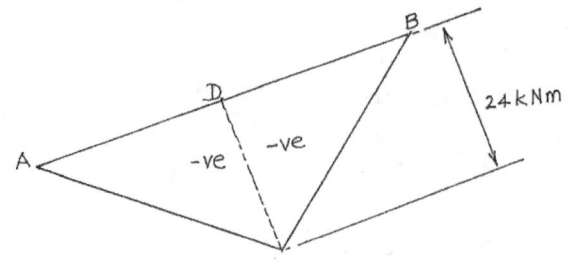

FIG 6. BENDING MOMENT DIAGRAM

Q10. Draw to scale the shearing force and bending moment diagrams stating their maximum values for a teak log floating horizontally in brackish water. The log is 10m long and has a square cross-section 20cm x 20cm. A mass 'M' is at the centre of the log such as to cause the depth of the log below water to be 18cm. Assume the log is impervious to water and take density of teak as 675kg/m³; density of brackish water = 1000 kg/m³, and g = 10 m/s².

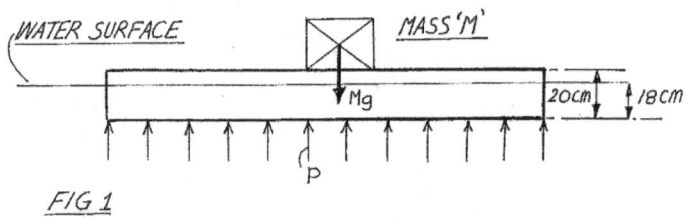

FIG 1

In Fig. 1, the log is in equilibrium in the water; the downward force W due to mass 'M' on the log plus downward force due to mass of log = upward force due to pressure P acting on underside of log.

Now, from the data provided mass of log

$$= \ 10 \times \frac{20}{100} \times \frac{20}{100} \times 675$$

$$= \ 270\text{kg}$$

Therefore, volume of brackish water displaced by the log of mass 270kg and mass M together

$$= \ \left(\frac{270 + M}{1000}\right) m^3$$

Horizontal cross-section of log

$$= \ 10 \times \frac{20}{100}$$

$$= \ 2\text{m}^2$$

$\therefore$　　Depth of immersion $= \left(\dfrac{270+M}{1000 \times 2} \right) m$

so that　　$\dfrac{270+M}{2000} = \dfrac{18}{100}$

or　　　　$270 + M = 360$

from which　　$M = 90$kg

Checking pressure P

From hydrostatics

$$P = 1000 \times \dfrac{18}{100}$$
$$= 180 \text{kg/m}^2$$

Taken over the bottom of the log of area $20 \times \dfrac{10}{100} m^3 = 2m^2$

$\therefore$　　Upward force due to P $= 180 \times 2$kg
$$= 360 \text{kg}$$

Summarising therefore, taking g = 10m/s²,

Weight of log per unit metre　　$= 270$N, i.e. $270 \times \dfrac{10}{10} N$

Similarly, upward force on log due to P $= 360$N/metre. Refer to Fig. 2

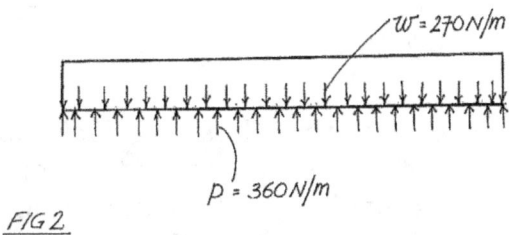

FIG 2.

The distribution of the nett upward force of 90N/m is shown in Fig. 3.

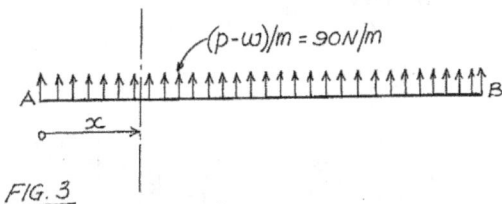

FIG. 3

Shearing Force at 'XX', which is x metres from A = (p-w)x = 90x Newtons. Consider a section 'AM' x metres from A. The nett force of 90x will cause the section to move upwards.

The convention we are using throughout in these examples of shearing-force and bending moment diagrams is as stated previously. They are here repeated for ease of reference.

Shearing Force	Bending Moment
Left upward : +ve	Left clockwise : +ve
Right downward : +ve	Right anticlockwise : +ve
Left downward : -ve	Left anticlockwise : -ve
Right upward : -ve	Right clockwise : -ve

The shearing force to the left of 'XX' is upward in direction and therefore positive; it is proportional to 'x' and the shearing force envelope from 'A' to the middle of the log is a straight line AC. With reference to Fig. 3, if we consider a section 'x' from B, then the shearing force will cause the section to move upward. 'Right upward' is negative and therefore the envelope is also a straight line as indicated.

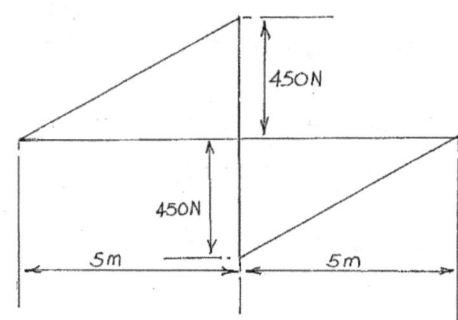

450N

450N

5m — 5m

FIG.4: SHEAR FORCE DIAGRAM   SCALE: 1cm = 90N

At the centre of the beam, the shearing force is W=900N, equal to the force due to the mass of 90kg located at the centre. Of course we could have continued the analysis of shearing-force distribution by merely extending our working from the left. For example, nett upward force due to (p-w) at C = 450N. But at 'C' there is a force of 900N due to mass 'M'. This results in a downward shearing force $OC' = 450N$ as shown in Fig. 4, i.e. +450N – 900N = -450N.

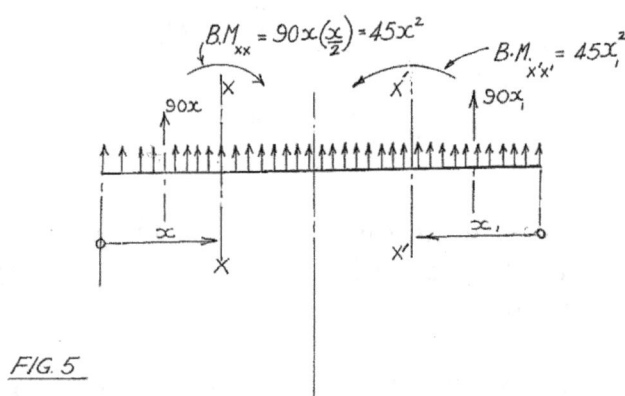

$$B.M_{xx} = 90x \left(\frac{x}{2} \right) = 45x^2$$

$$B.M_{x'x'} = 45x_1^2$$

90x   90x₁

FIG. 5

Referring to Fig. 5 and considering the left-hand section, the bending moment at XX is $90x\left(\dfrac{x}{2}\right) = 45x^2$. The bending moment envelope is therefore parabolic. Similarly, if we consider the right-hand portion of the log the bending moment at XX is $45x_1^2$, giving a parabolic variation.

The bending moment of the left-hand section is clockwise and that of the right-hand section anti-clockwise. In our convention, these are positive moments. The log therefore sags concave upwards.

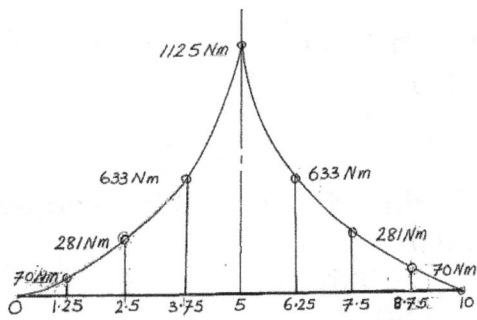

FIG 6. BENDING MOMENT DIAGRAM   SCALE: 1cm APPROX 180Nm

The complete Bending Moment diagram is shown in Fig. 6

Q11. The bending moment diagram for a beam simply supported at its ends is shown in Fig. 1. Draw the shear force and load diagrams.

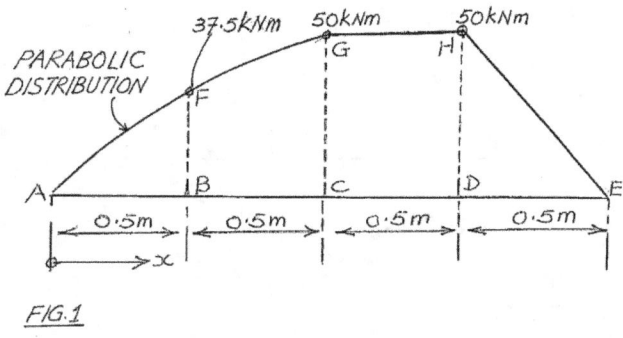

FIG.1

For the parabolic bending-moment distribution along section of beam AC, assume

$$M_{AC} = ax^2 + bx \qquad \ldots\ldots\ldots\ldots \text{(i)}$$

Considering this section

M = 50kNm at $x = 1m$, and

M = 37.5kNm at $x = 0.5m$

Substituting these values in (i), we have

50   = a + b   . . . .   (ii)

37.5 = 0.25a + 0.5b   . . . .   (iii)

12.5 = 0.25a + 0.25b

37.5 = 0.25a + 0.5b

∴   25 = 0.25b

∴   b = 100

from which a = -50

This means that B.M distribution along section AC is given by :

M_{AC} = $-50x^2 + 100x$ . . . .   (iv)

Now $\dfrac{dM}{dx}$ = Shear Force

so that shear force distribution over AC is given by

SF_{AC} = $-100x + 100$ . . . .   (v)

At $x = 0$;   SF = 100kN

At $x = 0.5m$;   SF = 50kN

" $x = 1m$;   SF = 0

Next we consider section of beam CD where

M_{CD} = Constant = 50kNm

Clearly $\dfrac{dM_{CD}}{dx} = 0$ . . . .   (vi)

    i.e. SF_{CD} = 0

∴ Shear force on beam from $x = 1m$ to $x = 1.5m$ is zero.

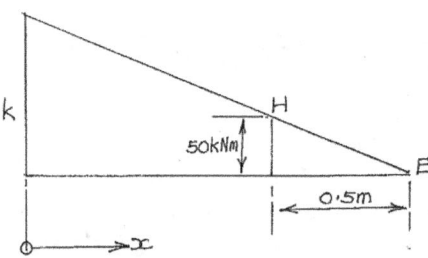

FIG. 2

To obtain equation HE, reference is made to Fig. 2. We know that the equation of HE is of the form:

M_{HE} = $-\dfrac{50}{0.5}x + k$ . . . .   (vii)

~ 56 ~

Constant k is obtained from
$$\frac{50}{k} = \frac{0.5}{2}$$
i.e.   k = 200
∴ M_{HE} = -100x + 200   . . . . (viii)
This is valid from $x = 1.5m$ to $x = 2m$.
Differentiating (viii) w.r.t. x, we have
$$\frac{d}{dx}(M_{HE}) = -100$$
i.e. SF_{HE} = -100   . . . . (ix)
∴   SF = -100kN from $x = 1.5m$ to $x = 2m$

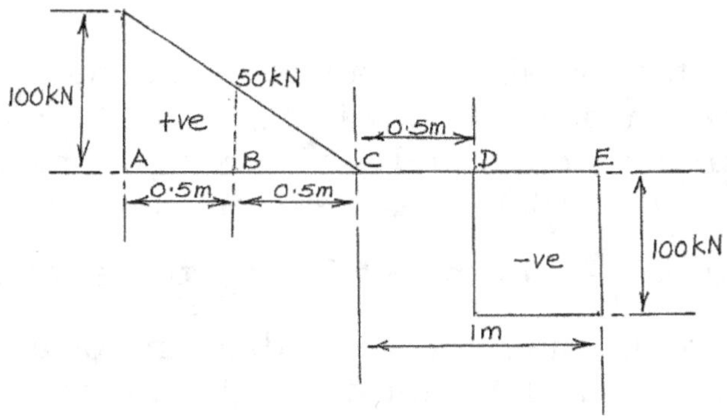

FIG.3 : SHEAR FORCE DIAGRAM

The S.F. diagram is plotted at Fig. 3.
Now for the load diagram.
Collecting our results for S.F. distribution over the sections AC, CD and DE :

SF_{AC} = $-100x + 100$ . . . . (viii)
SF_{CD} = 0
and SF_{HE} = -100   . . . . (ix)
We know
$$\frac{d}{sx}(SF) = -w$$ . . . . (x)
where   w = nett downward load at x

From an inspection of Fig. 3 we may infer that there is a uniformly distributed load, say wkN/m over the section AC

Now $\frac{d}{dx}(SF_{AC})$ = -100

∴   -w = -100
∴   w = 100kN/m

~ 57 ~

Evidently the reaction at A is upwards and = 100kN. It is also evident there is no load on CD, that the reaction at E is 100kN; and because SF_{HE} = -100kN over whole of section HE, there is a point load of 100kN at D. The load diagram is shown at Fig. 4.

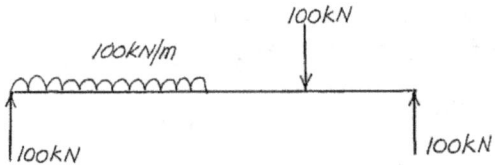

FIG.4 : _LOAD DIAGRAM_

Q12. The depth of water contained in a water catchment area held by timber planks and wooden piles, shown in Fig. 1 is 10 metres. The piles are driven deep into the ground and are braced by struts 10 metres above it. The piles are placed 10 metres between centres. Assuming that the thrust in a strut is $\frac{3}{10}th$ of the total water thrust against each pile, draw the shear force and bending moment diagram for a pile. Determine also the bending moment at the point where the pile emerges from the ground and the position along the pile where the bending moment is zero. Take the density of water = 1000kg/m³, and g = 10m/s².

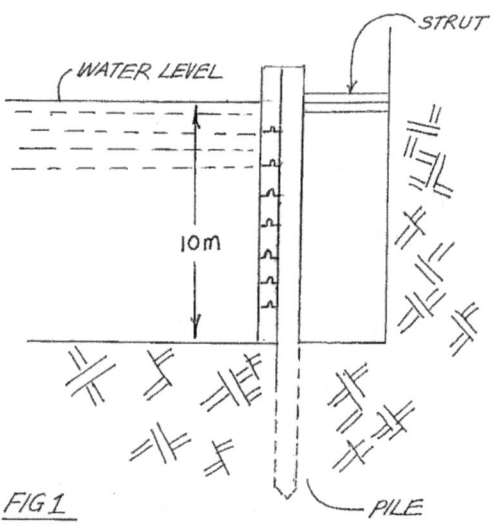

FIG 1

It is assumed that each pile carries the force distributed over 10 metres of dam width, as depicted in Fig. 2. Thus, the pile marked 'B' supports the total load due to water pressure against the area marked 'abcd'.

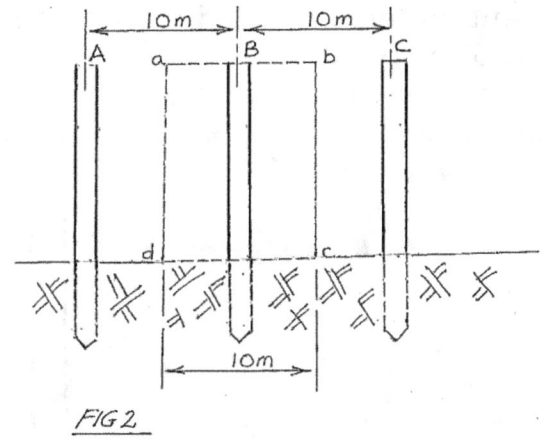

FIG 2

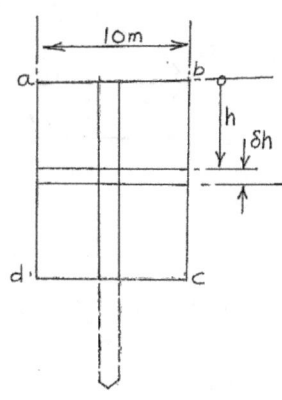

FIG 3

Considering the infinitesimal strip 10m wide and 'δh' long in Fig. 3, the infinitesimal force 'δF' on it is

$\qquad \delta F$ = pressure × area $\qquad\qquad$ . . . . (i)

i.e. $\quad \delta F = \rho \, gh \times 10\delta h \qquad\qquad$ . . . . (ii)

$\therefore \qquad$ Total force 'F' on area $\quad = 10\rho g \int_0^{10} h \, dh$

$$= 10\rho g \left[\frac{h^2}{2}\right]_0^{10}$$

$$= 5\rho g[100 - 0]$$

$\therefore \qquad F = 500\rho g$ Newton

Note that from equation (i)

Force $= \dfrac{kg}{m^3} \cdot \dfrac{m}{s^2} \cdot m \times m.m$

$\therefore \qquad$ Force $= kg \cdot \dfrac{m}{s^2}$

which are the dimensional units of Newton.

Substituting, ρ = 1000kg/m³ and

$\qquad$ g = 10 m/s² approx.

We see that the resultant force resisted by each pile is

$$F = 500 \times 1000 \times 10$$
$$= 5000000N$$
or $\quad F = 5000kN$ or $5MN$

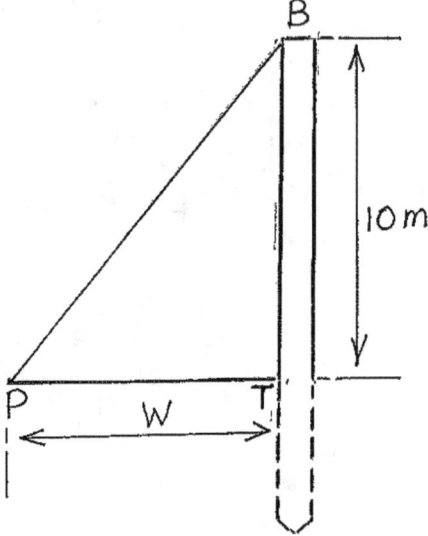

FIG 4

This force of 5000kN on pile B may be considered to vary from 0 at B to W at T, where

$$W \times \frac{10}{2} = 5000kN$$
i.e. $\quad W = 1000kN$

We may now draw a free-body diagram of a loaded pile. This is shown in Fig. 5.

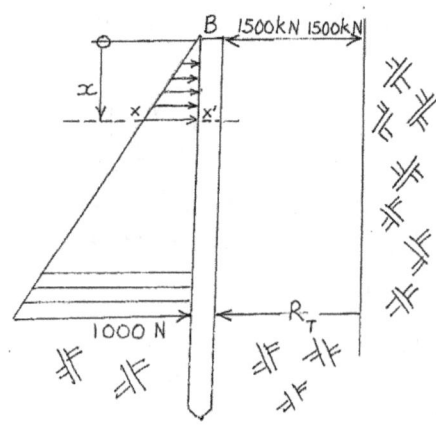

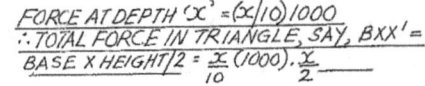

FORCE AT DEPTH 'x' $= (x/10)1000$
∴TOTAL FORCE IN TRIANGLE, SAY, BXX' $=$
BASE X HEIGHT/2 $= \frac{x}{10}(1000) \cdot \frac{x}{2}$

FIG. 5

~ 60 ~

Total thrust against pile = 5000kN

$\therefore$ Thrust on strut $= \dfrac{3}{10}(5000)$

                               $= 1500kN$

$\therefore$ Reaction at R_T $= 3500kN$, i.e. (5000-1500)kN

Let us now work out values of shear force starting at B, selecting points x metres from B, where $x = 0$, 2m, 4m, 6m, 8m and 10m.

At $x = 0$; SF = 1500kN

If we turn Fig. 4, clockwise through 90°, this S.F is 'right upwards and therefore negative in our sign convention for S.F.

At $x = 2m$; SF $= 1500 - \dfrac{2}{10}(1000) \cdot \dfrac{2}{2} = 1500 - 200$

i.e. SF_2 = 1300kN (also negative)

At $x = 4m$;SF $= 1500 - \dfrac{4}{10}(1000) \cdot \dfrac{4}{2} = 700kN$, also negative

At $x = 6m$; SF = 1500kN $- \dfrac{6}{10}(1000) \cdot \dfrac{6}{2} = 300kN$, positive

At $x = 8m$, SF = 1500kN $- \dfrac{8}{10}(1000) \cdot \dfrac{8}{2} = 1700kN$, positive; and

At a point just marginally above T;

SF $= 1500 - \dfrac{10}{1}(1000) \cdot \dfrac{10}{2} = 3500kN$; positive.

These values of shear force are plotted to produce the S.F Diagram at Fig. 6.

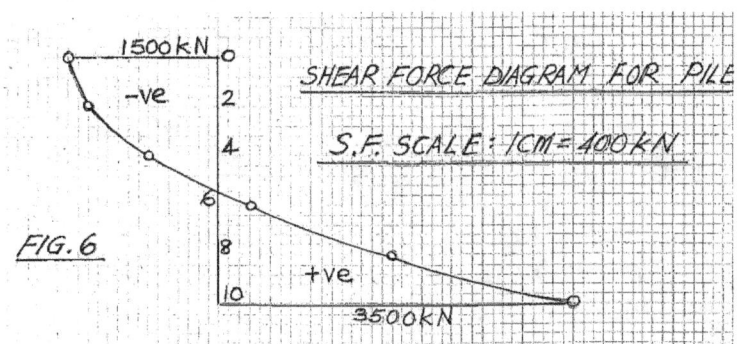

Let us now calculate bending moments for the same sections we choose for S.F. determinations.

At $x = 0$;   B.M. = 0

At $x = 2$; $B.M_2 = + 1500(2) - \dfrac{2}{10}(1000)\dfrac{2}{2} \cdot \dfrac{2}{3}$

~ 61 ~

$$= +3000 - 133.3$$
$$= +2866.7 \text{ kNm}$$

At $\quad x = 4$; $\text{B.M}_4 = +1500(4) - \dfrac{4}{10}(1000) \cdot \dfrac{4}{2} \cdot \dfrac{4}{3}$

$$= +6000 - 533.3$$
$$= 5466.7 \text{kNm}$$

At $\quad x = 6$; $\text{B.M}_6 = +1500(6) - \dfrac{6}{10}(1000) \cdot \dfrac{6}{2} \cdot \dfrac{6}{3}$

$$= +9000 - 3600$$
$$= +5400 \text{kNm}$$

At $\quad x = 8$; $\text{B.M}_8 = +1500(8) - \dfrac{8}{10}(1000) \cdot \dfrac{8}{2} \cdot \dfrac{8}{3}$

$$= +12000 - 8533.3$$
$$= +3466.7 \text{kNm}$$

At $\quad x = 0$; $\text{B.M}_{10} = +1500(10) - \dfrac{10}{10}(1000) \cdot \dfrac{10}{2} \cdot \dfrac{10}{3}$

$$= +15000 - 16333.3$$
$$= -1333.3 \text{kNm}$$

Where is S.F. = 0?

Let this section be distant 'd' metres from the top of the pile. Then

$$1500 - \dfrac{d}{10}(1000) \cdot \dfrac{d}{2} = 0$$

i.e. $\quad 50d^2 = 1500$

or $\qquad d = \sqrt{30}$

$\therefore \qquad d = \approx 5.5m$

Where is B.M. = 0?

Let this section be distant 'z' metres from the top of the pile. Then

At $x = z$; $BM_z = +1500(z) - \dfrac{z}{10}(1000) \cdot \dfrac{z}{z} \cdot \dfrac{z}{3}$

For $BM_z = 0$

$$1500z = \dfrac{50z}{3}z^2 , \text{ i.e. } 1500z = \dfrac{50z^3}{3}$$

i.e. $z^2 = 90$

or $z \approx 9.4m$

The computed values of B.M. are plotted at Fig. 7. (q.v.).

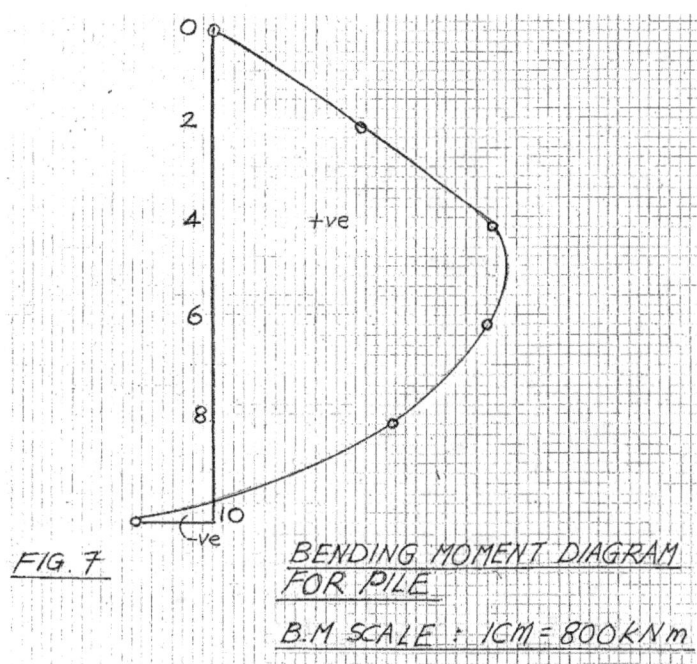

FIG. 7   BENDING MOMENT DIAGRAM
FOR PILE

B.M SCALE : 1CM = 800KN m

Q13. A ship 100 metres in length has a total load distribution characteristic shown by the solid lines in Fig. 1. The upward water-thrust distribution is given by the dotted lines in the same diagram. Plot the nett force distribution diagram to a scale of 1cm = 10m and 1cm = 50kN and shear force diagram to a scale 1cm = 4m and 1cm = 160kN. Calculate the positions of zero shear force.

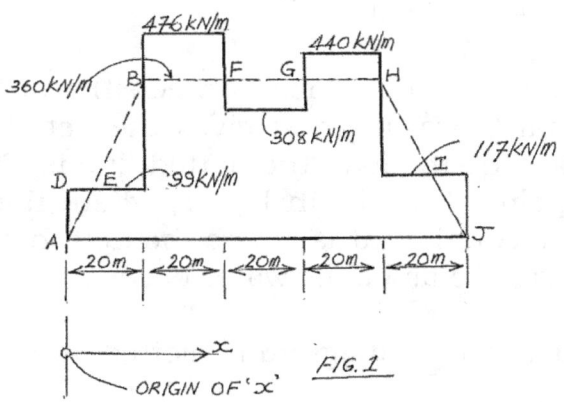

FIG. 1

Water thrust 'p' is upwards and of opposite sign to load 'w' which is downwards.

In Fig. 1, p goes from 0 to 360kN/m in 20m. Considering the origin of 'x' and 'p' at A, the equation of the thrust line AB is therefore given by:

~ 63 ~

$$p = \frac{360}{20}x$$

i.e. $p = 18x$ . . . .   (i)

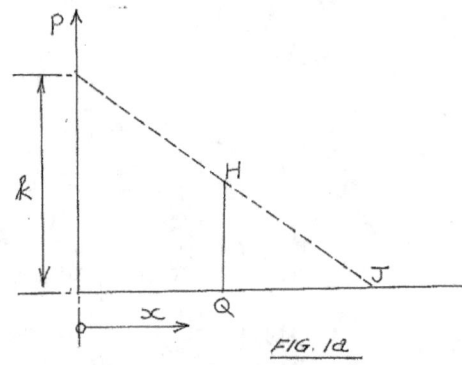

FIG. 1a

Similarly, equation of line HJ in Fig. 1a is given by

$$p = -18x + k$$

where 'k' is the intercept in the positive direction of the 'p' axis. From similar triangles

$$\frac{HQ}{k} = \frac{QJ}{100}$$

i..e $\dfrac{360}{k} = \dfrac{20}{100}$

∴ $k = 1800$

so that equation of HJ is

$$p = -18x + 1800$$ . . . .   (ii)

Referring to Fig. 1, the difference between the ordinates of total load of the ship and total upward thrust gives the nett force intensity. Where the buoyancy exceeds the load the dotted line is above the solid line and vice versa. At points E and I in Fig. 1, where the upward water thrust and total load are equal, that is where the nett force intensity is zero; the sections are referred to as being 'water borne'.

The nett force distribution diagram is plotted at Fig. 2.

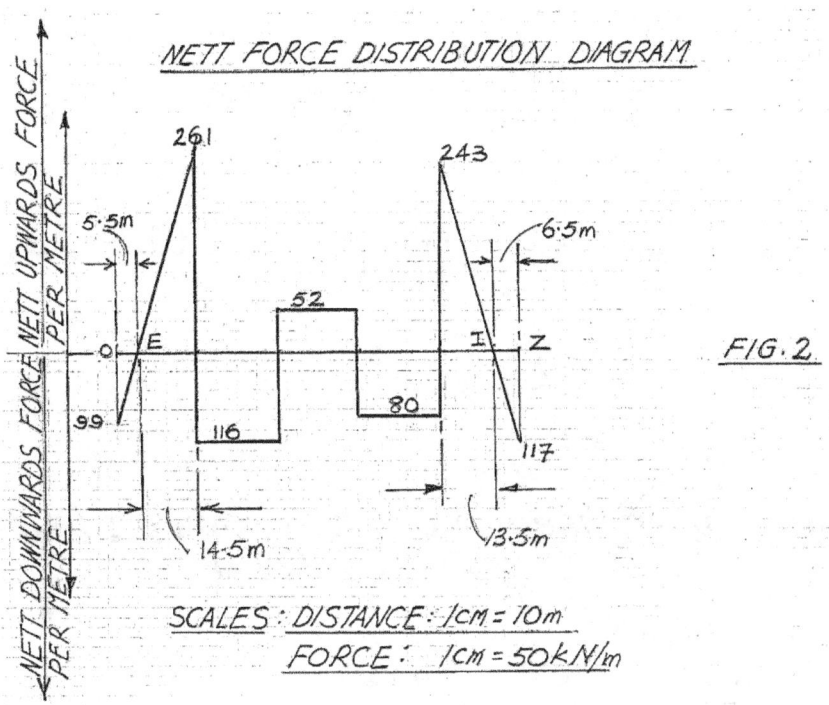

NETT FORCE DISTRIBUTION DIAGRAM

FIG. 2

SCALES : DISTANCE : 1cm = 10m
FORCE : 1cm = 50kN/m

At E,

$$18x = 99$$
$$\therefore OE = 5.5m$$

Similarly, at I

$$-18x + 1800 = 117$$
$$\text{or} \quad 18x = 1683$$
$$\therefore x = 93.5m$$

measured from 0. See Fig. 2

The student must note that because the total weight of the ship must equal the total upthrust, the forces corresponding to the areas above OZ in Fig. 2 must also equal the forces corresponding to the areas below OZ. Let us check this out. First we repeat Fig. 2, but this time not to scale. We make the diagram large enough to be able to write in the relevant forces.

With reference to Fig. 2 therefore,

Fig. 3 : (not drawn to scale)

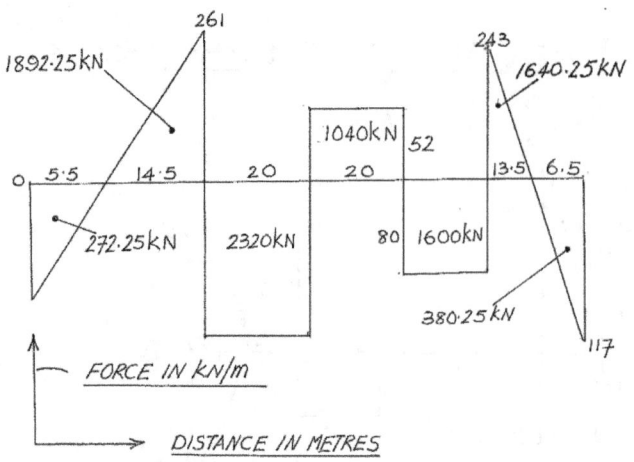

FIG. 3. _DOWNWARD LOAD AND UPWARDS THRUST_
(NOT DRAWN TO SCALE)

Total upward thrust
$$= \quad 1892.25 + 1040 + 1640.25$$
$$= \quad 4572.5\text{kN}$$
Total downward load
$$= \quad 272.25 + 2320 + 1600 + 380.25$$
$$= \quad 4572.5\text{kN}, \quad \text{Check}$$

The forces corresponding to the areas in Fig. 3 act through their centroids and the student should note that it is also necessary that the moments of these forces about any point in the axis of the ship must balance. We leave the demonstration of this for the problem immediately following.

Referring again to Fig. 1, we observe that between D and E, total downward load 'w' is greater than upward thrust 'p'

$$\therefore \qquad \text{Nett downward force} \quad = \quad w - \text{p}$$
$$= \quad 99x - 18x$$

Now, we know that the first differential of the shear force (SF) with respect to x = -(nett downward force)

i.e. $\quad \dfrac{d(SF)}{dx} = -(99 - 18x)$ $\qquad$ . . . . (iii)

$\therefore \qquad SF \;=\; 9x^2 - 99x + C_1$

But shear force = 0 at x = 0

$\qquad \therefore \quad C_1 \;=\; 0$

So that S.F distribution between D and E is given by

$$SF \;=\; 9x^2 - 99x \qquad \qquad \text{. . . . . (iv)}$$

If SF = 0

$$9x^2 - 99 = 0$$

from which, $x = 0$; $x = 11$

so, SF is also zero at x = 11m

Between E and B, nett downward force = $-(18x - 99)$

$$\therefore \quad -(\text{nett downward force})$$
$$= -\{-(18x - 99)\}$$

$$\therefore \quad \frac{d(SF)}{sx} = 18x - 99$$

for section EB, so that

$$SF = 9x^2 - 99x + C_2$$

Now at $x = 5.5m$, we may find out what the value of SF is by putting $x = 5.5m$ in equation (iv)

$$\text{i.e.} \quad SF = 9(5.5^2) - 99(5.5)$$
$$= -272.25\text{kNm}$$

Therefore to find C_2, we may say S.F = -272.25kN at $x = 5.5$

i.e.   -272.25 = 272.25 – 544.5 + C_2

$$\therefore \quad C_2 = 0$$

So, for the first 20-metre section of the ship, S.F distribution is given by

$$SF = 9x^2 - 99x \qquad \cdots\cdots\cdots \quad \text{(iv)}$$

We now draw up a table of SF values within the range $x = 0$ to $x = 20m$.

x(in metres) :	1	2	3	4	5	5.5
SF(in kN) :	-90	-162	-216	-252	-270	-272.25

x(in metres) :	6	10	11	12	15	18
SF(in kN) :	-270	-90	0	108	540	1134

x(in metres) :	20
SF(in kN) :	1620

Next, we consider portion of the ship BF.

There, nett downward force = 476 – 360

$$\therefore \quad \frac{d}{dx}(SF) = -(476 - 360)$$

$$= -116$$

$$\therefore \quad SF = -116x + C_3$$

But at x = 20, SF = +1620
$$\therefore \qquad 1620 = -116(20) + C_3$$
or $\qquad 1620 = 2320 + C_3$
$$\therefore \qquad C_3 = 3940$$
Accordingly
$$SF_{BF} = -116x + 3940 \qquad \qquad \ldots\ldots\ldots \text{(v)}$$

Again, we calculate values of SF at 5m intervals. The table of values is as follows:

x (in metres) :	25	30	35	40
SF(in kN) :	1040	460	-120	-700

Also, when SF = 0
$$116x = 3940$$
$$x = 33.96 \text{ (say) } 34\text{m}$$
For portion of the ship FG
$$\text{Net upward force} = 360 - 308$$
$$\therefore \quad \text{Nett downward force} = -(52)$$
so that −(net downward force)
$$= -(-52)$$
$$= 52$$
i.e. $\quad \dfrac{d}{dx}(SF) = 52$
or $\quad SF = 52x + C_4$

But, at $x = 40$, SF = -700
$$\therefore \quad 700 = 52(40) + C_4$$
or $\quad C_4 = -2780$
$\therefore \quad$ Between F and G
$$SF = 52x - 2780 \qquad \qquad \ldots\ldots\ldots \text{(vi)}$$

We draw up another table, this time using equation (vi).

x (in metres) :	45	50	55	60
SF(in kN) :	-440	-180	+80	+340

We inspect equation (vi) to determine where SF = 0; from the above table we observe that the position is somewhere between 50m and 55m.

Where
$$52x = 2780$$
$$x = 53.46\text{m}, \text{ say } 53.5\text{m}$$

Next, we examine potion of the ship GH. There, nett downward force = 440 – 360

$$\therefore \quad \frac{d}{dx}(SF) = -80$$

From which $SF = -80x + C_5$

But $\qquad SF = +340\text{kN}$ at $x = 60m$

$$\therefore \quad 340 = -4800 + C_5$$
$$\therefore \quad C_5 = 5140$$

Accordingly for this section
$$SF_{GH} = -80x + 5140 \qquad \ldots\ldots\ldots\ldots \text{(vii)}$$

$SF = 0$, where $x = 64.25\text{m}$

As before, we calculate values of SF at 5m intervals, employing (vii) for this purpose. We tabulate the results as follows :

x (in metres) :	65	70	75	80
SF (in kN) :	-60	-460	-860	-1260

From earlier working we determine the equation of HJ to be
$$p = 18x + 1800 \qquad \ldots\ldots\ldots\ldots \text{(ii)}$$

Between H and I, nett downward force
$$= -\{-(p-2)\}$$
$$= -18x + 1800 - 117$$
$$= -18x + 1683$$

Therefore, for this section of the ship
$$\frac{d}{dx}(SF) = 18x + 1683$$
$$\text{or} \quad SF = -9x^2 + 1683x + C_6$$

But $SF = -1260\text{kN}$ at $x = 80$
$$\therefore \quad -1260 = 9(80^2) + 1683(80) + C_6$$
$$\text{or} \quad -1260 = 57600 + 134640 + C_6$$

from which $\quad C_6 = -78300$
$$\therefore \quad SF = -9x^2 + 1683x - 78300 \qquad \ldots\ldots\ldots\ldots \text{(viii)}$$

The student should determine for himself or herself that the S.F distribution between I and J is given by the same equation as in (viii).

We are going to use (viii) to determine SF from $x = 80$ to $x = 100$ using 5-m intervals.

We tabulate the results as follows :

x (in metres) :	85	90	95	100
SF (in kN) :	-270	+270	360	0

For SF = 0 in this part of the ship

$$-9x^2 + 1683x - 78300 = 0$$

i.e.
$$x = \frac{1683 \pm \sqrt{2832489 - 2818800}}{18}$$

$$= \frac{1683 \pm \sqrt{13689}}{18}$$

$$= \frac{1683 \pm 117}{18}$$

$$x_{1,2} = \frac{1566}{18}, \frac{1800}{18}$$

$$= 87m, \quad 100m$$

i.e. 87m from A and as we would expect also at the very end of the ship. The tabulated values of SF are plotted to produce Fig. 4.

So you want to be a professional in international shipping; appreciate then that the distribution of cargo on a ship is highly important; after all, in a sense, a ship may be likened to a floating beam.

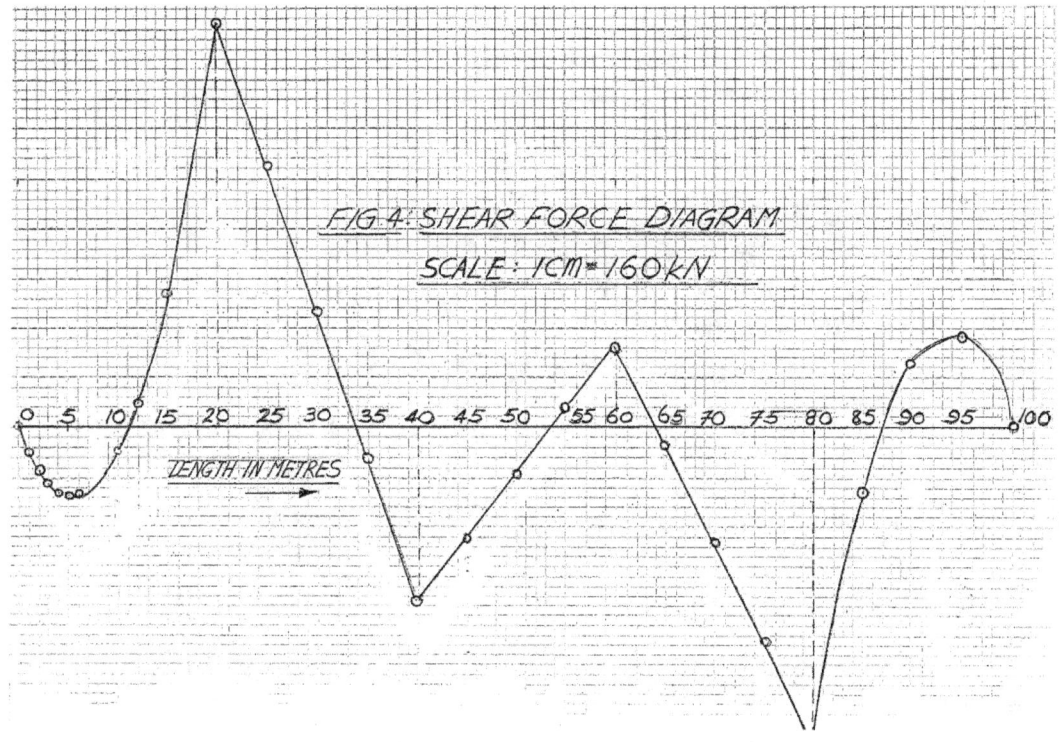

FIG 4: SHEAR FORCE DIAGRAM
SCALE: 1CM = 160 kN

LENGTH IN METRES

Q.14 A simply supported beam ABCD shown in Fig. 1 is 18 metres long and has a concentrated load of 5kN at end A. From a point C, 8 metres from A and along the beam to end D a total load of 25kN attributed to a varying linear loading from 'w'kN/m to 'z'kN/m, is carried. What must be the values of 'w' and 'z' in order that the reactions at B and D be equal to each other? Determine also the position of any point of contraflexure.

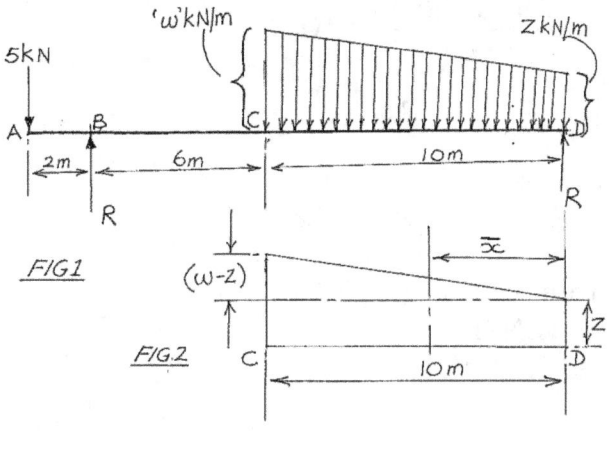

FIG1

FIG.2

~ 71 ~

The total load attributed to the varying load distribution is 25kN. Dividing the varying linear load into rectangular and triangular components, we have : $10z + (w-z)\dfrac{10}{2} = 25$

$$\text{i.e.} \quad 10z + 5w - 5z = 25$$
$$5z + 5w = 25$$
$$\text{or} \quad w+z = 5 \qquad \ldots\ldots\ldots\ldots \text{(i)}$$

For vertical equilibrium :
$$2R = 5 + 25$$
$$\text{i.e.} \quad R = 15kN$$

Taking moments about D
$$5(18) - R(16) + 25\,(\bar{x}) = 0 \qquad \ldots\ldots\ldots\ldots \text{(ii)}$$

where $\bar{x}$ distance of load centre (due to the varying load) from end D

i.e.
$$\left\{10(z) + \frac{(w-z)}{2}\cdot 10\right\}\bar{x} = 10z\cdot 5 + \left(\frac{w-z}{2}\right)\cdot 10\cdot\frac{2}{3}\cdot 10$$

$$\therefore \quad 5(z+w)\bar{x} = 50z + 100\frac{(w-z)}{3}$$

$$\text{or} \quad (z+w)\bar{x} = 10z + \frac{20}{3}(w-z)$$

$$3(z+w)\bar{x} = 30z + 20w - 20z$$

$$3(z+w)\bar{x} = 10z + 20w$$

$$\therefore \quad \bar{x} = \frac{10(z+2w)}{3(z+w)} \qquad \ldots\ldots\ldots\ldots \text{(iii)}$$

Substituting this result in (ii)
$$5(18) - R(16) + 25\left\{\frac{10(z+2w)}{3(z+w)}\right\} = 0$$

Putting R = 15
$$90 - 240 + \frac{250}{3}\frac{(z+2w)}{z+w} = 0$$

$$\text{or} \quad \frac{15}{3}\left(\frac{z+2w}{z+w}\right) = 15$$

$$\text{i.e.} \quad 5z + 10w = 9w + 9z$$
$$w = 4z \qquad \ldots\ldots\ldots\ldots \text{(iv)}$$

Putting w = 4z in equation (i) gives 5z = 5
$$\therefore \quad z = 1kN/m$$
$$\text{and} \quad w = 4kN/m$$

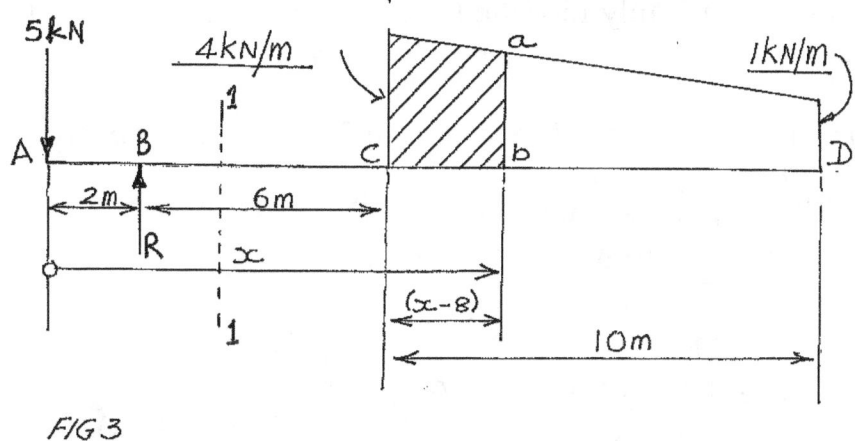

FIG 3

By inspection it is evident that no point of contraflexure exists between A and B. Let us inspect that part of the beam BC, say, at section 1-1, when 'x' is between 'B' and 'C'

Nx from A.

$$\begin{aligned}
BM_{1,1x} &= -5(x) + R(x-2) \\
&= -5x + 15(x-2)) \\
&= -5x + 15x - 30 \\
&= 10x - 30
\end{aligned}$$

When we put $10x - 30 = 0$, (i.e. $x = 3m$ we see that there is a point of contraflexure 1 metre to right of B.

Let us now extend x into the regime of the varying load, up to, say, section marked a-b. This is shown in Fig. 3.

$BM_{a,b,x} = -5(x) + R(x-2)$ – load due to shaded portion of diagram at Fig 3 times distance of load centre from section ab   . . . . (v)

Now, load due to shaded portion of diagram is found as follows :

$$\begin{aligned}
ab &= 4-(x-8)0.3 = 4-0.3x+2.4 \\
&= 6.4-0.3x
\end{aligned}$$

$\therefore$ Area of shaded portion $= \left(\dfrac{4+6.4-0.3x}{2}\right)(x-8)$

$$= (10.4-0.3x)\dfrac{(x-8)}{2}$$

To obtain the load centre for the shaded portion we employ the expression (iii) suitably modified, i.e. in this case $z = ab = (6.4 - 0.3x)$; w = 4

Therefore if $\bar{u}$ = distance of load centre from section ab, then

$$\bar{u} = \frac{10\{6.4 - 0.3x) + 2(4)\}}{3\{6.4 - 0.3x) + 4}$$

$$= \frac{10}{3} \frac{(14.4 - 0.3x)}{(10.4 - 0.3x)}$$

Substituting this result in equation v we get

$$BM_{a,b,x} = -5x + 15(x-2) - (10.4 - 0.3x)\left(\frac{x-8}{2}\right)\frac{10}{3}\frac{(14.4 - 0.3x)}{10.4 - 0.3x}$$

$$BM_{abx} = -5x + 15x - 30 - (10.4 - 0.3x)\left(\frac{x-8}{2}\right)\frac{10}{3}\frac{(14.4 - 0.3x}{10.4 - 0.3x)}$$

$$= -5x + 15x - 30 - \left(\frac{x-8}{2}\right)\frac{10}{3}(14.4 - 0.3x)$$

$$= 10x - 30 - \frac{5(x-8)(14.4 - 0.3x)}{3}$$

For point of contraflexture $BM_{ab,x} = 0$

$$30x - 90 = 5(x-8)(14.4 - 0.3x)$$

$$6x - 18 = 14.4x - -0.3x - 115.2 + 2.4x$$

i.e. $0.3x^2 - 10.8x + 97.2 = 0, \quad x = 18m$

or $x^2 - 36 + 324 = 0$

i.e. $(x-18)^2 = 0, \quad x = 18m$

that is to say 'D' is a point of contraflexure, which is evident anyway.

Q15. An iron frame ABCD made from a bar weighing 100N per metre is fixed horizontally as shown in Fig. 1. A concentrated load equivalent to 1000N is carried at point 'D', 1.2m from C. Draw the bending moment diagrams for the three parts of the frame giving maximum values.

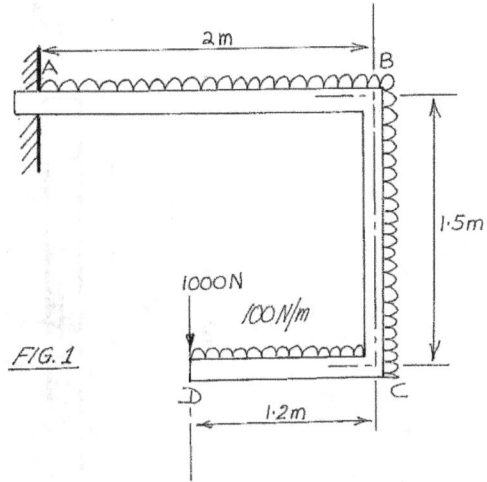

Commencement of the analysis should be done by considering the forces and moments required for maintaining the equilibrium of each section. Let us start with DC. Refer to Free-Body Diagram (FBD) at Fig. 2.

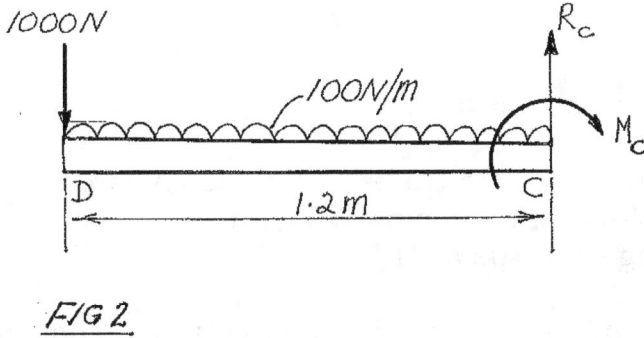

To establish equilibrium we apply M_c as shown, a (hogging moment in our B.M convention). We also apply an upward reaction R_c at C to balance the 1000N at D and the udl.

Taking moments about 'C'

$$1000(1.2) + 100(1.2)\frac{(1.2)}{2} - M_c = 0$$

$$\therefore \quad M_c = 1272 \text{ Nm, clockwise}$$

For vertical equilibrium of DC

$$1000 + 100(1.2) - R_c = 0$$

$$\therefore \quad R_c = 1120N$$

Note that since M_c and R_c are positive, it means that the direction or sense of application as shown in the diagram is correct i.e. M_c clockwise and R_c upwards. Let us consider next the section BC, for which purpose, reference is made to the FBD at Fig. 3.

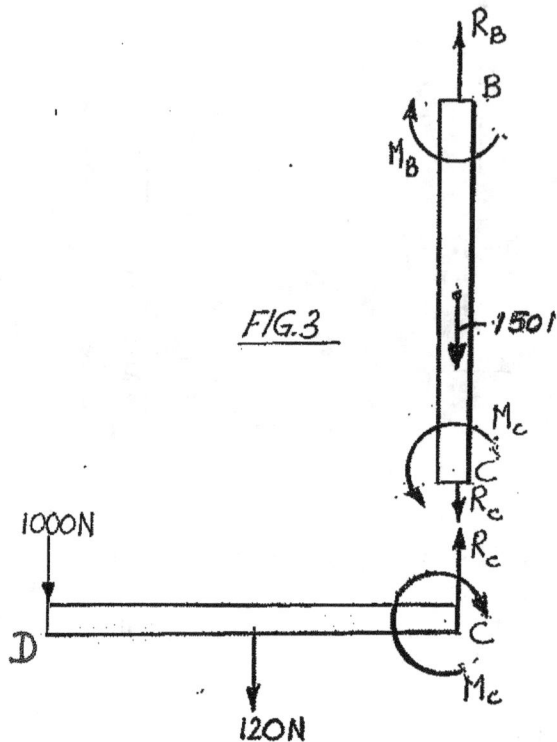

FIG.3

The forces acting on CB are :

(i)   its weight = 100 x 1.5 = 150N, downwards
(ii) R_c = 1120N acting downwards; and
(iii) R_B acting upwards at B

Bending moment 'M_c' on BC must act counter-clockwise as shown in the diagram. Why? Because when 'BC' is joined to 'CD' the M_c on BC and the M_c on DC must neutralize each other. Accordingly, M_B at 'B' must act in a clockwise sense. Therefore BC is in a state of pure bending. BC would therefore bend with its center of curvature to the right. In our convention this is positive bending. Now for vertical equilibrium:

$$R_B = 150 + R_c = (150 + 1120)N, \text{ i.e. } R_B = 1270N$$

Finally, we consider section AB. The relevant FBD is shown as Fig. 4.

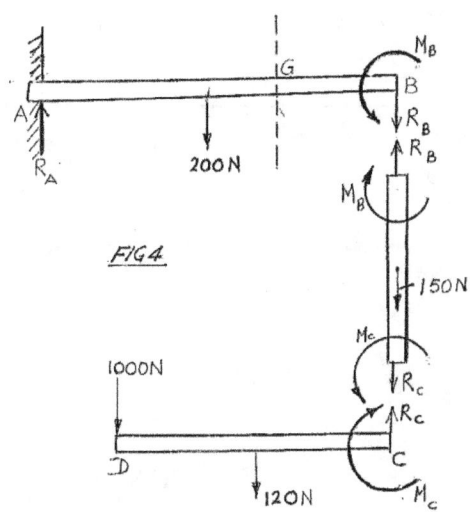

FIG 4

The forces on AB are:

(i)   vertical reaction at support A : R_A
(ii)   weight of section AB : 200N
(iii)   downward vertical reaction at B

For vertical equilibrium
$$R_A - 200 - R_B = 0$$
i.e. $$R_A - 200 - 1270 = 0$$
∴ $$R_A = 1470N$$

Bending Moment Diagrams

We shall show individual diagrams for each member.

For DC
(i)   Due to point load of 1000N at D

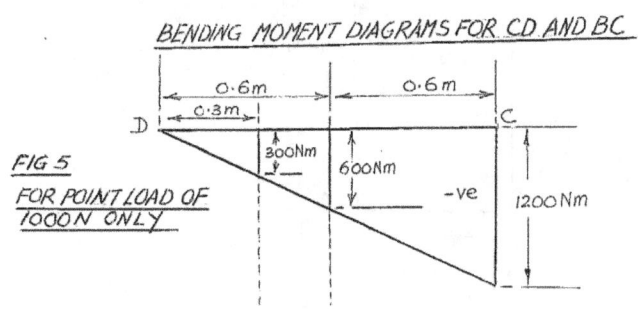

BENDING MOMENT DIAGRAMS FOR CD AND BC

FIG 5

FOR POINT LOAD OF
1000N ONLY

(ii)   Due to udl only

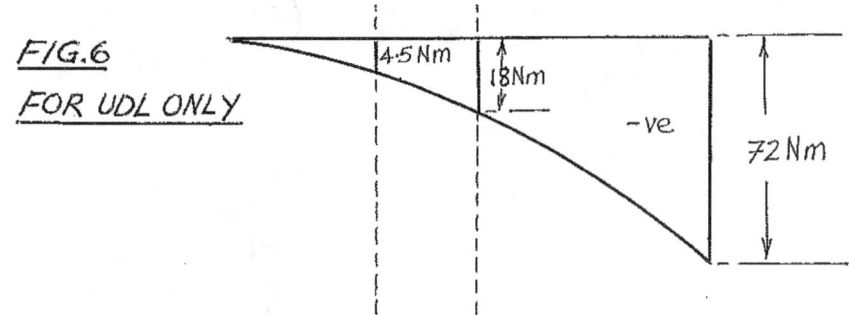

FIG.6
FOR UDL ONLY

4·5 Nm   18 Nm   −ve   72 Nm

(iii)   Due to point load of 1000N at D and udl on DC, (i) and (ii) combined)

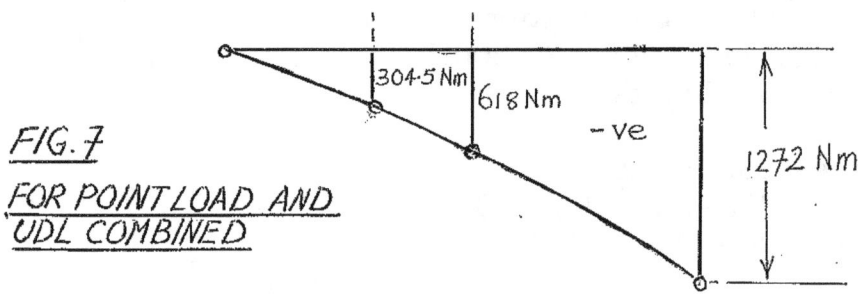

FIG.7
FOR POINT LOAD AND
UDL COMBINED

304·5 Nm   618 Nm   −ve   1272 Nm

(iv)   Due to sagging bending moment M_C and M_B on BC

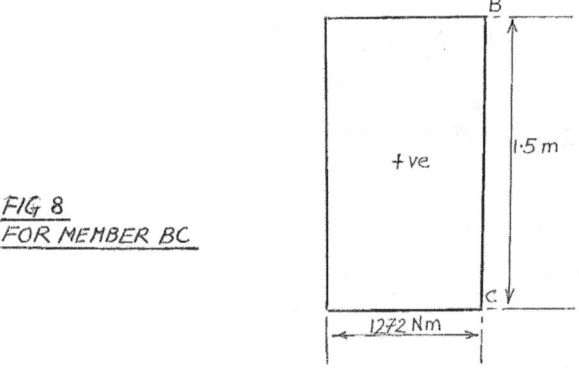

FIG 8
FOR MEMBER BC

B   +ve   1·5 m   C   1272 Nm

(v)   <u>Bending Moment Diagrams for AB</u>

Referring to Fig. 4, M_B is anti-clockwise on the right and therefore a sagging moment and therefore positive. The

~ 78 ~

moments due to 'R_B' and the udl about 'A' are clockwise and therefore negative. The bending moment diagrams for R_B and the udl are shown respectively as Figs. 9 and 10. Fig. 11 is the B.M diagram for R_B and the udl combined. In Fig. 12 the sagging moment M_B of magnitude +1272Nm is 'added' to the B.M diagram at Fig. 11.

For RB

BENDING MOMENT DIAGRAMS FOR AB

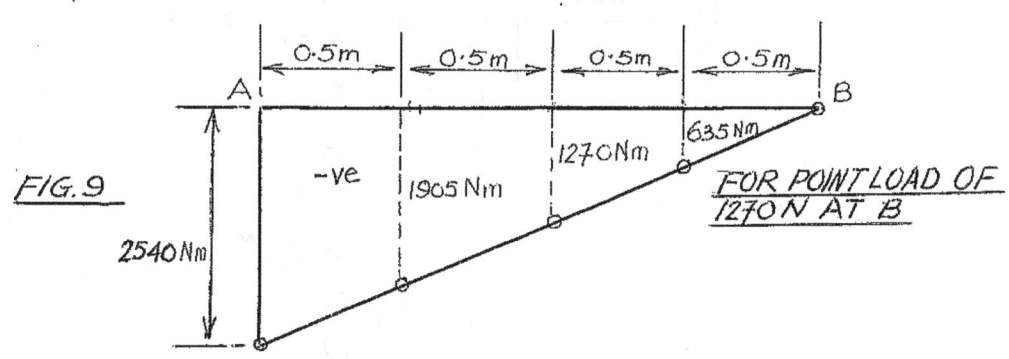

FIG.9 — FOR POINT LOAD OF 1270N AT B

For udl

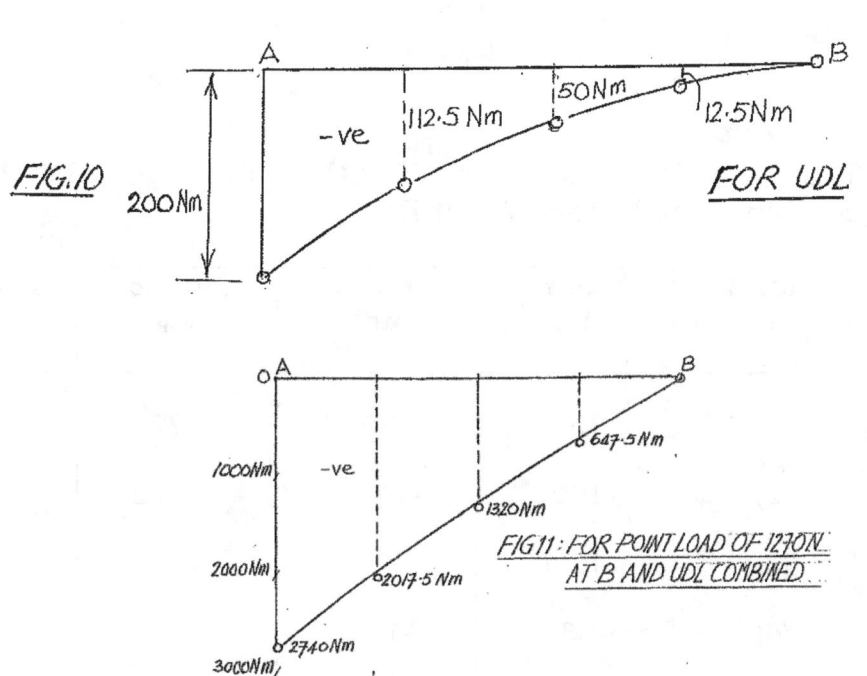

FIG.10 — FOR UDL

FIG.11: FOR POINT LOAD OF 1270N AT B AND UDL COMBINED

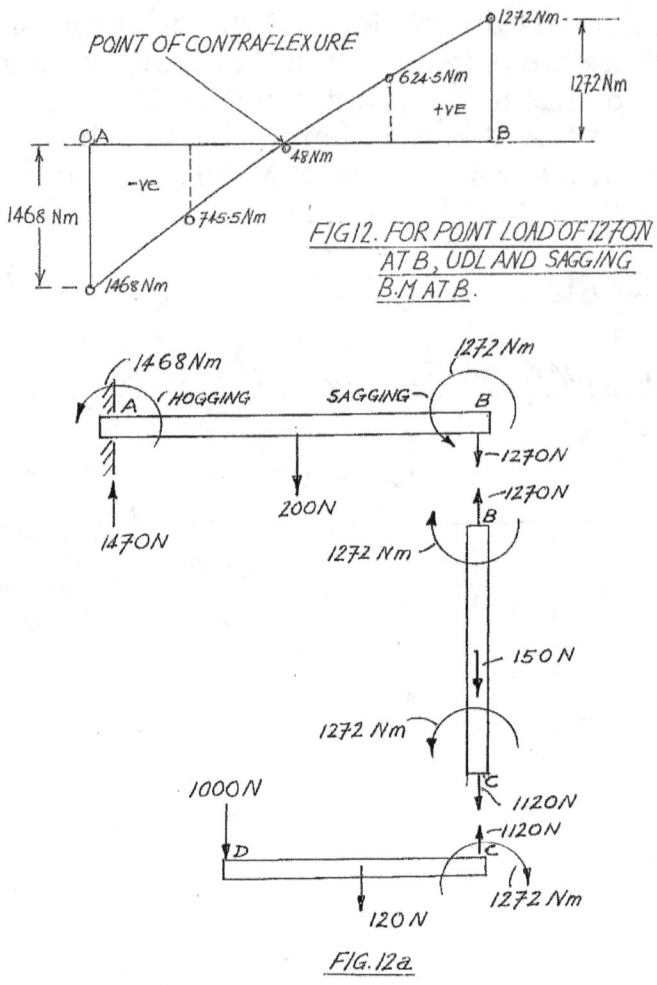

FIG 12. FOR POINT LOAD OF 1270N
AT B, UDL AND SAGGING
B.M AT B.

FIG. 12a.

The following table provides the values of BM for the combined loads and 'M_B' at B.

Station No.	Due to R_B	Due to udl	Due to M_B	Total BM in Nm
(1)	-2540	-200	+1272	-1468
(2)	-1905	-112.5	+1272	-745.5
(3)	-1270	-50	+1272	-48
(4)	-635	-12.5	+1272	+624.5
(5)	0	0	+1272	+1272

It is interesting to observe at Fig. 12 and from the above table of values that there is a point of contraflexure on AB at a point slightly beyond station (2) from A. This means that as we move along the beam from B to A, B.M. changes sign from positive, i.e. sagging at B to hogging at 'A'. At A the hogging moment is 1468Nm. All forces and bending moments are shown in Fig. 12a.

Q16. A crane is lifting a uniform concrete pile of length 7.5m and mass 707kg by means of cables as shown in Fig. 1, the inclination of the pile to the horizontal being 45°. What are the maximum shear force and bending moment on the pile? ; and, calculate where the latter occurs. Also, draw the shear force and bending moment diagrams. Take $g \approx 10\text{m/s}^2$

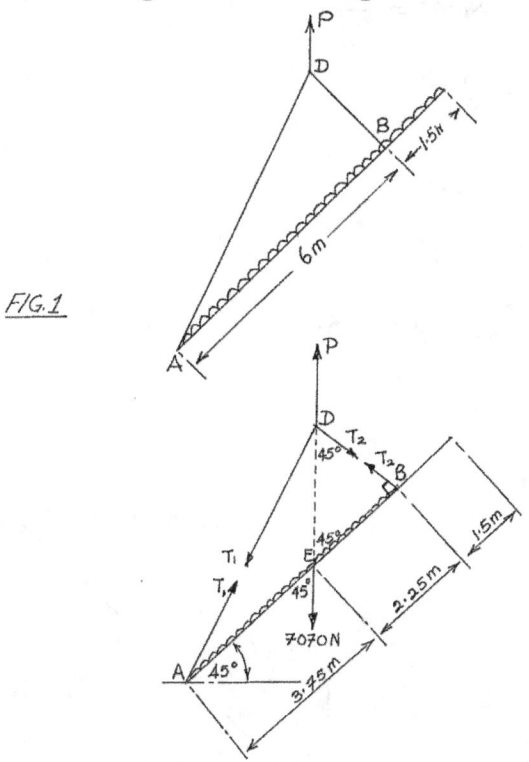

FIG. 1

FIG 2: FREE BODY DIAGRAM

Evidently, the line of action of the force of 7070N passes through the middle of the concrete pile E. i.e. through the pile's centre of mass. This line of action is the same as that of P through D.

By inspection we infer that angles DEB and EDB are equal, each being 45°.

$$\therefore \quad BD = 2.25\text{m}$$
so that $\quad AD^2 = 2.25 + 6^2$
or $\quad AD \approx 6.4\text{m}$

Now in the right-angle triangle ABD

$$SinADB = \frac{6}{6.4}$$

$$= 0.9363$$

∴ $\angle ADB = 69.44°$

and $\angle ADE = 69.44° - 45°$

$$= 24.44°$$

Therefore $\angle EAD = 180 - (135 + 24.44)$

$$= 20.56°$$

Now, taking moments about A

$$7070(AF) = T_2(6)$$

But $\frac{AF}{3.75} = Sin45 = \frac{1}{\sqrt{2}}$

$$AF = \frac{3.75}{\sqrt{2}}$$

$$T_2 = 7070 \times \frac{3.75}{6\sqrt{2}}$$

i.e. $T_2 = 3125N$

Considering vertical equilibrium

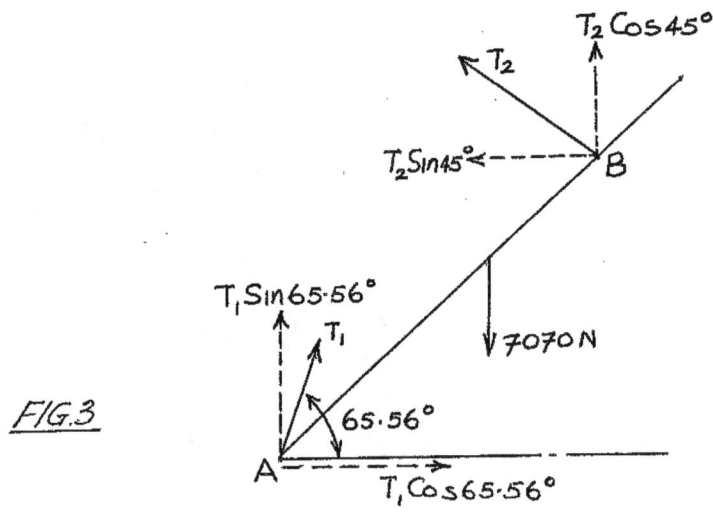

FIG.3

$$T_1Sin65.56° + T_2Cos45 = 7070$$

i.e. $0.9104T_1 + \frac{3125}{\sqrt{2}} = 7070$

$$0.9104T_1 + 2210 = 7070$$

$$0.9104T_1 = 4860$$

∴ $T_1 = 5338N$

We may now insert these forces in a new diagram, Fig. 4. Let us now proceed to resolve forces perpendicular to the pile.

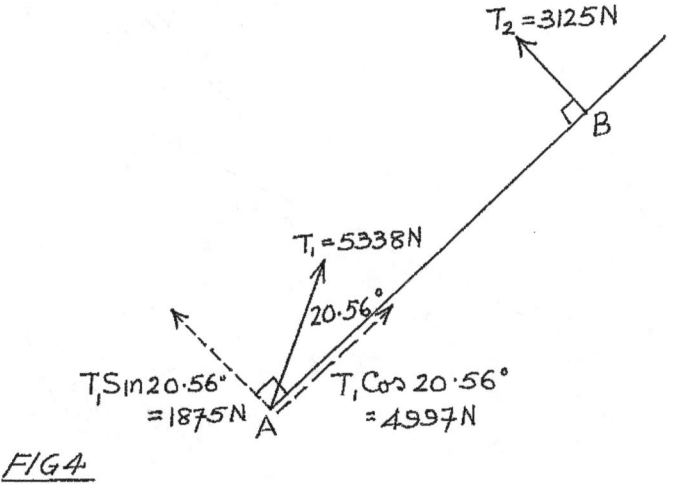

FIG 4

Force per unit length along the pile due to its mass = 7070/7.5m = 943N/m. This downwards force translates into $943 Cos 45°$ parallel to the pile and a force of the same magnitude perpendicular to the pile as shown in Fig. 5.

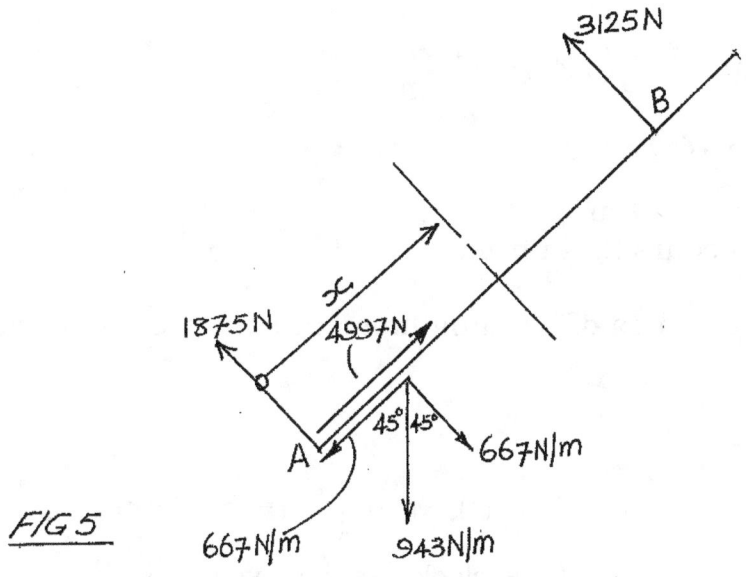

FIG 5

In Fig. 6, S.F at A = 1875N, left upwards, +ve; S.F. 1m from A = 1875 – 667 = 1208N; SF 2m from A = 1875 – 2(667) = 541N; SF at mid-point M = 1875 – 3.75(667) = -626N. Considering S.F from C, SF at B = 1.5(67) = 1000N, right downwards, i.e. +ve so that BU in Fig. 6 = 1000 and BV = 1000 – 3125 = -2125N. SF at M = -2125 + 2.25(667) = -624.25N. This is almost equal to the value of –626N obtained when we started from the point A. The error or difference between the two values is caused by rounding.

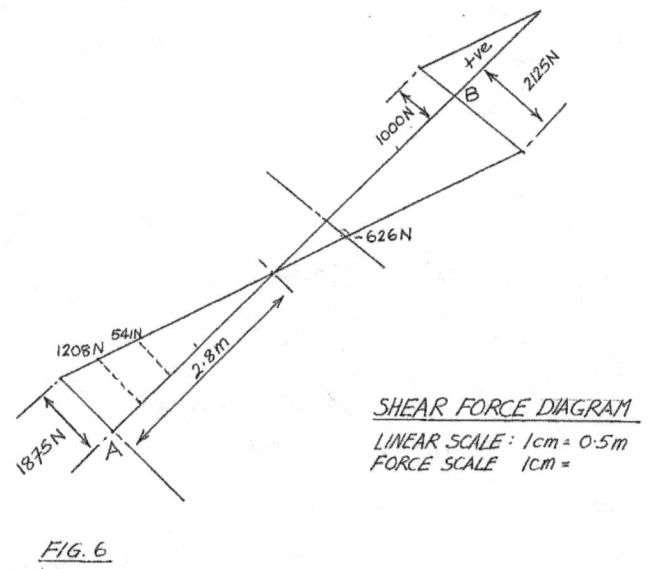

SHEAR FORCE DIAGRAM
LINEAR SCALE : 1cm = 0·5m
FORCE SCALE   1cm =

FIG. 6

Referring to Fig. 5, BM at section 'x' from A is :

$$1875(x) - 667\left(\frac{x^2}{2}\right)$$

For

maximum or minimum $\frac{d}{dx}(BM_x) = 0$

i.e. $\frac{d}{dx}(BM_x) = 1875 - 667x = 0$

or $x = 2.8m$

∴   Maximum BM occurs 2.8m from A.

Let us work out some values of bending moment as a prelude to drawing the B.M diagram.

B.M at A = 0
B.M. 1m from A

$$= \quad 1875(1) - 667\left(\frac{1^2}{2}\right) = 1541.5Nm,$$

left – clockwise, i.e. +ve

B.M., 2m from A $= \quad 1875(2) - 667\left(\frac{2^2}{2}\right) = 2416Nm,$

left – clockwise, i.e. +ve

B.M. 2.8m from A $= \quad 1875(2.8 \quad - \quad 667\left(\frac{2.8^2}{2}\right) =$

2636Nm, left – clockwise, + ve

~ 84 ~

B.M 3.7m from A	=	$1875(3.75) - 667\left(\dfrac{3.75^2}{2}\right)$
	=	234Nm, left – clockwise, +ve
B.M. 6m from A	=	$1875(5) - 667\left(\dfrac{5^2}{2}\right)$
	=	- 1037.5Nm
B.M. 6m from A	=	$1875(6) - 667\left(\dfrac{6^2}{2}\right)$
	=	-756Nm

The B.M diagram is plotted at Fig. 7, the maximum value occurring 2.8m from A being 2636Nm

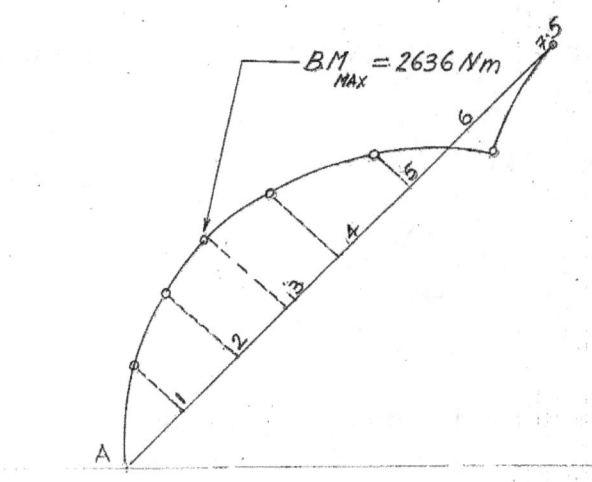

FIG 7 : BENDING MOMENT DIAGRAM

B.M. SCALE : 1 CM = 1000 Nm

LINEAR SCALE: 1CM = 0.5m

Q17. Along a freely-supported girder AB of length 5L moves a robotic device of length 'L'. The device may be approximated to a uniformly distributed load of w per unit length. Determine (i) the shearing force at a point distant 'L' from A and (ii) the maximum bending moment at that point.

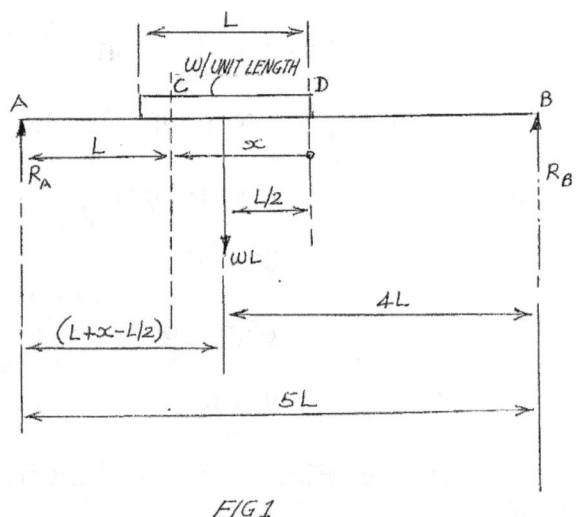

FIG 1

In Fig. 1, let C be the point where the maximum bending moment occurs distant 'L' from A.

Let x = distance of C from D
Taking moment about A :

$$R_B(5L) - wL\left(L + x - \frac{L}{2}\right) = 0$$

$$\text{i.e.} \quad R_B = \frac{w}{5}\left(\frac{L}{2} + x\right)/5$$

Bending moment at a distance L from A is :

$$M_C = R_B(4L) - \frac{wx^2}{2}$$

Substituting the value earlier obtained for R_B

$$M_C = \frac{wL}{5}\left(\frac{L}{2} + x\right) \cdot 4 - \frac{wx^2}{2}$$

$$M_C = w\left(\frac{4L^2}{10} + \frac{4Lx}{5} - \frac{x^2}{2}\right)$$

and

$$dM_c / dx = 4L/5 - x$$

For maximum M_C, $\dfrac{dM_c}{dx} = 0$, i.e. $x = \dfrac{4L}{5}$

$$\therefore \quad M_{max} = w\left\{4\frac{L^2}{10} + \frac{16L}{5} \cdot \frac{4L}{5} - \left(\frac{4L}{5}\right)^2 \cdot \frac{1}{2}\right\}$$

$$= w\left(\frac{4L^2}{10} + \frac{16L^2}{25} - \frac{8L^2}{25}\right)$$

$$= \frac{2wL^2}{5} + \frac{8wL^2}{25}$$

$$M_{max} = \frac{18wL^2}{25}$$

Taking moments about B

$$R_A(4L) = wL\{5L - (L + x - L/2)\}$$
$$= wL(5L - L - x + L/2)$$
$$= wL(4L - x + \frac{L}{2})$$

$$R_A = \frac{wL}{4L}\left(\frac{7L}{2} - x\right)$$

$$= \frac{w}{4}\left(\frac{7L}{2} - x\right)$$

Let V_c = shearing force at 'C' which is distant 'L' from A

But $V_c = R_A - w(L - x) = \frac{w}{4}\left(\frac{7L}{2} - x\right) - wL + wx$

i.e. $V_c = \frac{7wL}{8} - \frac{wx}{4} - wL + wx = \frac{3wx}{4} - \frac{wL}{8}$

$$= w\left(\frac{3x}{4} - \frac{L}{8}\right)$$

We are asked to find the shearing force at a point distant 'L' from 'A', i.e. when $x = L$. Refer to Fig. 1.

$$\therefore \quad V_{x=L} = w\left(\frac{3L}{4} - \frac{L}{8}\right)$$

or $\quad V_{x=L} = \frac{5wL}{8}$

$$\therefore \quad V_c max = w\left(\frac{3L}{4} - \frac{L}{8}\right)$$

$$V_c max = 5wL/8$$

Q18. The shear force diagram of a simply supported beam is the parabolic distribution shown in Fig. 1. From it derive the load distribution and bending moment diagrams.

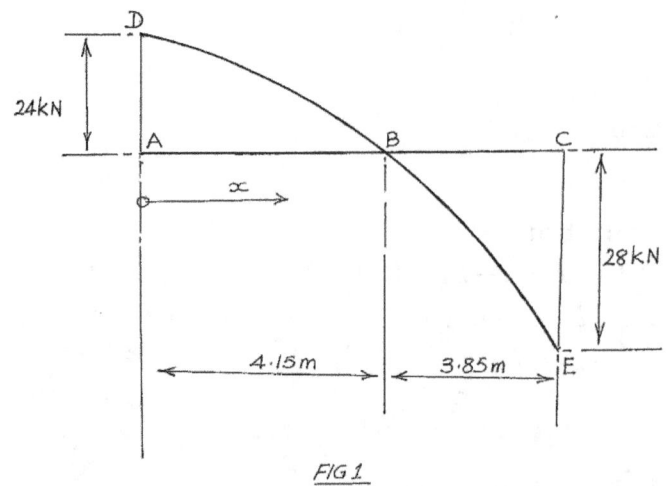

FIG 1

Because the shear force characteristic is parabolic and originates at D, it is reasonable to assume that the relationship between shear force SF_x and distance x measured from origin A, is of the form :

$$SF_x = ax^2 + bx + c \qquad \cdots\cdots\cdots \text{(i)}$$

and a, b, and c are constants.

At A where $x = 0$, S.F. $= +24$kN

$$\therefore \quad 24 = a(0^2) + b(0) + c \qquad \cdots\cdots\cdots \text{(ii)}$$

which means that

$$c = 24$$

Further, at $x = 4.15$m, S.F $= 0$

$$\therefore \quad 0 = a(4.15^2) + b(4.15) + 24$$

from which

$$17.22a + 4.15b = -24 \qquad \cdots\cdots\cdots \text{(iii)}$$

Finally at $x = 8$m, shear force $= -28$kN

$$\therefore \quad 64a + 8b + 24 = -28$$

$$\text{or} \quad 64a + 8b = -52$$

Multiplying equation (iii) by 8 and equation (iv) by 4.15 we obtain

$$137.78a + 33.2b = -192$$

and

$$265.6a + 33.2b = -215.8$$

$$\therefore \quad 127.82a = -23.8$$

$$\text{or} \quad a = -0.186$$

Substituting this value for a in (iv) we get

$$-64(0.186) + 8b = -52$$

$$-11.92 + 8b = -52$$

$$\therefore 8b = -40.08 \quad \text{i.e.} \quad b \approx 5$$

Substituting the values of these constants in equation 1 produces the equation of S.F distribution along the beam:

$$SF_x = -0.186x^2 - 5x + 24 \qquad \dots\dots\dots\dots \text{(v)}$$

Now, we know that $\dfrac{d}{dx}(SF_x) = -w \qquad \dots\dots\dots\dots \text{(vi)}$

i.e. $\quad -0.372x - 5 = -w \qquad \dots\dots\dots\dots \text{(vii)}$

Evidently the load distribution 'w' varies linearly along the beam.
At x = 0
$-w = -5$
At A therefore w = 5kN/m
At x = 8
$\qquad -w = -0.372(8) - 5$
$\qquad -w = -2.98 - 5$
$\qquad\quad = -7.98kN/m$
or $\quad w \approx 8kN/m$

Evidently the load on the beam varies linearly from 5kN/m at one end to 8 kN/m at the other as illustrated in Fig. 2.

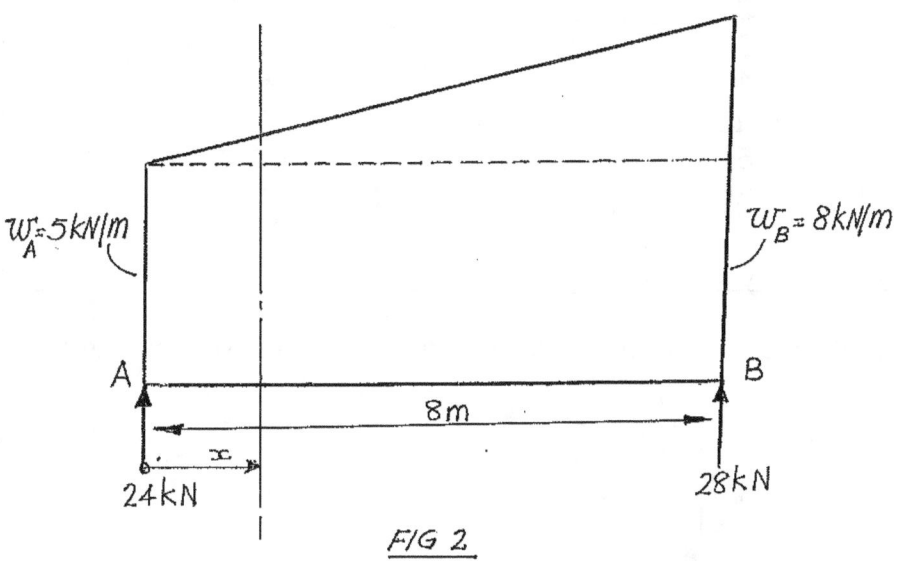

FIG 2

To obtain the bending moment diagram we may proceed from first principles and at sections, say, 'x' from A, consider bending moment from the following expression:

$$BM_x = R_A(x) - 5(x)\left(\frac{x}{2}\right) - x\left(\frac{8-3}{8}\right)\frac{x}{2}\cdot\frac{x}{3} \qquad \dots\dots\dots\dots \text{(viii)}$$

or alternatively because we know that

$$\frac{d}{dx}(BM_x) = SF_x$$

which in this case is :

$$\frac{d}{dx}(BM_x) = -0.186x^2 - 5x + 24$$

Integrating we obtain :

$$BM_x = -0.062x^3 - \frac{5x^2}{2} + 24x + K$$

But at $x = 0$, $BM_x = 0$, $\therefore$ K = 0
Accordingly,

$$BM_x = -0.062x^3 - \frac{5x^2}{2} + 24x \qquad \dots\dots\dots\dots \text{(ix)}$$

Using expression (ix) let us draw up a table of values of BM for different xs on the beam

x (in metres)	B.M. (in kNm)
0	0
1	21.4
2	37.5
3	47.8
4	52
4.15	52.12
5	49.8
6	40.6
7	24.2
8	0

These values are plotted on squared paper to produce Fig. 3.

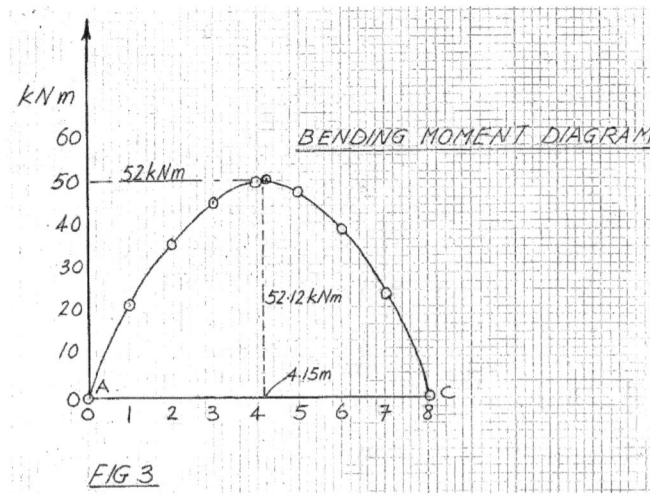

Q19. The end supports A and B of the structure shown in Fig. 1 are restrained not only in position but also in direction. Pins are at D, E and F. Draw the shear force and bending moment diagrams for AG, GF and FE.

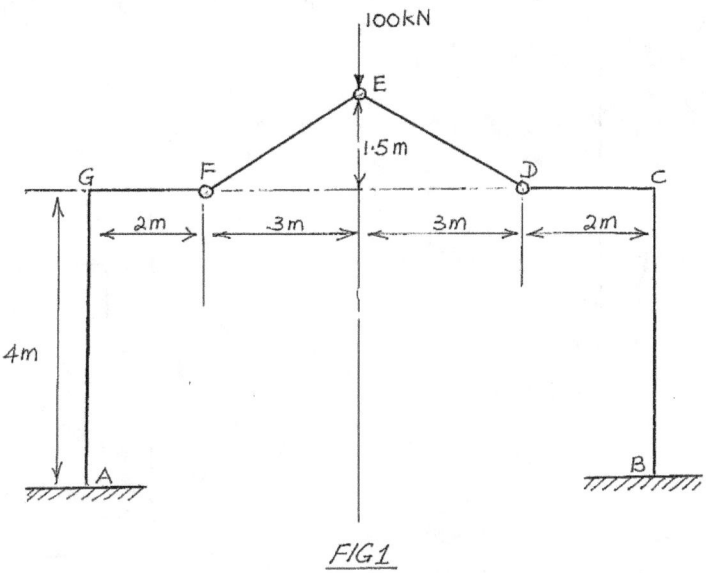

FIG 1

Fig. 1 not drawn to scale

We infer from what we are told that the reactions at A and B consist of vertical and horizontal forces and fixing moments.

Let us proceed to construct free-body diagrams (FBDs). In Fig. 2 we have drawn the 2 halves of the structure which are identical except for the labelling. The student should note that because the crown E is pinned, there cannot be any resisting moment there, as indeed there cannot be at pins F and D either. Observe also that as the two points of E come together, the two HE components vanish, and the only force at E then is 100kN downwards.

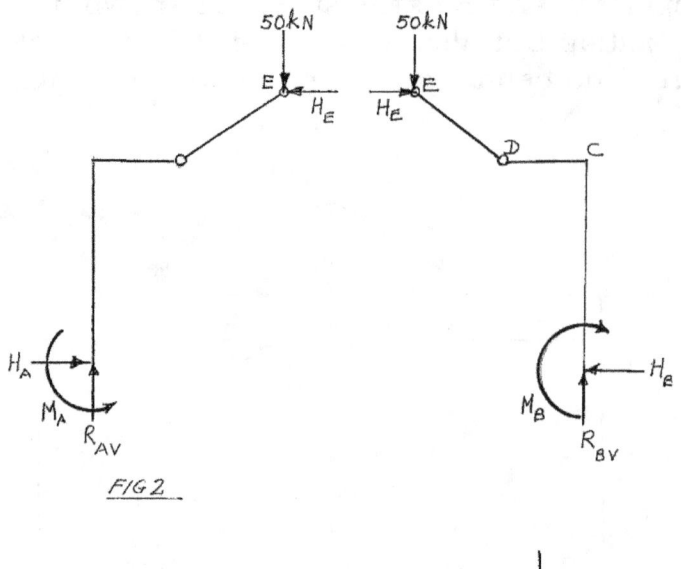

FIG 2

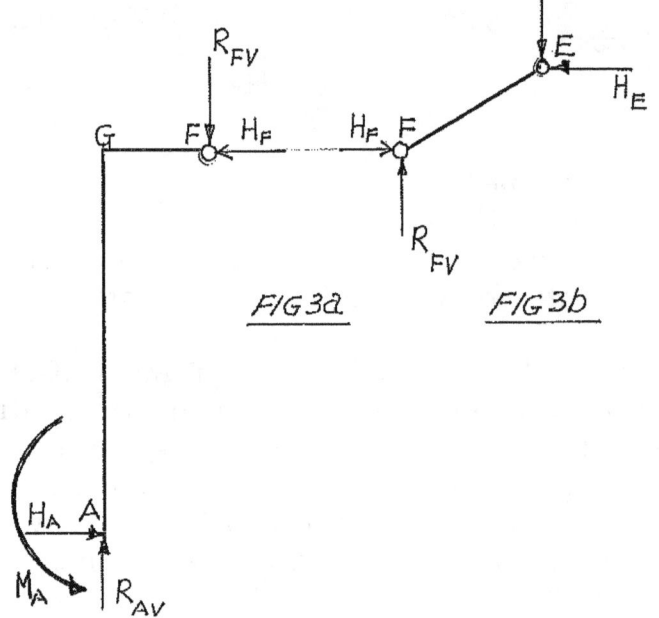

FIG 3a   FIG 3b

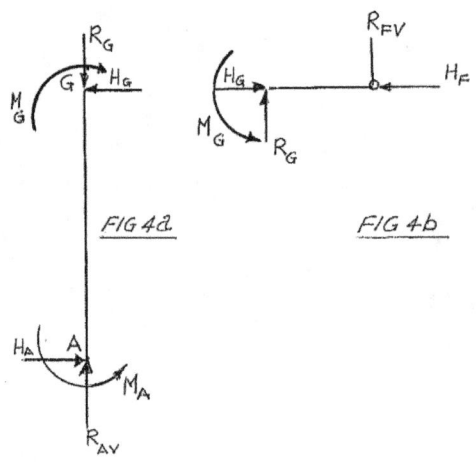

FIG 4a   FIG 4b

Considering equilibrium of FBD in Fig. 3b it is evident that:
$$R_{FV} = 50kN$$

Also, taking moments about E
$$R_{FV}(3) = H_F(1.5)$$
$$\therefore \quad H_F = 100kN$$
Further $\quad H_E = 100kN$

Considering equilibrium of FBD in Fig. 4b
$$R_G = 50kN$$
$$H_F = H_G = 100kN$$

Taking moments about G
$$R_{FV}(2) - M_G = 0$$
$$\therefore \quad M_G = 50(2)$$
$$\text{i.e. } M_G = 100kNm$$

Considering equilibrium of element AG of Fig. 4a
$$R_{AV} = R_G = 50kN$$
$$H_A = H_G = 100kN$$

Taking moments about A
$$M_G - H_G(4) - M_A = 0$$
i.e. $\quad 100 - 100(4) - M_A = 0$
or $\quad M_A = -300kNm$

Let us consider B.M on component FE.

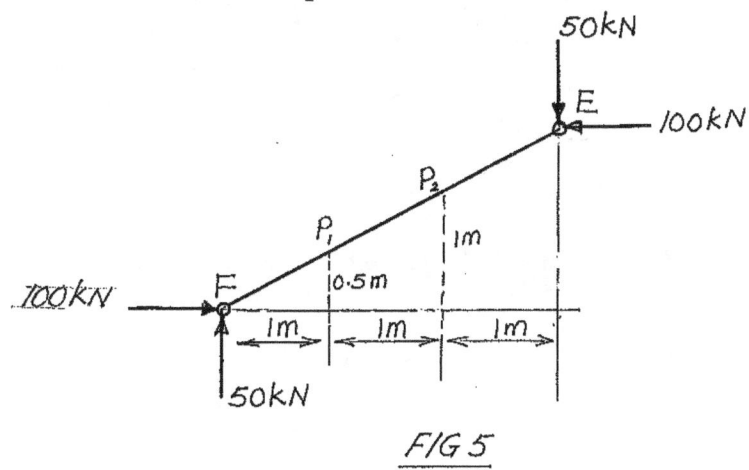

FIG 5

B.M. at $P_1 = +50(1) - 100(0.5) = 0$
B.M at $P_2 = 50(2) - 100(1) = 0$
Also $M_F = M_E = 0$
We may now draw our shear force and bending moment diagrams

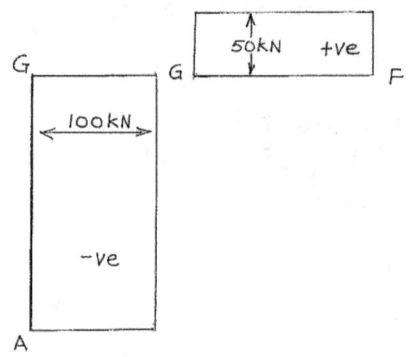

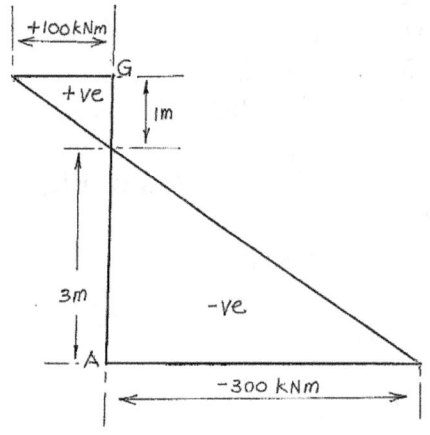

FIG.6 SHEAR FORCE DIAGRAMS FOR MEMBERS AG AND GF

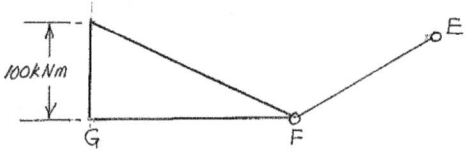

FIG.7. BENDING MOMENT DIAGRAMS FOR MEMBERS AG,GF
AND FE

Q20. A bar bent in the form of a circular segment is acted upon by the forces 'Q' as indicated in Fig. 1. Derive expressions for shear force and bending moment at a section 'x' metres from the crown as indicated. Express your answers in terms of 'Q' 'a', 'x' and 'ℓ'.

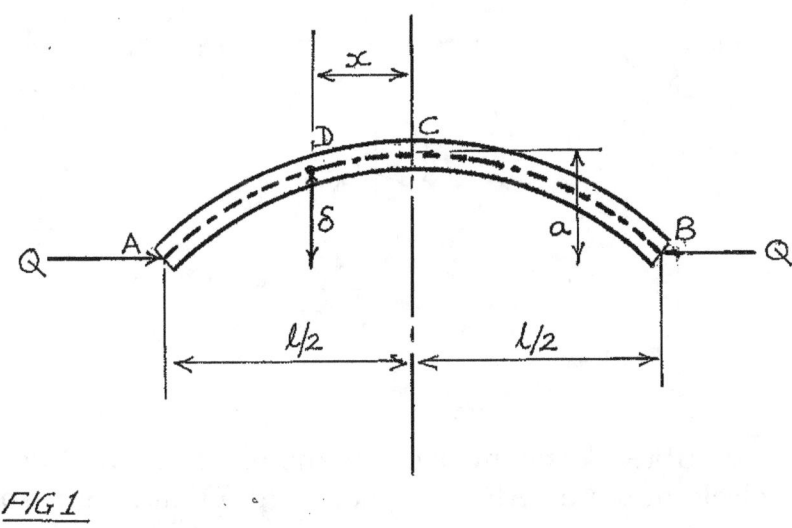

FIG 1

The first thing we must recall is that by definition shear force at a section acts at right angles to the axis of that section. Therefore in Fig. 2

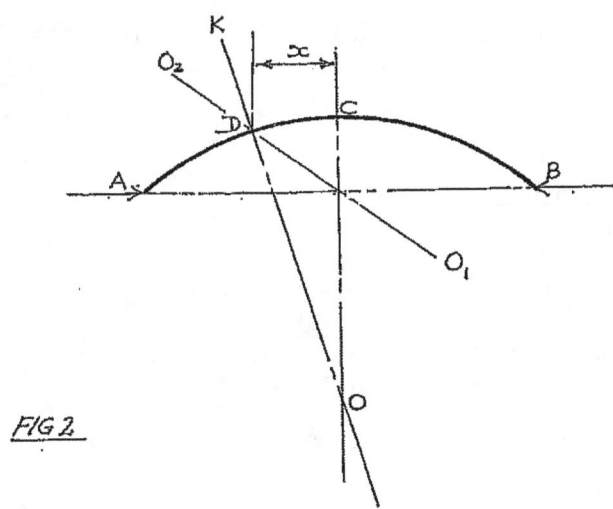

FIG 2

the force at point 'D' cannot act along O_1, O_2. Rather it has to act along OK, OD being the radius of a circle of which the segment is a part, because all radial planes cut a circumference at right angles to the axis of the circumference.

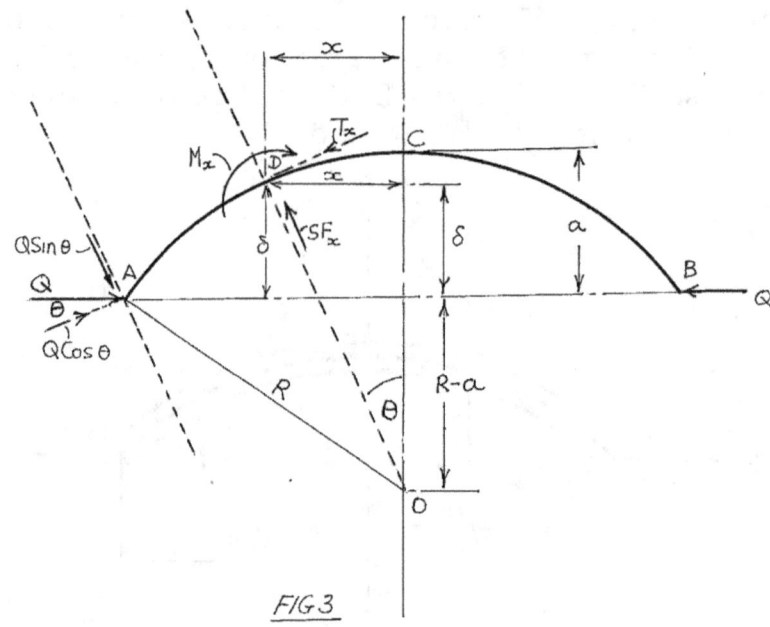

FIG 3

Let angle 'θ' subtend the horizontal distance 'x' in Fig. 3; and let R = radius of circle of which ABC is a segment 'Q' at A is resolved into $QSin\theta$ in a direction parallel to radial plane OD and $QCos\theta$ perpendicular to the same plane.

SFx is the shear force at D and M_x the bending moment, also at D.

Considering equilibrium of the section AD assumed disconnected from AC, we have

$$QSin\theta = SF_x \qquad \qquad \text{..........} \quad (i)$$
$$QCos\theta = T_x \qquad \qquad \text{..........} \quad (ii)$$

and taking moments about D

$$Q(\delta)) - M_x = 0 \qquad \qquad \text{..........} \quad (iii)$$

Considering Fig. 3

$$Sin\,\theta = \frac{x}{R}$$

Also $R^2 = \left(\frac{\ell}{2}\right)^2 + (R-a)^2$

i.e. $R^2 = \dfrac{\ell^2}{4} + R^2 - 2aR + a^2$

or $R = \dfrac{\left(\dfrac{\ell^2}{4} + a^2\right)}{2a} \qquad \qquad \text{..........} \quad (iv)$

so that $\quad Sin\theta \;=\; \dfrac{x \cdot 2a}{\left(\dfrac{\ell^2}{4} + a^2\right)}$

$\quad$ SF$_x$(Shear Force$_x$) $\;=\; \dfrac{2axQ}{\left(\dfrac{\ell^2}{4} + a^2\right)}$ $\qquad$ . . . . (v)

By inspection of Fig 3 we see that

$\quad Tan\theta \;=\; \dfrac{x}{(R - a + \delta)}$ $\qquad$ . . . . (vi)

Now because $Sin\,\theta \;=\; \dfrac{x}{R}$

$Tan\theta = \dfrac{x}{\sqrt{(R^2 - x^2)}}$

Substituting this result in (vi)

$$\dfrac{x}{\sqrt{(R^2 - x^2)}} = \dfrac{x}{(R - a + \delta)}$$

$\therefore \qquad \sqrt{(R^2 - x^2)} = R - a + \delta$

i.e. $\quad \delta \;=\; \left\{\sqrt{(R^2 - x^2)}\right\} - R + a$ $\qquad$ . . . . (viii)

Using the relation (iv)

$\quad$ radius, R $\;=\; \left\{\dfrac{\left(\dfrac{\ell^2}{4} + a^2\right)}{2a}\right\}$

equation (vii) becomes

$$\delta = \left\{\sqrt{\left(\dfrac{\dfrac{\ell^2}{4} + a^2}{2a}\right) - x^2}\right\} - \left(\dfrac{\dfrac{\ell^2}{4} + a^2}{2a}\right) + a$$

Let us simplify - $\left(\dfrac{\dfrac{\ell^2}{4} + a^2}{2a}\right) + a \;-\; \left(\dfrac{\dfrac{\ell^2}{4} + a^2}{2a}\right) + a$

$$= -\frac{\ell^2 - 4a^2}{8a} + a$$

$$= -\frac{\ell^2 - 4a^2 + 8a^2}{8a}$$

$$= -\frac{\ell^2 + 4a}{8a}$$

$$\therefore \quad \delta = \left\{ \sqrt{\left(\frac{\frac{\ell^2}{4} + a^2}{2a}\right)^2 - x^2} \right\} - \frac{\ell^2 + 4a^2}{8a}$$

so that

$$M_x = Q\left\{ \sqrt{\left(\frac{\ell^2 + 4a^2}{8a}\right)^2 - x^2} \right\} - \left(\frac{\ell^2 - 4a^2}{8a}\right)$$

Although we were not asked to obtain the thrust in the bar it is evident that

$$T_x = P Cos\theta$$

and because

$$Cos\theta = \frac{\sqrt{(R^2 - x^2)}}{R}$$

$$= \frac{\sqrt{\left(\frac{\ell^2}{4} + a^2\right) - x^2}}{\dfrac{\left(\dfrac{\ell^2}{4} + a^2\right)}{2a}}$$

$$= \frac{2a\sqrt{\left(\frac{\ell^2}{4} + a^2\right) - 4a^2 x^2}}{2a\left(\frac{\ell^2}{4} + a^2\right)}$$

$$\therefore \quad T_x = \frac{\sqrt{\left(\frac{\ell^2}{4} + a^2\right) - 4a^2 x^2}}{\left(\frac{\ell^2}{4} + a^2\right)}$$

Q21.

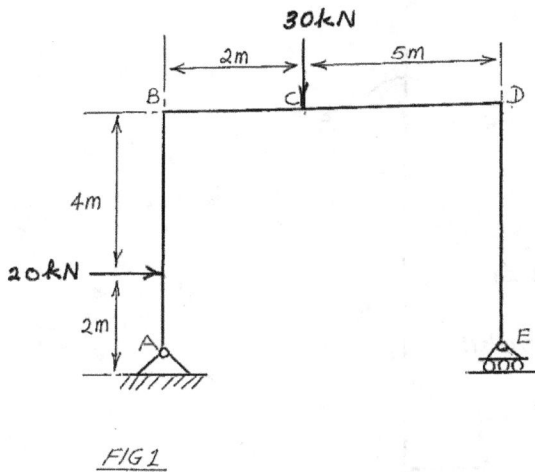

FIG 1

Draw the shear force, thrust and bending moment diagrams for the portal frame shown in Fig. 1 above.

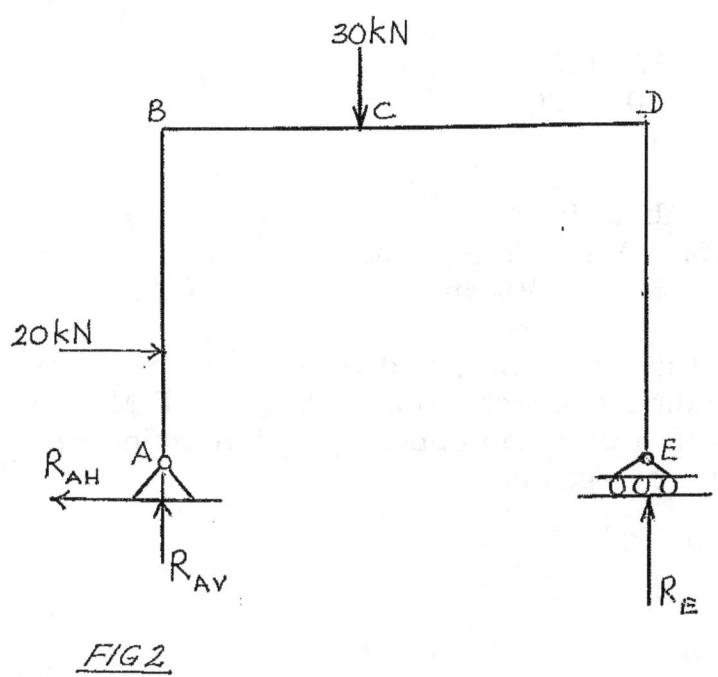

FIG 2

Let us proceed to determine reactions R_{AV}, R_{AH} and R_E.

Taking moments about E

$$R_{AV}(7) + 20(2) - 30(5) = 0$$
$$\therefore \qquad 7R_{AV} = 110$$
$$\text{Or} \qquad R_{AV} = 15.7kN$$

Evidently, $R_E = 14.3kN$; $R_{AH} = 20kN$
Consider AB separately

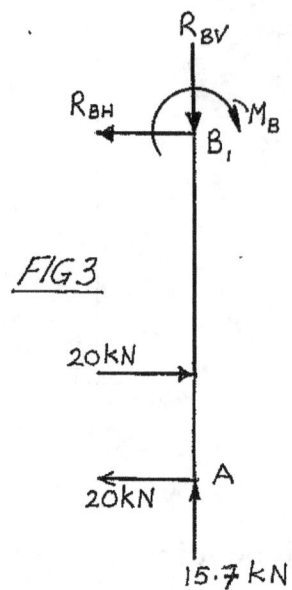

FIG 3

Assuming the forces and moments at B and considering equilibrium of the member

$$R_{BV} = 15.7kN$$
$$R_{BH} + 20 = 20$$
$$\therefore \quad R_{BH} = 0$$

Taking moments about B

$$20(6) + M_B - 20(4) = 0$$
$$\text{i.e.} \quad M_B = -40kNm$$

Note that as we have said time and time again, pin jointed supports do not sustain bending moments; that is why no B.M. was applied at A. Evidently M_B is in a direction opposite to that which we assumed. It is sagging rather than hogging.

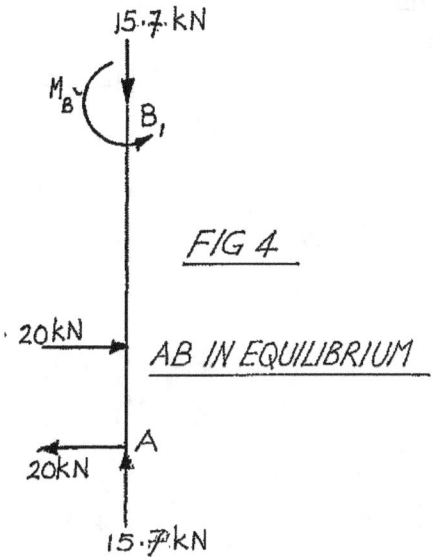

FIG 4

AB IN EQUILIBRIUM

AB in equilibrium
Now, let us consider BD

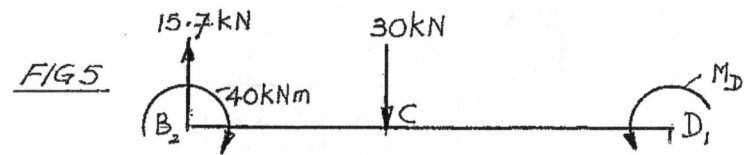

FIG 5

There are no horizontal forces at B and D; $R_{BH} = 0$; therefore $R_{DH} = 0$.
For vertical equilibrium
$$R_{DV} + 30 - 15.7 = 0$$
$$\therefore \quad R_{DV} = -14.3 kN$$
Considering moments
$$40 + 15.7(7) - 30(5) - M_D = 0$$
$$40 + 110 - 150 - M_D = 0$$

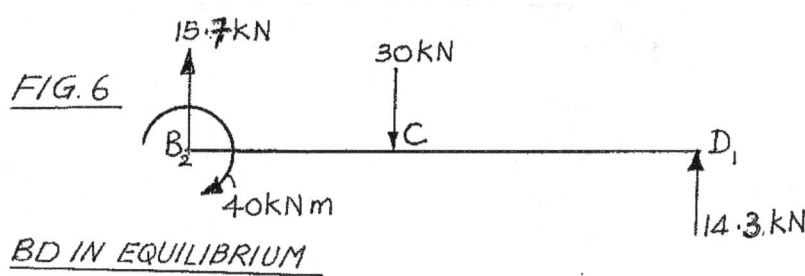

FIG. 6

BD IN EQUILIBRIUM

BD in equilibrium

Finally we come to DE

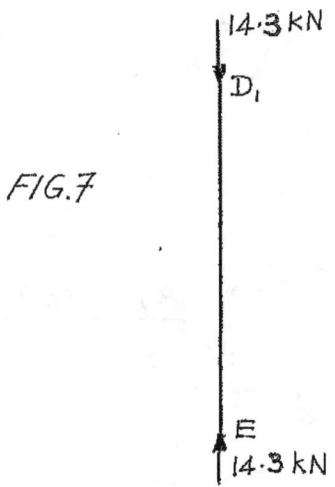

FIG.7

14·3 kN

D₁

E
14·3 kN

DE is in equilibrium under the action of 2 forces : 14.3kN downwards at D₂ and 14.3kN upwards at E. There is no B.M. on member DE.

The shear force and bending moment diagrams are plotted at Figs. 8 and 9.

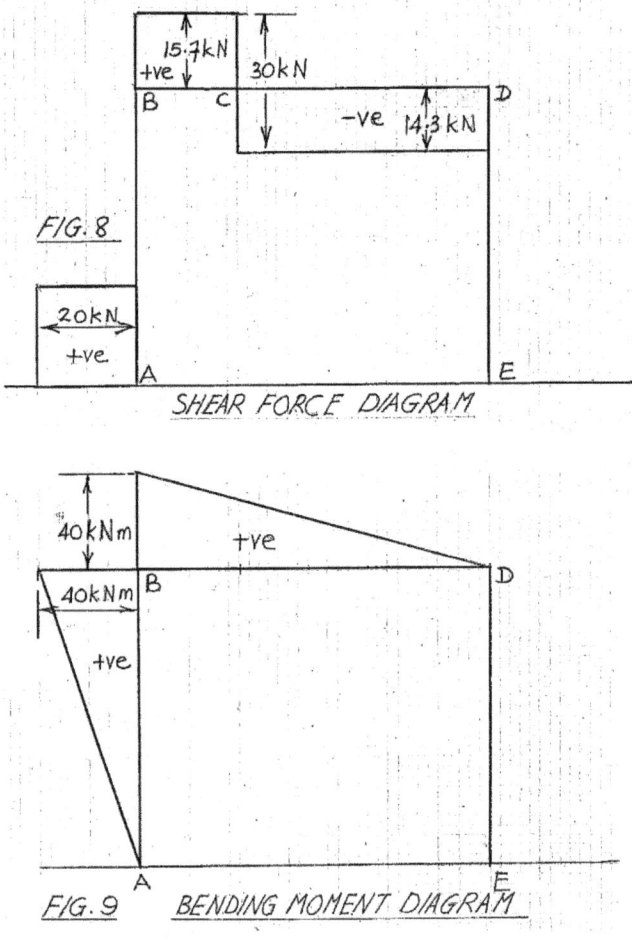

FIG.8

15·7kN
+ve

30kN

B   C   -ve 14·3 kN   D

20kN
+ve

A   E

SHEAR FORCE DIAGRAM

40kNm   +ve

40kNm B   D

+ve

A   BENDING MOMENT DIAGRAM   E

FIG.9

Q22. Find expressions for shear and bending moment for the loaded semi-circular ring shown as in Fig. 1, stating them in terms of W, R and θ. Evaluate the bending moment at $\theta = 0°$ and $180°$ and shear force at $\theta = 90°$. What is the maximum bending moment and where does it occur? Draw the shear force and bending moment diagrams.

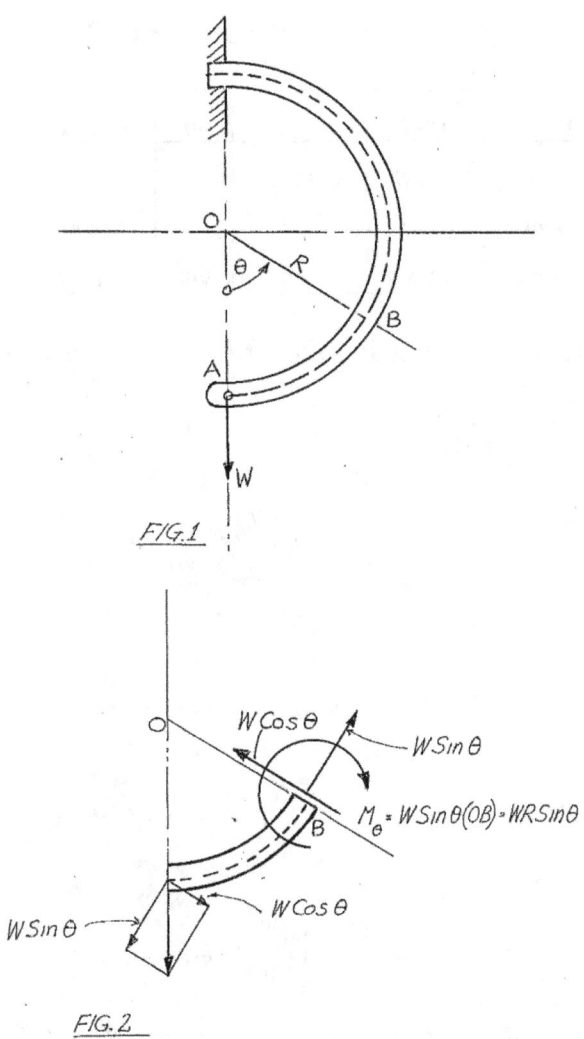

FIG.1

FIG.2

To begin the analysis we cut the ring at B. See Fig. 2. We then resolve the force W into 2 components parallel and perpendicular to the cut, viz, $W Sin\theta$ and $W Cos\theta$. Now, equal and opposite forces are applied at B as indicated. 'W' alone acting at A produces an anti-clockwise moment about B. Therefore, the balancing moment Mθ at B must be clockwise. We add Mθ and the system is in static equilibrium.

Forces W and $W Cos\theta$ pass through the center 0 therefore $\sum M_o = W Sin\theta \times R$ i.e. $\quad$ Mθ = $W R Sin\theta$

Evidently shear force at section 'θ' = $WCos\theta$ and Thrust = $WSin\theta$.
These results may now be summarized thus:

Shear Force at section 'θ' = $WCos\theta$
       Thrust Force at section 'θ' = $WSin\theta$
       Bending Moment at section θ = $WRSin\theta$

We now tabulate values of shear, thrust and B.M. for various values of θ

	0	22 ½ °	45°	67 ½ °	90°	112 ½ °	135°	157½ °	180
S.F :	W	0.924W	0.707W	0.383W	0	-0.383W	-0.707W	-0.924W	-W
Thrust :	0	0.383W	0.707W	0.924W	W	0.924W	0.707W	0.383W	0
B.M :	0	0.383W	0.707WR	0.92WR	WR	0.924W	0.707W	0.383W	0

Maximum bending moment is WR and occurs where θ = 30°.

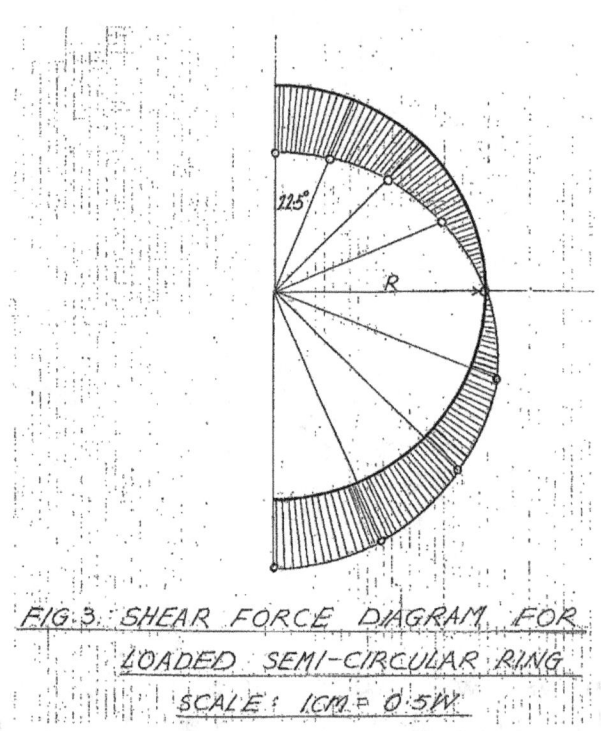

FIG 3 SHEAR FORCE DIAGRAM FOR
LOADED SEMI-CIRCULAR RING
SCALE : 1CM = 0.5W

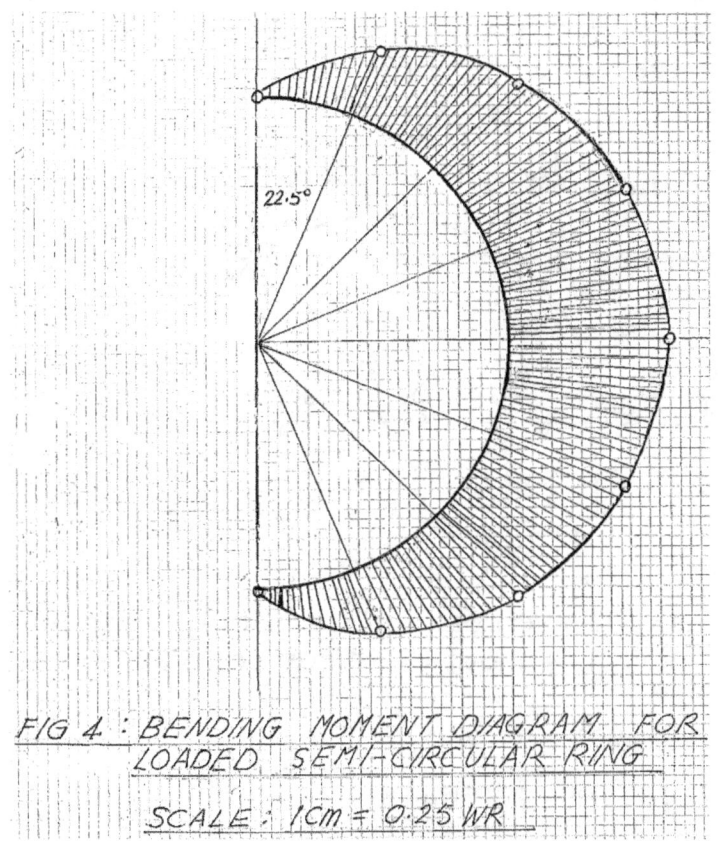

FIG 4 : BENDING MOMENT DIAGRAM FOR
LOADED SEMI-CIRCULAR RING

SCALE : 1CM = 0.25 WR

Q23. A semi-circular beam ABCDE is pin-jointed at support A and rests
on rollers at support B. The beam carries three radial loads each
equivalent to 100kN at the points indicated in Fig. 1. Derive expressions
for shear force, thrust and bending moment at sections of the beam in
the portions AE, ED, DC and CB.

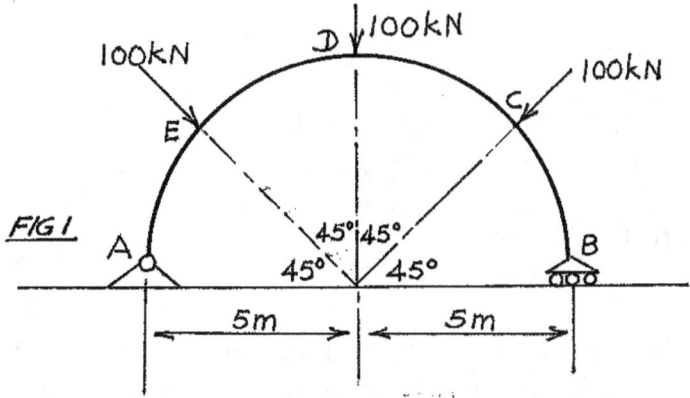

FIG 1

In Fig. 2, the horizontal and vertical components of the forces at E are
entered to replace the loads there. At pin-support A, the horizontal force
H_A and vertical force R_{AV} which are respectively the horizontal and
vertical components of the reaction at A, are also entered. The roller

support B cannot resist any horizontal force. Hence at B in Fig. 2, only the vertical force R_{BV} is present.

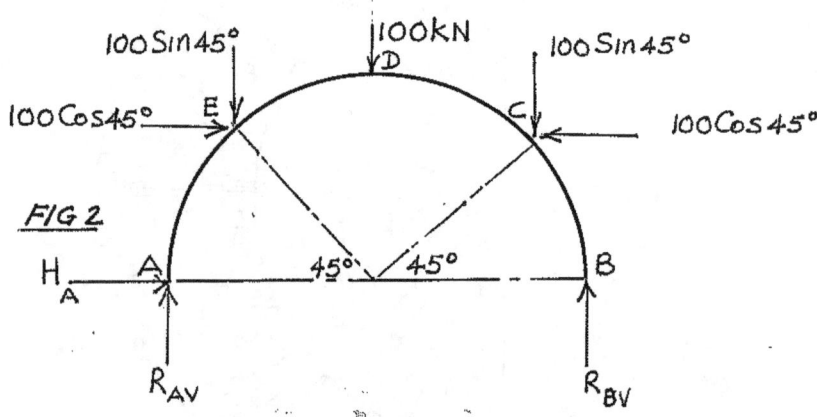

FIG 2

The equations of the vertical and horizontal forces on the beam and the equations of moments about B are now considered.

Vertical Equilibrium

R_{AV} – 100Sin45 – 100- 100Sin45 + R_{BV} = 0

i.e. R_{AV} + R_{BV} - 200$\cdot\dfrac{1}{\sqrt{2}}$ -100 = 0

∴ R_{AV} + R_{BV} = 241.4kN .... (i)

Horizontal Equilibrium

H_A + 100Cos45 – 100Cos45 = 0
∴ H_A = 0 .... (ii)

Moments about B

10R_{AV} – 1000Sin45 – 500 = 0
i.e. 10R_{AV} = 120.
∴ R_{AV} = 120. .... (iii)

According to (i) therefore
R_{BV} = 120.7

In Fig. 3 we show all the external loads as indicated in Fig. 1 and the values of the reactions just computed. Also shown is a radial plane *OS'S* which cuts the beam perpendicularly. *OS'S* is θ^o measured anti-clockwise from AO.

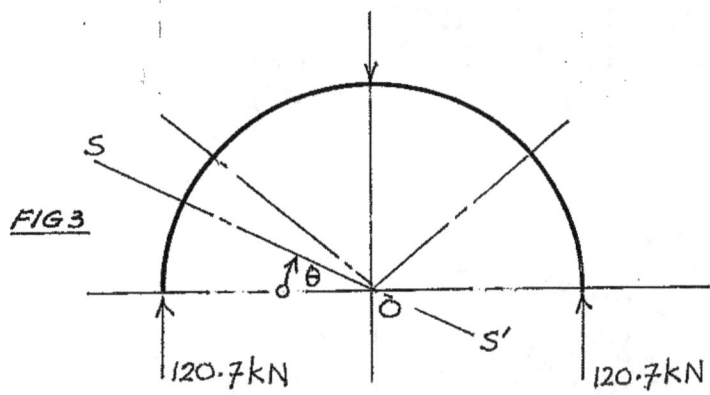

FIG3

We have now to determine shear force, thrust and bending moment for:

AE, i.e. when θ is between 0° and 45° ; ED, .
   i.e. when θ is between 45° and 90°; DC, .
   i.e. when θ is between 90° and 135° and CB, .
   i.e. when θ is between 135°; and 180°. We consider each section of the beam in turn.

Section AE

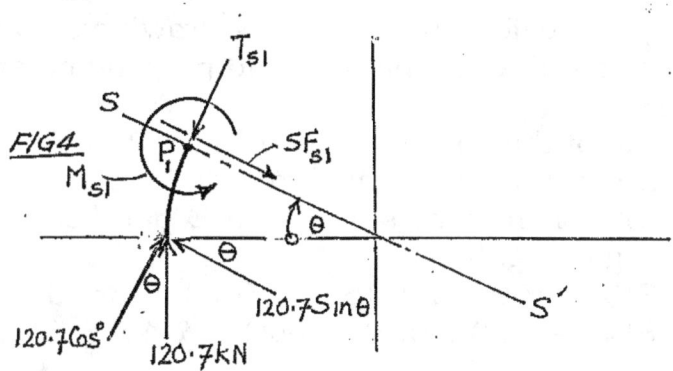

FIG4

Here 'θ' is between 0° and 45°. The radial plane OS'S defines the position we have selected. The force $R_{AV}(=120.7\text{kN})$ is now resolved into components parallel and perpendicular to OS'S. For equilibrium of the section, SF_s at P_1 must act in the direction as shown; T_s must act opposite to $120.7Cos\theta$ at A; and the anti-clockwise moment M_s at P must equal the clockwise moment $R_{AV}(r)$ of R_{AV} about 0, and the anti-clockwise moment of T_s about 0. Our results for S.F, T and M for portion of the beam AE are therefore:

$$SF_{s1} = 120.7Sin\theta \qquad \ldots\ldots\ldots\ldots \text{(iv)}$$
$$T_{s1} = 120.7Cos\theta \qquad \ldots\ldots\ldots\ldots \text{(v)}$$

$$M_{s1} + T_{s1}(r) = 120.7(r)$$
$$\text{or} \quad M_{s1} = 120.7r(1-Cos\theta)$$
$$\therefore \quad M_{s1} = 603.5(1-Cos\theta) \qquad \ldots\ldots\ldots \text{(vi)}$$

Section ED

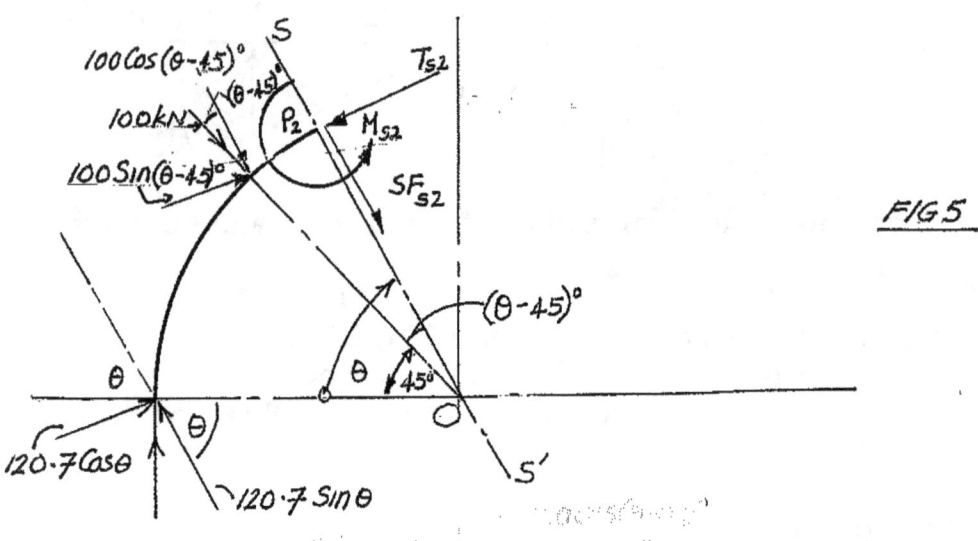

FIG 5

In Fig. 5, the plane $OS'S$ has now moved beyond E and angle OEP_2, is now $(\theta - 45°)$. As before, when $OS'S$ is drawn we obtain the components of force at A and E that are parallel and perpendicular to $OS'S$.

For equilibrium of the portion of the beam AP_2, SF_{s2} at P_2 must equal the sum of all the other forces parallel to $OS'S$. We have also shown in Fig. 5, the components of the forces at A and E parallel and perpendicular to $OS'S$. Referring to Fig. 5

$$SF_{s2} + 100Cos(\theta-45) - 120.7Sin\theta = 0$$
$$\therefore \quad SF_{s2} = 120.7Sin\theta - 100Cos(\theta-45) \qquad \ldots\ldots \text{(vii)}$$

Similarly for thrust T_{s2}
$$T_{s2} = 120.7Cos\theta + 100Sin(\theta-45) \qquad \ldots\ldots \text{(viii)}$$

Taking moments about 0
$$R_{AV}(r) - T_k(r) - M_s$$
$$\text{i.e.} \quad 120.7(r) - 120.7(r)Cos\theta - 100rSin(\theta-45) - M_s = 0$$
from which
$$M_{s2} = 603.5(1-Cos\theta) - 500Sin(\theta - 45)$$

Collecting our results for portion of beam ED:
$$SF_{s2} = 120.7Sin\theta - 100Cos(\theta - 45) \qquad \ldots\ldots\ldots \text{(vii)}$$
$$T_{s2} = 120.7Cos\theta + 100Sin(\theta - 45) \qquad \ldots\ldots\ldots \text{(viii)}$$

~ 108 ~

$$M_{s2} = 603.5(1-Cos) - 500Sin(\theta - 45) \qquad \ldots\ldots\ldots\ldots \text{(ix)}$$

We come now to section DC. As in the two previous cases considered, plane $OS'S$ cuts the beam at a point P_3 between D and C. 'θ' is now the obtuse angle AOP_3.

Section DC

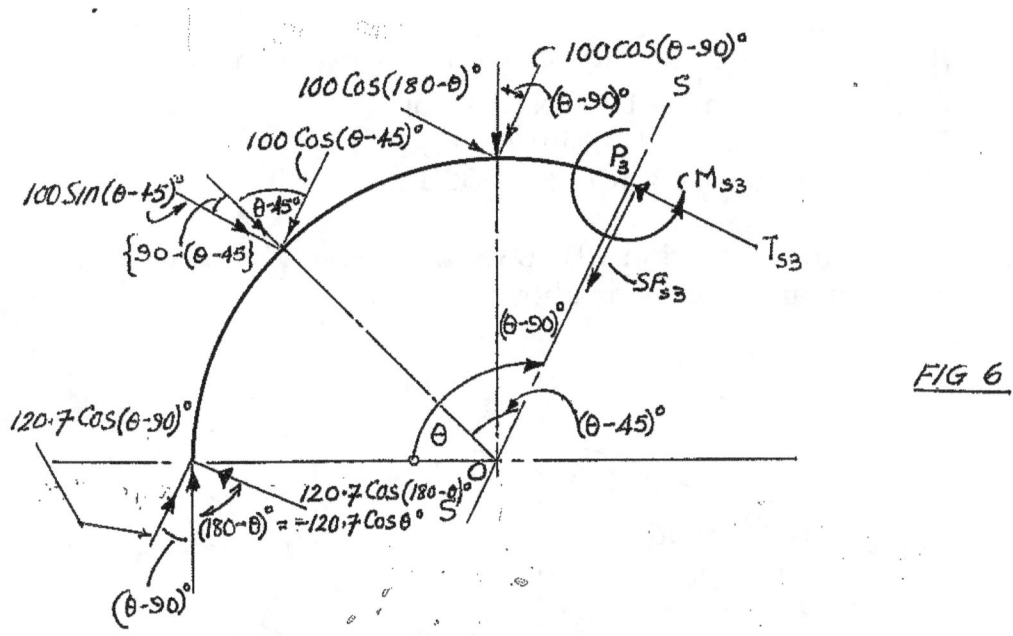

FIG 6

At A, E and D the forces there are resolved into components parallel and perpendicular to plane $OS'S$. SF_s, T_s and M_s are applied at P_3 to maintain equilibrium. By inspection we see

$SF_{s3} + 100Cos(\theta-90) + 100Cos(\theta-45) - 120.7Cos(\theta-90)= 0$
Noting that $Cos(\theta-90) = Cos\{-(90-\theta)\}$ and remembering that $Cos(-G) = CosG$ we may rewrite the SF_{s3} equation as
$SF_{s3} + 100Cos(90-\theta) + 100Cos(\theta-45) - 120.7(\theta-90) = 0$
But $Cos(\theta-90) = Sin\theta$
$\therefore \qquad SF_{s3} + 100Sin\theta + 100Cos(\theta-45) - 120.7Sin\theta = 0$
$\therefore \qquad SF_{s3} = 120.7Sin\theta - 100Sin\theta - 100Cos(\theta - 45)$
or $\qquad SF_{s3} = 20.7Sin\theta - 100Cos(\theta - 45)$

Again, referring to Fig. 6 and considering equilibrium perpendicular to $OS'S$ we have
$T_{s3} - 100Cos(180-\theta) - 100Sin(\theta - 45) \qquad +120.7Cos(180-\theta) = 0$
$\qquad \therefore \qquad T_s + 100Cos\theta - 100Sin(\theta - 45)$

$$- 120.7\cos\theta = 0$$

i.e. $T_{s3} - 20.7\cos\theta - 100\sin(\theta - 45) = 0$

∴ $T_{s3} = 20.7\cos\theta + 100\sin(\theta - 45)$

Considering moments about 0

$$120.7(r) - T_{s3}(r) - M_s = 0$$

r = 5

∴ $603.5 - \{103.5\cos\theta + 500\sin(\theta - 45)\} = -M_s = 0$

i.e. $M_{s3} = 603.5 - 103.5\cos\theta - 500\sin(\theta - 45)$

Collecting our results for this part, θ going from 90° to 135°.

$SF_{s3} = 20.7\sin\theta - 100\cos(\theta - 45)$ . . . . (x)

$T_{s3} = 20.7\cos\theta + 100\sin(\theta - 45)$ . . . . (xi)

$M_{s3} = 603.5 - 103.5\cos\theta - 500\sin(\theta - 45)$ . . . . (x)

Finally, we come to section CB; *OS'S* has moved beyond the point load at C and the new angle 'θ' is as shown in Fig. 7

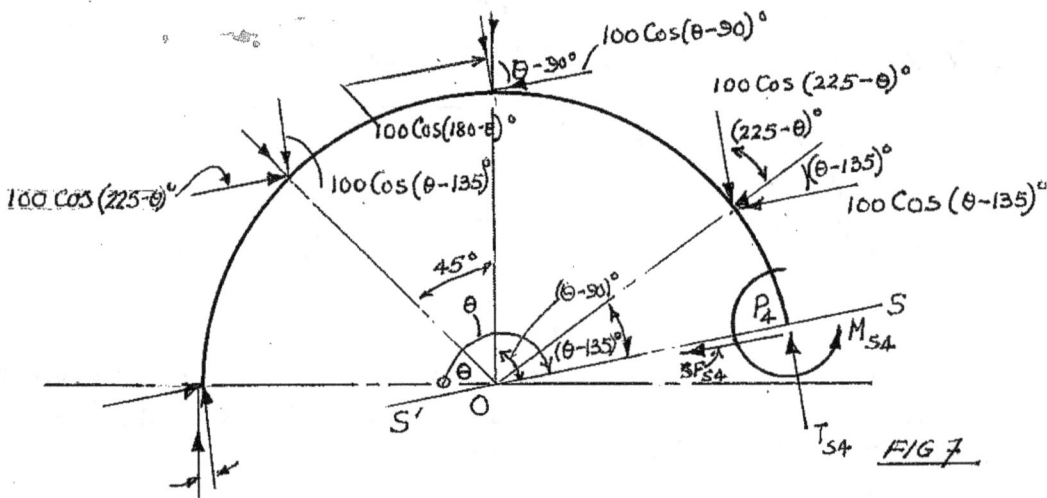

By inspecting Fig. 7 we see that for equilibrium in direction of SF_{s4} we have:

$SF_{s4} + 100\cos(\theta - 135) + 100\cos(\theta - 90) -$
$\quad 100\cos\{90 - (\theta - 135)\} - 120.7\cos(\theta - 90) = 0$

i.e. $SF_{s4} + 100\cos(\theta - 135) + 100\cos(\theta - 90) -$
$\quad 100\sin(\theta - 135)\} - 120.7\cos(\theta - 90) = 0$

We may rewrite this last expression as:

$SF_{s4} + 100\cos - 90 - (\theta - 45)\} + 100\cos\{-90 - \theta)\}$
$\quad -100\sin - \{90 - (\theta - 45)\} - 120.7\cos\{-(90 - \theta)\} = 0$

i.e. $SF_{s4} + 100\sin(\theta - 45) + 100\cos(90 - \theta) +$
$\quad +100\cos(\theta - 45) - 120.7\cos(90 - \theta) = 0$

or $SF_{s4} + 100\sin(\theta - 45) + 100\sin\theta + 100\cos(\theta - 45)$

~ 110 ~

$$-120.7 \sin\theta = 0$$
$$\therefore \quad SF_{s4} = 120.7\sin\theta - 100\sin(\theta - 45) - 100\sin\theta - 100\cos(\theta - 45)$$
$$\qquad\qquad\qquad \cdots\cdots\cdots\cdots \text{(xiii)}$$

For equilibrium perpendicular to plane $OS'S$
$$T_{s4} - 100\cos\{90 - (\theta - 135)\} - 100\cos(180 - \theta) -$$
$$-100\cos(\theta - 135) + 120.7\cos(180 - \theta) = 0$$
i.e. $\quad T_{s4} - 100\sin(\theta - 135) + 100\cos[\{90 - (\theta - 45)\}]$
$$-127.7\cos\theta) = 0$$
$\therefore \quad T_{s4} = 100\sin(\theta - 135) - 100\cos + 100\sin(\theta - 45)$
$$+ 120.7\cos\theta)$$
$\therefore \quad T_{s4} = 100\sin - \{90 - (\theta - 45)\} - 100\cos\theta + 100\sin(\theta - 45)$
$$+ 120.7\cos\theta)$$
or $\quad T_{s4} = 100\sin\{90 - (\theta - 45)\} - 100\cos\theta$
$$+ 100\sin(\theta - 45) + 120.7\cos\theta$$
$$= -100\cos(\theta - 45) - 100\cos\theta$$
$+ 100\sin(\theta - 45) + 120.7\cos\theta$
$T_{s4} = \quad 100\sin(\theta - 45) - 100\cos(\theta - 45) -$
$$100\cos\theta + 120.7\cos\theta) + 120.7\cos\theta \qquad \cdots\cdots \text{(xiv)}$$
$$= \quad 100\sin(\theta - 45) - 100\cos(\theta - 45) + 20.7\cos\theta$$

We come now to the bending moment at the section of the beam at P_4.

For moment equilibrium of the section, we consider as was done before, moments about 0 :
$$R_{AV}(r) - T_{s4}(r) - M_s = 0$$
i.e. $\quad 120.7(r) - T_{s4}(r) - M_s = 0$
Using the expression at (xiv) for T_s we get
$M_{S4} = 120.7(r) - 100(r)\sin(\theta - 45) + 100r\cos(\theta - 45)$
$$+ 100\cos\theta - 120.7r\cos\theta$$
$$= 120.7r(1 - \cos\theta) - 100r\sin(\theta - 45) +$$
$$100r\cos(\theta - 45) + 100r\cos\theta$$
For $r = 5m$
$M_s = 603.5(1 - \cos\theta) - 500\sin(\theta - 45)$
$$+ 500\cos(\theta - 45) + 500\cos\theta \qquad \cdots\cdots\cdots \text{(xv)}$$

We may now summarise our results as follows
Section AE : from θ to $45°$
$$SF_{s1} = 120.7\sin\theta$$
$$T_{s1} = 120.7\cos\theta$$
$$M_{s1} = 603.5(1 - \cos\theta)$$

At E, D, C there are concentrated radial loads each of magnitude 100kN

Section ED : θ from just to right of E and just to left of D

$$SF_{s2} = 120.7Sin\theta - 100Cos(\theta - 45)$$
$$T_{s2} = 120.7Cos\theta + Sin(\theta - 45)$$
$$M_{s2} = 603.5(1 - Cos\theta) - 500Sin(\theta - 45)$$

Section DC : θ from just to right of D and just to left of C
$$SF_{s3} = 20.7Sin\theta - 100Cos(\theta - 45)$$
$$T_{s3} = 20.7Cos\theta + 100 Sin(\theta - 45)$$
$$M_{s3} = 603.5(103.5Cos\theta - 500Sin(\theta - 45)$$

Section CB : θ from just to right of C down to B
$$SF_{s4} = 20.7Sin\theta - 100Sin(\theta - 45) - 100Cos(\theta - 45)$$
$$T_{s4} = 100Sin(\theta - 45) - 100Cos(\theta - 45) + 20.7Cos\theta$$
$$M_{s4} = 603.5(1-Cos\theta) - 500Sin(\theta - 45)$$
$$+ 500Cos(\theta.45) + 500Cos\theta$$
$$= 603.5 - 103.5Cos\theta - 500Sin(\theta - 45) +$$
$$500Cos(\theta - 45)$$

Q24. For the semi-circular beam of Question 23, draw the shear force, thrust and bending moment diagrams for $\theta = 180°$.

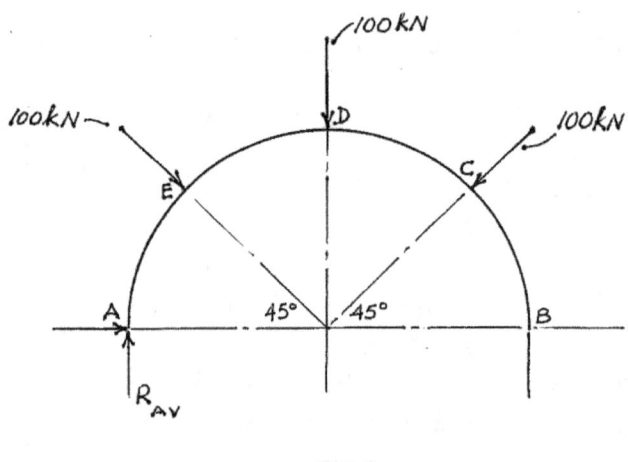

FIG 1

In Question 23 we showed that the shear force, thrust and bending moment at sections '$\theta°$' from AO where, $0 \le \theta \le 45$; $45 \le \theta \le 90$; $90 \le \theta \le 135$; and $135 \le \theta \le 180$ are given by the following expressions:

$0 \le \theta \le 45$: Section AE
$$S.F = 120.7Sin\theta$$
$$T = 120.7Cos\theta$$
$$BM = 603.5(1-Cos\theta)$$

$45 \le \theta \le 90$: Section ED

$$\text{S.F} = 120.7\sin\theta - 100\cos(\theta - 45)$$
$$\text{T} = 120.7\cos\theta + 100\sin(\theta - 45)$$
$$\text{B.M} = 603.5(1-\cos\theta) - 500\sin(\theta - 45)$$

$90 \le \theta \le 135$: <u>Section DC</u>
$$\text{S.F} = 20.7\sin\theta - 100\cos(\theta - 45)$$
$$\text{T} = 20.7\cos\theta + 100\sin(\theta - 45)$$
$$\text{B.M} = 603.5(103.5\cos\theta - 500\sin(\theta - 45)$$

$135 \le \theta 180$: <u>Section CB</u>
$$\text{S.F} = 20.7\sin\theta - 100\sin(\theta - 45) - 100\cos(\theta - 45)$$
$$\text{T} = 100\sin(\theta - 45) - 100\cos(\theta - 45) + 20.7\cos\theta$$
$$\text{B.M} = 603.5 - 103.5\cos\theta - 500\sin(\theta - 45)$$
$$+ 500\cos(\theta - 45)$$

Using these equations, values of SF, T and M were computed.

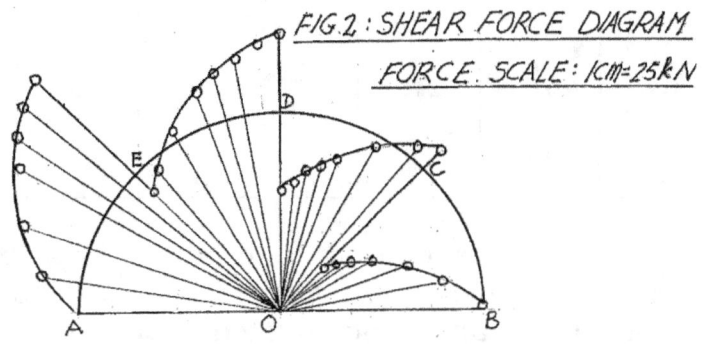

FIG.2 : SHEAR FORCE DIAGRAM

FORCE. SCALE : 1cm=25kN

These are tabulated below.

Table of θ, SF, T and BM for arch.

θ degrees	SF kN	T kN	B.M. kNm	θ degrees	SF kN	T kN	B.M kNm
0	0	+120.7	0	100	-37	78	212
10	+21	+118.8	9	110	-23	84	186
20	+41	+113	36	120	-8	86	172
30	+50	+104	81	130	+7	86	172
40	+78	+93	141	135	+14.7 /-85.3	85	177
45	+85.3/ -14.7	+85	177	140	-78	93	141
50	-7	86	172	150	-60	104	81
60	+8	86	172	160	-41	113	36
70	+23	84	186	170	-21	119	9
80	+37	78	212	180	0	121	0
90	+50/- 50	71	250				

The tabulated values were plotted to produce Figs. 2, 3 and 4. The student's attention is drawn to the entries for SF at 45°, 90° and 135° at which points on the beam concentrated loads, of value 100kN are located. Hence there are changes in the sign of the shear force at these points. Both values were entered against the relevant angles in the Table.

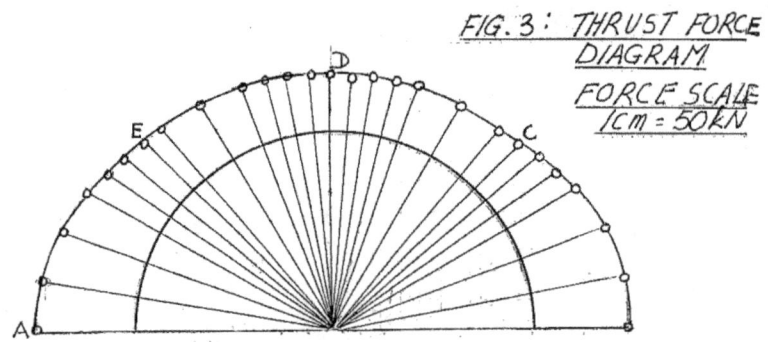

FIG. 3: THRUST FORCE DIAGRAM
FORCE SCALE
1cm = 50kN

Apart from the S.F diagram where values on one half of the beam are of the opposite sign but of identical magnitude as those at corresponding radial points on the other side, both the thrust and bending moment diagrams are symmetrical.

It is also evident from the analysis conducted in Problem No. 23 that the thrust T is of a compressive nature.

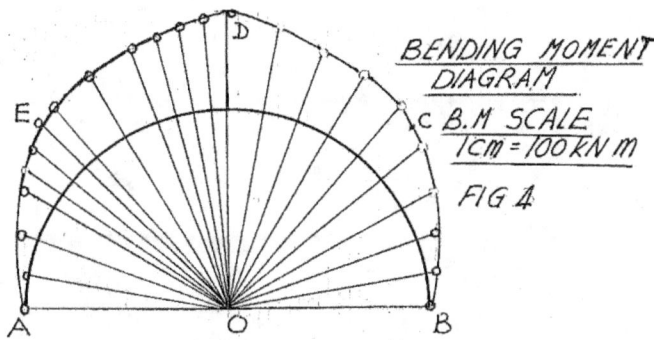

BENDING MOMENT DIAGRAM

B.M SCALE
1CM = 100 KN m

FIG 4

This page is intentionally left blank.

CHAPTER 7

ELASTIC BENDING STRESSES: THE FLEXURE FORMULA FOR STRAIGHT BEAMS; PLASTIC MODULUS

As I have pointed out many times elsewhere in this textbook, it is important to understand the foundations upon which any theory is mounted, especially so if one expects to apply it with confidence or to conduct research so as to advance knowledge of the subject.

In this Chapter I shall outline the fundamental theory and therefrom deduce a relation connecting elastic bending stress, 'σ'; bending moment, 'M'; radius of curvature, 'R' (assumed to have the profile of the arc of a circle) of a straight beam subjected to the bending moment 'M'; 'I' the second moment of area of a section of the beam about an axis called the neutral axis about which more later; and, 'y', the distance from the neutral axis to the place where the value of bending stress 'σ' is required.

The relation is generally expressed thus: $\dfrac{M}{I} = \dfrac{\sigma}{y} = \dfrac{E}{R}$

in which E is the elastic modulus of the material of the beam, assumed constant throughout. This relation is of such fundamental importance in the theory and practice of mechanics of materials that the student of engineering should, as Anselm O'Callaghan, Presentation Brother, my Soul Brother, polymath and mentor of revered memory, used to say: "Frame it and place it over your bed-head."

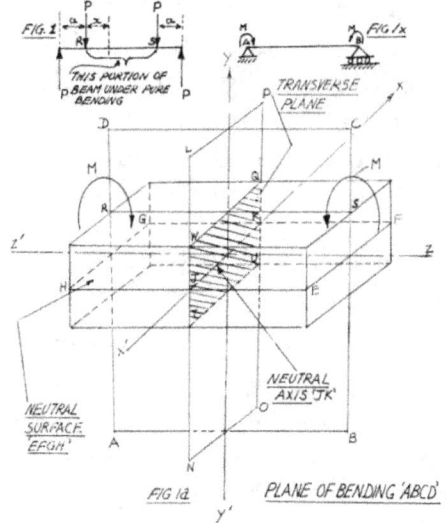

In Fig. 1, a simply supported beam is shown with equal loads 'P' equidistant from each end. Accordingly no shear force acts on the portion of the beam between 'R' and 'S' and bending moment on that portion is constant, being equal to Pa (from $M_x = P(a + x) - Px$). The portion of the beam RS is said to be in a state of pure bending or pure flexure[2]. Characteristics required for this condition? : no shear and constant bending moment. The portion of the beam 'RS' is shown in Fig.1a with the constant bending moment designated 'M' acting on it. It does not require much imagination to visualize that portion of the beam taking up the shape shown in exaggerated fashion for emphasis in Fig. 1b; only part of the portion of the beam RS is shown there.

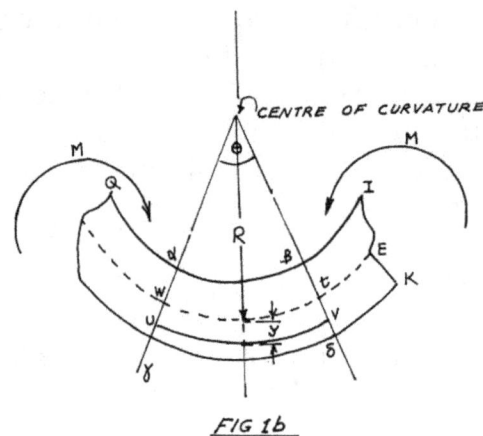

FIG 1b

Clearly the fibres of the beam on the outer surface, of which the trace is JK, are stretched, whereas those on QI are compressed. We shall treat tensile stresses as positive; compressive stresses as negative in accordance with our convention. Thus, as we move from the outermost fibres to the innermost, there must be a surface between JK and *QI* where there is zero stress. This occurs at the layer, say, 'EFGH' in Fig. 1a; at 'HwtE' in Fig. 1b. This layer is called the neutral surface; and the line JK in Fig.1a which is the trace of the intersection of a transverse section of the beam and the neutral surface, which as we just said is the surface with zero stress, is called the neutral axis.

Before proceeding any further it is well to note the assumptions undergirding the foundation of the flexure formula. These are:

---

[2]In some texts e.g. *Strength of Materials* by Shanley, McGraw- Hill Inc., 1957, at page 235, the state of pure bending is erroneously illustrated by a beam resting on end supports at which constant bending moments are applied, e.g. as shown in **Fig. 1x** in that text. Evidently there must be support forces at A and B even if not shown in the diagram. These support reactions would definitely cause shear on the beam and bending moment would vary as a consequence. Pure Bending or Flexure is a state or condition where there is constant bending moment and no shear, i.e. shearing force $'V' = \frac{dM}{dx} = 0$.

(i)   bending stresses and strains are within the elastic range of the material. Accordingly, Hooke's law is applicable; and Young's modulus 'E' is constant throughout the beam, assumed to be of isotropic material.

(ii)   plane transverse sections such as 'WTUQ' in Fig.1a remain plane under the effect of bending, as before bending; at distance y in Fig. 1b, below the neutral axis; σ_{max} is in fact the maximum tensile stress At the neutral axis JK, in Fig. 1a, by definition, the stress there is zero.

(iii) the radius of curvature of the beam in bending is much greater in comparison with the depth and width of the beam; and,

(iv)   the sides of the beam are unloaded. Thus, there is no resultant force on any transverse section of the beam.

Back to Fig.1b: the beam is shown bent. Let R = radius of curvature of the neutral surface, the trace of which is HE. Consider the portion of the beam '$\delta\beta \propto \gamma$' and let '$uv$' be a fibre at distance 'y' from the neutral surface. Accordingly, the radius of curvature at this same fibre 'uv' is $(R + y)$; "wt" is that part of the neutral surface in the slice '$\delta\beta \propto \gamma$'.

From Fig.1b,

$$\frac{uv}{wt} = \left(\frac{R+y}{R\theta}\right)\theta = \frac{(R+y)}{R}$$

Also, strain in the fibre distant 'y' from the neutral surface:

$$= \frac{uv - wt}{wt}$$

$$= \frac{uv}{wt} - 1$$

$$= \frac{(R+y)}{R} - 1$$

$$= \frac{y}{R}$$

If we denote this strain by 'e', then by definition: $e = \frac{y}{R}$

Applying Hooke's law,

$$E = \frac{stress}{strain}$$

$$E = \frac{\sigma}{e} = \frac{\sigma}{y/R}, \sigma \text{ being the stress in the fibre at '}y\text{'.}$$

therefore $\dfrac{\sigma}{y} = \dfrac{E}{R}$ or $\sigma = \dfrac{Ey}{R}$ (ii)

Thus, for constant 'E' and 'R', the bending stress 'σ' in any fibre distant 'y' from the neutral surface is directly proportional to this distance, which of course is another way of saying that stress 'σ' varies linearly with 'y' from top to bottom of a section of the beam. Accordingly, we write $\sigma = Cy$ where 'C' is an arbitrary constant. This linear variation of stress over a typical cross-section of the beam is shown in Fig.2. The maximum bending stress σ_{max} occurs at distance y_2 below the neutral axis JK, at the outermost fibres of the beam; σ_{max} is a tensile bending stress. By a similar token, the minimum stress σ_{min} occurs at distance y_1 above the neutral axis; σ_{min} is in fact the maximum compressive stress, i.e. the greatest negative stress. By definition at the neutral axis JK, the stress there is zero.

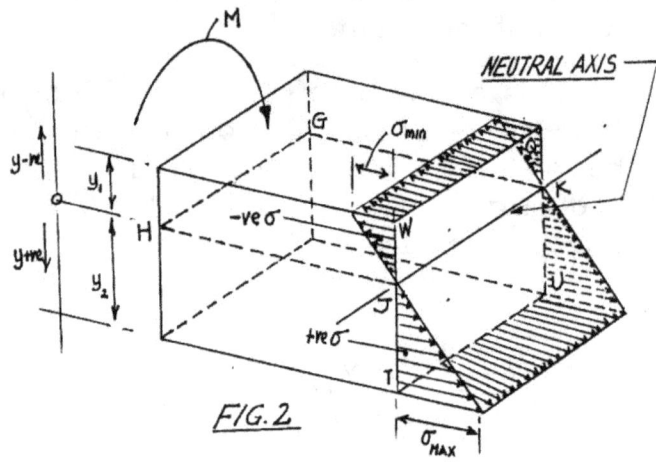

FIG.2

Referring again to Fig.2, observe that 'HJKG' is part of the neutral surface and if we were to consider a fibre distant 'y' below or above JK, then we may write,

$$\pm\sigma = \frac{E}{R}y \text{ or alternatively } \sigma = \frac{E}{R}(\pm y)$$

Clearly, with the beam being of constant cross-sectional dimensions, any 'y' would be a constant for a particular fibre surface and so for constant 'E', the radius of curvature of the beam under the influence of a uniform bending moment would also be constant.

Conclusion: In pure bending the elastic line of a beam has the configuration of a circular arc.

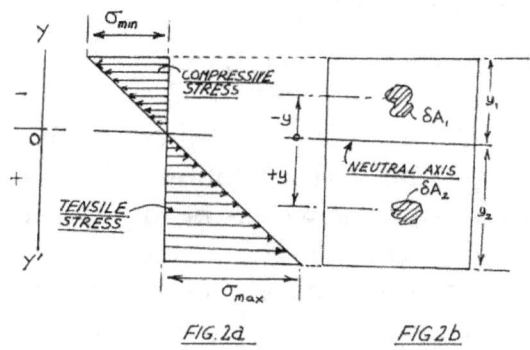

FIG. 2a   FIG 2b

In Fig.2a I have shown a view of the stress distribution in a typical longitudinal plane at a typical cross-section. The face of a typical cross-section is at Fig.2b. The neutral axis is at distant 'y' from the top of the section; and, 'y' from the bottom. Accordingly, at any section above the neutral axis: $\sigma_y = -\sigma_{min}\left(\frac{-y}{y_1}\right)$ a compressive stress; and, at section 'y' below the neutral axis $\sigma_y = \sigma_{max}\frac{y}{y_2}$ (a tensile stress).

If we now consider the forces acting on the infinitesimal areas δA in Fig. 2b above and below the neutral axis, then for the former, $\delta F = -\sigma_{min}\left(\frac{-y}{-y_1}\right)\delta A$; and $\delta F = +\sigma_{max}\frac{(+y)}{+y_2}\delta A$, for the latter. Therefore, the total tension on the section below the neutral axis $= +\frac{\sigma_{max}}{+y_2}\int +ydA$; and the total compressive force above the neutral axis is $\frac{-\sigma_{min}}{y_1}\int ydA$.

Since there can be no resultant force across the section, we have for $\Sigma F_x = 0$:

$$\underset{\substack{y_2 \text{ below N.A}}}{\left[\sigma_{max}\int ydA_2\right]} + (-)\underset{\substack{y_1 \text{ below N.A}}}{\left[\sigma_{min}\int -ydA_2\right]} = 0$$

Referring to similar triangles OAB and ODC in Fig 25a: $\frac{\sigma_{max}}{y_2} = \frac{\sigma_{min}}{y_1}$

Therefore $\int ydA$ for that part of the beam below the neutral axis is equal *to* $\int ydA$ for the part above the neutral axis, but the signs are different. Therefore the sum of the moments 'δA' by distance 'y' for the portions below and above the neutral axis cancel each other, that is to say, $\int ydA$ for the entire section = 0.

Alternatively, from the earlier result $\sigma = \frac{E}{R}$ we could write, for each of the infinitesimal areas:

Nett axial force = $\int\sigma\delta A = 0$

$$= \int \frac{E}{R} y \delta A = \frac{E}{R} \int y dA = 0$$

which leads to the same conclusion viz. $\int y dA = 0$.

By definition for any set of areas δA and distances 'y'from, say, the X- axis ydA = Aÿ; and because 'A' cannot equal zero, it follows that a result such as ydA = 0, means that ÿ = 0. Evidently, since ÿ, which is the distance from a datum axis – in this case the neutral axis which is the X- axis, the result ÿ = 0 means that the neutral axis and the centroid axis are collinear, or to put it another way, the neutral axis passes through the centroid of the cross-section; a very important result to note in cases of simple bending. Frame that one too!

In the foregoing analysis we applied one of the laws of statical equilibrium, viz $\Sigma F_{x, y, z} = 0$ to the bending stress distribution. We shall now apply another law of statical equilibrium viz. $\Sigma M_{x, y, z} = 0$, in considering the moments of the forces due to bending stresses about the neutral axis. Referring again to Fig. 2 bending moment 'M' is a sagging moment in the X –Y plane.

Accordingly, $\Sigma M_z = 0$ which means that the total moment of the forces due to 'σ' acting on all the infinitesimal elemental areas summed up across the cross-section 'WTUQ' in Fig. 2b, about the neutral axis must equate to 'M_z'. The moment of the compressive forces above the neutral axis and also that of thetensile forces below it constitute a couple.

Therefore, $M = \sigma_{max} \cdot \frac{1}{y_2} \int y^2 \, dA + (-\sigma_{min}) \frac{1}{(-y_2)} \int (-y)^2 dA$

As was explained in the chapter on Second Moment of Area $\int y^2 dA$ is a function of only the geometrical properties of a cross-sectional area. It is in fact that well-known property of a section: the moment of inertia, 'I'. Here, 'I' is with reference to the centroidal axis of the section. Again, noting that $\frac{\sigma_{min}}{y_2} = \frac{\sigma_{max}}{y_1}$ our earlier expression for 'M' may be rewritten, substituting:

$$\sigma_{min} = \sigma_{max} \frac{y_1}{y_2} \text{ as } M = \frac{\sigma_{max}}{y_2} \int y^2 \, dA + \sigma_{max.} \frac{-y_1}{y_2} \frac{(1) \int (-y)^2}{-y_1} \, dA$$

which reduces to

$$M = \frac{\sigma_{max}}{y_2} \; [(\int y^2 dA)_{\text{below N.A}} + \{(-y)^2 dA \}_{\text{aboveN.A}}]$$

Referring to Fig.2b we observe that the quantity $(y^2 dA)_{\text{below N.A}} + \{(-y)^2 dA\}_{\text{above N.A}}$ is in fact the second moment of inertia of the section QRST about the neutral axis.

Therefore we may write,

$$M = \frac{\sigma_{max}}{y_2} \; x \, I$$

or $\quad \dfrac{M}{I} = \dfrac{\sigma_{max}}{y_2}$

If in the equation for bending moment 'M' we substituted $\sigma_{max} = \sigma_{min}\left(\dfrac{y_2}{y_1}\right)$, then we would have obtained the result $\dfrac{M}{I} = \dfrac{\sigma_{min}}{y_1}$.

Try it and see!

In fact, generally $\qquad \dfrac{M}{I} = \dfrac{\sigma_{max}}{y_2} = \dfrac{\sigma_{min}}{y_1} = \dfrac{\sigma}{y}$ $\qquad\qquad$ ....(iii)

Recalling earlier result $\dfrac{\sigma}{y} = \dfrac{E}{R}$ $\qquad\qquad\qquad\qquad\qquad\qquad$ ....(ii),
its combination with equation (iii) gives the Flexure Formula for simple bending in straight beams:

$$\frac{M}{I} = \frac{E}{R} = \frac{\sigma}{y}$$

Bear in mind when using the Flexure formula that 'y' is the distance of a fibre from the neutral axis: plus 'y' (i.e. $+y$) if on the side of tensile stress and minus 'y' (i.e. $-y$) if on the side of compressive stress. Hence we may write:

$$\pm\sigma = \frac{My}{I} = \frac{M}{I_y} = \frac{M}{Z}$$

The quantity I/y is called the section **ELASTIC MODULUS** and it is generally denoted by the letter 'Z'. Thus, if the tensile stress 'σ_t' in a fibre is distant y_t from the neutral axis, then

$$+\sigma_t = \frac{M}{I_{yt}} = \frac{M}{Z_t}$$

Similarly, if the compressive stress $-\sigma_c$ is in a fibre y_c from the neutral axis, then

$$-\sigma_c = \frac{M}{Z_c}$$

Z_t is the elastic tension modulus of the section and Z_c the elastic compression modulus.

Make a vignette of these results and add to the lot over your bedhead.

Illustrative Example 1

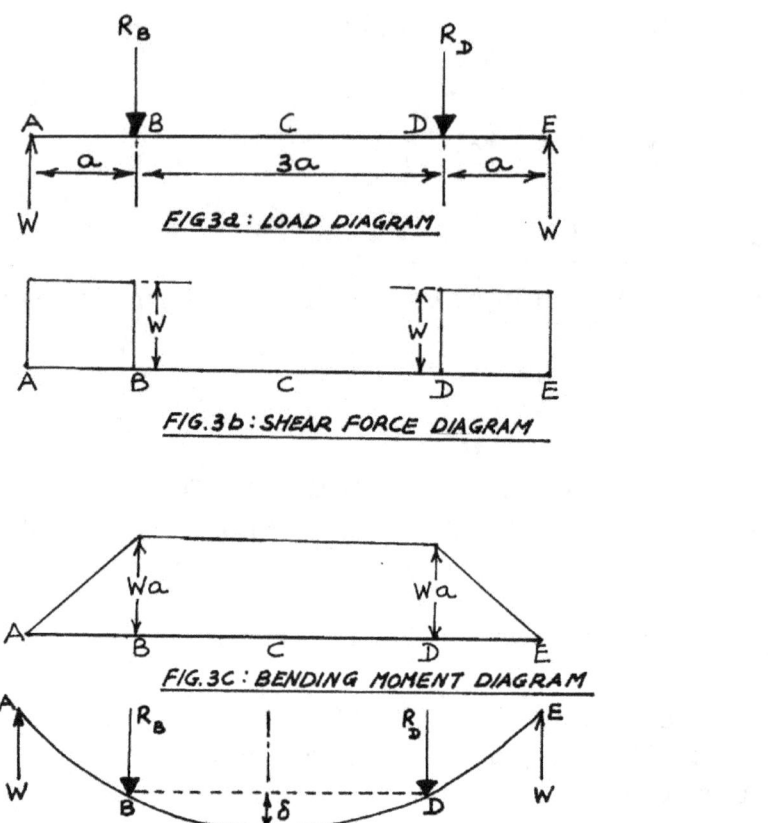

FIG.3a : LOAD DIAGRAM

FIG.3b : SHEAR FORCE DIAGRAM

FIG.3c : BENDING MOMENT DIAGRAM

FIG.3d : BENDING MOMENT ON PORTION BCD IS CONSTANT AND SHEAR FORCE IS ZERO

In Fig. 3a I have shown a beam ABCDE with knife-edge supports at B and D and with equal point loads *W* acting at *A* and *E*. Bending moment over the portion of the beam BD ' is constant and equal to Wa; also there is no shear force on this portion of the beam. Refer to Fig. 3b and 3c. Consequently, a state of pure bending exists on BD, and the elastic curve for this section is an arc of a circle of radius 'R'.

Let us evaluate the sag 'δ' of the beam at mid-span. As represented in Fig. 3e the straight line BD is a chord of a circle of radius 'R' and the elastic curve BCD is part of the circumference of this circle.

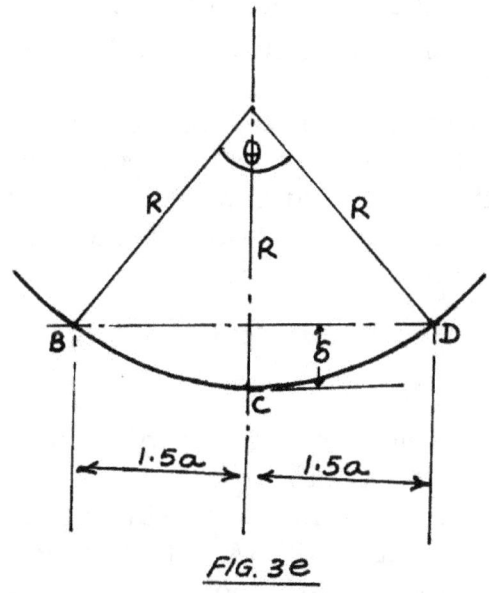

FIG. 3e

From geometrical considerations,

$(R - \delta)^2 + (1.5a)^2 = R^2$

which may be reduced to:

$2R\delta - \delta^2 = 2.25a^2$

Because δ is small in comparison with R, we may discard δ^2.

$\therefore \qquad\qquad 2R\delta = 2.25a^2$

or $\qquad\qquad R = \dfrac{2.25a^2}{2\delta}$

Substituting this value of 'R' in the Flexure formula, we have

$$\frac{M}{I} = \frac{E}{R} = \frac{E}{2.25a^2/2\delta}$$

from which $M = \dfrac{2EI\delta}{2.25a^2}$ or $M = \dfrac{k\delta}{a^2}$, in which 'k' is a constant $= EI/1.125$.

Referring to Fig.3d, it is evident that by varying 'W' for a beam of constant cross-section, and therefore constant moment of inertia 'I', we could by writing *Wa* for bending moment 'M' restate the last result as,

$$Wa = \frac{k\delta}{a^2} \quad \text{or} \quad W = \frac{k\delta}{a^3}$$

~ 125 ~

from which it is seen that a graph of 'W' (as ordinates) against deflection 'δ' as (abscissae) should be a straight line passing through the origin of the 'W versus δ' axes. The slope of such a characteristic would be equal to $\dfrac{EI}{1.125a^3}$.

Hence an estimate of Young's modulus can be obtained in this way once we know the values of 'I' and the distance between the supports. Now, go and build an apparatus in the engineering workshop and design an appropriate experiment to determine the value of Young's modulus based on this model. Here is a project to test your skill in applying the Flexure formula to straight isotropic beams.

Illustrative Example 2

The plan view of a short stable column is that of a symmetrical I- section as shown in Fig. 4a. The cross-sectional area of the column is uniform, being $40 \times 10^3 (mm)^2$ throughout its length, and its maximum radius of gyration $k_{xx} = 50$ mm. A vertical concentrated load of $25\ kN$ acts through the axis of the column and another vertical load PkN acts at a point $100\ mm$ away from this load. See Figs. 4b and 4c. If the maximum tensile stress is not to exceed $8\ MN/m^2$, then determine the value of P.

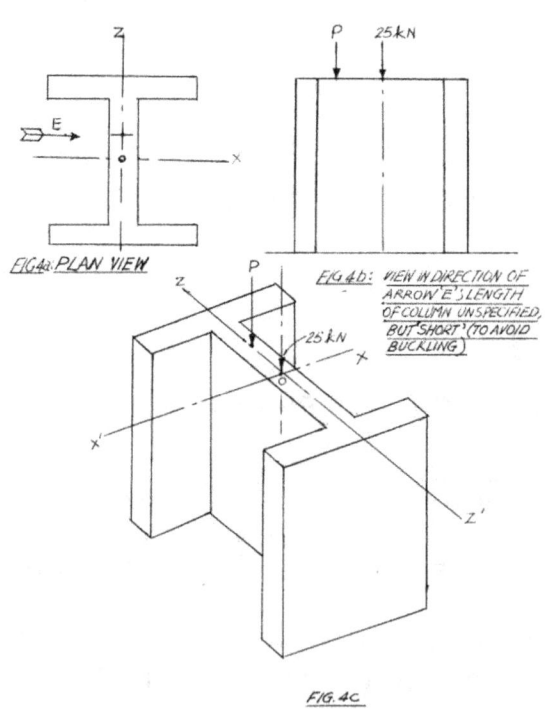

FIG 4a: PLAN VIEW

FIG 4b: VIEW IN DIRECTION OF ARROW 'E', LENGTH OF COLUMN UNSPECIFIED, BUT 'SHORT' (TO AVOID BUCKLING)

FIG. 4c

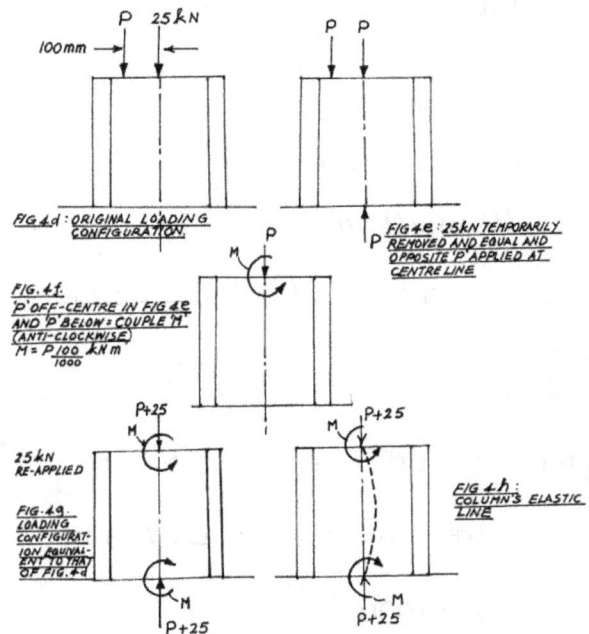

Fig.4.d: ORIGINAL LOADING CONFIGURATION.

FIG 4.e: 25kN TEMPORARILY REMOVED AND EQUAL AND OPPOSITE 'P' APPLIED AT CENTRE LINE

FIG.4.f: 'P' OFF-CENTRE IN FIG 4.e AND 'P' BELOW = COUPLE 'M' (ANTI-CLOCKWISE) $M = P \frac{100}{1000}$ kN m

25kN RE-APPLIED

FIG.4.g LOADING CONFIGURAT-ION EQUIVAL-ENT TO THAT OF FIG.4.d

FIG 4.h: COLUMN'S ELASTIC LINE

I have gone into considerable detail to show how the loading configuration on the column may be reduced to a compressive load of *(P + 25) kN* acting at the centroid of cross-section, and a bending moment $'M''M' = P\left(\frac{100}{1000}\right)$ *kNm* causing the elastic line of the column to take shape shown at Fig.4h: the loading configuration as stated in the question is represented in the series of drawings Fig. 4a through Fig. 4h. In Figs. 4a, 4b and 4c, the direct compressive load of *25 kN* is acting through *'O'*, the centroid of the cross-section. In Fig. 4d the *25 kN* is temporarily removed, and in Fig. 4e, two equal and opposite forces *'P'* act through *'O'*. The net effect is shown in Fig. 4f viz. a moment or couple *'M' = P(100/1000) kNm* acting anti-clockwise about the neutral axis *'XX'* of the cross-section. If your ambition is to be a professional engineering designer then the foregoing is representation of the kind of analysis you would be expected to perform. Finally, we reinstate the direct compressive load of *25 kN* in Fig. 4h. The total compressive force acting through the centroid of the area of column is *(P + 25) kN;* and the given cross-sectional area is *40 x 10³ (mm)²*. Therefore the direct compressive stress, say, σ_c on this cross- sectional area is:

$$\sigma_c = \frac{(P+25)10^3}{\frac{40 \times 10^3}{10^6}}$$

$$= \frac{(P+25)}{40}\, 10^6\ N/m^2$$

Remembering that compressive stresses are treated in this text as negative, we may accordingly write:

$$\sigma_c = -\frac{(P+25)}{40}\, 10^6\ N/m^2$$

Maximum moment of inertia is about the $XX-$ axis. I_{xx} is obtained from

$$I_{xx} = A \; (k_{xx})^2$$

i.e.

$$I_{xx} = \frac{40 \times 10^3}{10^6} \; x \; \frac{(50)^2}{1000}$$

$$= 100 \times 10^{-6} \; m^4$$

To determine the bending stress due to the bending moment of $P\left(\frac{100}{1000}\right) kNm$,

recall

$$\frac{M}{I} = \frac{\sigma}{y}$$

Clearly the maximum tensile bending stress σ_t occurs where $y_t = \frac{400}{2} = 200 \; mm$

Working in newtons and metres and noting that P is in kN, we have

$$\frac{P \; x \; 1000 \; x \; \frac{100}{1000}}{100 \; x \; 10^{-6}} = \frac{\sigma_t}{\frac{2}{1000}}$$

from which $\sigma_t = 0.2P \; x \; 10^6 \; N/m^2$.

Now maximum tensile stress we are told is not to exceed 8 MN/m². This means that the maximum tensile bending stress: $\sigma_t = + 0.2P \; x \; 10^6 \; N/m^2$ plus the direct compressive: $\sigma_c = -\left(\frac{P+25}{40}\right) 10^6 \; N/m^2$ must not exceed 8 MN/m².

Accordingly, maximum value of P is obtained from:

$$-\left(\frac{P+25}{40}\right) 10^6 + 0.2P \; x \; 10^6 = 8 \; x \; 10^6$$

$$-0.025P - 0.625 + 0.2P = 8$$

therefore, $0.175P = 8.625$ or $P = 49.3 \; kN$

RADIUS OF CURVATURE AND BEAM DEFLECTION

For pure bending with constant moment of inertia *(I)* and constant Young's modulus *(E)* it should be clear that the elastic line or elastic curve of the beam is part of an arc of a circle of radius *'R'* i.e. curvature is constant. Now, consider a beam similar to that in Fig. 1a, but this time subjected to transverse loading thereby causing bending moment *'M'* to vary along its length. When the effect of shear stress is neglected the relationship $\frac{M}{I} = \frac{E}{R}$ may be used to determine the beam's deflection due to bending, which we demonstrated in an earlier example.

Evidently, if I and E are constant and $'M'$ is varying, then the radius of curvature $'R'$ or alternatively curvature $(1/R)$, must be varying.

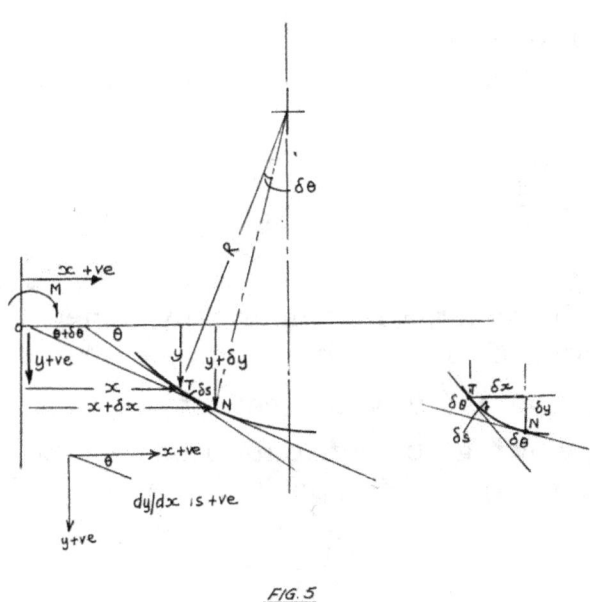

FIG. 5

In Fig.5, points $'T'$ and $'N'$ are on the elastic curve of a beam of varying bending moment $'M'$ and varying $'R'$. Point $'T'$ has coordinates '' and $'y'$ and point $'N', +$ δx; $y+\delta y$; the tangent to the curve at T makes an angle $'\theta'$ with the positive directions of X- axis; the tangent to the curve at N, an angle $(\theta + \delta\theta)$. Evidently, the angle between these two tangents is $\delta\theta$ which is the same as the angle between the two perpendiculars erected at T and N and meeting at O. If $\delta\theta$ tends to zero these two perpendiculars 'merge' to become R, the radius of curvature at T. It is important to note that the y- axis is positive downwards.

Referring again to Fig. 5, $\delta s = R\delta\theta$

or $$\frac{1}{R} = \frac{\delta\theta}{\delta s} \approx \frac{\delta\theta}{\delta x}$$

In the limit $$\frac{1}{R} = \frac{\delta\theta}{\delta s} = \frac{\delta\theta}{\delta x}$$

Also, $$Tan\theta = \frac{\delta y}{\delta x}$$

But we also know that $Lt \frac{Tan\,\theta}{\theta} = 1$ as θ approaches zero.

Therefore since $$\theta = \frac{dy}{dx}$$

$$\frac{d(\theta)}{dx} = \frac{d^2y}{dx^2}$$

i.e.
$$\frac{1}{R} = \frac{d^2y}{dx^2}$$

But from the relationship $\frac{M}{I} = \frac{E}{R}$

or,
$$\frac{I}{R} = \frac{M}{EI}$$

and so
$$\frac{M}{EI} = \frac{d^2y}{dx^2} = \frac{d\theta}{dx}$$

As may be observed by reference to Fig.5, y increases as x increases; slope θ is falling as x increases. In fact, $\delta\theta$ is numerically a negative quantity: external angle $TUS = \theta$ is the sum of the two interior angles NLS $(\theta+\delta\theta)$ and LVU $(= -\delta\theta)$. So the rate of change of dy/dx with respect to 'x' is decreasing i.e. $\frac{d^2y}{dx^2}$ is negative. Therefore, following our convention for clockwise moment on the left (which is a sagging moment) as being positive, we may write thusly:

$$\frac{d^2y}{dx^2} = \frac{M}{EI}$$

Established thereby is the mathematical relationship between bending moment 'M', flexural rigidity (EI) and deflection 'y' of the elastic curve. Another one for the bed-head.

I shall be referring to it shortly in the chapter dealing with beam deflection. However it is well to note at this juncture that in some textbooks the fundamental mathematical relationship just derived is instead given by:

$$\frac{d^2y}{dx^2} = +\frac{M}{EI}$$

This is simply due to the fact that the sign convention employed in the derivation of the latter is different from the one we employed. For whereas in our analysis the 'y' coordinate was considered positive downwards, in the other 'y' is taken as positive upwards. Otherwise, the other criteria are the same: 'x' is positive from left to right; angle of rotation θ is positive clockwise from the x-axis; bending moment 'M' is sagging (positive); and curvature is positive when the beam bends with concavity upwards.

In Chapter 17 elastic deflection of straight beams and some other engineering structures shall be considered using the foregoing analytical foundation.

CLOSURE ON THE FLEXURE FORMULA

As important as it is in the analysis of mechanics of materials one must always bear in mind a theory's underlying assumptions especially the one about pure bending, in this case. The state of pure bending is a condition that is most uncommon in practical engineering problems. One may well ask, "What happens when the bending moment is not constant, but varying along a beam, say?"

Well, according to the Mathematical theory of Elasticity[3], the relationship codified as it were, in the Flexure formula, is a sufficiently good approximation when bending moment M changes along the beam, provided the cross-sectional area of the beam is constant. In fact in the mathematical Theory of Elasticity, the relations $e = y/R$; $\sigma = Ey/R = MI/y$ are referred to as being of the Approximate Theory of Flexure.

Let us now apply it to a common problem of practical importance.

Illustrative Example 3

A mechanical-engineering technician is supervising the off-loading of cement-lined ductile-iron pipes from the MV "Ernie" at Point Lisas. She must first determine the positioning of the slings which are to be equidistant from each end of a typical pipe. She does this to ensure that the bending stress in the pipe is a minimum during lifting. See Fig. 6.

A typical cement-lined ductile-iron pipe of nominal diameter (ND) 1000 mm, has a length of *8m* and its total mass is 2980 kg. Assuming that the weight of the pipe is uniformly distributed over its length and that it is being lifted in a horizontal position, determine where the technician should order that the slings be positioned to induce minimum bending stress in the pipe as each length of pipe is being hoisted. Assume $g = 9.80$ m/s².

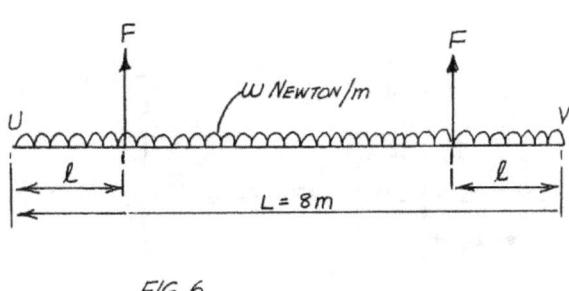

FIG. 6

[3]See for example: Southwell, R.V. An Introduction to the Theory of Elasticity for Engineers and Physicists. OUP, 2nd Edn,p. 157 ff.

'w' newton/metre $= \dfrac{2980 \times 9.80}{8}$

$$= 3650.5 \ N/m$$

$$= 3.65 \ kN/m$$

For minimum stress, a condition is required such that a minimum bending moment occurs on the pipe.

Let l be the distance of each sling from each end of the pipe for minimum bending stress. Therefore, *B.M* at each sling $= \dfrac{wl^2}{2}$ for this condition. This means the maximum B.M on the pipe must not exceed this valve.

Now, let x = distance from U where maximum B.M occurs. See Figs. 7 and 7a.

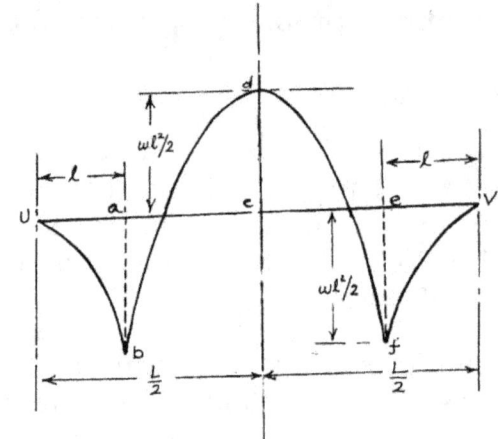

FIG 7. BENDING MOMENT DIAGRAM

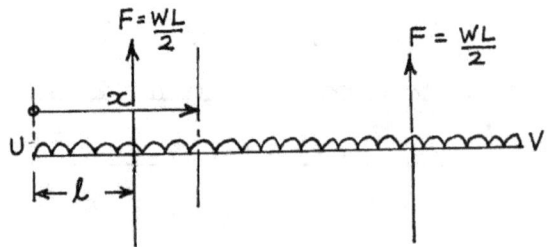

FIG 7a

$$BM_x = F\,(x - l) - \frac{wx^2}{2}$$

$$BM_x = \frac{wL\,(x-l)}{2} - \frac{wx^2}{2} \qquad \text{(i)}$$

i.e. $\quad BM_x = \frac{wLx}{2} - \frac{wLl}{2} - \frac{wx^2}{2} \qquad \text{(ii)}$

For maximum or minimum $BM, \frac{d}{dx,}(B.M_x) = 0.$ Accordingly

$$\frac{d}{d_x}(BM_x) = \frac{wL}{2} - wx = 0$$

therefore $x = \frac{L}{2}$

This means that the maximum bending moment occurs at mid-span; its value is from (i)

$$\text{BM}_{\text{max}} = \frac{wL}{2}\left(\frac{L}{2} - l\right) - \frac{wL^2}{8}$$

Evidently, in order to satisfy the conditions of minimum $B.M$ on the pipe, BM_{max} must be equal to the value of BM at the position where each sling is fitted, because whatever the value of 'ℓ', the bending moment at 'a' and 'e' in Fig.7 must always be: $\frac{wl^2}{2}$ a movable minimum value, if you will. Therefore BM_{max} must equal $\frac{wl^2}{2}$ for minimum bending stress, i.e. cd must equal ab in Fig. 7.

i.e. $\frac{wL^2}{4} - \frac{wLl}{2} - \frac{wL^2}{8} = \frac{wl^2}{2}$

or $\qquad L^2 - 4\,Ll = 4l^2$

We must now solve the quadratic equation:

$$4l^2 + 4Ll - L^2 = 0$$

to obtain the value of 'l'.

Accordingly, $\quad l = \dfrac{-4L \mp \sqrt{16L^2 + 16L^{\,2}}}{8}$

$$= \frac{-4L \mp (4\sqrt{2})\,L}{8}$$

$$= \frac{-4L \pm 5.65L}{8}$$

Only possible solution is:

$$l = \frac{1.656L}{8}$$

The length 'L' of the pipe = *8m*

therefore $l = 1.656m$, say, $1.66m$.

Accordingly, the slings must be placed *1.66m* from each end of the pipe.

The minimum bending moment $= \dfrac{wl^2}{2} = \dfrac{3.65 \, x \, (1.66)^2}{2} = 5.03 \, kNm$, say $5000Nm$

or $5kNm$.

To determine the stress 'σ' in the pipe as it is being lifted, we recall

$$\sigma = \pm \frac{M}{Z} = \pm \frac{5 \, kNm}{Z}$$

Where Z = section modulus of the pipe in appropriate units, in this case M^4 because we are working in metres. This value may be obtained from the manufacturer's handbooks. All you have to do is to substitute it in the last equation taking care to ensure conformity of units.

PLASTIC MODULUS

Structural steels are typical elastic-plastic construction materials and their behaviour under stress is reflected in the idealized stress vs. strain diagram, Fig. 8a, derived from the typical characteristic shown in Fig. 8b.

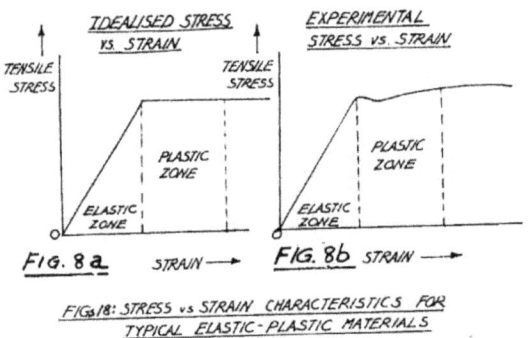

FIGs 18: STRESS vs STRAIN CHARACTERISTICS FOR TYPICAL ELASTIC-PLASTIC MATERIALS

In elastic bending, the stresses $+\sigma$, tensile, and $-\sigma$, compressive due to bending moment 'M' are determined from the truncated Flexure Formula:

$$\frac{M}{I} = \frac{\sigma}{y} \quad or \quad M = \sigma Z$$

'Z' being the Elastic Modulus = I/y, and y = distance from the neutral axis to the fibres where the bending stress = $\pm\delta_y$. See Fig. 9.

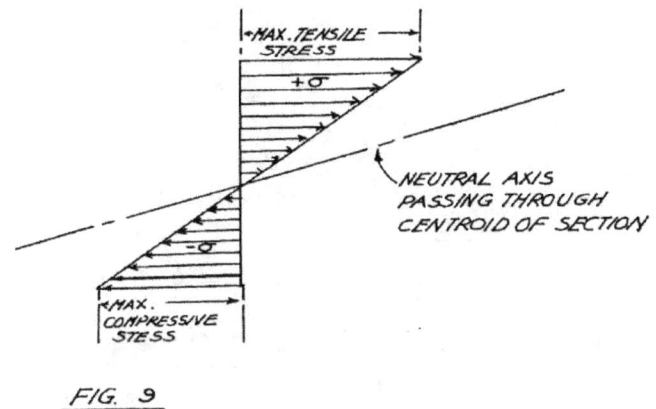

FIG. 9

If the bending moment '*M*' on, say, a beam of uniform cross-section throughout, is increased to a value such that the maximum stress in the outermost fibres on each side of the neutral axis is equal to the yield stress of the material, say, σ_y then the bending moment at which this occurs is the yield moment designated M_y. Accordingly at this stage:

$$\sigma_y = \frac{M_y c}{I} \ or \ M_y = \frac{\sigma_y I}{c}$$

in which C = <u>furthest distance of fibres from the neutral axis</u>. Evidently for a beam with a square cross-section of side '*b*'

$$\sigma_y = \frac{6M_y}{b^3}$$

$$M_y = \frac{\sigma_y b^3}{6}$$

When the bending moment '*M*' is increased, the strain in the fibres of the beam will increase and as shown in Fig. 10 the outer fibres will yield plastically. At this stage, fibres close to the neutral axis will still be elastic, the idealized stress distribution in the beam being somewhat like that also shown in Fig. 10.

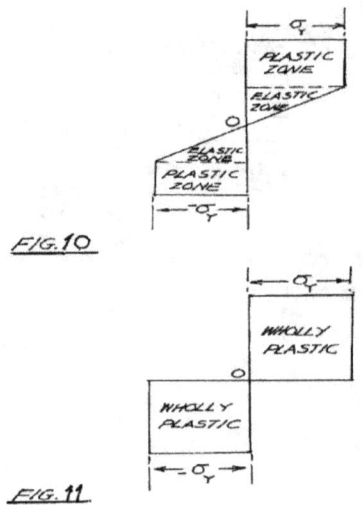

FIG.10

FIG.11

With further increase in bending moment, the plastic zone eventually extends all the way to the neutral axis. The stress distribution in this condition is shown in Fig. 11, the neutral axis being perpendicular to the plane of this sheet and through 'O'.

The bending moment for this condition is called the plastic moment, designated M_p. In Fig. 12a, the distribution of σ_y across the width of the beam is shown. In Fig. 12b, the entire cross section 'abcd' is also shown with the neutral axis separating the cross-sectional area into two parts: A_1 above the neutral axis and A_2 below it as indicated. Evidently:

$$\sigma_y A_1 = \sigma_y A_2, i.e. \; A_1 = A_2$$

and because $A_1 + A_2 = A$, it follows that $A_1 = \frac{A}{2} = A_2$

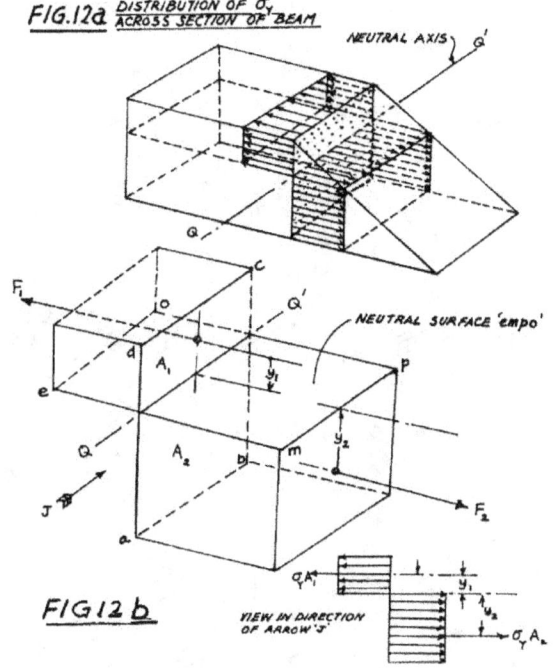

FIG.12a DISTRIBUTION OF σ_y ACROSS SECTION OF BEAM

FIG 12 b

VIEW IN DIRECTION OF ARROW 'J'

With reference to Fig. 12b the moment of resistance, M_R at the cross-section abcd is:

$$M_R = F_1(y_1) + F_2(y_2)$$

and for equilibrium at the section $M_R = M_P$. See Figs. 12c and 12d.

Also equilibrium at abcd in a direction perpendicular to the neutral axis, requires that $F_1 = F_2$. Accordingly:

$$M_R = M_P = F_1(y_1) + F_2(y_2)$$

$$= F_1 (y_1 + y_2)$$

and because $\quad F_1 = \sigma_y A_1 = \sigma_y \dfrac{A}{2}$

the plastic moment may be expressed as:

$$M_p = \sigma_y \frac{A}{2} (y_1 + y_2)$$

Recalling the regular Flexure Formula

$$\frac{\sigma}{y} = \frac{M}{I} \quad or \quad M = \sigma \, \frac{I}{y} = \sigma Z$$

we may write: $\qquad\qquad M_p = \sigma_Y Z_p$

in which
$$Z_p = \frac{A}{2}(y_1 + y_2).$$

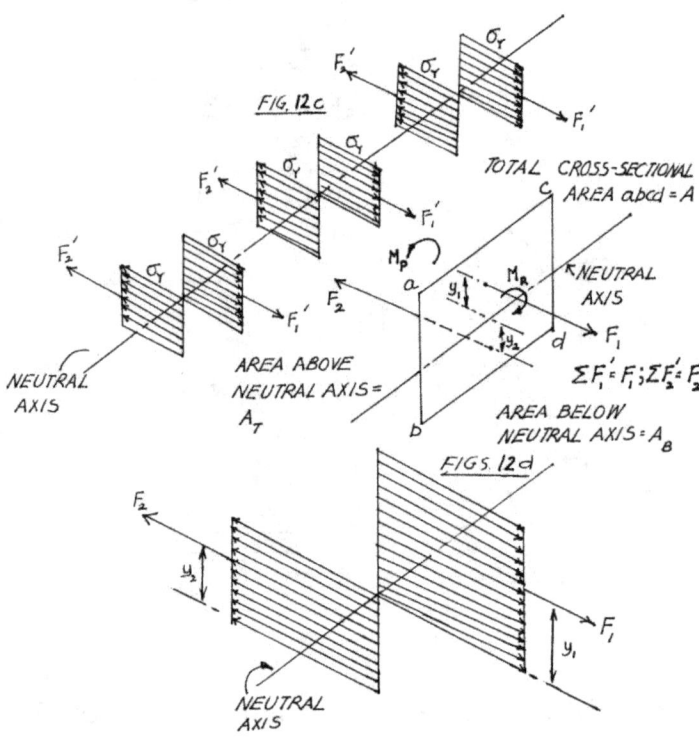

FIG. 12c

TOTAL CROSS-SECTIONAL AREA abcd = A

NEUTRAL AXIS

AREA ABOVE NEUTRAL AXIS = A_T

AREA BELOW NEUTRAL AXIS = A_B

$\Sigma F'_i = F_i ; \Sigma F'_2 = F_2$

NEUTRAL AXIS

FIGS. 12d

NEUTRAL AXIS

This latter expression is referred to as the **plastic modulus** of the section. The ratio $\frac{M_p}{M_y}$ is known as the Shape Factor, designated 'f'. Given a beam of a particular cross-section, in order to calculate the plastic modulus, the first order of business is to determine the centroid of the cross-sectional area. Having thus divided the area, the next step is to determine the distance of the centroid of each part from the neutral axis. Hence y_1 and y_2 and y_3 In at least one Table of the dimensions and properties of sections shown in the Appendix, values of plastic moduli are tabulated.

Illustrative Example 4

The cross-section of a hypothetical I-beam is shown in Fig. 13. Determine the plastic modulus of the section.

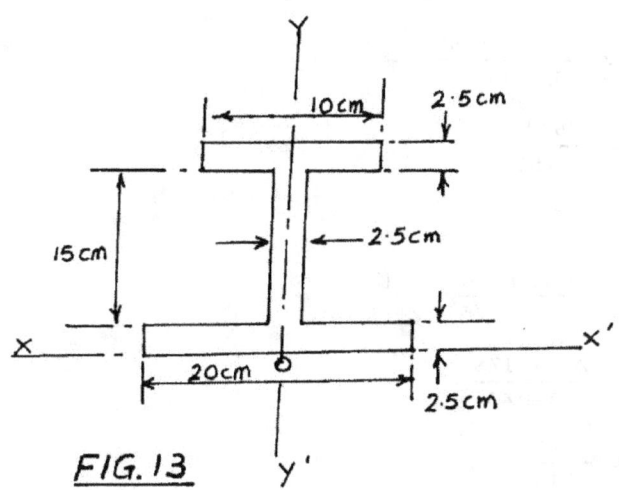

FIG. 13

Determine first the centroid of the section. Let the distance of the centroid be y from the X-X' axis.

Therefore:

$$\bar{y} = 20(2.5)(1.25) + 15(2.5)(10) + 10(2.5)(18.75)$$

$$= \frac{62.5 + 375 + 468.75}{50 + 37.5 + 25}$$

$$= \frac{706.25}{112.5} = 6.28 \; cm$$

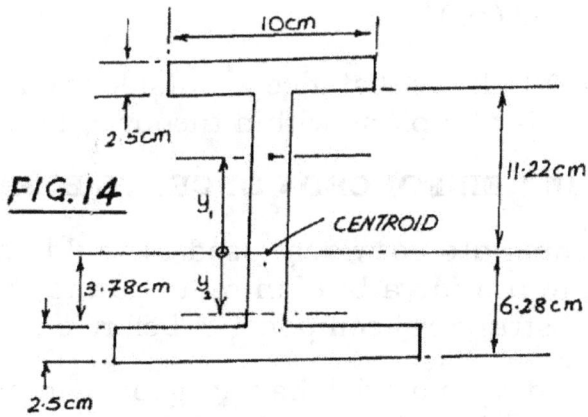

FIG. 14

In Fig. 14, let y_1 and y_2 be the distances of the areas respectively above and below the line X'-X' passing through the centroid of the cross-section.

Considering the upper area first:

$$y_1 = \frac{10(2.5)(13.72 - 1.25) + 11.22(2.5)(5.61)}{10(2.5) + 2.5(3.78)}$$

$$= \frac{311.75 + 157.36}{25(2.5) + 2.5(3.78)}$$

$$= \frac{469.11}{53.05}$$

$$\therefore \qquad y_1 = 8.84 \; cm$$

and the lower: $\qquad y_2 = \dfrac{20(2.5)(5.03) + 2.5(3.78)(1.89)}{20(2.5) + 2.5(3.78)}$

$$= \frac{251.5 + 17.9}{50 + 9.45}$$

$$= \frac{269.4}{59.45}$$

$$y_2 = 4.53 \; cm$$

Let Plastic Modulus $= Z_p$

$$Z_p = \frac{A(y_1 + y_2)}{2}$$

$$A = 10(2.5) + 15(2.5) + 20(2.5)$$

or $\qquad A = 112.5 \; (cm)^2$

$\therefore \qquad Z_p = 112.5 \dfrac{(8.84 + 4.53)}{2}$

i.e. $\qquad Z_p = 752 \; (cm)^2$

Observe that while Fig. 14 shows distance y_2 outside the flange, the calculation proves otherwise, i.e. '$y_2{}'$ is at a plane within the lower flange of the section.

BENDING STRESSES IN REINFORCED CONCRETE BEAMS

Reinforced concrete beams are so widely used in building construction that I thought it appropriate to provide a brief introduction to the subject of bending stress in these common structural components before ending this Chapter.

Plain concrete is a building material having good compressive strength but relatively poor tensile behavioural properties. In contrast, steel possesses both good tensile and compressive strength. It follows therefore that a plain concrete beam subject to an external bending moment or even bending under its own self weight would fail as a direct consequence of induced tensile stress. In order to overcome this inherent deficiency, steel bars are installed in the matrix of the plain concrete at a position such that the bars which now constitute a component of the composite material consisting of concrete and steel, carry the tensile forces. In essence the steel takes care of the tensile stresses and the concrete, the compressive stresses in a manner described in what is called "reinforced concrete". Generally speaking the amount of steel used in an optimum design, i.e. where both materials of the composite are stressed to the maximum, is referred to as the "economic percentage".

It is well to note at the outset that the following theory called the "No tension Theory" is just one of the many available. It is widely accepted however.

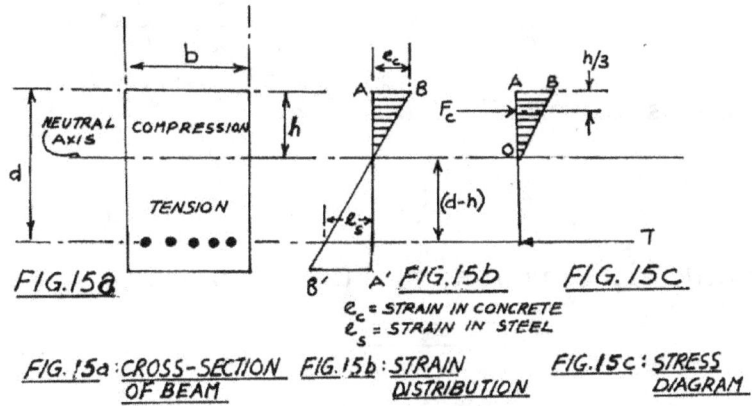

FIG. 15a: CROSS-SECTION OF BEAM   FIG. 15b: STRAIN DISTRIBUTION   FIG. 15c: STRESS DIAGRAM

The underlying assumptions of the "no tension theory" are summarized thus:

(i)   no relative motion between concrete and steel reinforcement;

(ii)   the steel bars carry all the tension;

(iii)   the tensile stress induced in each steel bar is evenly distributed over each bar's cross-sectional area;

(iv)   the stresses in the concrete and steel are elastic; and,

(v)   plane sections before bending remain plane after bending.

Consider the cross-section of a reinforced beam shown in Fig. 15a and note that its breadth is 'b'; depth at which the steel bars are placed is 'd' from the top surface of the beam; and 'h' the distance of the neutral axis from the top. Note in particular that 'h' does not pass through the centroid of the section. It is assumed that the bars are placed in the tension face of the beam.

In the strain diagram Fig. 15b, the plane through 'AA' remains a plane under the influence of a bending moment and its trace takes up the position 'BB'. This means that strains are proportional to their distance from the neutral axis, i.e.

$$\frac{e_c}{e_s} = \frac{h}{d-h}$$ .... (1)

denoting, σ_c = stress in concrete at AB, and

σ_s = stress in steel

then $e_s = \dfrac{\sigma_s}{e_s}$ and $E_c = \dfrac{\sigma_c}{e_c}$ or alternatively

$$\frac{e_c}{e_s} = \frac{\sigma_c}{E_c} \div \frac{\sigma_s}{E_s}$$

Therefore by (i) $\quad\dfrac{\sigma_c}{E_c} \div \dfrac{\sigma_s}{E_s} = \dfrac{h}{d-h} \quad or \quad \dfrac{\sigma_c}{E_c} \cdot \dfrac{E_s}{\sigma_s} = \dfrac{h}{d-h}$

The modular ratio $\dfrac{E_s}{E_c}$ is universally denoted 'm'

therefore, $\qquad\qquad \dfrac{\sigma_c}{\sigma_s} = \dfrac{l}{m}\left(\dfrac{h}{d-h}\right)$ .... (2)

from which by a little algebraic manipulation we obtain

$$h = \frac{\sigma_c\,md}{(\sigma_s + \sigma_c\,m)}$$ ....(3)

Assuming that the total cross-sectional area of all the steel bars is A_s, it is evident that the tensile force designated 'T' in Fig. 15c is given by:

$$T = \sigma_s A_s$$ .... (4)

The distribution of compressive stress being as represented by the right-angled triangle AOB in Fig. 15c, the total compressive force F_c is given by:

$$F_c = \left(\tfrac{1}{2}bh\right)\sigma_c$$ .... (5)

And F_c acts through the centroid of said triangular area, i.e. $\dfrac{h}{3}$ from the top surface of the beam.

The compressive force F_c and T must be equal and therefore constitute a couple. This couple is what balances the applied bending moment, say, M, at any cross-section. Accordingly,

$$M = \sigma_s A_s \left(d - \tfrac{h}{3}\right) \quad \text{for steel}$$ .... (6)

and $\qquad\qquad M = \dfrac{1}{2}bh\sigma_c \dfrac{(d-h)}{3} \quad$ for concrete .... (7)

The distance $\dfrac{(d-h)}{3}$ is the lever arm of the couple

$$\frac{1}{M}\frac{(h)}{d-h} = \frac{2A_s}{bh}$$

therefore
$$A_s = \frac{bh}{2m}\frac{(h)}{d-h} \quad \dots\dots\dots\dots\dots(8)$$

which gives rise to the following quadratic in 'h' viz.

$bh^2 + 2mA_s h - 2mdA_s = 0$

the solution of which produces

$$h = \frac{-2mA_s \pm \sqrt{4m^2 A_s + 8bmd\,A_s 4ac}}{2b}$$

$$= \frac{-2mA_s \mp 2\sqrt{m^2 A_s + 2bmd\,A_s}}{2b}$$

i.e. $\quad h = \dfrac{-mA_s \pm \sqrt{m^2 A_s^2 + 2bmdA_s}}{b} \quad \dots\dots\dots\dots\dots(9)$

Designers ensure that at the section where the maximum bending moment occurs on the beam, concrete and steel each achieve maximum permissible level. Accordingly, by giving appropriate values to σ_t and σ_s in equations (6) and (7) feasible dimensions of a beam can be determined.

Equations (6) and (7) are fundamental in design:

$$\sigma_s A_s \left(d - \frac{h}{3}\right) = \frac{1}{2}\sigma_c\, bh \left(d - \frac{h}{3}\right)$$

Clearly, if A_s is too small, the permissible stress in the steel will be exceeded and the beam will fail in tension. If on the other hand A_t is too large, the steel will be under-stressed. By a similar token the dimensions 'b' and 'h' in a concrete beam should be such that σ_c is at its maximum permissible value.

Illustrative Example 5

The flange of a singly-reinforced concrete Tee-beam is 150 cm wide and 10 cm thick. The rib is 30 cm thick and the steel reinforcement bars in it have a total cross-sectional area of 20 (cm)^2 and are placed 40 cm below the top surface of the flange. Determine (i) the position of the neutral axis; (ii) moment of resistance of the beam; (iii) actual maximum stresses in the steel and concrete, given that the limits of stress are: (a) for concrete 4 MN/m^2 and (b) for steel 110 MN/m^2.

A cross-section of the Tee-beam is shown in Fig. 16.

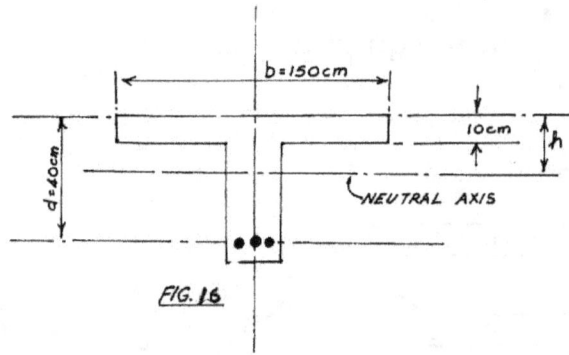

FIG. 16

Recalling the fundamental equalities for B.M. at any cross-section,

$$M = \sigma_s A_s \left(d - \frac{h}{3}\right) \text{ for steel reinforcement}$$

and $$M = \frac{1}{2} bh\sigma_c \left(d - \frac{h}{3}\right)$$

in which 'A'_s is the total cross-sectional area of the steel reinforcement and the other symbols refer to the dimensions shown in Fig. 16. From these 2 equalities:

$$A_s = \frac{bh}{2m} \frac{(h)}{(d-h)}$$

i.e. $bh^2 + 2mA_s h - 2mdA_s = 0$

'm' being the modular ratio.

and so $$h = \frac{-m A_s \pm \sqrt{m^2 A_s^2 + 2bmd A_s}}{b}$$

Here, $A_s = 20(cm)^2$; $m = 15$, $b = 150cm$

therefore,

$$150h^2 + 600h - 24000 = 0$$

or $$h^2 + 4h - 160 = 0$$

i.e. $$h = \frac{-4 \pm \sqrt{16(1+40)}}{2}$$

$$= -2 \pm 2\sqrt{41}$$

$$h = -2 \pm 12.8$$

Hence, only possible solution is h = 10.8 cm.

We are given: Maximum permissible stress in steel is 110 MN/m².

Remember that Moment of Resistance (M.R.) at any section must equal the applied bending moment at the same section.

Let us consider first the reinforcement steel at its maximum value. Accordingly when we apply

$$M.R_s = \sigma_s A_s \left(d - \frac{h}{3}\right)$$

we obtain, working in newtons and metres,

$$M.R_s = 110 \times 10^6 \times \frac{20}{10^4}\left(\frac{40-10.8}{100} \quad \frac{}{300}\right)$$

$$= 110 \times 10^6 \times \left(40 - \frac{10.8}{3}\right)\frac{1}{10^2}$$

$$M.R_s = 110\,(20)\,(36.4)\ \text{Nm}$$

$$= 80080\ \text{Nm}.$$

Consider next the concrete

$$M.R_c = \frac{1}{2}\sigma_c bh \left(d - \frac{h}{3}\right)$$

$$= 1.4^2 \times 10^6 \times \frac{150}{100}\cdot\frac{10.8}{100}\frac{(36.4)}{10^2}$$

Therefore

$$M.R_c = 300 \times 10.8 \times 36.4\ Nm = 117936\ Nm.$$

We note at once that MR_c exceeds MR_s. With MR_c = 117936 Nm, the value of σ_s becomes:

$$117936 = \sigma_s \times \frac{20}{10^4}\left(\frac{36.4}{10^2}\right)$$

from which $\qquad \sigma = 162 \times 10^6\ N/m$

This value exceeds the maximum allowable stress for steel and the reinforcement would therefore fail. Clearly, this means that the moment of resistance is the value calculated earlier for the steel, viz. *80080 Nm*. With such a value we can now find the maximum stress in the concrete.

Therefore, $\qquad 80080 = \frac{1}{2}\sigma_c\frac{150}{100} \times \frac{10.8}{100}\left(\frac{36.4}{10^2}\right)$

$$2 \times 80080 \times 10^6 = 150(10.8)(36.4)\sigma_c$$

therefore, $\qquad \sigma_c = 2.72 \times 10^6\ or\ \sigma = 2.72\ MN/m^2$

Therefore the actual maximum stresses are (a) in the steel reinforcement $110MN/m^2$ and (b) in the concrete $2.72 MN/m^2$.

Let us now derive relationships for moments of resistance when concrete and steel are stressed to their permissible working stresses. Recalling equation (2)

$$\frac{\sigma_c}{\sigma_s} = \frac{1}{m}\left(\frac{h}{d-h}\right)$$

Rearranging this we write

$$\frac{\sigma_s}{m\sigma_c} = \frac{d-h}{h}$$

Assuming permissible stresses in steel and concrete to be $110MN/m^2$ and $5MN/m^2$ respectively with modular ratio 'm' = 15.

$$\frac{110 \; x \; 10^6}{15 \; x \; 10 \; x \; 10^6} = \frac{d}{h} - 1$$

i.e

$$\frac{22}{15} = \frac{d}{h} - 1$$

from which

$$h = \frac{15}{37}d, \text{ say } 0.41d$$

Now, the lever arm

$$= \left(\frac{d-h}{3}\right) = d - \frac{0.41d}{3}$$

$$= 0.863d$$

Now the moment of resistance due to the compressive force in concrete

$$= \frac{1}{2}\sigma_c hb \left(\frac{d-h}{3}\right)$$

$$= \frac{1}{2} \; x \; 5 \; x \; 10^6 \; x \; 0.41d \; x \; b \; x \; 0.863d$$

$$= 885 \; x \; 10^3 \; bd^2 \; Nm$$

Let us now consider moment of resistance due to steel in tension. This moment

of resistance $= \sigma_s A_s \; x \left(\frac{d-h}{3}\right)$.

In any reinforced concrete beam, the tensile steel ratio 'r' is the ratio of the total cross-sectional area of the steel, A_s to the cross-sectional area of the beam, i.e. 'bd' where 'b' = breadth of beam and 'd' = depth from the top surface of the beam or slab to the centroid of the steel reinforcement. Refer again to Fig. 15a. The dimension 'd' is referred to as the effective depth. Thus $r = A_s/bd$. When however at the section of the reinforced beam the steel (in tension) and the concrete (in compression) are at their permissible-stress limits at the same time, the proportion of steel that satisfies this criterion is known as the Equal

Strength Ratio; generally designated ESR. Thus there is a specific ESR for a particular grade of steel in a particular grade of concrete. Therefore,

$$\sigma_s A_s \left(\frac{d-h}{3}\right) = \sigma_s \, rbd \, (0.863d)$$

$$= 110 \times 10^6 \, rbd \times 0.863d$$

Going back to the equality

$$\sigma_s A_s = c \, hb$$

i.e $$\frac{\sigma_s A_s}{bd} = \frac{1}{2} \, \sigma_c \left(\frac{h}{d}\right)$$

Therefore $$\frac{A_s}{bd} = r = \frac{5 \times 10^6}{2 \times 110 \times 10^6} (0.41)$$

i.e. $$\text{ESR} = 0.009318$$

Accordingly moment of resistance due to steel in tension

$$= 110 \times 10^6 bd \times 0.009318 \times 0.863d$$

$$\text{M.R}_s = 885000 \, bd^2 \, \text{Nm}$$

SOLVED AND OTHER PROBLEMS

Q1. The C-clamp shown in Fig.1 is exerting a force 'F' of 2000 newtons on the work specimen in its grip. Determine the maximum stress in the clamp at *AA'* where the cross-section may be treated as a rectangle of dimensions 75 mm x 25 mm.

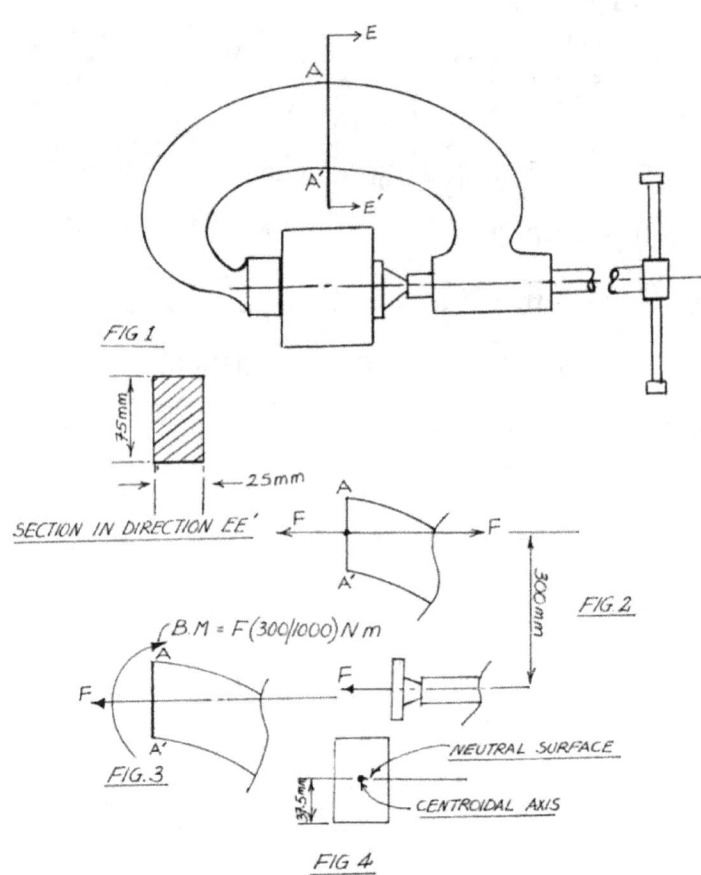

FIG 1

SECTION IN DIRECTION EE'

FIG 2

B.M = F (300/1000) N m

FIG.3

NEUTRAL SURFACE

CENTROIDAL AXIS

FIG 4

In Fig. 2, I have taken a section at *AA'* and applied two equal and opposite forces '*F'* on it as shown. In Fig. 3 the two forces forming the couple have been replaced by a bending moment equal to F times $\left(\frac{300}{1000}\right) = \frac{3F}{10}$ newton metres, *F* being in newtons; and in addition there is a direct tensile force "*F''* on the cross-section.

Evidently the maximum stress at section '*AA'* is the combination of the maximum bending stress due to the bending moment plus the direct tensile stress due to *F*.

Direct tensile stress due to $F = \dfrac{F}{Csa}$

$$= \frac{2000}{\frac{25}{1000} \, x \, \frac{75}{1000}}$$

$$= 1.067 \, \text{MN/m}^2$$

Bending stress σ is obtained from

$$\frac{\sigma_{AA}}{y_{AA}} = \frac{M}{I}$$

Referring to Fig. 4.

$$I = \left(\frac{25}{1000}\right) \, x \, \left(\frac{75}{1000}\right)^3 \, x \, \frac{1}{12}$$

$$= \frac{25 \, x \, (75)^3 \, x \, 10^{-12}}{12} \, m^4$$

$$= 878906.25 \, x \, 10^{-12} \, m^4$$

so that

$$\frac{\sigma_{AA}}{\left(\frac{37.5}{1000}\right)} = \frac{M}{878906.25 \, x \, 10^{-12}} = F \, x \, \frac{\frac{300}{1000}}{878906.25 \, x \, 10^{-12}}$$

$$\frac{\sigma_{AA} \, x \, 1000}{37.5} = 2000 \, x \, \frac{300}{1000} \, x \, \frac{1}{878906.25} \, x \, 10^{12}$$

$$\sigma_{AA} = \frac{37.5 \, x \, 3 \, x \, 10^{12}}{5 \, x \, 878906.25}$$

$$= \frac{37.5 \, x \, 3 \, x \, 10^6}{5 \, x \, 0.8789}$$

$$= 25.6 \, MN/m^2$$

Therefore Maximum stress $= 1.067 \, \text{MN/m}^2 + 25.6 \, \text{MN/m}^2$

$$= 26.7 \, \text{MN/m}^2$$

Q2. A short piece of a mild steel Tee 15 cm x 10 cm and 3 cm thickness throughout is used as a column. A bracket is fixed to the 15cm face of the tee and carries a vertical load 'W' whose line of action is 5cm from the face on the centre-line of the Tee. See Fig.1. Determine the maximum value of 'W' if the tensile stress induced in the section is not to exceed 4 MN/m².

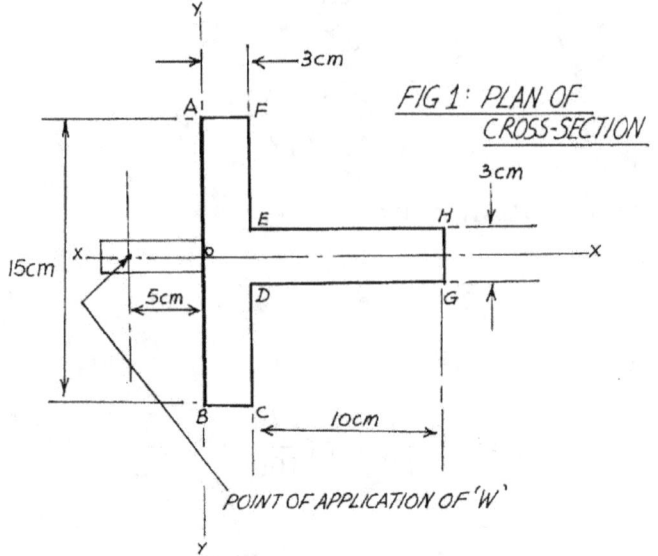

FIG 1: PLAN OF CROSS-SECTION

POINT OF APPLICATION OF 'W'

It is given that the Tee is a short column so that the effect of buckling may be ignored. Let the face *AB* be on the *Y-axis* and the line *XX* the *X-axis*. The first thing to be done is finding the location of the centroid of the cross-section of the Tee.

Let its distance from face $A\bar{B} = x$; Evidently $y = 0$.

Therefore

$$\bar{x} = \frac{15(3)(1.5) + 10(3)(5 + 3)}{15(3) + 10(3)}$$

$$= 4.1\ cm$$

The position of the centroid is shown in Fig.2.

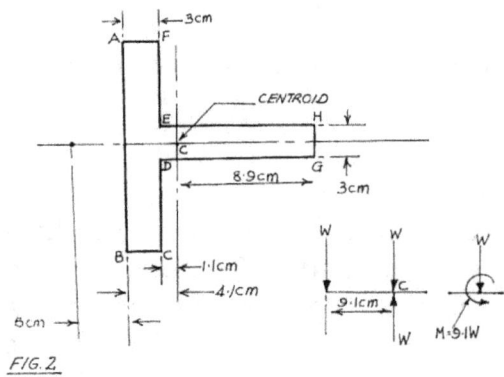

FIG. 2

~ 150 ~

Now,

$$I_{yy}(cm)^4 = \frac{15 \times (3)^3}{12} + 15(3)(4.1 - 1.5)^2$$

$$+ \frac{3(1.1)^3}{3} + \frac{3(8.9)^3}{3}$$

$$= 1044.25, \text{ say}$$

$$= 1044 \ (cm)^4$$

The Tee being symmetrical about its centre line *XX* we have found its centroid to be *4.1cm* from face *AB*: this is the position of the neutral axis, perpendicular to the plane of the sheet.

The bending moment about the neutral axis is accordingly *W (4.1 + 5) = 9.1W* newton centimetres assuming *W* is in newton.

Under the influence of this bending moment it should be evident that the maximum tensile stress will be at edge '*HG*', distant *8.9cm* from the neutral axis.

Relying on the Flexure formula $M = \sigma Z$, we have, working in metres:

$$M = \frac{W(9.1)}{100} \text{ newton metres}$$

$$Z = \frac{I}{y}$$

$$I = \frac{1044}{10^8} m^4$$

$$y = \frac{8.9}{100}$$

therefore,
$$\frac{9.1W}{100} = \sigma \times \frac{1044 \times 100}{10^8 \times 8 \times 9}$$

'σ' being in newton/m^2

i.e.
$$\sigma = \frac{9.1W \times 10^8 \times 8.9}{1044 \times 10^4}$$

so that
$$\sigma = 776 \ W/m^2$$

But, it is important to take into account the direct compressive stress on the section due to load *W*. The cross-sectional area of the Tee is:

$$\{(15 \times 3) + 10(3)\} \ cm^2 = 75(cm)^2$$

Therefore the compressive stress on the section is: $\dfrac{W}{\frac{75}{10^4}} = \dfrac{10^4}{75}\,W/m^2$

Being a compressive stress, we may state that the total tensile stress at HG is: $776\,W - 133\,W = 643\,W/m^2$

And since the maximum tensile stress must not exceed 4MN/m², i.e.:

$4 \times 10^6\,N/m^2$ we may write

$643W/m^2 = 4 \times 10^6\,N/m^2$ from which $W = 622\,/N$ or $W = 6.2$ kN

Q3. Using the data of Q. No.2, what is the compressive bending stress of face AB? Determine also the total compressive stress on this face.

Compressive bending stress, 'σ_c' may be obtained using the Flexure formula.

Therefore $\dfrac{9.1W}{10^8} = \sigma_c' \times \dfrac{1044}{4.1} \times \dfrac{100}{4.1}$

i.e. $\sigma_c' = 776\,W/m^2 \times \dfrac{4.1}{8.9}$

from which $\sigma_c' = 359\,W/m^2$

$\qquad\qquad = 359(6.2)\ kN/m^2$

$\qquad\qquad = 2226\ kN/m^2$

or $\qquad \sigma_c' = 2.2\ MN/m^2$

Total compressive stress σ_c'' on AC is given by

$\qquad \sigma_c'' = (359 + 133)\ W/m^2$

$\qquad\qquad = 492\ W/m^2$

Knowing W to be *6.2kN*, makes total compressive stress on AC

$\qquad \sigma_c'' = 492(6.2)\ kN/m^2$

$\qquad\qquad = 3050\ kN/m^2$

Therefore $\sigma_c'' = 3\ MN/m^2$

Q4. A purpleheart beam 3m long has the cross-section shown in Fig.1. A semi- circular groove 10cm diameter is milled out along the mid-line of each side for the beam's total length. The beam is simply supported at its ends and carries a uniformly distributed load of 15 kg/metre on its top surface, and also a point load of 100 kg at a distance of *60cm* from the left-hand end as shown in Fig.2. Calculate the position of the maximum bending moment and the maximum bending stress in the beam. Take $g = 9.81 \text{ m/s}^2$ and neglect weight of the beam.

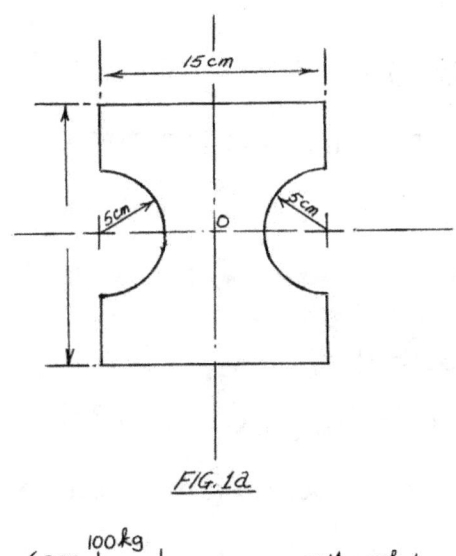

FIG.1a

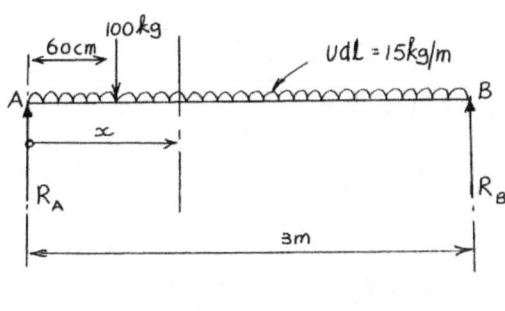

FIG.1b

The cross-section of the beam is symmetrical about the X- and Y-axes; the centroid of the section is therefore at *'O'*.

$$I_{xx} \ (cm)^4 = \frac{15}{12}(18)^3 - \frac{\pi (5)^4}{4}$$

$$= 7290 - 491$$

$$= 6799$$

Taking moments about *'A'*

$$R_B(3) - 3\,(15)(9.81)\frac{3}{2} - 100\,(9.81)\frac{(60)}{100} = 0$$

$$3R_B - 662 - 588.6 = 0$$

i.e. $$R_B = 416.6N$$

and $R_A = (3 \times 15 \times 9.1 + 100 - 416.6)N = 125.5N$

Let section of the beam at which maximum bending moment, say, M_{max} occurs, be distant 'x' from 'A'.

Therefore $B.M_x = R_A(x) - 100(x - 0.6) - 15(9.81)(x)\dfrac{x}{2}$

$$= 125.5x - 100x + 60 - 73.75x^2$$

For max.B.M

$$\frac{d}{dx}(B.M_x) = 0$$

i.e. $125.5 - 100 - 147.5x = 0$

from which $x = 0.1728$ m or approximately 17¼ cm from 'A'.

Accordingly, the maximum bending moment 'M'_{max} is given by:

$M_{max} = 125.5(0.1728) - 100(0.1728) + 60 - 73.75(0.1728)^2$

$= 21.7 - 17.3 + 60 - 2.2 = 62.2Nm.$

Using $\quad \dfrac{M}{I} = \dfrac{\sigma}{y}$

$$\frac{62.2}{6799/10^8} = \frac{\sigma}{9/100}$$

i.e. $\quad \dfrac{62.2 \times 10^8}{6799} = \dfrac{100\sigma}{9}$

therefore $\quad \sigma = \dfrac{9 \times 62.2 \times 10^6}{6799}$

i.e. $\quad \sigma = \dfrac{82336N}{m^2}$ or $82.3 \ kN/m^2$

Q5. An asymmetrical girder has the cross-sectional dimensions shown in Fig. 1. If it is simply supported at its ends then determine the maximum span for which it can be used assuming the total uniformly distributed lead including the self weight of the beam is equivalent to 12 kN/m and the maximum tensile bending stress is limited to 75 MN/m².

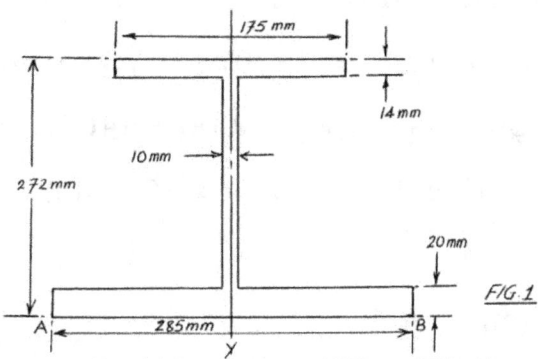

FIG. 1

The first order of business is determination of the position of the centroid. Let $\bar{y}$ = distance of centroid from bottom of flange AB.

$$\bar{y} = \frac{175\,(14)(265) + 10\,(238)(139) + 285\,(201)(10)}{175\,(14) + 10\,(238) + 285\,(20)}$$

   = 98.48mm, say,

$\bar{y} \approx 98.5\ mm.$

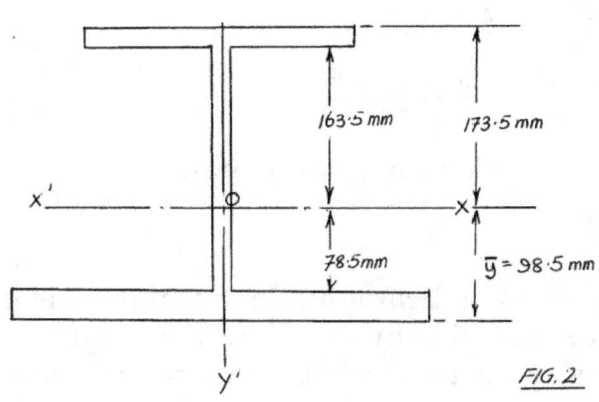

FIG. 2

Next we determine I_{xx}, for which purpose see Fig.2. Working in mm units:

$$I_{xx} = \left\{ \frac{175 \times (14)^3}{12} + 175\,(14)(167.5)^2 \right\} + \left\{ \frac{10\,(163.5)^3}{12} + 10\,(163.5)(81.75)^2 \right\}$$

$$I_{xx^1} = + \left\{ \frac{10\,(78.5)^3}{12} + 10\,(78.5)(49.25)^2 \right\}$$

This expression for I_{xx} eventually reduces to:

I_{xx}= {40016.7 + 68737812.5} + {43707228.75 +10926807.19}

+{4837366.25 + 1891938.75} + {190000 + 44643825}

= 174974995.1 (mm)^4 or 17497995.1 x 10^{-12} (m)^4

i.e.

I_{xx} = 0.175 x 10^{-3} (m)^4.

Maximum B.M = $\frac{wl^2}{8}$; w = udl (N/m); l = length in metres.

$$= \frac{12000\,l^2}{8} = 1500\,l^2$$

No $\qquad \dfrac{\sigma}{y} = \dfrac{M}{I}$

Given $\sigma = 75 MN/m^2$

Working in newtons and metres

Therefore $\qquad \dfrac{75 \times 10^6}{\frac{78.5}{1000}} = \dfrac{1500\,l^2}{0.175 \times 10^{-3}}$

i.e. $\qquad \dfrac{75 \times 10^9}{78.5} = \dfrac{1500 \times 10^3 \times l^2}{0.175}$

i.e. l^2 = 47.7, from which maximum span

l = 6.9m.

Q6. The cross-section of a beam made from an inverted steel channel is approximated to that shown in Fig.1. The beam 3m in length is simply supported at its ends and carries two point loads of magnitude 'W'kN distant one–half-metre from each end. See Fig 2. Determine the value of 'W' if maximum bending stress is not to exceed 80 MN/m².

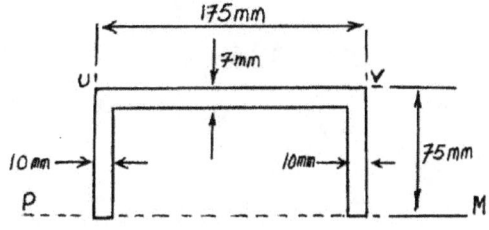

FIG. 1

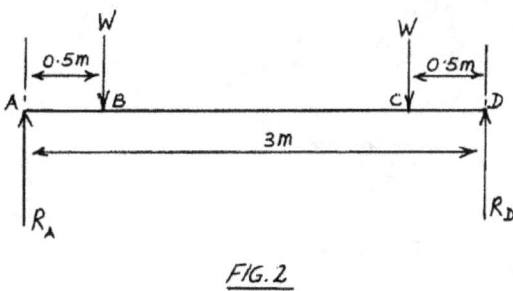

FIG. 2

Maximum B.M occurs over portion BC. It is equal to R_A (0.5),i.e. B.M$_{max}$= 0.5WkN

How would you describe the state of bending over BC?

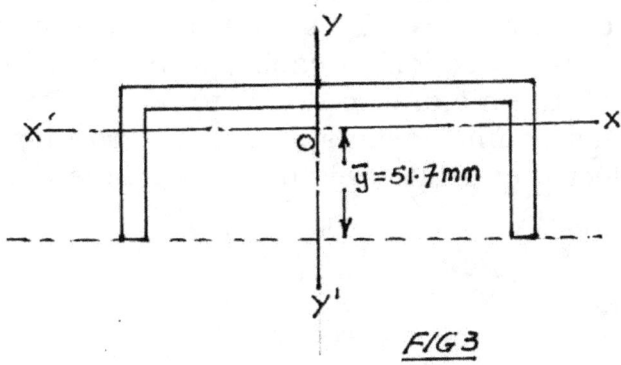

FIG 3

$$\bar{y} = \frac{(2)75\,(10)(37.5) + 155\,(7)(71.5)}{2\,(75)(10) + 155\,(7)}$$

$$= (56250 + 77577.5)/(1500 + 1085)$$

$$= 133827.5 \,/\, 2585$$

$$\bar{y} = 51.7 \text{mm}$$

$$I_{xx'} = \frac{175(23.3)^3}{3} - \frac{155(16.3)^3}{3} + 2\left\{10\,\frac{(51.7)^3}{3}\right\}$$

$$= 514135 + 921260$$

$$I_{xx'} = 1435395 \text{ (mm)}^4 \text{ or } 1435395 \times 10^{-12} \text{ m}^4$$

$$I_{xx'} = 0.1435 \times 10^{-5} \text{ m}^4$$

$$\frac{\sigma}{y} = \frac{M}{I}$$

Therefore $\dfrac{80 \times 10^6}{\frac{51.7}{1000}} = \dfrac{0.5 \; x \; 1000W}{0.1435 \; x \; 10^{-5}}$

$$\frac{80 \times 10^{9}{}^{10}}{51.7} = \frac{0.5 \times 1000 \times 10^5 \; W}{0.1435}$$

$$W = \frac{80 \times 100 \times 0.1435}{51.7 \times 0.5}$$

$$W = 4.4 \text{ kN}$$

Q7. A steel tank 1.5 m x 1.5 m x 1m (depth) is to be used for storage of diesel fuel. The tank is to be supported symmetrically on two cantilevers which are essentially two structured Tees split from a universal beam. See Figs. 1a and 1b. The cross-section of each Tee has the dimensions shown in Fig. 2. Determine the maximum depth to which the tanks may be filled, assuming the maximum tensile stress in the cantilevers is not to exceed 65 MN/m². Neglect the tank's weight and assume density of diesel fuel = 900kg /m³.

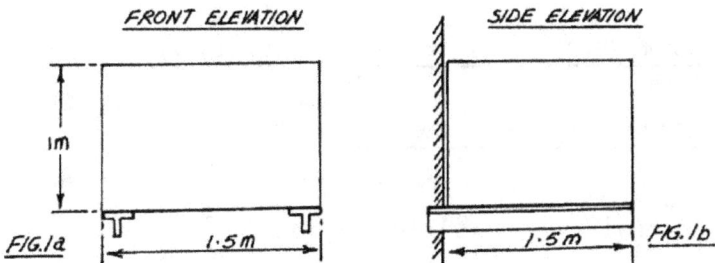

FRONT ELEVATION   SIDE ELEVATION

FIG.1a   1·5m   1·5m   FIG.1b

The first step is to find the position of the centroid of each cantilever section. Taking moments about EF

$$(152.4 \times 10.9)(224.8 - 5.45) + (224.8 - 10.9)(7.6)x \, \frac{(224.8 - 10.9)}{2}$$

$$= \{152.4\,(10.9) + 7.6\,(224.8 - 10.9)\}\bar{y}$$

i.e. $3286.8\bar{y} = 538156$

therefore $\bar{y} = 163.7 \, mm$

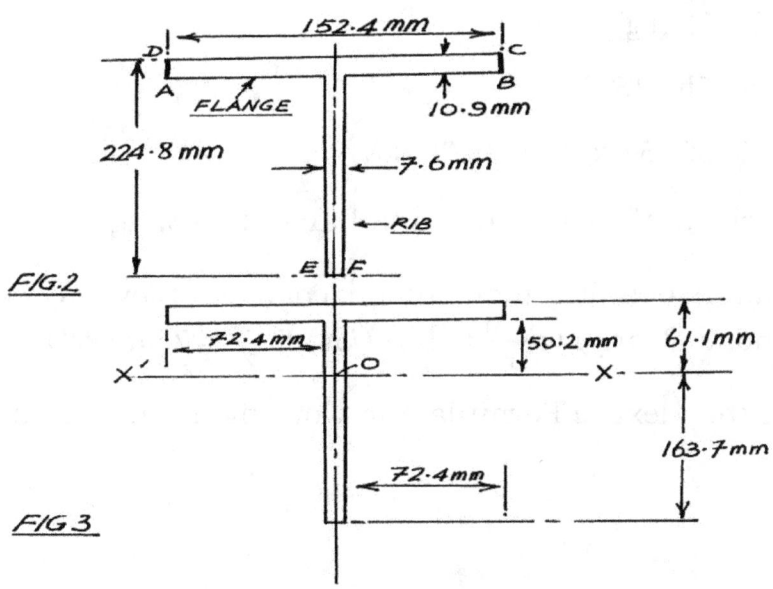

FIG.2

FIG 3

Referring to Fig. 3 in which the position of the centroid of the section is shown,

$$I_{xx}^1 \ (mm)^2 = \frac{152.4 \times (61.1)^3}{3} - \frac{144.8 \times (50.2)^3}{3} + \frac{7.6 \times (163.7)^3}{3}$$

$$= \frac{152.4 \times 228099}{3} - \frac{144.8 \times 126506}{3} + \frac{7.6 \times 4386781}{3}$$

$$= \frac{1}{3}(3461704 - 18216864 + 33339535)$$

$$= \frac{18589775}{3}$$

$$I_{xx}^1 \ (mm)^2 = 6196592$$

Continuing,

Let h = maximum depth of oil in metres

  V' = volume of oil of density 900kg /m³ at depth 'h'

  W = weight of volume 'V'

   = 1.5 x 1.5 x h x 900kg

   = 2025h (kg)

   = 2025h *(9.81)* N

i.e. W = 19865.25 hN, say, 19.9h (kN)

Assume each cantilever carries one-half of this weight, i.e. $\frac{w}{2} \approx 10H$ *(kN)*.

The maximum bending moment on each cantilever occurs at the wall support and is clearly: $w\frac{(1.5)}{2} = 10H$ *(kN)* $X \ \frac{1.5}{2} = 7.5H$ *(kN)*.

Employing the Flexure Formula and working in metres and newtons:

$$\frac{M}{I} = \frac{\sigma}{y}$$

We get

$$\frac{7.5h \times 10^3 N}{\frac{6196592 \ (m^4)}{10^{12}}} = \frac{\sigma}{y} \qquad \ldots\ldots\ldots\ldots (i)$$

Evidently each cantilever will, under the loading deflect in such a way that the fibres in their flanges will stretch, i.e. will be subject to tensile stress. This means that the value of 'y' must be the distance from the neutral plane to the outermost fibres of the flange, which in this case is 61.1 mm.

Accordingly, with stress 'σ' restricted to 65 MN/m² equation (i) may now be re- written as:

$$\frac{7.5h \times 10^3 \times 10^{12}}{6196592} = \frac{65 \times 10^6}{\frac{61.1}{1000}}$$

$$\text{or } \frac{7.5 \times 10^6 \times h}{6.2} = \frac{65}{61.1}$$

$$\text{i.e.} \qquad h = \frac{65 \times 6.2}{7.5 \times 61.1}$$

$$= 0.879 \, m, \text{ say,}$$

$h \approx 0.9 \, m$ (I suggest $\frac{7}{8}$ m to be on the safe side)

Q8.   A vertical lamp standard 10 m long and rigidly fixed at ground level is of square-section throughout, but its sides taper uniformly from 15 cm at bottom to 7.5 cm at the top. A horizontal pull of 100 N is applied at the top of the standard, in a direction along a diagonal of the section. Determine the distance from the top where the maximum bending stress occurs and also calculate its magnitude.

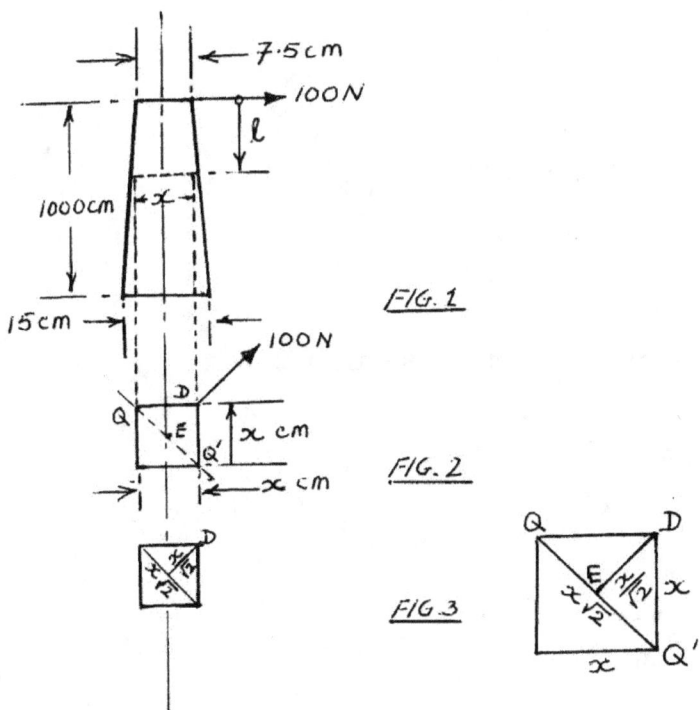

A sketch of the lamp standard is shown at Fig.1. Fig. 2 is a cross-section x by $x(cm)$ at a distance $l\,cm$ from top of standard.

Therefore at $l\,cm$ from top, side 'x' is $\left\{7.5 + \frac{l}{1000}(7.5)\right\}\,cm$

i.e. $\qquad\qquad\qquad\qquad x = \frac{7500 + 7.5l}{1000}$ ....(i)

From (i) $\qquad\qquad 1000x = 7500 + 7.5l$

or $\qquad\qquad\qquad\qquad l = \frac{(1000x - 7500)}{7.5}\,cm$ .... (ii)

Employing the Flexure Formula: $\frac{M}{I} = \frac{\sigma}{y}$ or

$$\sigma = \frac{M_y}{I} = \frac{M}{Z}$$ ....(iii)

Neutral plane is QQ'.

To find $I_{QQ'}$ consider second moment of area of triangle QDQ' (in Fig.3) about QQ'; and double the value obtained to get $I_{QQ'}$ for the entire cross section at distance 'l' from top.

Now I for triangle QDQ' about its centroid is:

$$\frac{QQ'}{36}(ED)^3 + \text{Area of } QDQ'\left(\frac{1}{3}ED\right)^2$$

$$= \frac{x\sqrt{2}}{36}\left(\frac{x}{\sqrt{2}}\right)^3 + \frac{x\sqrt{2}}{2}\cdot\frac{x}{\sqrt{2}}\left(\frac{1x}{3\sqrt{2}}\right)^2$$

$$= \frac{x^4}{72} + \frac{x^4}{36} = \frac{x^4}{24}$$

Therefore $I_{QQ'}$ for entire cross-section x by x

$$= 2\frac{(x^4)}{24} = \frac{x^4}{12}(cm)^4$$

Accordingly,

$$Z_{QQ'} = \frac{I_{QQ'}}{ED} = \frac{x^4\sqrt{2}}{12x}$$

$$= \frac{x^3 \sqrt{2}}{12}$$

$$\therefore \quad \sigma = \frac{M}{Z} = \frac{100l(12)}{x^3 \sqrt{2}}$$

Substituting the value of 'y' given by (ii)

$$\sigma = 100 . \frac{1000}{7.5} \frac{(x-7.5) .(12)}{x^3 \sqrt{2}}$$

$$= 100 . \frac{1000}{7.5} \frac{(x-7.5) . \quad (12)}{x^3 \sqrt{2}}$$

For maximum/minimum σ

$$\frac{d\sigma}{dx} = \frac{100 (1000) (12)}{7.5 \sqrt{2}} (-2x^{-3}) + 22.5x^{-4}) = 0$$

i.e. $\quad -2x^{-3} + 22.5x^{-4} = 0$, or $2x = 22.5$

from which $x = 11.25 \, cm$

But $\quad l = \frac{1000x - 7500}{7.5}$

So that for $\quad x = 11.25 \, cm$

$$l = \frac{1000(11.25) - 7500}{7.5} = \frac{3750}{7.5} cm = 500 \, cm = 5 \, m$$

Therefore maximum stress occurs at a section 11.25 cm x 11.25 cm, 5 m from top of standard.

$$\sigma_{max} = \frac{100 (1000) (12)(11.25 - 7.5)}{7.5(11.25)^3 \sqrt{2}}$$

$$= 298 \, N / cm^2 = 298N / \frac{1}{10^4} m^2$$

$$= 298 \times 10^4 \, N / m^2 = 2.98 \times 10^6 \, N/m^2$$

i.e. $\quad \sigma_{max} \approx 3 \, MN / m^2$

Q9. A vertical flagpole is of circular section throughout and its diameter varies uniformly from *40cm* at ground level to *10cm* at its top. A horizontal force of *150N* is applied to the flagpole at its top end. Calculate the maximum stress due to bending.

Ans: σ_{max} = *71kN/m²*, distance *10m* from top of flagpole.

Q10. A long 3 metre-high cavity wall is made of 2 sets of concrete blockwork separated by a gap of 10 cm. See Fig.1. Each set of blockwork is tied to the other by wall ties so that the entire structure may be regarded as one in resisting any overturning moment. Calculate the intensity of wind load in N/m² on one side of the wall which acting normal to it would be sufficient to induce tension in the block work and cause the structure to fail. Assume 1 m³ brickwork = 2000 kg and take g = 9.81 m/s²

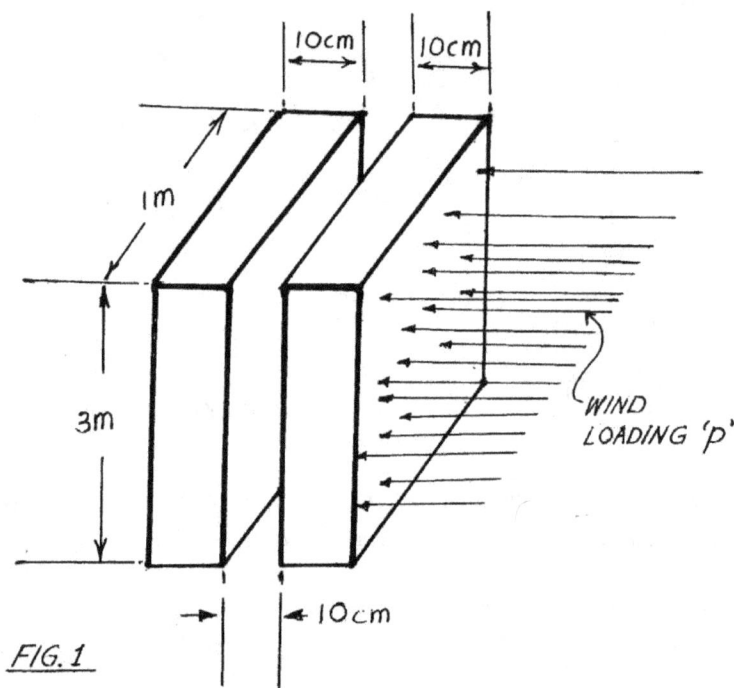

FIG. 1

Consider a 1 metre length of one part of the wall. Its weight, say, Wkg = length x thickness x height x 2000.

$$W = 1m \ x \ \frac{10}{100}m \ x \ 3 \ m \ x \ 2000 \ kg$$

$$= 600 \ kg$$

$$= 600 \ kg \ x \ 9.81 \ m/sec^2$$

i.e. $W = 5886 \ N$

cross-sectional area of wall $= \frac{10 \ m}{100} \ x \ 1 \ m = 0.1 \ m^2$.

Because the two parts of the cavity wall are to be regarded as a single structure, the uniform compressive stress $'\sigma_c'$ at the base of the wall is obtained from either:

$$\sigma_c = 2 \ \frac{(5886)}{(0.1)m^2} N \ or \ \frac{5886N}{0.1 \ m^2}$$

$$= 58860 \ \frac{N}{m^2} \ or \ \sigma_c = 58,86 \ kN/m^2$$

Due to pressure 'p acting on the face of the wall, the force on the wall caused by it is given by: p x Area of face of wall.

For the one metre length of wall under consideration, this force, say, F is given by: F = p x 1 m x 3 m = 3 pN in which 'p' is in N/m².

The overturning moment at base of wall due to F, say, M$_B$ is:

M$_B$ = Fx moment arm.

Accordingly, $\qquad M_B = 3 \ pN \ x \ \frac{3}{2}m = \frac{9}{2}p$

To determine the bending stresses $\pm\sigma_B$ due to M$_B$, reliance is placed on the Flexure formula; in this case:

$$\frac{M_B}{I} = \frac{\sigma_B}{y} \ or \ M_B = \pm\sigma_B.Z$$

For the combined structure $I'(m^4) = \left\{ 1\left(\frac{30}{100}\right)^3 - 1x\left(\frac{10}{100}\right)^3 \right\}/12$

i.e. $\qquad I'(m^4) = 0.013/6$

Here, y = (10 cm + 10 cm + 10 cm)/2 = 15 cm = 0.15 m

$$\therefore \qquad \frac{9}{2}p = \pm\sigma_B x \frac{0.013}{6} \ x \frac{1}{0.15}$$

from which $\qquad \pm\sigma_B = \frac{9p}{2} \ x \frac{6}{0.013} \ x \ 0.15 = 3.115 \ pN/M^2$

$\pm\sigma_B$ is the tensile bending stress in the base of the wall.

For σ_B (tensile) to overcome the uniform compressive stress σ_C in the base of the wall, $\pm\sigma_B$ must first equal 58.86 kN/m². Equating $\pm\sigma_B$ and σ_C.

$$\therefore 311.5.p(N/m^2) = 58860 \ N/m^2$$

i.e. $\qquad\qquad p \cong 189 \ N/m^2.$

Uniform pressure on the wall greater than 189 N/m² would induce tension in its base and result in failure.

Q11. A universal beam 127 mm x 76 mm and 5 mm long is simply supported at its ends. A cross-section of the beam is shown in Fig.1. Excluded from the diagram are the fillet radii. The beam, self-weight of which is 13 kg per metre, carries a mass of 1 tonne at mid-span. Calculate the maximum bending stresses in the beam. Take g = 9.81 m/s² and E = 207 GN/m².

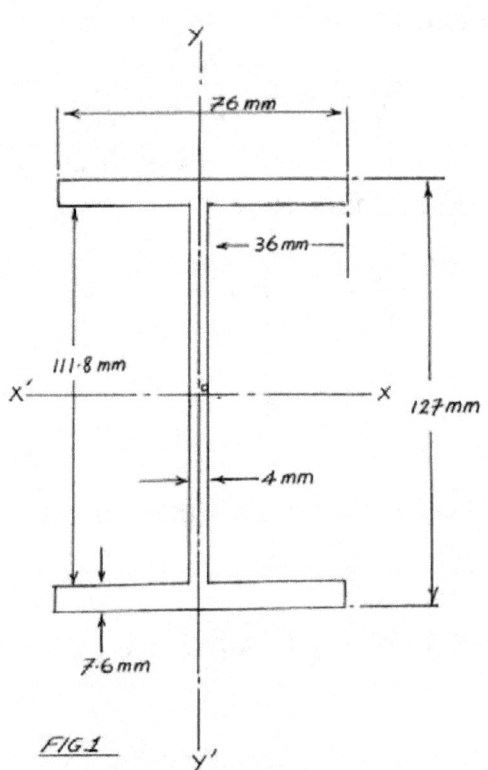

FIG 1

UNIVERSAL BEAM 127mm x 76mm X 13
(FILLET RADII NOT SHOWN).

Working second moment of area in centimetres

$$\frac{1}{12}\left\{\frac{76}{10}x\left(\frac{127}{10}\right)^3\right\}-\frac{1}{12}\left\{\frac{72}{10}x\left(\frac{111.8}{10}\right)^3\right\}$$

$$\approx 459(cm)^4 \text{ or } \frac{459}{10^8}(m)^4$$

Mass of 1 tonne is equivalent to 1000 kg x 9.81 m/s² = 9810 kg m/s² = 9810 newton.

Accordingly, bending moment 'M' due to 1–tonne load at mid- span is, from $\frac{WL}{4}$ in which W = mid-span load and L = total span $= \frac{9810\,x\,5}{4} =$ 12262.5 N m.

Employing the Flexure Formula $\frac{M}{I}=\frac{\sigma}{y}$ and making the relevant substitutions (working in metres):

$$\frac{12262.5}{\dfrac{459}{10^8}}=\frac{\sigma}{\dfrac{63.5}{1000}}$$

i.e.

$$\frac{12262.5\,x\,10^8}{459}=\frac{\sigma\,x\,1000}{63.5}$$

therefore

$$\sigma=\frac{12262.5\,x\,10^5\,x\,63.5}{459}$$

$$= 1696.4\,x\,10^5 \text{ or } 169.6\,x\,10^6 N/m^2$$

i.e.

$$= 169.6\,MN/m^2$$

For the uniformly distributed load due to the beam's self weight

Let w = beam's udl. = 13 kg/m = 13 x 9.81 N/m = 127.5 N/m

therefore bending moment due to 'w' $= \frac{wL^2}{8} = \frac{127.5}{8} x\,25 = 398.5$ Nm.

Using the Flexure Formula as before $\frac{M}{I}=\frac{\sigma}{y}$

Therefore $\qquad \dfrac{398.5}{\dfrac{459}{10^8}}=\dfrac{\sigma}{\dfrac{63.5}{1000}}$

Therefore $\quad \sigma = \dfrac{398.5\,x\,63.5\,x\,10^5}{459}$

$$= 55.1\,x\,10^5\,N/m^2$$

$$\approx 5.5 \text{ MN/m}^2$$

Therefore bending stresses are = Tensile = 169.6 + 5.5 = 175.1 MN/m² ;

Compressive = -175.1 MN/m².

Q12. A twin-chamber-brickwork-flue 1.2m high was found to fail under the influence of gusty winds, the pressure of which was estimated at *1500N/m²*. A cross- section of the flue is shown in Fig. 1 Assume that the wind pressure acted normal to the side AB. Determine the reason for failure of the flue.

Take density of brick work = 18.8 kN/m³

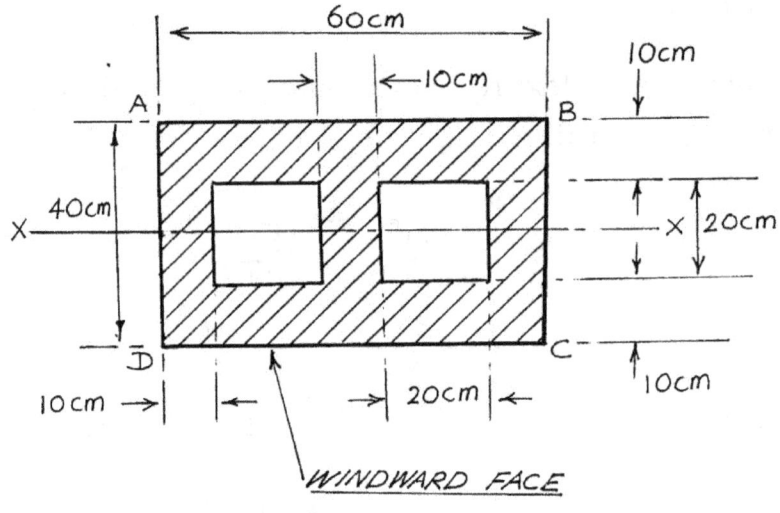

FIG 1 : SECTION THROUGH FLUE

Let CSA = shaded cross-sectional area (as in Fig. 1)

Therefore volume of brick work = 1.2(CSA)

so that

Weight of brick work = 18800(1.2) (CSA) newton which means that Compressive stress at base of brick work = $\dfrac{18800\ (1.2)(CSA)}{CSA}$ = 22560 N/m²

[Observe, density in kN/m³; therefore kN/m³ x height (in m) = kN/m²]

Continuing,

$$I\ (m^4)_{xx} = \frac{1}{12}\left\{\frac{60}{100} \ x \ \frac{40^3}{100} - \frac{20}{100} \ x \ \frac{20^3}{100}\right\}$$

~ 168 ~

$$= \frac{1}{12} \frac{(368)}{10000} \frac{100}{20}$$

$$z(\text{m}^3) = \frac{I}{y}$$

$$= \frac{1}{12} \frac{368}{10000} \frac{100}{20}$$

$$Z = \frac{184}{12000} = 0.0153 \ m^3$$

Let us now determine the bending moment on the base of the chimney.

Area of windward face = $\left(1.2 \ x \ \frac{60}{100}\right) m^2$

Wind pressure of $1500 N/m^2$ acting on this area contributes a force of

$1500 \ x \ 1.2 \ x \ \frac{60}{100}$ newton

This force acts at the centroid of the windward face, i.e. 60 cm from base.

Accordingly, the bending moment 'M' is given by

$1500 \ x \ 1.2 \ x \ \frac{60}{100} \ x \ \frac{60}{100} = 648 Nm$

Therefore bending stress, say, $\sigma_B = \pm \frac{M}{Z}$

$$= \frac{648}{0.0153}$$

$$= 42353 \ \text{N/m}^2$$

But, the compressive stress due to the weight of the brick work = 22560 N/m²

Therefore σ_B (tensile) on windward side

= +42353 N/m² – 22560 N/m² =+ 19793 N/m²

Also, σ_B (compressive) on leeward side =

- 42353 N/m² – 22560 N/m² = - 64913 N/m²

Evidently the chimney flue failed because the wind load induced nett tension in the brick work.

Q13. Determine the second moment of area of a square about its diagonal. A short hollow column *1.2m* square and *10cm* wall thickness on all sides carries a vertical load at a point on the diagonal distant *28cm* from the vertical axis of the column. The load is equivalent to a force of *850kN*.

Ignoring the weight of the column, determine the normal stresses at the four outside corners on a horizontal section of the column. Refer to Fig. 1.

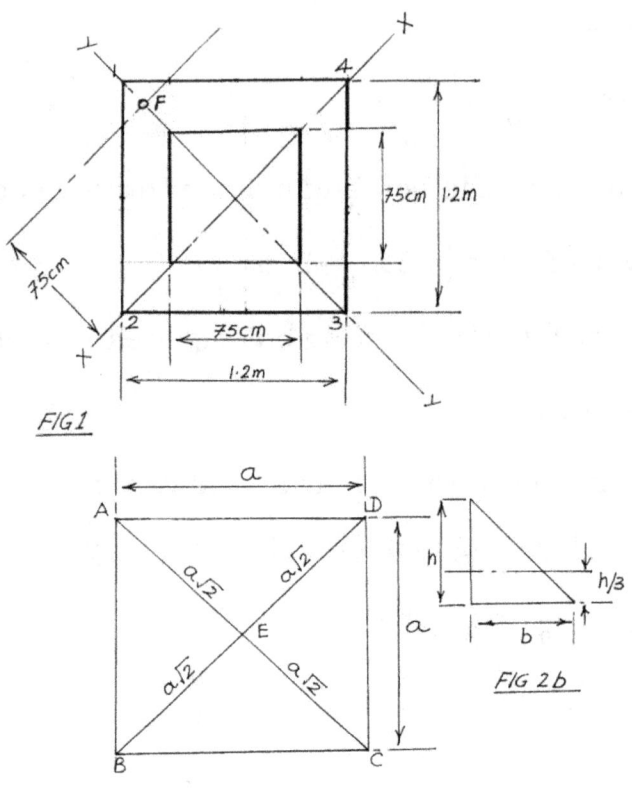

FIG 1

FIG 2a

FIG 2b

ABCD is a square of side 'a' and the diagonal about which we are to determine the second moment of area is "BD".

You would recall that the second moment of area of the triangle shown in Fig. 2b about an axis passing through its centroid which is a distance $h/3$ from the base is $\frac{1}{36} bh^3$ Now consider triangle ADE.

Designating second moment of area about its centroid as I_c, then

$$I_c = \frac{1}{36} \cdot \frac{a}{\sqrt{2}} - \frac{(a)^3}{\sqrt{2}}$$

i.e. $I_c = \frac{a^4}{144}$

We must now use the Parallel – Axis theorem to find the value of *'I'* about *ED*. It is

$$I_{ED} = \frac{a^4}{144} + Area\ of\ Triangle\ AED\ (distance\ of\ its\ centroid\ from\ ED^2)$$

$$= \frac{a^4}{144} + \left\{\frac{1}{2}\ \frac{a}{\sqrt{2}}\ \frac{a}{\sqrt{2}}\right\} \cdot \left(\frac{1}{3}\ \frac{a}{\sqrt{2}}\right)^2$$

$$I_{ED} = \frac{a^4}{144} + \frac{a^4}{72} = \frac{a^4}{48}$$

and because there are 4 such triangles in the square

$$I_{diagonal} = \frac{4a^4}{48}$$

$$= \frac{a^4}{12}$$

Applying this result to the short column

$$I_{xx} = \frac{1}{12}\left\{(1.2)^4 - \frac{(75)^4}{100}\right\} m^4$$

$$= \frac{1}{12}\ \{(1.2)^4 - (0.75)^4\}$$

$$= \frac{1}{12}\ \{(1.2)^2 - (0.75)^2\}\ \{(1.2)^2 + (0.75)^2\}$$

$$= \frac{1}{12}\ (1.44 - 0.5625)\ (1.44 + 0.5625)$$

$$= \frac{1}{12}\ (0.8775)\ (2.0025)$$

$$I_{xx} = 0.146\ m^4$$

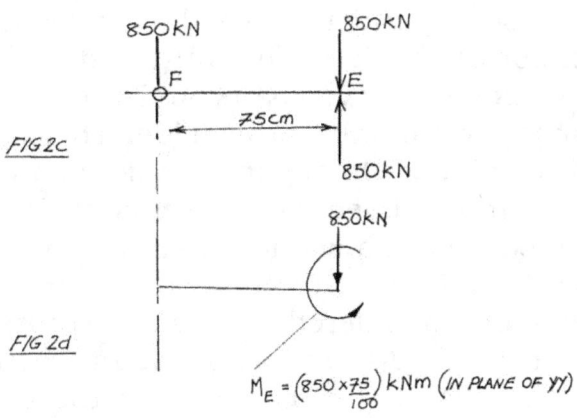

850kN
850kN
F
E
75cm
850kN

FIG 2c

850kN
M_E = (850 x 75/100) kNm (IN PLANE OF YY)

FIG 2d

The loading configuration reduces to a direct (compressive) load of *1200kN* on the column and an anti-clockwise moment $= 850 \times \frac{75}{100}$ *kNm* = 637.5 *kNm*.

Refer to Figs. 2c and 2d.

Therefore bending stresses σ_b are obtained from

$$\sigma_b = \pm \frac{M}{Z}$$

$$= \pm \frac{637.5}{I} \cdot 1.2 \sqrt{2}$$

$$= \frac{637.5}{0.1464} \cdot 1.2 \sqrt{2}$$

$$= 7389 \ kN/m^2$$

Now, the direct compressive stress $\dfrac{Load}{Cross-sectional\ are\ of\ column}$.

$$= \frac{850 \ kN}{(1.2)^2 - (0.75)^2}$$

$$= \frac{850}{1.44 - 0.5625}$$

$$= 968 \frac{kN}{m^2}$$

At point labeled '1' in Fig.1, the total stress is $(7389 + 968)$ kN /m², i.e. 8357 kN /m²

At points '2' and '4', the total stress is the direct compressive stress = 968 kN /m², and at point 3' where the bending stress is tensile, the total stress = $(7389 - 968)$ kN /m², i.e. 6421 kN /m².

Q14. A number of dressed cylindrical wooden logs closely packed together make up a temporary bridge. The butt-ends of the logs rest on an abutment on one side of a river bank and the other ends on a concrete plinth at the same level on the other side. The length of each log is *12* metres and their combined diameter considered as a unit at the butt-end is *50cm*. If *'Z'* the combined section modulus at the butt-end is related to *'Z_x'* the section modulus at a section distant *'x'* from the butt-end by the expression $Z_x = Z - 0.5x$, in which *'x'* is in centimetres, then find the position on the span, considered as a single unit, where the maximum bending stress occurs. Assume the span to be uniformly loaded throughout its length and the reactions at the ends of the span to be equal.

Let *'r'* = radius of combined unit at butt end

Therefore $Z = \dfrac{l}{y} = \dfrac{\pi r^2}{4.r} = \dfrac{\pi r^3}{4}$

For $r = 25$ cm

$Z = \dfrac{\pi\,(25)^3}{4} = \dfrac{15625\pi}{4} = 3906.25\pi$, say

$Z = 3906\pi\ (cm)^3$

Accordingly we may now write

$Z_x = 3906\pi - 0.5\,x$

Let $\qquad M_x$ = bending moment at section 'x' from butt end

and $\qquad \sigma_x$ = bending stress at section 'x' from butt end

therefore $\quad \sigma_x = \dfrac{M_x}{Z_x}$

Because we are told to assume that the reactions at each end of the span are equal

$M_x = \dfrac{wlx}{2} - \dfrac{wx^2}{2}$

where l = span

Accordingly

$\sigma_x = \dfrac{M_x}{Z_x} = \dfrac{wlx}{2} - \dfrac{wx^2}{2}$

i.e. $\quad \sigma_x = \dfrac{M_x}{(3906\pi - 0.5x)} = \dfrac{wlx}{2} - \dfrac{wx^2}{2}$

For max, min $\quad \dfrac{d}{dx}(\sigma_x) = 0$

Differentiating our expression for σ_x, we have

$\dfrac{d\sigma_x}{dx} = \dfrac{(3906\pi - 0.5x)\left(\frac{wl}{2} - wx\right) - \left(\frac{wlx}{2} - \frac{wx^2}{2}\right)}{(3906\pi - 0.5x)^2}$

For $\quad \dfrac{d\sigma_x}{dx} = 0 \qquad \pi$

we have $1953\pi wl - 3906\pi wx - 0.25wlx + 0.5wx^2 + 0.25wlx - 0.25wx^2 = 0$

or $\qquad 1953\pi l - 3906\pi x + 0.25x^2 = 0$

Substituting 'ℓ' = 12 m = 1200 cm, the latter equation becomes

$7361716 - 12270x + 0.25x^2 = 0$

or $x^2 - 49080x + 29446864 = 0$

Solving this quadratic

$x = \dfrac{49080 + \sqrt{(49080)^2 - 4(29446864)}}{2}$

$= \dfrac{49080 + 47865}{2} = 607.5 \ cm \ or \ 4872.5 \ cm$

or x = 6.1 m or 48.7 m

Only possible solution is x = 6.1cm; and this is the position of the section with maximum bending stress.

Q15. Explain by illustrating your answer with reference to the properties stated below of concrete and steel, why in a simply supported reinforced concrete beam loaded on top and with reinforcement steel bars at the bottom, the concrete around the bars will crack. Assume there is no slip of the bars in the concrete and that the steel bars are stressed to 120 MN/m². Take E_s = 208 GN/m²; E_c = 18.5 GN/m².

What will be the estimated computed stress in the concrete close to the bar?

Q16. A hollow circular cylindrical concrete pier has an external diameter of 100cm and inside diameter of 75cm. Determine the maximum distance from the centre of the pier of the point through which the vertical line of action of the resultant thrust must pass so that there be no tension in the section.

Let x = distance from central axis of pier to point of application of vertical load 'P' in cm.

Therefore bending moment due to 'P' = Px Ncm, assuming P in say newtons.

Let us determine the bending stress due to Px Ncm.

Using the Flexure formula, taking σ, to be the bending stress

$\dfrac{\sigma_1}{50} = \dfrac{Px}{I}$ ....(i)

For the pier which is annular in cross-section, I with reference to any diameter is given by

$$I = \frac{\pi d_o^4}{64} - \frac{\pi d_i^4}{64}$$

Here $\quad d_o = 100\ cm;\ d_i = 75\ cm$

therefore $\quad I = \frac{\pi}{64}(100 + 75)\{(100)^2 + (75)\}$

$$= \frac{\pi}{4}(175)(25)(15625$$

i.e. $\quad I = 3355164\ cm^4$

Employing (i) we have

$$\frac{\sigma_1}{50} = \frac{Px}{3355164}$$

or $\quad \sigma_1 = \pm\frac{50Px}{3355164}\ N/cm^2$

The direct compressive stress 'σ_2' due to 'P' is $\frac{P}{csa}$, csa being the cross-sectional area of the pier.

Here $\quad csa = \frac{\pi}{4}d_0^2 - \frac{\pi}{4}d_1^2 = \frac{\pi}{4}\{(100)^2 - (75)^2\} = \frac{\pi}{4}(175)(75)$

$$= 3436\ cm^2$$

$$\sigma_2 = \frac{P}{3436}\ N/cm^2$$

This is a compressive stress, i.e. $\sigma_2 = \frac{-P}{3436}\ (N/cm^2)$

For zero tension in the section: $+\frac{50\ Px}{3355164}\ \frac{-P}{3436} = 0$

i.e. $\quad 50(3436)Px = 3355164P \qquad$ or $\qquad x = 19.5\ cm$

Q17. An aluminium plate 10 cm wide and 2 cm thick is 1 metre long and has pin-jointed ends. The plate is oriented with the 2 cm side horizontal and is subjected to an eccentric pull of 120 kN which is 1 cm above the longitudinal axis of the plate. What load must be placed at centre-span so that the distribution of longitudinal bending stress at the centre of the plate is uniform?

The eccentric load causes a bending moment = 120 kN x 1 cm = 120 kN cm

'I' for the plate about a horizontal axis through its centroid = $\frac{2x(10)^3}{12}$ cm

= 166.7 cm^4

If σ_1 = bending stress due to the eccentric loading, then $\frac{\sigma_1}{5} = \frac{120}{166.7}$

or $\sigma_1 = \frac{600}{166.7} kN/cm^2 = 3.6\, kN/cm^2$

As concerns mid-span loading, let the required load = $W\,kN$

Therefore the bending moment 'M' at centre-span due to $W = \frac{WL}{4}$

Let σ_1 = bending stress due to centre-span loading

Therefore $\frac{\sigma_2}{5} = \frac{WL}{4} \cdot \frac{1}{166.7}$

Substituting $L = 1\,m = 100\,cm$

i.e. $\sigma_2 = \frac{5W(100)}{4(166.7)} kN/cm^2 = 3.6\, kN/cm^2$

$\sigma_2 = 0.75\,W\ kN/cm^2$

For uniformity of bending stress

$0.75 = 3.6\, kN/cm^2$

from which $W = 4.8\,kN$

Q18. Solve problem Q 8 .taking into account the direct tensile stress due to the pull.

Ans: 12.8 kN.

Q19. Determine the dimensions of a rectangular beam to be fashioned from a cylindrical log *36cm* in diameter such that it would be capable of withstanding the greatest bending stress.

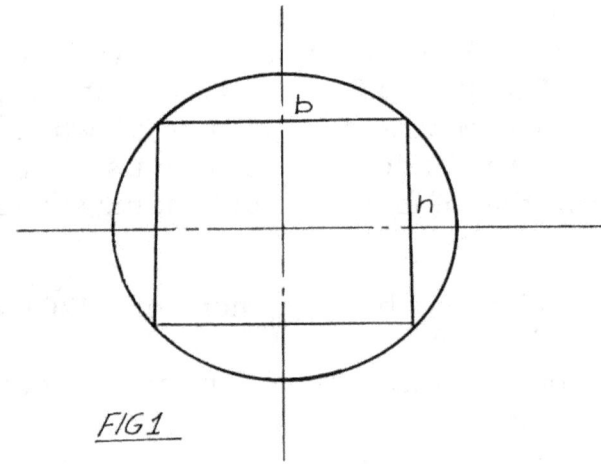

FIG 1

Begin by letting the required dimensions be:

b = breadth of beam; h = depth of beam, as in Fig. 1.

Evidently $b^2 + h^2 = (36)^2$

Let the constant bending moment to produce bending stress = M

Second moment of area about centroid = $\dfrac{bh^3}{12}$

Putting σ = bending stress, I would write

$$\frac{\sigma}{\frac{h}{2}} = \frac{M}{\frac{bh^3}{12}} \quad \text{or} \quad \frac{2\sigma}{h} = \frac{12M}{bh^3}$$

from which

$$\sigma = \frac{6M}{bh^2}$$

but $\quad h^2 = (36)^2 - b^2 = 1296 - b^2$

therefore $\quad \sigma = \dfrac{6M}{b(1296-b^2)} = \dfrac{6M}{(1296-b^3)}$

Noting that M is a constant. Therefore, for maximum $'\sigma'$, $\dfrac{d\sigma}{db} = 0$.

$$\frac{d\sigma}{db} = \frac{(1296b-b^2)\,0 - 6M\left(1296-3b^2\right)}{(1296b-b^3)^3}$$

$$\frac{d\sigma}{db} = 0$$

$$1296 - 3b^2 = 0$$

or $\qquad\qquad b^2 = 432$

from which $\quad b = 20.78\ cm$

so that $\qquad h^2 = 1296 - 432 = 864$

or $\qquad\qquad h = 29.39\ cm$

The dimensions of the strongest beam are: width 20.78 cm; depth 29.39 cm, say 20.8 cm by 29.4 cm.

Q20. A purple heart beam of cross-section 30 cm x 30 cm and 5 cm long is reinforced by two mild-steel plates 30 cm x 5 cm fixed rigidly to its top and bottom surfaces of which a sketch is shown in Fig. 1. The composite beam is simply supported at its ends and carries a uniformly distributed load of 8 kN/metre.

Determine the maximum bending stresses in the purple heart and in the steel at mid-span. Take E_{STEEL} = 205 GN/m² and $E_{PURPLE\ HEART}$ = 18.4 GN/M².

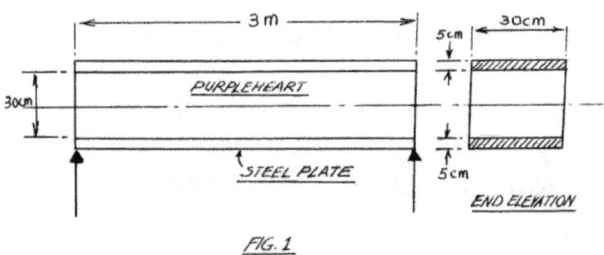

FIG. 1

Let M_p = maximum bending moment in purple heart at mid-span;

M_s = maximum bending moment in steel also at mid-span

therefore $M_p + M_s = \dfrac{wl^2}{8} = \dfrac{8000}{8} \times (5)^2 = 25000 Nm$

but by the Flexure relationship

$$\frac{M_p}{I_p} = \frac{\sigma_p}{y_p} \quad and \quad \frac{M_s}{I_s} = \frac{\sigma_s}{y_s}$$

or, alternatively, denoting Z as section modulus

$M_p = \sigma_p Z_p; \quad M_s = \sigma_s Z_s$ so that

$\sigma_p Z_p + \sigma_s Z_s = 25000$

Now $\quad I_p = \dfrac{30}{12} \times (30)^3 = 67500\ (cm)^4 = 67500 \times 10^{-8}\ (m)^4$

$y_p = 15cm \quad therefore\ Z_p = \dfrac{67500 \times 10^{-8}}{15 \times 10^{-2}} = 4500 \times 10^{-6}\ (m)^3$

Also, $I_s = \dfrac{30\,(40)^3}{12} - \dfrac{30\,(30)^3}{12} = 30\,\dfrac{(64000-27000)}{12}$

$\qquad = \dfrac{30}{12}\,(37000) = 92500 \times 10^{-8}\ m^4$

$y_s = 20cm\ therefore\ Z_s = \dfrac{92500 \times 10^{-8}}{20 \times 10^{-2}} = 4625 \times 10^{-6}\ m^3$

so that $\sigma_p Z_p + \sigma_s Z_s = 25000$

becomes $\sigma_p\,(4500 \times 10^{-6}) + \sigma_s\,(4625 \times 10^{-6})$ ....(i)

Because the reinforcement and the purple heart bend to the same radius of curvature

$$\frac{\sigma_p}{y_p} = \frac{M_p}{I_p} = \frac{E_p}{R}; \frac{\sigma_s}{y_s} = \frac{M_s}{I_s} = \frac{E_s}{R}$$

R, being radius of curvature

For purple heart $\quad R = \frac{E_p \, y_p}{\sigma_p}$

and for steel $\quad\quad R = \frac{E_s \, y_s}{\sigma_s}$

therefore $\quad\quad\quad \frac{E_p y_p}{\sigma_p} = \frac{E_s y_s}{\sigma_s}$

or $\quad\quad\quad\quad\quad \frac{\sigma_p}{\sigma_s} = \frac{E_p y_p}{E_s y_s}$

i.e. $\quad \frac{\sigma_p}{\sigma_s} = \frac{18.4 \times 10^9}{205 \times 10^9 \times 20} \quad \frac{15}{100} \times 100 = \frac{55.2}{820}$

or $\quad \frac{\sigma_s}{\sigma_p} = 14.85$

Substituting this result in (i)

$$\sigma_p \, (4500 \times 10^6) + 14.85\sigma_p \, (4625 \times 10^6) = 25000$$

$$4500\sigma_p + 68581.25\sigma_p = 25 \times 10^9$$

$$73181.25\sigma_p = 25 \times 10^9$$

Therefore $\quad\quad \sigma_p = 342.4 \, kN/m^2$

and $\quad\quad\quad\quad \sigma_s = 14.85\sigma_p$

i.e. $\quad\quad\quad\quad \sigma_s = 5085.6 \, kN/m^2$

Therefore the maximum bending stresses in purple heart and steel at mid- span are 342.4 kN/m² and 5085.6 kN/m², respectively.

Q21. A symmetric hollow cylindrical steel beam with external radius 'R$_0$' and internal radius 'R$_i$' is shown in Fig. 1. Show that the plastic modulus Z_p is given by the expression

$$Z_p = \frac{2}{3} \, (R_o - R) \, (R_o^2 - R_o R_1 + R^2)$$

[Assume that the position of the centroid of a semi-circle is distant $\frac{4r}{3\pi}$ from its diameter].

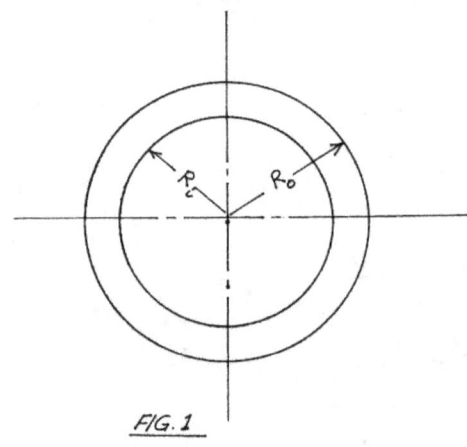

A line through the centroid of the whole cross-sectional area divides it into two (2) semi-circular areas. The distances of the centroids of these two areas from this original centroid must now be determined. Because of symmetry we shall calculate the position for one-half of the area. Let y $= y_1 = y_2$ be the vertical distance of this position from the diameter.

Accordingly: $\quad y = y_1 = y_2 = \left\{ \pi R_o^2 \left(\frac{4R_o}{3\pi} \right) - \pi R_I^2 \left(\frac{4R}{3\pi} \right) \right\} / \frac{\pi}{2} (R_0^2 - R_1^2)$

$$= \frac{4}{3\pi} \frac{(R_0^3 - R_1^3)}{(R_0^2 - R_1)^2} = \frac{4}{3\pi} \frac{(R_0^2 - R_1^2 - R_0 R_1)}{(R_0 + R_1)}$$

Plastic Modulus, $Z_p = \frac{A}{2} (y_1 + y_2)$

$$\therefore Z_p = \frac{(\pi R_o^2 - \pi R_1^2)}{2} \cdot 2 \left\{ \frac{4}{3\pi} \frac{(R_o^2 + R^2 - R_o R_1)}{(R_o + R_1)} \right\}$$

$$Z_p = \frac{4}{3} (R_0 - R_1) (R_0^2 + R_1^2 - R_0 R_1)$$

Q22. If the yield stress of the beam in Q is 280MN/M^2, determine the plastic moment of 'M_p' and shape factor 'f' given that $R_1 = 400$ mm; and thickness of beam = 25 mm.

Ans. 3.4 MNm; f = 0.02

Q23. The steel T-beam shown in Fig. 1 has a yield stress of 250 MNm^2. Determine (i) the position of the beam's neutral axis; (ii) moment of inertia of section with reference to the neutral axis; (iii) Yield Moment; (iv) Plastic Modulus; (v) Plastic Moment; and (vi) Shape Factor

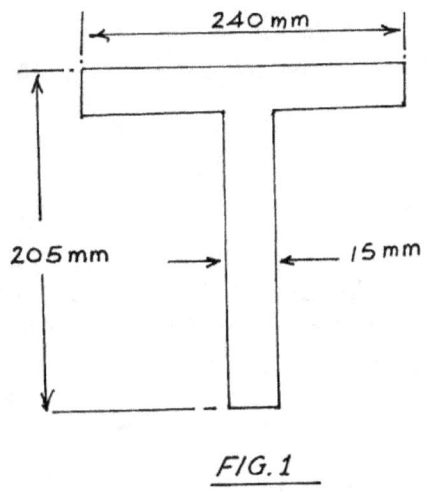

FIG. 1

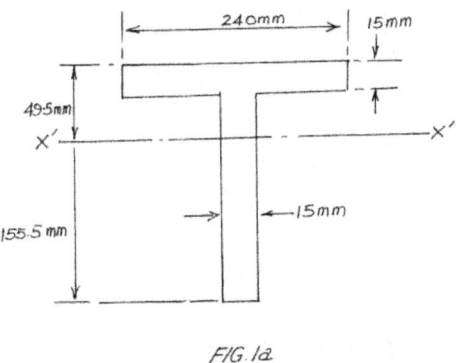

FIG. 1a

Let distance of neutral axis from x – x = $\bar{y}$. Referring to Fig. 1,

$$\bar{y} = \frac{15\,(240)(7.5)\quad(190)(15)\,(95 \div 5)}{15(240)\ +\ 190\,(15)} = 52.8\ mm$$

This distance from axis X–X positions the line through the neutral axis $x' - x'$. Referring now to Fig. 1a,

$$Ix' - x' = \frac{15\ x\ (52.2)^3}{3} + \frac{240\ x\ (52.8)^3}{3} - 2\ \left\{112.5\ x\ \frac{(37.8)^3}{3}\right\}$$

$$= 25353518\ mm^4$$

or $\quad Ix' - x' = 2523\ x\ 10^{-8}m^4$

Let 'C' be the furthest distance from axis $x' - x'$ of the fibres of the beam, in this case = 155 mm or 0.155 m. The Yield Moment M_Y is computed from:

~ 181 ~

$$M_\gamma = \frac{\sigma_Y I_{x'-x'}}{c}$$

$$= \frac{250 \times 10^{-6} \times 25.4 \times 10^{-6}}{\frac{152.2}{10^3}} \quad \text{(working in metres)}$$

$$= 41.7 \times 10^3 \; Nm$$

say $M_\gamma = 41.7 \; kNm$

To determine the Plastic Modulus 'Z_0" it is necessary to find first of all the distances y_1 and y_2 of the centroids of the areas of the section above and below the neutral axis, respectively.

With reference to Fig. 1a,

$$y_1 = \frac{37.8 \times 15 \times 18.9 + 240 \times 15 \, (45.3)}{37.8 \, (15) + 240 \, (15)}$$

$$= 173796.3/41.67$$

i.e. $y_1 = 41.7 \; mm$

Evidently, $\qquad\qquad y_2 = 76.1 \; mm$

Accordingly, $\qquad\qquad Z_p = \frac{A \, (y_1 + y_2)}{2}$

A = area of section

$\therefore \quad A = (15 \, (240) + 15 \, (190)$

$$= 6450 \; (mm)^2$$

i.e. $A = 6450/10^6 \; (m)^2$

$$Z_p = \frac{6450}{10^6} \cdot \frac{1}{2} \left(\frac{41.7 + 76.1}{1000} \right)$$

$$= 57405/10^3 \; m^3$$

or $Z_p = 57405 \; (cm)^3$

from which Plastic Moment M_p

$$= \frac{\sigma}{\gamma} \cdot Z_p$$

i.e. $M_p = 250 \times \frac{57405}{10^3}$

$$= 14.4 \; kNm$$

or $M_p \cong 14.4 kNm$

and Shape Factor $'f' = M_p/M_y$ is given by

$$f = \frac{14.4 KNm}{41.7 kNm}$$

i. e. $f = 0.345, say, 0.35$

Q24. Determine how many reinforcing steel bars of diameter 20 mm are required in the design of a floor slab which is to carry a load equivalent to 2000 N/m^2 including the estimated weight of the slab itself over a 3 metre span. Assume permissible stress of 110 MN/m^2 in each steel bar.

Assume the relationship for Moment of Resistance (MR): MR = 885000 bd^2 NM in which b= breadth or width of slab and d = depth of slab, both in metres. Also, take the lever arm for computing resisting force in a steel bar = 0.45d.

Considering a 1 metre width of steel slab parallel to the span
$$885000\ bd^2 = \frac{wl^2}{8} = \frac{2000\ x\ 9}{8}$$

i.e. $885000\ x\ 1\ x\ d^2 = 2250\ Nm$

from which $d^2 = 0.00025\ or\ d - 0.05\ m\ or\ 5\ cm.$

Now with σ_s = permissible stress in steel bar and A_s = total cross-sectional area of all steel.

$\sigma_s A_s\ x\ lever\ arm = 2250\ Nm$

i.e. $\sigma_s A_s\ x\ 0.45\ (0.05) = 2250$

$\therefore\ A_s(m)^2 = \dfrac{2250}{0.45\ x\ 0.05\ x\ 110\ x\ 10^6}$

or $A_s(mm^2) = \dfrac{2250}{0.45\ x\ 0.05\ x\ 110\ x\ 10^6} = 909$

One 20 mm diameter steel bar has a cross-sectional area = 314 mm². Therefore three (3) 20 mm Ø bars per metre width of slab should suffice.

Q25. If the yield stress of the beam in Q. 21 is 280 MN/m^2, determine the plastic moment 'M_p' and shape factor 'f' given that $R_1 = 400mm;$ and thickness of beam = 25mm.

Ans: 5.1 MNm

This page is intentionally left blank.

CHAPTER 8

IMPORTANT PROPERTIES OF MATERIALS: TENSILE AND COMPRESSIVE STRENGTH; HARDNESS; IMPACT STRENGTH; AND, FRACTURE TOUGHNESS. STRESS CONCENTRATION; CREEP; STRESS RELAXATION AND VISCOELASTICITY; AND RECOVERY

As pointed out in Chapter 1 it was during the Renaissance, the period of enlightenment which followed the Dark Ages, that Galileo Galelei recognized that engineering structures could neither be designed properly nor their behaviour predicted accurately by a reliance solely on theoretical analysis based on geometrical considerations. He posited correctly that such considerations had to be supplemented by empirical knowledge of the properties of the engineering materials used to design and build a structure. Thus, was born the engineering science of Strength of Materials whose theorems have strong phenomenological connections.

In the preceding chapter the reader was administered a soupcon of the fundamental theory of engineering stress and strain. But let me state at once that the treatment there was not altogether devoid of matters related to the mechanical and physical properties of materials because, for example, we assumed Hooke's law for uni-axial stress which has definite connections to ductility and strength; and, we also dealt with thermal properties, when stress in combination with thermal expansion was considered.

Nowadays engineers have at hand an astonishing array of materials from which to formulate their designs and manufacture products to practically any shape and size imaginable. A wide variety of ferrous materials is available: from carbon and alloy steels to structural steels and stainless steels. In the non-ferrous category are included aluminium, zinc, magnesium, chromium and their alloys, so too copper and copper-based alloys such as brass and bronze; and, also the so-called space-age metals like titanium, niobium, tantalum and vanadium and their alloys.

An important group of compounds made up of metal and non-metallic constituents comprises the ceramics. Clay is the vital component of such products. Widely known examples of these are porcelain and firebrick. Glass, Portland cement and other specialty cements used extensively in oil-well exploration and production fall within this category.

In modern-day manufacturing the raw materials derived from the processing of crude oil, condensate, natural gas and LNG (liquefied natural gas) comprise a highly economically-significant group of non-metallics. From these are produced products such as lubricants including the synthetic variety made from polymers, greases, asphalt, bitumen, naphthenic oils, and plastics, synthetic rubber, resins and epoxy resins and an assortment of gaseous hydrocarbons such as methane, ethylene, acetylene, and isobutane which are used for the production of other petrochemicals in so-called downstream industries.

Notwithstanding the many alternative materials available, wood continues to be preferred over many modern-day substitutes because of its ease of working, machineability, heat-and sound-insulating value and load bearing capacity. Wood is still very widely used in furniture manufacturing and building construction. Some of the more popular woods are cedar, mahogany, balata, laurier, teak, greenheart, purpleheart and even bamboo. These are fairly widely distributed throughout the Caribbean region.

The properties of several of the materials referred to in the foregoing are so unique that they have influenced directly the development of a whole range of new manufacturing methods, in a positive way generally speaking. Consider as an example of this the case of the ubiquitous 'plastics'. In what is called Integrated Product Design (IPD for short) of this material, the use of moulding techniques whereby a component of unusual and complicated shape, say, is generated in a single operation or process, has all but replaced completely the traditional manufacturing method of producing separate parts and subsequently assembling or fabricating them to form the finished product.

This improvement does have its downside however in that whereas the engineer designer's imagination may in IPD produce a remarkable specimen of technological ingenuity and complexity of form, stress analysis of a representative model of such mental lucubration might turn out to be virtually impossible, even with much mathematical weariness of the flesh and the aid of the electronic computer. IPD in plastics is based principally on rules rather than on conventional stress analysis.

In this chapter we shall spend some time examining among other matters the empirical relationship between stress and strain and by such examination also identify those properties which broadly speaking are characteristic of engineering materials; the properties with which engineers must be familiar when choosing materials for particular applications. This will be done by reporting, in some cases on actual experimental observations and determinations made in connection with a number of tests conducted under engineering–laboratory conditions. These include the following: Static Tensile and Compressive; Fatigue, and Hardness Tests.

It is well to note at this juncture that the properties of engineering materials are intimately bound up with their microstructure. It is therefore important for engineers to be aware of factors that can control and influence changes in the properties of the materials they specify for particular applications. Thus it is important to teach the subject of Strength of Materials in conjunction with the science and technology that undergird it, metallurgy being one such.

TENSILE AND COMPRESSIVE STRENGTH

Static Tensile Test on Mild Steel

The apparatus consisted of a Riehle testing machine fitted with an autographic recorder. In the arrangement the specimen was stretched and the recorder plotted the load vs extension characteristic as the experiment proceeded. The specimen was made from structural steel plate *(C : 025%; Mn = 1.5%; P : 0.045%; S : 0.05%; Si ; 0.20%; Cu; 0.30% Ni : 0.6%).* It was rectangular in cross-section with csa = 367(mm)^2. The gauge length of the specimen was 200 mm, the width and thickness along this length being respectively 38mm and 9.6mm. Refer to Fig. 1a.

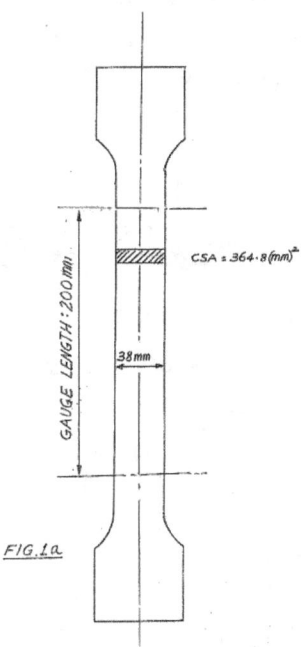

FIG. 1a.

Along the centre-line of the specimen a number of light centre-punch marks were made distant 10mm apart symmetrically about the middle. See Fig. 1b.

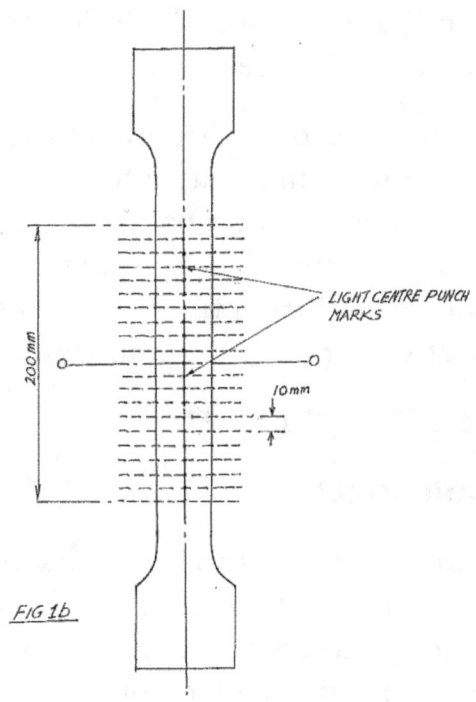

FIG 1b

LIGHT CENTRE PUNCH MARKS

200 mm

10 mm

The purpose of this was to examine the relationship between elongation and the 10mm gauge lengths. A travelling microscope was used to measure the actual distances between the centre-punch marks as the specimen elongated, and a micrometer was used to measure the variation in thickness of the specimen under load. In reporting on the experiment appropriate comments shall be made to elucidate observation.

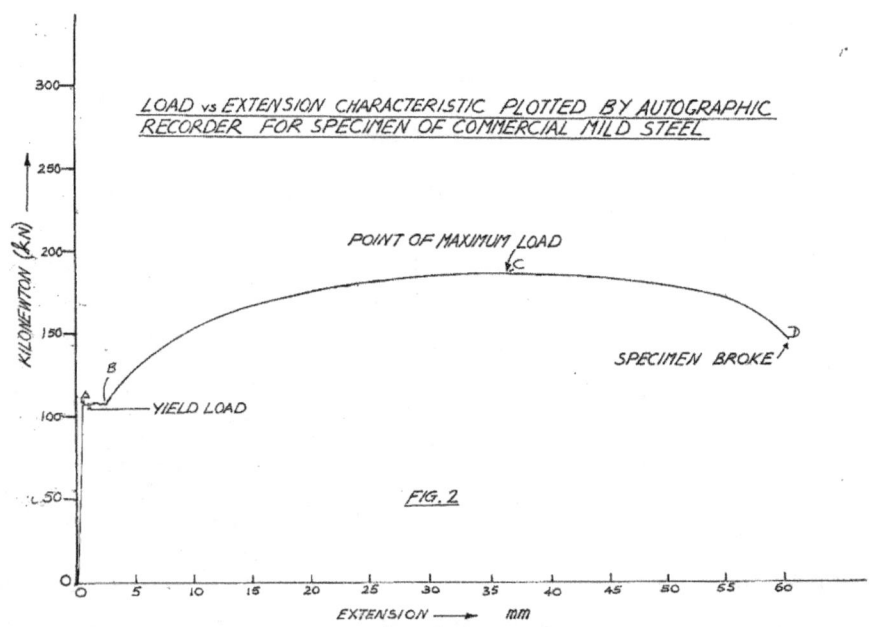

LOAD vs EXTENSION CHARACTERISTIC PLOTTED BY AUTOGRAPHIC RECORDER FOR SPECIMEN OF COMMERCIAL MILD STEEL

POINT OF MAXIMUM LOAD

SPECIMEN BROKE

YIELD LOAD

FIG. 2

KILONEWTON (kN)

EXTENSION ——▶ mm

The graph obtained from the autographic recorder is shown as Fig. 2. The first portion of it is a straight line passing through the origin of the axes. Hence it was concluded that load is directly proportional to extension for this portion, the value of stress at its end being the proportional limit, i.e. where the characteristic ceased to be a straight line. Thereafter, the specimen extended significantly for little or no additional load. The magnitude of the load which initiated the onset of this phenomenon called yielding, is commonly referred to as the yield load. Yielding is an important phase in the atomic restructuring of a ductile metal. In the straight line portion OA in Fig. 2 which defines the elastic phase any extension is recoverable once the load causing it is removed. At yielding however, a significant dislocation of the atomic structure of the metal occurred, in short a deformation which was not totally recoverable. The yield load from which the yield stress was easily determined by dividing the former by the original cross-sectional area of the specimen, marked the separation of the elastic phase from the plastic phase. The yield stress commonly referred to as yield strength is an important property required of structural components. Service loads should be such as to avoid plastic deformation. This is not to say that loads causing plastic deformation are of no interest to the engineer. A knowledge of the loads that would cause such deformation is important especially in the metal-turning industry e.g. pressing of sheet steel into automobile fenders, hoods, extrusion of bars, cold-rolling and the like. Presses are designed to do just that.

After yielding at 'B' in Fig. 2, the specimen gained in strength because it required further increase in load to achieve more extension. This was caused by work-hardening, also called strain-hardening. At point 'C' on the graph work-hardening attained its limit and the maximum load was achieved.

Soon afterwards different cross-sections appeared to elongate at different rates; the cross-sectional area near the middle of the gauge length began to reduce significantly although the load was decreasing, thus heralding the phenomenon of 'necking', so called. The specimen suffered rupture shortly thereafter, breaking with a ballistic bang at 'D'. A diagram showing the physical appearance of the ruptured specimen with necking is shown as Fig. 3. The ultimate strength of the material was the value of the maximum load divided by the original cross-sectional area.

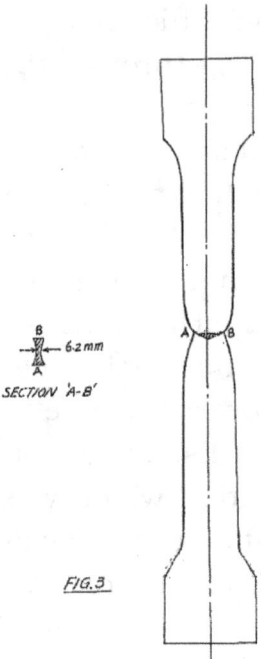

SECTION 'A-B'

FIG.3

In the course of the test, the elongation of each of the 10mm lengths marked off along the centre-line of the specimen was measured and a graph was plotted of these component gauge lengths against elongation for each for both parts of the specimen i.e. for sections above and below the mid-line of the specimen. The characteristic curve[4] is shown as Fig. 4. and by referring to it, it can be plainly seen that initially extension was gradual, almost uniformly distributed over the length of the specimen, after which with increasing load there was a gradual increase in extension and finally localized extension followed by fracture.

---

[4] The curve resembles the tractory or equitangential curve : "trakorie von Huygens." You may have come across it in your studies in pure mathematics where it is commonly referred to as the 'tractrix'

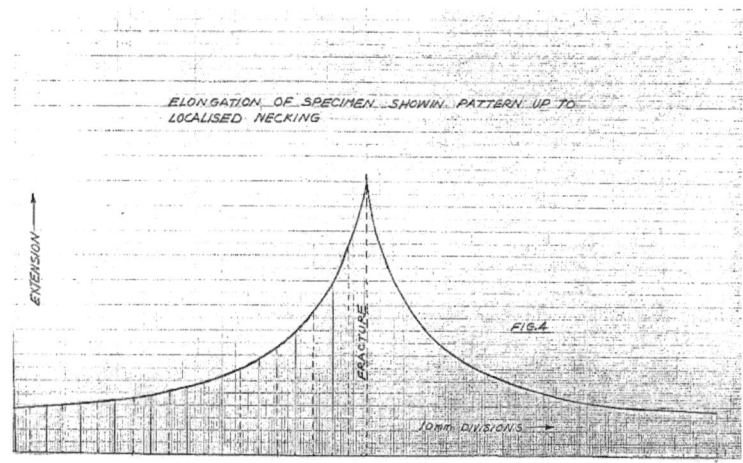

This behaviour is what led to the formulation of Barba's law which is generally expressed in the form. $\delta = c\sqrt{A} + B\ell$. That is to say, total extension 'δ' is a function both of gauge length 'ℓ' and cross-sectional area 'A'. 'B' and 'C' are often referred to as Unwin constants.

Employing Fig. 4, mid-ordinates were constructed to intersect the curve. The total elongation for specified gauge lengths was obtained by summing ordinates and another curve was plotted with values of elongation obtained by this method as ordinates against corresponding value of gauge lengths as abscissae. This curve is shown as Fig. 5.

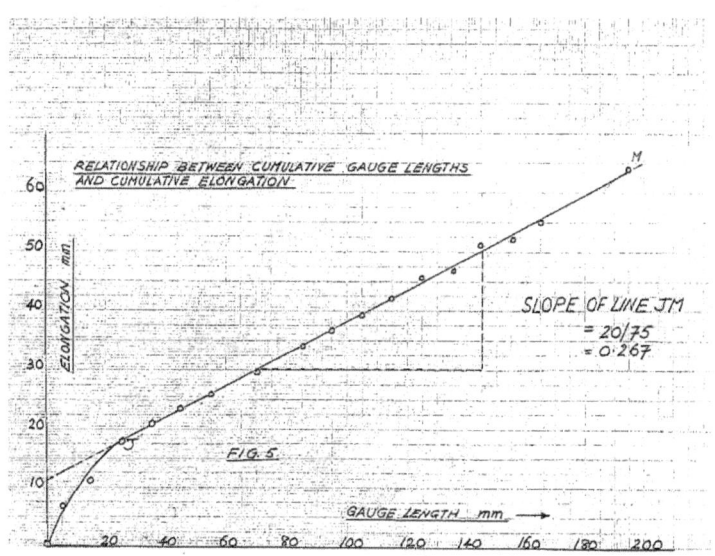

Determination of Properties

The following information was obtained from Fig. 2 and from measurements taken of the specimen at fracture :

Load at Yield Point = 107kN

Maximum Load = 173.5kN

Load at fracture = 142kN

Width at fracture = 27.8mm

Average thickness at fracture = 6.2mm

Original cross-sectional area = 364.8 (mm)^2

Based on these data the following were calculated :

(i)   Yield stress (or Yield Strength) $= \dfrac{Load\ at\ Yield\ Po\operatorname{int}}{Original\ csa}$

$$= \dfrac{107kN}{367/10^6}$$

$$= 291.5 MN/m^2$$

(ii)   Tensile Strength (also called Ultimate Tensile Strength)

$$= \dfrac{Maximum\ load}{original\ csa}$$

$$= \dfrac{173.5kN}{367/10^6}$$

$$= 472.8 MN.m^2$$

(iii)   Stress at Fracture (sometimes also called true fracture strength)

$$= \dfrac{Load\ at\ fracture}{Final\ csa}$$

$$= \dfrac{142kN}{172.4/10^6}$$

$$= 823.7 MN/m^2$$

(iv)   Percent reduction in area $\quad = \dfrac{100(original\ csa - Final\ csa)}{original}$

$$= \frac{100(367 - 172.4)}{367}$$

$$= 53\%$$

Referring to Fig. 5, it is seen that the portion of the curve JM is approximately a straight line. If this line is produced backwards to cut the elongation axis, the following are obtained:

Intercept on elongation axis $= 11.5mm$

Slope of characteristic JM $= 0.267$

$\therefore$ Elongation $'\delta' = 0.267\ell + 11.5$

Expressing elongation as per unit

$$\frac{\delta}{\ell_o}\% = 100\left(0.267 + \frac{11.5}{\ell}\right)$$

$$= 26.7 + 1150/\ell$$

Here $C\sqrt{a} = 1150$, in which, a = csa of bar

i.e. $C\sqrt{367} = 1150$ or $19.2C = 1150$ or $C = 60$

Unwin's constants are : *100C = 60*; and, *100B = 26.7*. The equation enables per cent elongation of geometrically similar bars of the same material to be determined. Based on it, the value of $\delta/\ell\%$ for the 200mm gauge length and 267(mm)^2 cross-sectional area in the experiment just described was 31%. The experimental determination based on $100(\ell - \ell_o)/\ell_o$, $'\ell_o'$ and $'\ell'$ being the original and final lengths respectively worked out to be $100\left(\frac{60}{200}\right) = 30\%$. It should be noted that the very basis on which Barba's law is grounded, that is, during straining a regime of uniform extension which is followed by localized 'necking', makes it important that in giving information on % δ/ℓ an indication should be given stating whether the gauge length includes the region in which fracture occurred. If the fracture occurred outside the gauge length Barba's law is inapplicable.

Percent elongation and percent reduction in area are generally the two most common means of specifying the ductility of materials. Ductility is the property which gives an indication of the amount of deformation a material can undergo before it fractures. Thus you are able to bend a piece of galvanized-iron wire back and forth quite a few times before it breaks, whereas a glass rod would fracture at once. Accordingly if there are circumstances in which an engineering component is overstressed, then its ductility would permit some deformation before it finally ruptures. This is exemplified in the long flat portion of the load vs extension characteristic in Fig. 2. It is customary for steel manufacturers to provide data on the properties $\%\left(\dfrac{d}{\ell_o}\right)$ and the reduction in area in their materials specifications.

Close observation of the specimen during application of load and after fracture resulted in identification of a particular phenomenon. Although the specimen was by no means highly polished, it was found that at yield point, the oxide coating on the steel in the region of the enlarged ends began flaking along certain 'lines' which made angles of approximately 45° to 60° with the axis of the specimen. These were identified as 'Luder-line traces' a manifestation of slip in the material. Fig. 6 shows the scaling effect produced on the specimen; the slip started at enlarged ends and worked inwards to the middle.

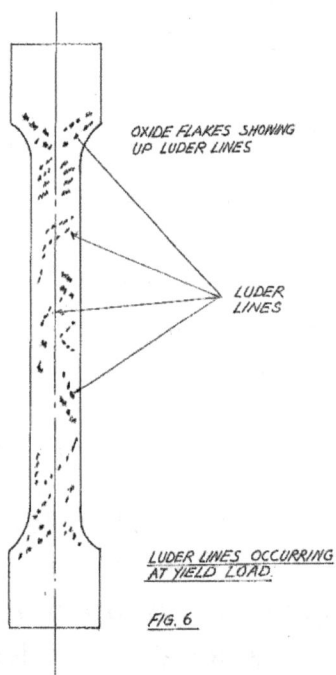

OXIDE FLAKES SHOWING UP LUDER LINES

LUDER LINES

LUDER LINES OCCURRING AT YIELD LOAD.

FIG. 6

It was not possible to determine the elongation of the specimen at the proportional limit and so neither an estimate of the stiffness of the specimen nor of Young's modulus was obtained. Referring again to the load vs extension graph, it would be seen that the specimen was initially almost rigid, there being very little extension up to a load of 100kN.

Static Tensile Test on Cast Iron

A circular cylindrical specimen of cast iron of composition Fe: 91.2%;

C : 19% and P = 69% and of gauge length 200mm and diameter 20.4mm was subjected to a tensile test using an Avery Universal Testing Machine. The machine was fitted with an autographic recorder which plotted the load extension characteristic redrawn here as Fig. 7. It is a straight line up to the fracture load of 57.5kN. Thus, instant fracture occurred at the maximum load. Because there is no deformation similar to that which occurred with the mild steel specimen, cast iron is regarded as a brittle material: cast iron is not normally regarded as a ductile material. We shall return to this subject in a trice.

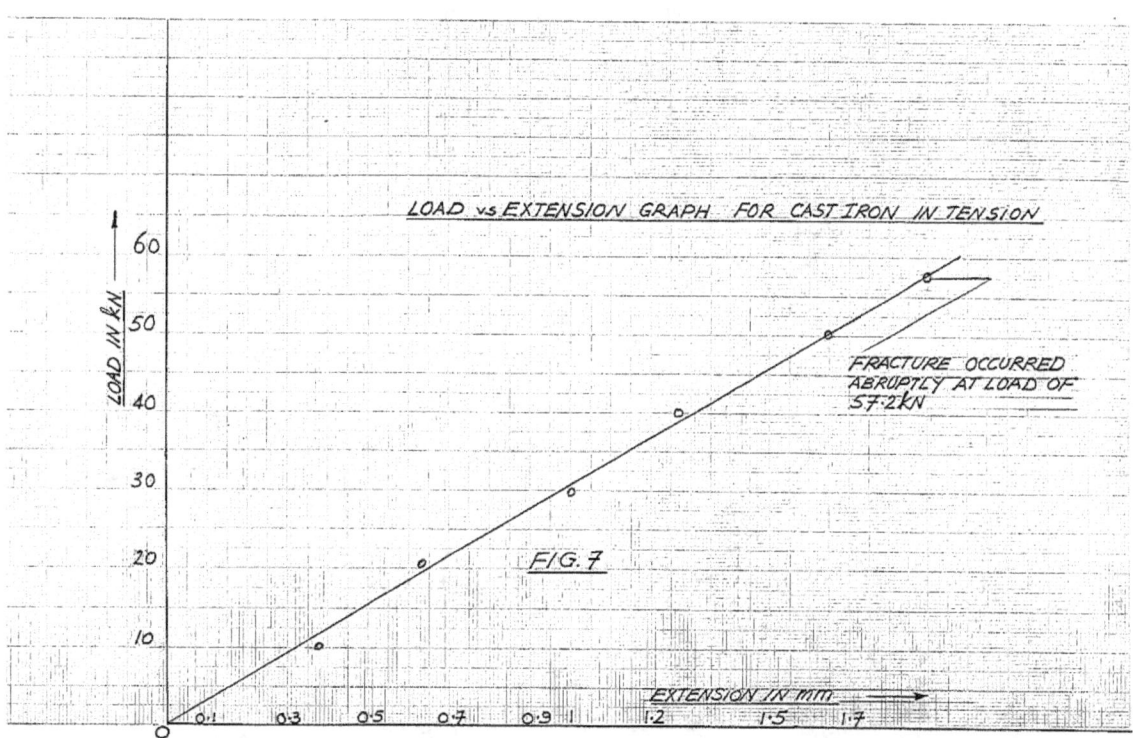

Based on the load at fracture a value of ultimate strength *(UTS)* calculated thus: Load at fracture = 57.5kN; Final extension (at fracture) = 1.905mm; and diameter of bar = 20.4mm. Therefore $UTS = 57.5kN / \frac{\pi}{4}(20.4)^2(10^{-6})m^2$

Percent elongation $= 100(1.91)/200 < 1\%$. The specimen broke outside the gauge length.

~ 195 ~

Static Compression Tests on Mild Steel and Cast Iron

To round off this section, the results of axial compression tests on mild steel and cast iron are given in order to compare them with some of the values obtained for the tensile tests.

The compression tests were performed on circular cylindrical specimens (Fig. 8) each 25.4mm long and diameter 12.7mm using a compression testing machine of 45kN capacity. During the test the speed of compression was extremely slow and readings were taken at the points of failure : barrelling in the case of the mild steel, and at the point of fracture in the case of the cast iron. Refer to Figs. 9 and 10.

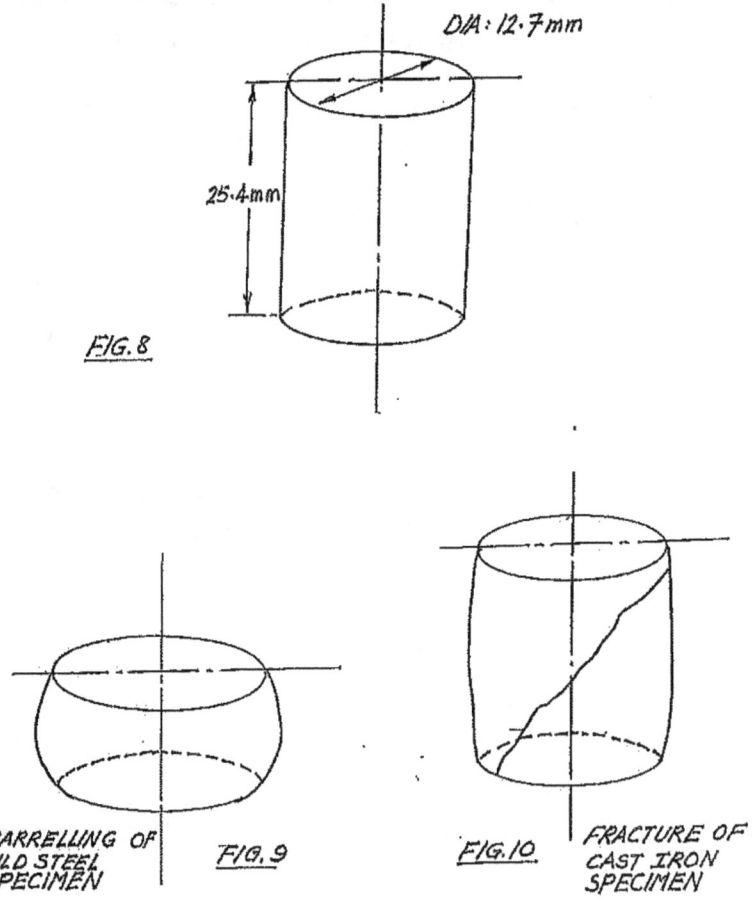

DIA: 12.7mm

25.4mm

FIG. 8

BARRELLING OF MILD STEEL SPECIMEN   FIG. 9   FIG. 10   FRACTURE OF CAST IRON SPECIMEN

The cast iron specimen broke at an ultimate stress of 772 MN/m², nearly four and one-half times the stress at which rupture occurred in the experiment to test that material's tensive qualities, thereby demonstrating unquestionably its superior compressive strength over its tensile strength. Although it was difficult to discern at what stage the mild steel specimen began to bulge, the stress at perceptible barrelling was calculated to be 941MN/m², a value higher than that at which the specimen tested in tension had fractured, viz. approximately 824MN/m². Without further experimental evidence it cannot be stated authoritatively that mild steel is stronger in compression than in tension but it may at least be fairly concluded that mild steel's compressive strength is as good as its tensile strength. Notwithstanding the results reported in the foregoing, cast iron should to be stereotyped as a material suitable only where compressive strength is required, as we shall learn presently.

There are many kinds of cast iron embracing a variety of Fe-C-Si-trace element alloy compositions. They are generally classified according to the manner in which carbon, in the form of graphite, is present in either a ferritic or pearlitic matrix. Cast iron owes its brittle nature to the distribution of carbon in the form of flakes of graphite throughout the matrix. Refer to Fig. 11. The key to solving the problem of improving the mechanical properties of cast iron lay in effecting a change not only in the size but also in the distribution of the carbon.

FIG.11. GRAPHITIC FLAKES IN GREY CAST IRON

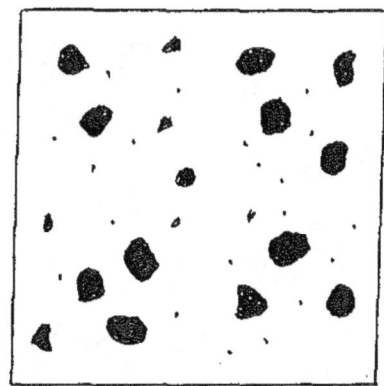

FIG.12. GRAPHITIC SPHEROIDS IN DUCTILE IRON

British and American research led ultimately in 1948 to the invention of a new form of cast iron called ductile iron, in which graphite is no longer present in the form of flakes, but instead as spheroids. Refer to Fig. 12. Ductile iron is sometimes referred to as S.G. iron. Its mechanical properties are far superior to those of ordinary cast iron. In Trinidad and Tobago ductile iron pipes are used extensively in potable – and waste-water systems.

Yield Stress (Yield Strength) vs Proof Stress (Proof Strength)

You would recall that the mild steel specimen in the tensile-test experiment reported earlier on, stretched in a rather distinctive fashion just after the proportional limit signalling a change from an elastic to a plastic phase of deformation. As Fig. 2 showed, the specimen was almost rigid during the start of straining, the line *OA* almost coinciding with the load axis up to about 110kN. The load at which yielding occurred was the yield load from which the yield strength of the material was calculated. Thus, design engineers regard yield strength as a very important criterion because it tells them whether a material would deform or not in service.

However not all materials exhibit a distinctive yield point, and by extension yield strength. This is well exemplified by the curve shown in Fig. 13 which is a typical load versus strain characteristic for a specimen of duralumin, 50mm gauge length and cross-sectional area equal to 8.84(mm)^2. A typical duralumin alloy has the following composition : 4% Cu; 0.5% Mn; 0.5%Mg; 0.5% Fe, 0.5%Si and 90%Aℓ.

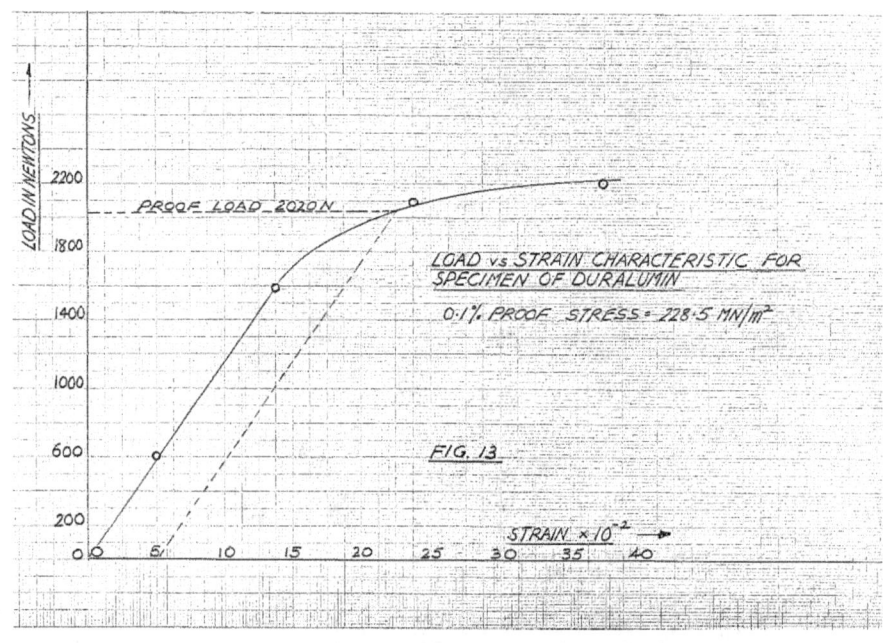

Because, as just stated, yield strength is such an important design criterion and because its determination cannot be made accurately using a graph such as that in Fig. 13, its determination being a matter more of judgment than anything else, it is generally agreed that the risk of a certain limited amount of permanent deformation can be tolerated in the design of a component whose material does not display the discernible yield strength. Thus, the device of a proof stress (or proof strength) was created to meet the need; not an ersatz mark you, but a more reliable determination than that of yield stress for certain materials.

Proof stress is the value of stress that causes a strain generally of either 0.1% or 0.01 for 0.1% proof stress; or 0.2% or 0.002 for 0.2 proof stress. Both values are found from the load vs extension (or strain) graph. This is achieved by constructing a line parallel to the elastic line of the graph and passing through the point marking the relevant abscissa on the strain axis : 0.001 in the case of 0.1% proof stress; and, 0.002 for 0.2% proof stress. Where the parallel line cuts the curve is the required proof stress or proof strength. In Fig. 13 the construction for 0.1% proof strength is shown, the value for the specimen of duralumin being $\{2020 \times 10^6 / 8.84\} N / m^2 \equiv 228.5 MN / m^2$.

It is necessary to point out that the load vs elongation graph obtained in the foregoing experiment did not reflect all the characteristics generally associated with typical mild steel. Because of this a typical textbook stress vs strain diagram for mild steel is shown in Fig. 14. It shows an elastic limit beyond the proportional limit. That is to say beyond 'A' and 'B' the direct proportional limit between stress and strain no longer exists. Point 'B' marks the elastic limit. If the stress is removed at that point the material returns to its original configuration. The graph obtained in the experiment described did not reflect that characteristic. Also, there are two yield points. It should also be noted that the slope of the line 'OA' in Fig. 14 is the modulus of elasticity of the material E being equal to the value of the elastic stress divided by the elastic strain.

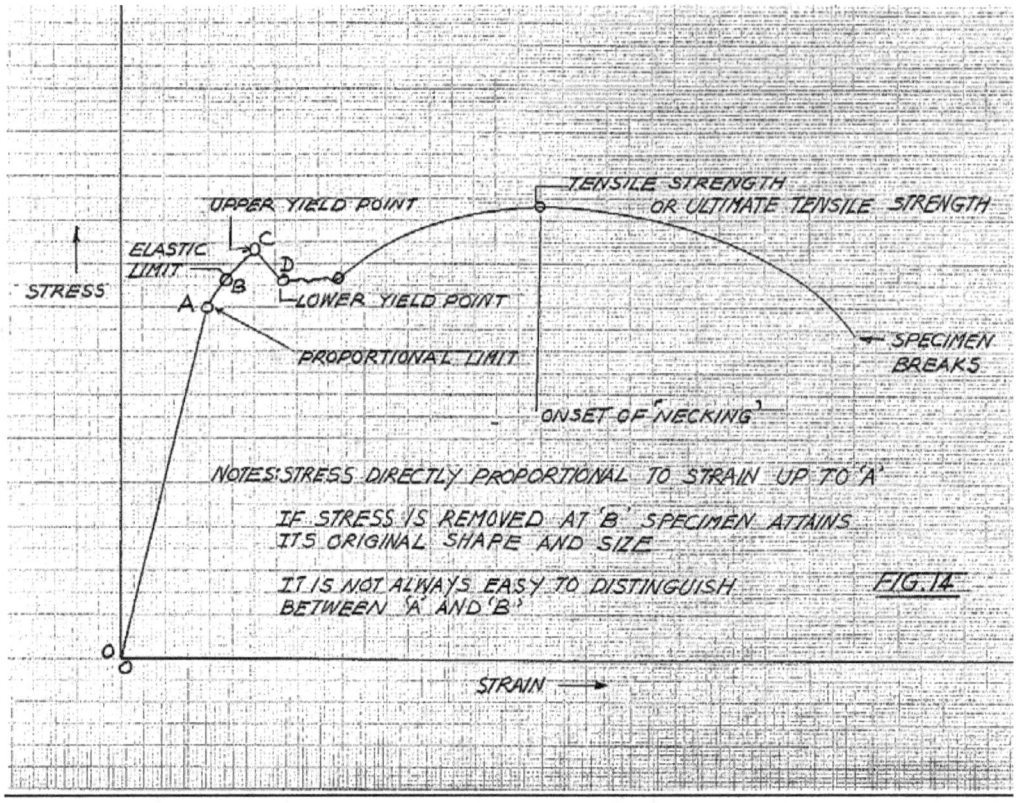

Figure showing stress-strain curve with labels: STRESS (vertical axis), STRAIN (horizontal axis), UPPER YIELD POINT, ELASTIC LIMIT, LOWER YIELD POINT, PROPORTIONAL LIMIT, points A, B, C, D, TENSILE STRENGTH OR ULTIMATE TENSILE STRENGTH, ONSET OF NECKING, SPECIMEN BREAKS.

NOTES: STRESS DIRECTLY PROPORTIONAL TO STRAIN UP TO 'A'

IF STRESS IS REMOVED AT 'B' SPECIMEN ATTAINS ITS ORIGINAL SHAPE AND SIZE

IT IS NOT ALWAYS EASY TO DISTINGUISH BETWEEN 'A' AND 'B'

FIG. 14

Hardness

When the term hardness is used in reference to a property of a material it is to convey an indication of the material's resistance to indentation and such and similar actions as for example surface scratching, say, with a nail, or abrasion as with a rasp, cutting as with a saw and machining as with a lathe or milling machine. The rotary-drill bit used in oil-well drilling operations is a good example of a mechanical component with hardened surfaces to enhance its resistance to wear by erosion and abrasion as the bit penetrates the earth. Hardness is also an important property in non-metallic materials such as the various kinds of timber such as teak, purple heart and greenheart used for manufacture of wooden flooring and furniture and, also in some polymers and ceramics.

Hardness tests are non-destructive procedures for checking properties of materials against specifications for compliance purposes. For example, a component made of quenched steel with 0.1%C should give a Vickers Pyramid Number (VPN) in the vicinity of 400, whereas one with 1.2%C should give one of about 900. Among the more popular tests are the Brinell (BHN) and the Vickers Diamond Pyramid (VPN or DPN). These consist of the pressing of a hardened steel ball, in the case of the former, and a pyramid-shaped diamond with an angle of 136º in the case of the latter, into the surface of the material, with a known force. For both, the Hardness number is derived from a relationship consisting of the load divided by the surface area of the indentation, a form of pressure measurement really.

In another procedure, the scleroscopic Hardness test, the measurement of hardness is based on the height to which a hammer with a rounded diamond tip rebounds after being dropped onto the surface of the specimen from a given height; the harder the specimen the greater the rebound; essentially an energy method. The height of rebound in the case of a fully-hardened tool steel is accorded a hardness number of 100 on the scale of hardness measurement as a standard, and based on the height of rebound of the hammer after striking a specimen under test with this standard the hardness number is determined on the basis of proportionality. The hardness of polymers is determined using a variation of this test.

With the Rockwell Hardness Tester (RHT), hardness number is based on the depth of the impression made by a loaded ball or diamond cone indenter applied to the surface of the material. One of the disadvantages with the RHT is the fact that there are several Rockwell scales embracing the range of values in metals and as a consequence the Rockwell values are not easily compared with VPNs, for example. One advantage of the Rockwell test however is that the pre-grinding and polishing of the surface of a material prior to BHN, VPN and Scleroscopic testing is not generally required. Hence the Rockwell Hardness Hardness Tester is the ersatz in situations where a quick quality-control determination is required : a rough and ready order of magnitude solution, as it were.

You may come across yet another scale of hardness measurement in the literature on mineralogy. It is called the Mohs hardness scale and is based on a manual scratch test. Diamond the hardest material is accorded a rank of 10 on this scale and the softest, talc a MHN of 1. By comparison, very hard steels are rated at MHN 7 and tungsten carbide used on some drill bits at a decimal rating of MHN 9.7; topaz is at 8; and calcite at 3. MHN ratings are very much a matter of judgment.

In timber technology, hardness is measured by means of a JANKA tool. The primary measurement is that of the resistance offered to penetration of a specially hardened steel tool consisting of a bar rounded to a diameter such that its projected area is 100(mm)^2.

During a test, the bar is embedded in the material a distance equal to one-half its diameter, an electrical contact indicating when the required depth is achieved. Thus the JANKA hardness number of a timber is measured by the value of the load required to achieve the given level of embedment of the tool[5]. Brief details relating to the computation of Brinell and Vickers Harness Numbers shall now be outlined. Remember that these indentation tests involve elastic and plastic deformation of the materials.

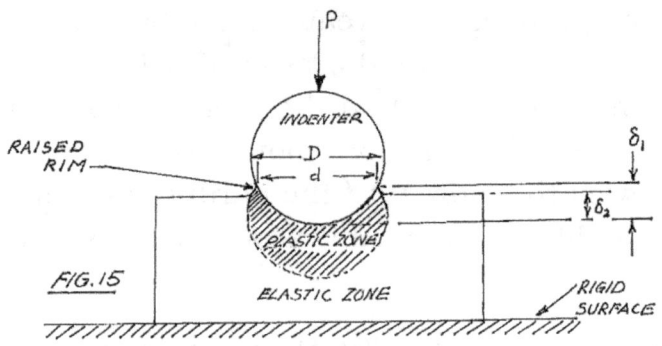

FIG. 15

INDENTATION TESTS FOR HARDNESS INVOLVE ELASTIC AND PLASTIC DEFORMATION OF THE MATERIAL. THE ABOVE DIAGRAM IS MEANT TO SHOW A PICTORIAL REPRESENTATION OF THE TWO ZONES IN A BRINELL TEST. (AFTER DOWLING AND BEAUMONT)

Brinell Hardness Number (BHN)

$$BHN = \frac{Load\ Applied}{Surface\ area\ of\ indentation\ (i.e.\ surface\ area\ of\ a\ segement\ of\ a\ sphere}$$

$$\frac{P}{\frac{\pi D}{2}\left\{ D - \sqrt{(D^2 - d^2)} \right\}}$$

in which, P = applied load in kg; D = Brinell ball diameter in mm and d = diameter of indentation also in mm. The depth of indentation is $\frac{1}{2}\left\{ D - \sqrt{D^2 - d^2}) \right\} mm$. One might be moved to consider that because the expression for BHN maybe expressed in the form : $P/\pi D$ (depth of indentation), determinations could be worked out simply by measuring the depths of

[5] Atrops, J.L. Strength Properties of Trinidadian Timbers, The University of the West Indies, 1970, Pp. 227.

indentation and substituting the relevant values in this expression. However, as shown in Fig. 15, the reading 'd' in the *BHN* expression is that corresponding to the depth 'δ_1' taken from the top of the raised rim which is different from depth 'δ_2' which would be the value recorded by the depth-measuring device. Thus resulting BHN determinations would be inaccurate. For different materials 'δ_1' and 'δ_2' would vary and there is no common constant by which the value of one could be multiplied to convert it to the other. Accordingly the diameter 'd' of an indentation must be measured experimentally.

In order to compare hardness determinations it is important that the shape of the indentation be of constant turn. Therefore if for a certain indenter-ball size, say, 'D', the indentation has a diameter 'd' with a certain load 'P', then for a rational comparison of the BHN thereby determined with that of , say, another material, then an appropriate load say 'P' might be required to produce an indentation of same diameter 'd', employing the same ball indenter of diameter 'D'. Such a method of comparing hardness is plainly impractical.

To obviate the need for a constant diameter 'd' the form of impression was allowed to vary within the limit implicit in the relation:

$$\frac{d}{D} = 0.25 - 0.50$$

the ideal form being the mean of the limits

$$\frac{d}{D} = 0.375$$

The British Standards Institution, relying on the principle of dimensional similarity, which may be expressed by the relationship : $P = K(D^2)$ i.e. Indenter load 'P' is directly proportional to indenter ball diameter squared, constant 'K' being such that d/D is within the range 0.25-0.50, has specified specific sizes of balls and unique values of the constant 'K' in each case. Accordingly for steels : $P/D^2 = 30$ so that for an indenter ball diameter of, say, 10mm, $P/(10)^2 = 30$ which gives an applied load of 3000kg; for copper and aluminium : $P/D^2 = 5$. For the same indenter-ball diameter of 10mm, $P = 500kg$.

In a real Brinell Hardness Test performed under engineering laboratory conditions the specimen chosen was a block of cold-drawn mild steel of dimensions 25mm x 25mm x 75mm. It was well polished. Various loads 'P' were applied in an increasing sequence and the diameters of the ball impressions measured by means of a traveling microscope and recorded. The results obtained are given in Table 1 below.

TABLE 1

RESULTS OF BRINELL HARDNESS TEST

ON SPECIMEN OF COLD-DRAWN MILD STEEL

TEST BALL DIAMETER, D = 9.525mm

LOAD P (kg)	Diameter D (mm)	d/D	Brinell Hardness Number (BHN) kg/(mm)^2
268.5	1.43	0.15	169
447.5	1.85	0.19	162
670	2.27	0.238	166
885	2.50	0.262	175
1342	3.15	0.330	169
1790	3.55	0.372	177
2240	3.93	0.412	179
2450	4.2	0.441	179
2680	4.3	0.445	174
2900	4.51	0.474	172
3120	4.75	0.500	165
3500	5.1	0.536	170
4020	5.38	0.563	157
4470	5.62	0.59	165

Vickers Diamond Pyramid Number (VPN):

This numeral which, like its Brinell counterpart, has the dimensions of pressure. It is given by the expression:

$$VPN = \frac{Load,\ (P)}{Area\ of\ indentation}$$

in which P = indenting load in kg, and the area of indentation is in (mm)^2. It is not generally realized that the Vickers Hardness Test was derived from considerations which arose in the Brinell Test. As stated previously, in order to obtain reliable results for the latter, the ratio : diameter of indentation/diameter of indenter, must be in the range 0.25 – 0.50; the average of 0.375 being the ideal. Fig. 16 shows such a configuration. Tangents drawn from the points of contact of the indentation with the indenter intersect at B, the included angle ABC being 136°. Hence the diamond indenter in the Vickers Hardness Test has its tip, that of a pyramid having an angle of 136° between opposite faces.

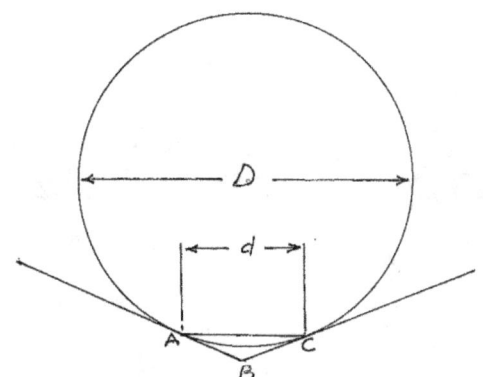

$d/D = 0.375$;   ANGLE $ABC = 136°$

FIG. 16. DIAGRAM SHOWING HOW THE ANGLE OF THE PYRAMID IN THE VICKERS HARDNESS TEST IS DERIVED FROM THE IDEAL BRINELL TEST RATIO $d/D = 0.375$.

Accordingly,

$$VPN = \frac{P}{\dfrac{d^2}{2} \cdot \dfrac{1}{Sin\,\theta/2}}$$

$$= \frac{2P \, Sin\theta/2}{d^2} = \frac{2P \, Sin68°}{d^2}$$

i.e. $$VPN = \frac{1.8544P}{d^2}$$

The beauty of Vickers Pyramid Test is at once abundantly clear. Evidently, for the same indentation, any multiple of the load applied is correspondingly reflected in the magnitude of the VPN. Hence it may be said that a material with VPN, say 184 is twice as hard as one with a VPN of 92.

The results obtained for real Vickers Diamond Pyramid tests on the same specimen of cold drawn mild steel for which Brinell Hardness test results were reported previously are given in Table 2. Like the earlier determinations, the VPN tests were done in an engineering laboratory. The results shown in Table 1 are plotted in Fig. 17. Fig. 18 is a plot of d/D against BHN for the results given in Table 1. The two plots demonstrate an approximate equality between the VPN and HBN in the range $d/D = 0.25 - 0.50$.

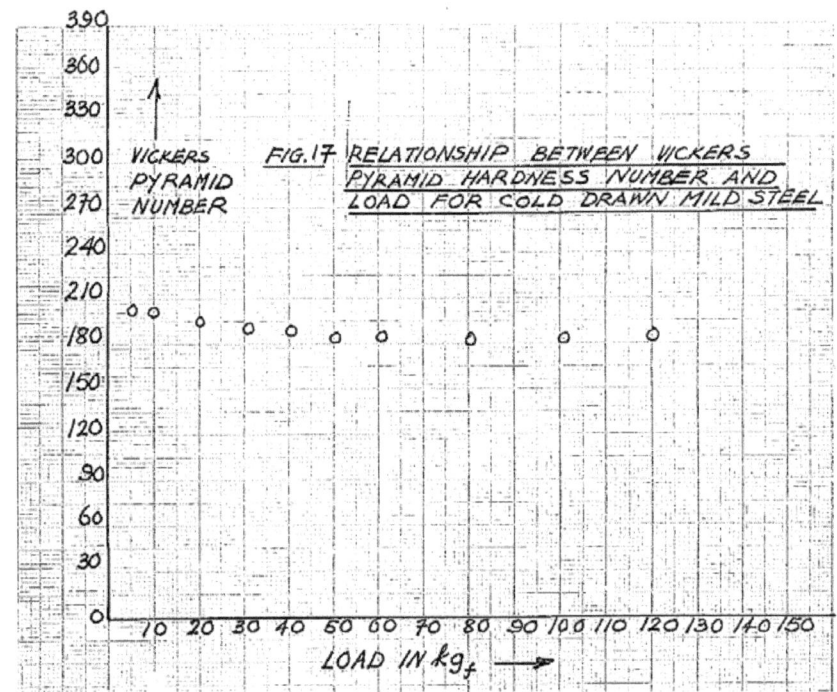

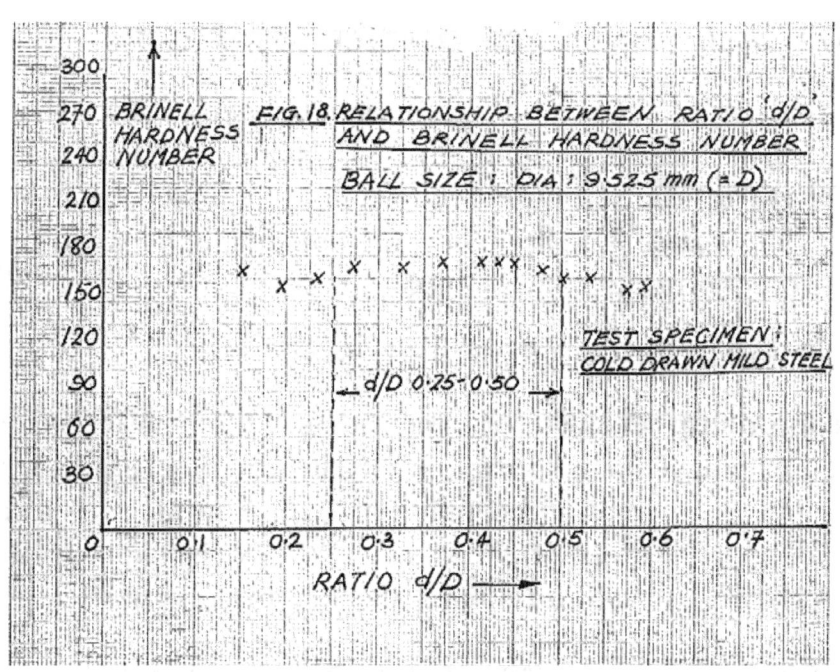

TABLE 2

RESULTS OF VICKERS DIAMOND PYRAMID TEST

LOAD IN kg	HARDNESS NUMBER kg/(mm)²
5	203
10	202
20	194
30	191
40	186
50	183
60	182
80	184
100	183
120	184

RELATIONSHIP OF HARDNESS TO TENSILE STRENGTH

Empirical relationships have been developed between VPN, HBN and tensile properties; notably ultimate tensile strength (UTS). Professor Dowling has reported the following approximate relationship for low and medium strength carbon steels, that is steels with a carbon content of less than 0.25%; and steels with a carbon content of 0.3% - 0.6%, respectively:

$$\text{UTS (in MN/m}^2) = 0.35 \text{ (HBN in MN/m}^2$$

This linear relationship derived on the basis of statistical analysis is not replicated for other classes of material, where it may be curvilinear for example. In the above equation the estimated value of UTS is in MN/m^2 and the value of HB must also be in MN/m^2. Note that when (HBN) is expressed as $kg/(mm)^2$ and UTS is in MN/m^2, the relationship stated in the foregoing is given by:

$$\text{UTS (in MN/m}^2) = 3.45 \text{ (HBN in kg/mm}^2) \text{ MN/m}^2$$

As you would imagine, and as we pointed out at the start when we referred to the hardened surfaces of oil-well drill bits, hardness is a property which is associated with good wear resistance.

IMPACT STRENGTH

Of what significance is the property generally referred to as the impact strength in a materials'–specifications, schedule? Reference to this characteristic as its name might suggest is not primarily to indicate the material's mechanical shock-resisting capability, but rather to point out whether the material was correctly heat-treated or not : in short, to evaluate its brittleness. Consider a practical situation. A client may require a supply of steel with a specified impact strength. In order to achieve this, it might be necessary for the manufacturer to treat the base stock in a certain way. For example, it may be necessary to oil-harden it at a particular temperature; and subsequently, after cooling, to temper it. i.e. to re-heat the material to a specific temperature so as to release the carbon as fine iron carbide, (Fe_3C) particles; and thereby obtain the requisite properties.

Modern-day methods of impact testing owe their origin to the work of Izod of whom it is said, in 1903 investigated the reasons for the rupture of a gun-barrel, in the course of which inquiry he clamped specimens of the material taken from the same barrel in a machinist's vice and applied hammer blows to them. He observed that defective specimens were brittle: breaking with a single blow, whereas ductile ones did not.

The two test methods employed nowadays for determining impact strength are (i) the Izod Test and (ii) the Charpy test (BS131). Both are based on the same principle inherent in an energy-absorbing process whereby a heavy pendulum of mass, say, 'M' falls through a vertical distance say, 'h_o, strikes the test specimen after which it, the pendulum that is, rises to a final height, say 'h_f'. Evidently the loss of potential energy of the pendulum $= Mg(h_o - h_f)$ must equal the energy absorbed by the specimen, assuming no losses due to air or bearing friction.

Evidently, in the case of a brittle material the value of 'h_f' would be greater than that of a ductile one. That is to say the energy absorbed by the brittle material would be less than that for the ductile material. The energy absorbed in an impact test is expressed in Joules (J), i.e. Newton metre (Nm) with dimensions kg m^2/s. Charpy's test has the advantage over Izod's in that clamping of the specimen is not required as in the other, in consequence of which many Charpy tests can be performed in short order. The Izod test is used nowadays mostly for non-metallic materials such as plastics.

In impact testing, the test specimen may either be notched or just simply plain. The Charpy V-notch test (ASTM, E23) is the one now almost universally applied. The drawing of a standard specimen used in both the Izod and Charpy tests is shown in Fig. 19 where other details concerning both test methods are shown. As it stands on the anvil of the machine in the Charpy test, the notch is vertical and on the side opposite that which is pendulum impacts. It should be noted that some test machines are fitted with strain gauges in order to obtain a "force-time read-out" during fracture of a specimen.

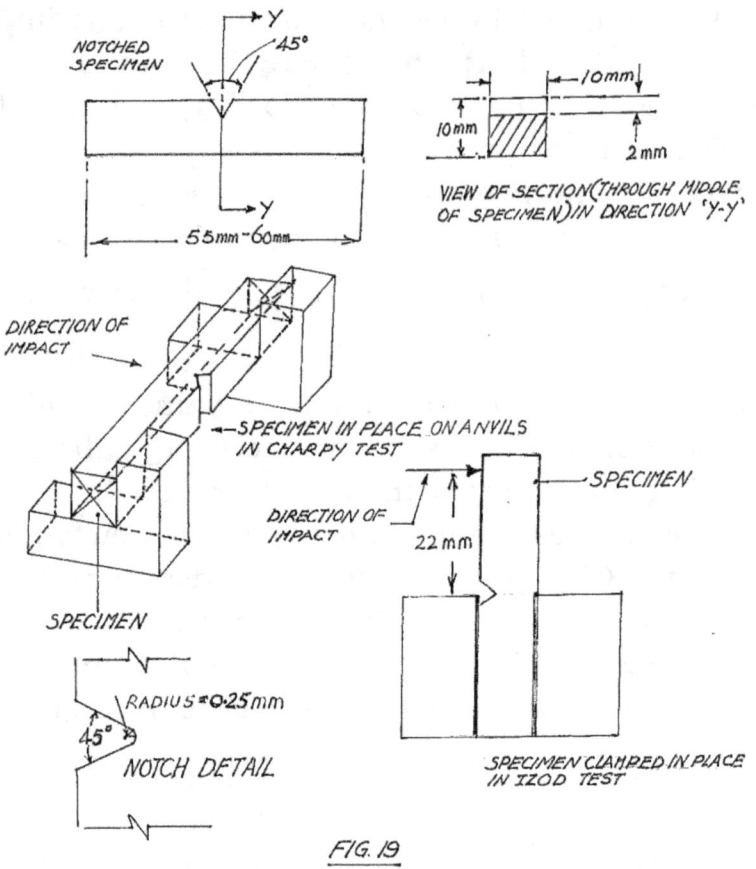

FIG. 19

The impact strength of metals varies with temperature. Fig 20 shows a typical graph for a low carbon steel i.e. in the 0.1% - 0.2% range. As can be seen, at low temperatures within a certain range, impact strength is relatively low and constant ; the material is brittle there. With increasing temperature impact strength rises significantly and almost proportionately with temperature and then more or less levels off ; the material is characteristically ductile at the higher temperatures.

The mid-point of the temperature range in which the transition from ductile to brittle behaviours occurs is called the ductile-to-brittle transition temperature (DBTT). This characteristic variation of impact strength with temperature is of great practical significance in steel components. This is so because the transition temperature of a component designed to withstand percussive blows in its service life should be lower than ambient temperature so as to avoid cracking and chipping of the steel.

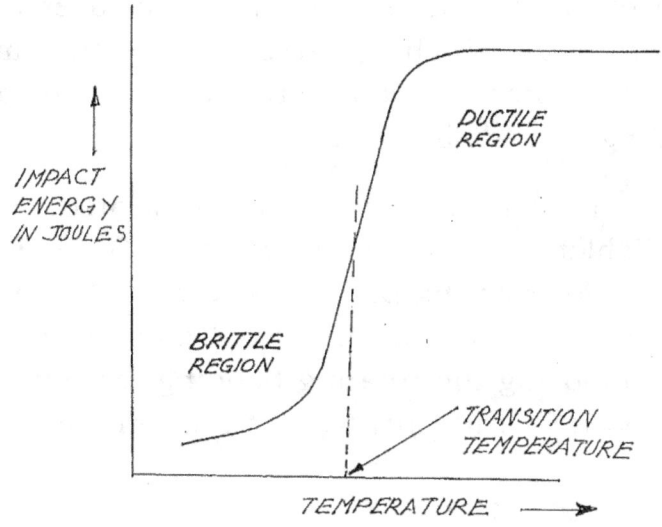

FIG.20 TYPICAL CURVE SHOWING VARIATION
OF IMPACT ENERGY WITH TEMPERATURE

METAL FATIGUE

Metal fatigue is an apt description for a state of "tiredness in a metal", to put it anthropomorphically, resulting from repeated cyclical variation of mechanical stress; and metaphorically, not unlike the mental deterioration of the human psyche due to continual psychological distress. The phenomenon ranks high in the study of mechanical failures of materials. In metal fatigue the stress required to cause rupture could occur at a level well below the yield strength or even the elastic limit. Crankshafts and truck axles are prime examples of components that receive complete reversals of high stress that lead to a high probability of progressive failure. On the other hand, the reversal of loading that occurs in some of the chords of roof trusses and similar structural members of which mention is made in Chapter 2 hardly qualify for consideration as a factor in metal fatigue. Metal fatigue is also caused by frequent variation in temperature which induces alternating expansion and contraction of parts, hence, the description Thermal Fatigue. Undoubtedly more tragically, metal fatigue has been responsible for a number of catastrophic failures that spawned human tragedy on a monumental scale, not to mention the considerable economic losses associated therewith. The high-altitude disintegration of two British De Havilland Comet airliners: G-ALYP on 10th January 1954 and G-ALYY on 8th April of the same year is a stark reminder of such spectacular failures[6]. In these two cases it was metal fatigue caused by cyclical variation of the hoop stress around the large windows in the fuselage that resulted in alternate contraction and expansion of the aircraft's skin, eventually leading to its disintegration.

The phenomenon of fatigue is readily demonstrated in the engineering laboratory using a Wöhler machine named after its German inventor. Wöhler machines are available for rotating bending and axial loading test procedures. A brief account will now be given of an actual fatigue test performed on two metallic specimens employing the rotating bending test method. A diagram of the Wöhler machine used for this purpose is shown in Fig. 21.

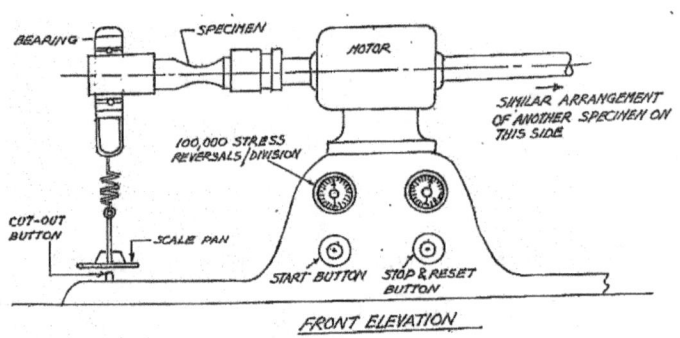

FIG.21 WHOLER ROTATING-BENDING FATIGUE TESTING MACHINE

In summary, the procedure involved the mounting of one end of the test specimen in the chuck of the motor. The other end was fitted into a special ball bearing attached to a hanger supporting a scale pan. During a test the specimen was in reality a rotating cantilever. At the start with a mass in the scale pan, the radius of curvature of the specimen was below its axis, i.e. its top surface was in tension; its bottom surface in compression. After rotating through 180° the surface that was initially in tension was subjected to compression. In Fig. 22 it is seen that the stress in the specimen went from a maximum tensile to a maximum compressive stress during 180°. Point 4, for example moving anticlockwise as shown in the diagram, went from 0 to maximum tensile stress after 90° rotation; then to zero stress after a further

---

[6] If one of the Comet's cruising altitude was 10kn, Cabin pressure 77kN/m^2 and atmospheric air-pressure 27kN/m^2, then for a fuselage diameter and skin thickness of 3.7m and 0.91mm respectively the circumferential or hoop stress would have been approximately 934 MN/m^2. Assuming a stress concentration factor of 3 arising from design considerations, the hoop stress in the skin of the aircraft made of a high strength aluminium alloy could have been anywhere in the vicinity of 3GN/m^2. [Note that the differential pressure the fuselage would have had to support was 50kN/m^2].

90° rotation; to maximum compressive stress for a further 90°; and finally back to zero stress after one complete revolution. Evidently, this occurred for every point on the surface of the specimen. Thus, for every rotation there were two reversals of stress.

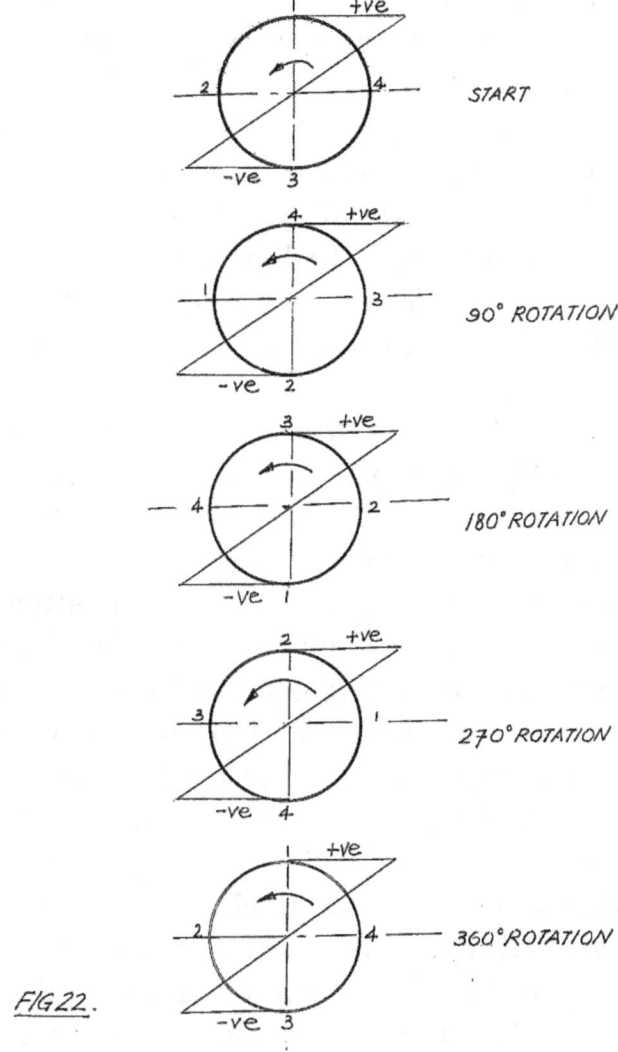

FIG 22.

Stress on a specimen was varied by changing the mass placed in the scale pan and the number of cycles of stress reversals to failure was automatically recorded for the value of stress that caused failure. If R = range of stress, then $R = \sigma_{max} - \sigma_{min}$, $+\sigma_{max}$ being the maximum tensile stress, positive in our convention; and $-\sigma_{min}$, the maximum compressive stress, negative in our convention. For the cylindrical rotating cantilever : $|\sigma_{max}| = |\sigma_{min}|$. Therefore $R = \sigma_{max} - (-\sigma_{max}) = 2\sigma_{max}$ and average stress : $\sigma_{avg} = (\sigma_{max} + \sigma_{min})/2 = 0$. Hence

$$\sigma_{max} = \sigma_{avg} + \frac{R}{2}; \quad \sigma_{min} = \sigma_{avg} - \frac{R}{2} \quad \text{or} \quad \sigma_{max,min} = \sigma_{avg} \pm \frac{R}{2}. \quad \text{Accordingly if}$$

~ 213 ~

$\sigma_{max} = +50MN/m^2$ and $\sigma_{min} = -50M/m^2$, then $\sigma_{max,min} = \pm 50MN/m^2$. Fig. 23 represents a reproduction of the actual graphical plots of stress *(S)* versus the logarithm of the number of cycles *(N)* to failure for a specimen of an aluminium alloy duralumin DTD 683 and another of rolled mild steel, *EN3B[1]*. Such graphs are generally referred to as 'Wöhler curves' or S/N curves. In Fig. 23 the magnitude of the stress below which failure by fatigue does not occur is in the vicinity of *254 MN/m²*. This stress is known as the fatigue limit of the specimen. On the other hand the curve obtained for the duralumin specimen does not reflect the characteristic knee: the curve slopes away downwards with increasing cycles. For such materials, instead of a fatigue limit, a fatigue strength or endurance limit is specified in terms of the stress at a certain number of cycles. For example, a fatigue strength of 125MN/m² at 100 million cycles could be specified for a Duralumin *683 DTD* component. Refer also to Fig. 23. It should be noted at this juncture that while a cyclical stress may be below either the fatigue limit or the endurance limit at which levels failure by fatigue would not normally be expected, the presence of incipient cracks or fatigue cracks or other internal defects arising in the course of material manufacture, or from stress raisers such as sharp re-entrant angles characteristic of some distortion, is likely to raise stress levels above safe-design limits. When therefore such incipient cracks or fatigue cracks or cracks due to stress concentration continue to be subjected to cyclical stresses these cracks and defect centres are propagated in the material and may lead to failure if the fracture toughness of the material is not sufficiently high. We shall refer to this property of fracture toughness later in this chapter and also in Chapter 19. The Paris equation which is sometimes referred to as the Paris-Erdogan Equation expresses a mathematical relationship between rate of crack growth with respect to load cycle and crack length and stress range. Other related terms you may come across in the engineering literature in connection with the phenomenon of fatigue are (i) fatigue ratio which is equal to fatigue limit divided by tensile strength, and (ii) fatigue life which is defined as the expected life of a component under repeated application of a certain stress. Is there an anthropological equivalent of this inanimate state?

---

[1] These tests were carried out by a group of students comprising members from the departments of Mechanical, Civil and Aeronautical engineering at City and Guilds College, Imperial College, London University. Note stresses which were calculated in tons (force) per sq. in were converted to MN/m².

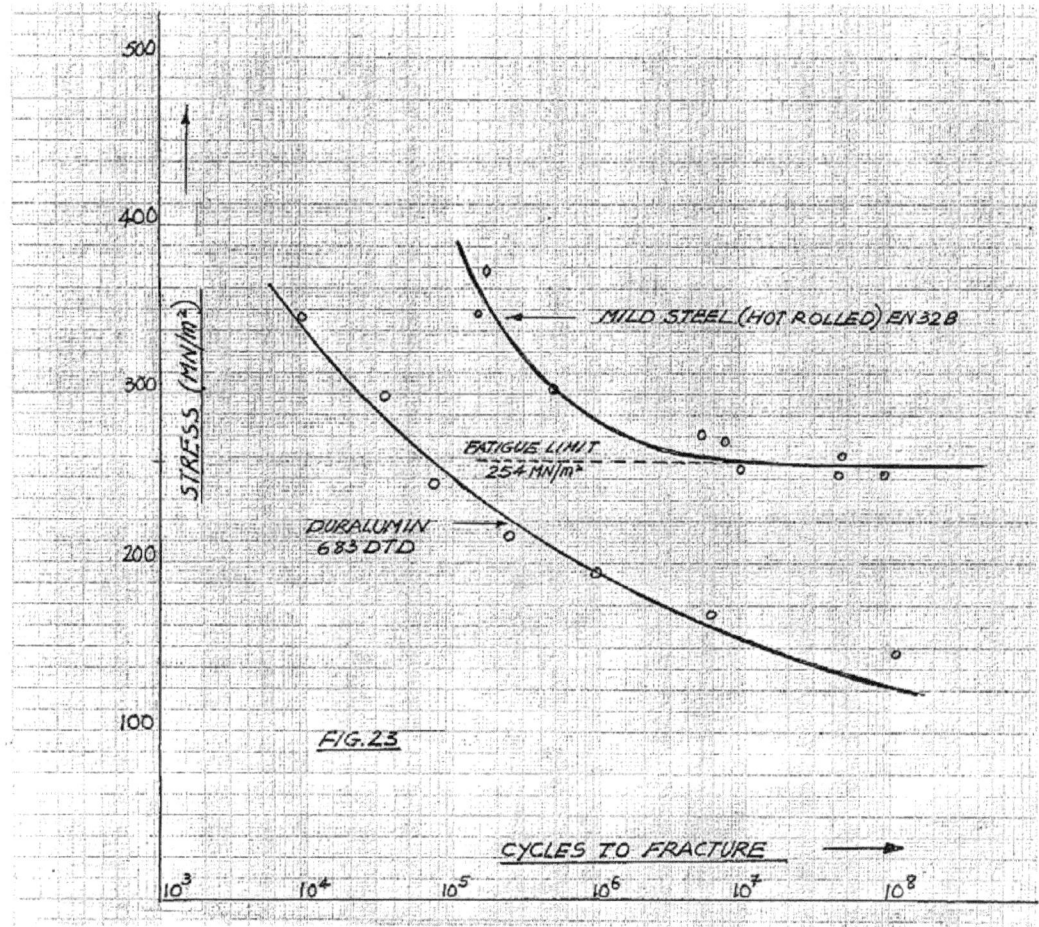

Axis labels: STRESS (MN/m²) (vertical); CYCLES TO FRACTURE (horizontal)

MILD STEEL (HOT ROLLED) EN 32B

FATIGUE LIMIT
254 MN/m²

DURALUMIN
683 DTD

FIG. 23

FRACTURE TOUGHNESS

When referring in Chapter 1 to the dynamic growth and development of the engineering science of Mechanics of Materials, a number of specialized fields were named as examples. Included among them was an offshoot of relatively recent sprouting called Fracture Mechanics.

Fracture Mechanics is a subject of no mean importance by virtue of the fact that it provides a basis for engineering design and choice of materials of construction while also paying due regard to imponderables such as manufacturing defects e.g. blowholes, cracks and other flaws in stress analysis. The property of the material that would be of interest here to the designer is called Fracture Toughness which in essence is a material's resistance to rapid crack growth.

Because of its significance a brief introduction to Linear Elastic Fracture mechanics is given in a Chapter 19 of this text. However for the purpose of this chapter an example of the application of fracture toughness is given to illustrate this important property of a material. There are published tables giving values of the plane strain fracture toughness of various representative materials at room temperature. The property fracture toughness is expressed in $MPA\sqrt{m}$ or $MNm^{-3/2}$. Consider now the case of a cylindrical pressure vessel made from a chromium-molylebdenum-vanadium steel. The internal diameter of the vessel is 1m and its wall thickness is 15mm. The fracture toughness of the steel is $60MPa\sqrt{m}$. The design engineer may wish to know what is the critical size of the crack in the wall of the vessel in comparison with its wall thickness. The internal pressure = 4MPa. Based on the data provided, the hoop stress σ_H which is the maximum stress in the wall of the vessel is calculated, i.e.

$$\sigma_H = \frac{pr}{t}$$

in which $p = 4MPa = 4 \times 10^6 N/m^2$; $t = 15mm$; and $r = 0.5m$.

$$\therefore \quad \sigma_H = \frac{4 \times 10^6 \times 0.5}{\dfrac{15}{1000}}$$

$$= 133 \times 10^6 N/m^2 N/m^2 \qquad \text{or} \qquad 133 NMPa$$

Plane strain fracture toughness K_{1c} is given by :

$$K_{1c} = \sigma\sqrt{\pi a_c}$$

in which σ =maximum stress and a_c = critical crack length. The latter expression may be rearranged as follows :

$$a_c = \frac{1}{\pi}\left(\frac{K_{1c}}{\sigma}\right)^2$$

so that

$$a_c = \frac{1}{\pi}\cdot\left(\frac{60MPa\ m^{1/2}}{133MPa}\right)^2 = 0.065m$$

i.e. $\quad a_c = 65mm$

This crack size is greater than the thickness of the vessel's wall. Consequently any crack will penetrate the wall and a leak will develop. See Fig. 24.

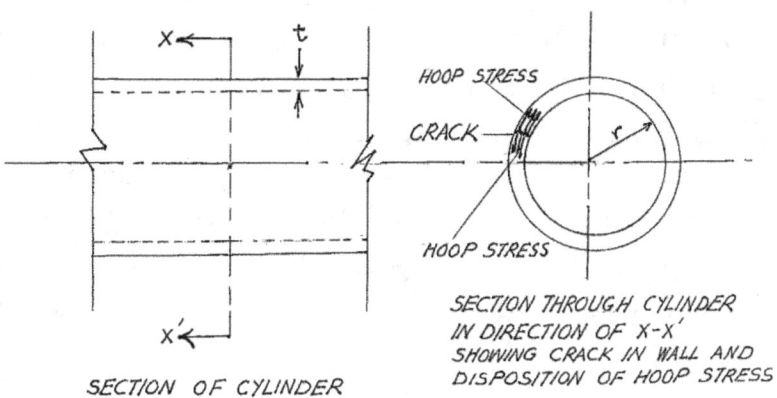

SECTION OF CYLINDER

SECTION THROUGH CYLINDER
IN DIRECTION OF X-X'
SHOWING CRACK IN WALL AND
DISPOSITION OF HOOP STRESS

FIG. 24

STRESS CONCENTRATION

While the phenomenon of stress concentration is of great importance in its own right, its interactive role in metal fatigue failure is of such overwhelming significance that it is considered appropriate to introduce the topic at this stage. Many engineering components fail precisely because insufficient attention is given to considerations of stress concentration in design especially in parts subjected to cyclical loading. I want to bring its attention to the student right from the start.

Evidence obtained in investigations of mechanical and structural failures indicate that in the vast majority of cases metal failure was the underlying cause (or was it the effect?). This was pointed out previously. Such failures accounted for between 80% and 90% of the total number. But what is also significant is that the failures almost always occurred in regions of high stress concentration.

Generally speaking any abrupt changes in the geometrical shape of a component due to, say, shoulder fillets, sharp re-entrant angles screw threads, keyways, splines, notches, blow holes formed during metal casting for example, oil holes for lubrication and the like, give rise while under loading, to localized stresses of much higher values than those determined by using the ordinary relationships of stress analysis. These localized irregularities in geometrical shape are commonly referred to as "stress raisers".

In order to explain the phenomenon, consider, as an example, a circular-cylindrical-bar component in which a notch has been machined and which is subjected to a tensile force 'P'. See Fig. 25a. With an un-notched bar of identical shape, size and loading as in that diagram, the average stress σ_{AVG} in this bar, at a section not less than two diameters away from the point of application of 'P', is P/A, according to the principle of St. Venant, previously discussed. However, the presence of the notch causes a different stress distribution pattern across the section of the bar. See Fig. 25b. Notice how the stress σ_{max} in the region of the shoulder fillet has increased quite markedly above the average value, P/A computed on the basis of the ordinary relationship.

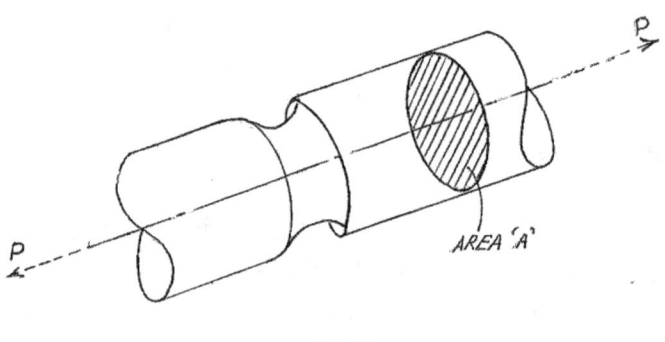

FIG.25a

The maximum stress σ_{max} acting alone or in combination with other stresses, when repeatedly applied can in time initiate a crack which would result inevitably in failure of the component. It is the build-up or spike if you will in the stress distribution that led to the coining of the term stress concentration. It is usually designated by a dimensionless number symbolically designated 'K_T' or just simply 'K' and is given by the following relationships, K_T being the stress concentration factor:

$$K_T = \frac{Maximum\ value\ of\ actual\ stress}{No\min al\ stress\ calulated\ on\ \min imum\ area}$$

Accordingly with reference to Fig. 25b

$$K_T = \frac{\sigma_{max}}{P/A'}$$

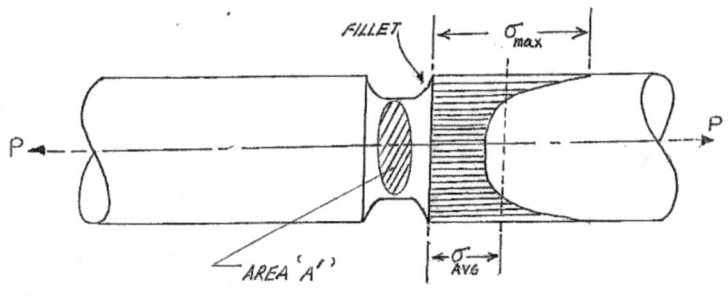

FIG.25b

Based on its origin as described in the foregoing, it is not surprising that stress concentration factors for a wide variety of geometrical discontinuities and different stress conditions have been determined empirically and expressed in several formulae.   See for example: "Formulas for Stress and Strain" by R.J. Roark.

Solely for illustrative and other purposes, curves showing the variation of K_T for (i) flat bars with circular holes and subjected to static elastic tension and (ii) flat bars with shoulder fillets and similarly loaded, are shown in Figs 26 and 27 respectively.

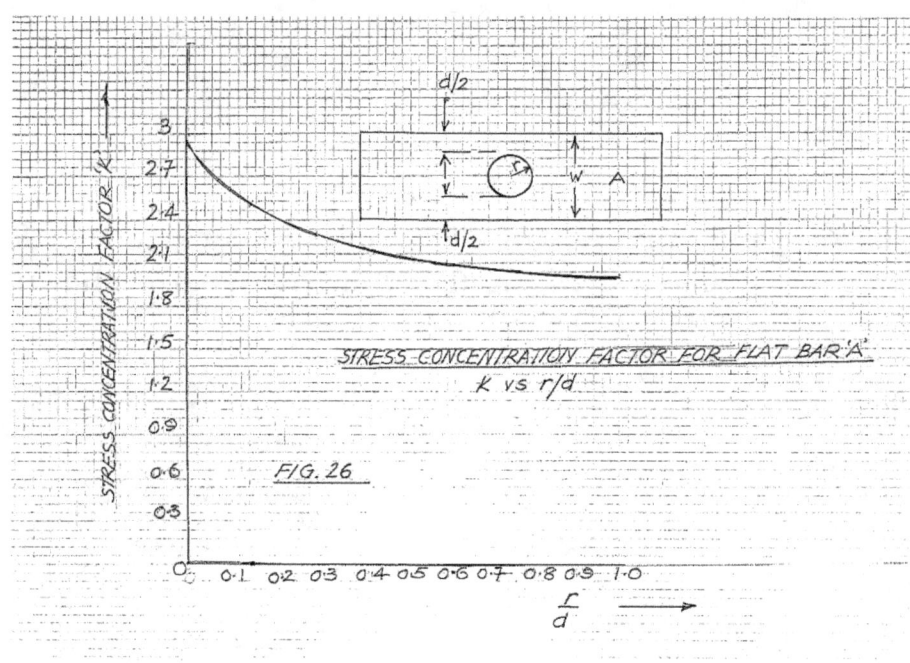

FIG. 26

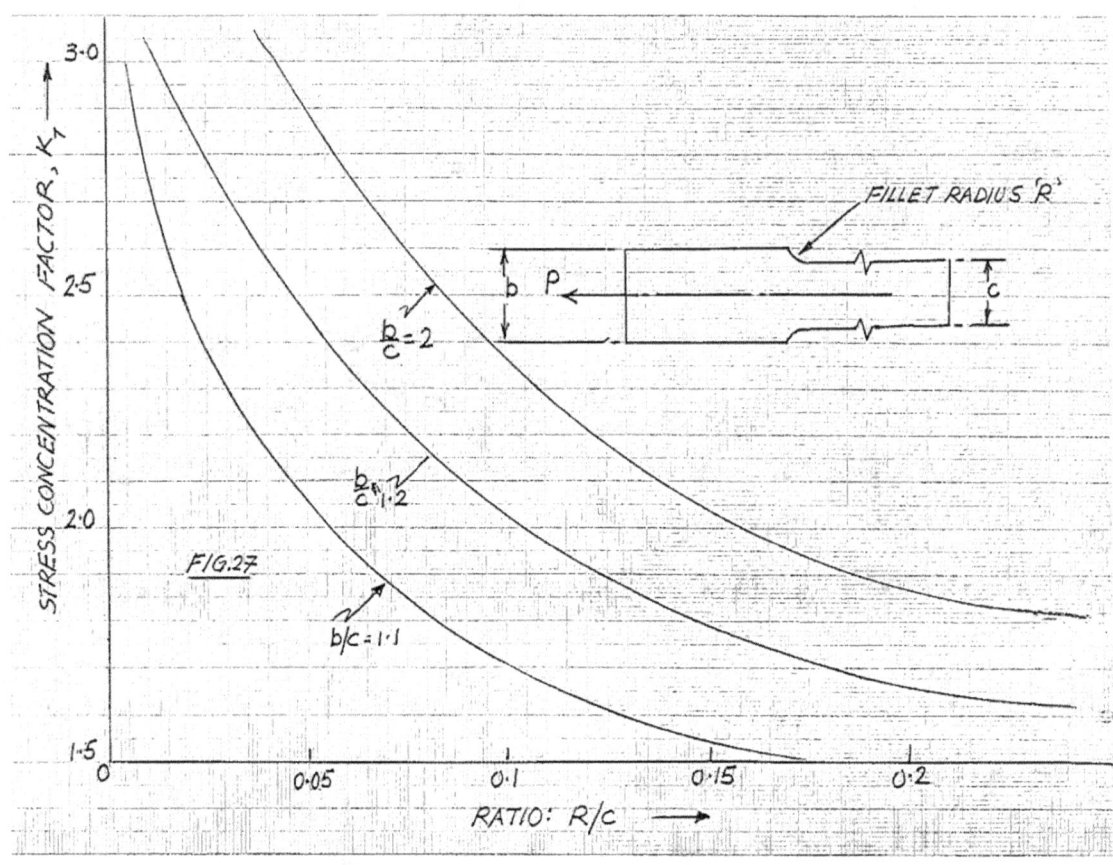

FIG.27

In a plot of K_T against the ratio of fillet radius to bar diameter for three values of the ratio of bar diameter for a circular cylindrical bar is shown in Fig. 28

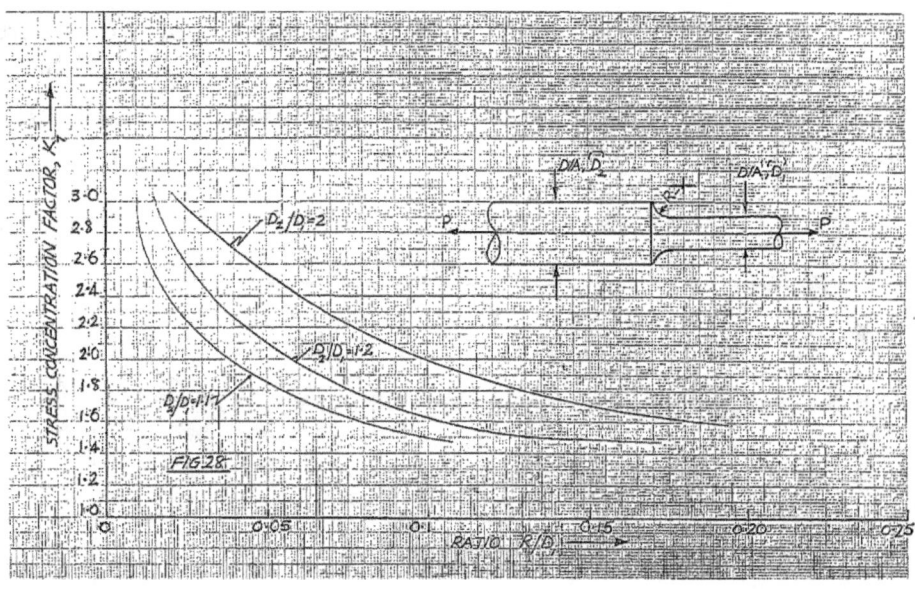

FIG 28

CREEP

Reference has already been made to elastic and plastic deformation in Chapter 8. In short, elastic deformation : recoverable ; plastic deformation : permanent. Neither of these deformations is of a time-dependent nature. But there is yet another kind of deformation : one that is time-dependent. It is called 'creep'; not to be confused with the term used to describe a boring specimen of 'homo sapiens' or the committee to re-elect the President (Nixon). How then is engineering creep manifested? It is a phenomenon displayed by a material such that strain increases with time when the material is subjected to a constant stress, elongation proceeding gradually until it eventually ruptures. This rupturing occurs even though the stress is less than its tensile strength. Worse! Even less than its yield strength sometimes. Creep occurs in metals, plastics and even in concrete which is a crystalline ceramic. It is to be noted that pure lead displays the phenomenon of creep in a marked manner at room temperature. One inter-metallic compound, a ceramic called nickel aluminide Ni_3 Al possesses the extraordinary and exceptional property of growing in strength with temperature increase up to about 830° Celsius.

In the design of mechanical components such as for example super- heater boiler tubes, high pressure steam lines, steam and aircraft turbine blades, the latter which operate at temperatures in excess of $0.4T_m$, T_m being the component material's melting point, creep resistance is a property of critical importance in the selection of materials for these particular applications. Maximum permissible creep is generally restricted to 1% during a component's lifetime.

For creep determination a specimen of the material is subjected to a constant stress while being heated at a constant temperature in a specially designed furnace. Measurements of strain and the time taken to attain each such level are recorded. Fig. 29 shows a typical creep characteristic: Creep Strain vs Time.

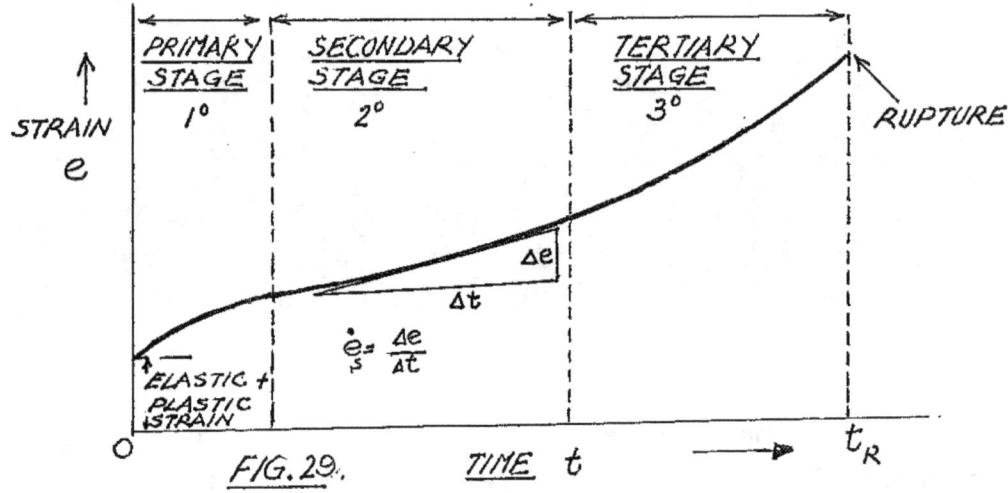

FIG. 29.

The characteristic shows three stages : (i) primary stage (1°) during which strain <u>rate</u> reflects a continuous decrease ; (ii) secondary stage (2°) when the strain rate is assumed to be as constant; and (iii) tertiary stage (3°) when strain rate increases rapidly, culminating in fracture. The secondary creep rate $\dot{e}_s$ is for many metals expressed as a power law, $\dot{e}_s = B\sigma^n$ in which σ = constant stress, n is of value between 3 and 4, and B is a constant.

A creep rate has to be specified at a particular constant stress and constant temperature most generally in units of per hour (h^{-1}) or per day (day^{-1}). As the characteristic just described clearly shows, creep strains are non-linear with respect to time. Remember each creep test is done under a constant stress at a constant temperature. Evidently for a variety of different constant stresses and different constant temperatures, a family of characteristics similar to that shown in Fig. 29 would be obtained. See Fig. 30.

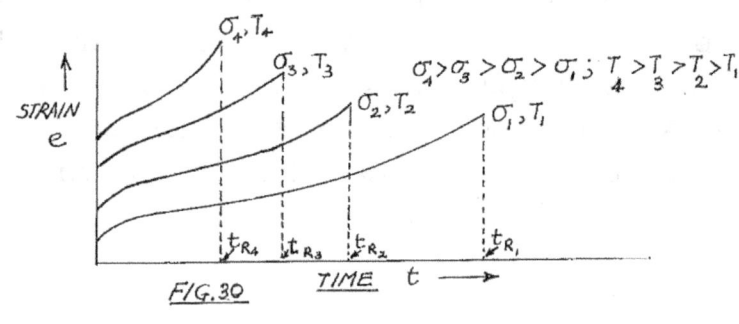

FIG. 30

Accordingly, it is perfectly reasonable to express strain rate $\Delta e/\Delta t = \dot{e}$ as a function of time 't' and also of temperature 'T', thus

$$\dot{e} = F(T,t)$$

Those of you who studied chemistry would recall that generally speaking a chemical reaction is speeded up by temperature increase, the process being regarded as one of thermal activation. One of the architects of the modern science of physical chemistry was an outstanding Swedish man of science and Nobel Prize Winner by name Svante August Arrhenius. He propounded a law, widely accepted, which explains the normal activation phenomenon. It is expressed thus :

$$\frac{d}{dT}(In\ k) = \frac{E_A}{RT^2}$$

in which T is absolute temperature, 'k' is a velocity constant, E_A is the energy activation, and R = universal gas constant = $8.314J$ per mole per degree. Integrating this expression results in the following:

$$In\ k = \frac{E_A}{R}\int \frac{dT}{T^2} = Constant - \frac{E_A}{RT}$$

which result may be rewritten as :

$$k = e^{(Constant - E_A/RT)}$$

or $\quad k = We^{-E_A/RT}$

W being $e^{(Constant)}$ which is of course another constant. Because engineering creep may be regarded as a speeding up of an atomic dislocation process in a material due to increasing temperature the Arrhenian phenomenon provides a convenient springboard as it were for the development of a relationship based on the parameters of stress, creep rate, temperature and rupture time.

Two such parametric equations are due to : (a) Sherby and Dorn and (b) Larson and Miller.

Let us take the case of the Larson–Miller parameter which is perhaps the more widely known of the two. Its foundation is an Arrhenius-type equation of the form:

$$\dot{e} = Ae^{-E/RT}$$

in which '$\dot{e}$' is the steady-state creep rate; A = constant; E = activation energy for creep which is assumed to be a function of stress 'σ'; and T is absolute temperature. Accordingly, by taking the logarithm of both sides:

$$In\,\dot{e} = In\,A - E/RT$$

which may be rewritten as $In\,\dot{e} = In\,A - Q(\sigma)RT$; E here is not Young's Modulus; it is activation energy for creep.

E being replaced by $Q(\sigma)$ as a function of stress

Hence, $\qquad T(In\,\dot{e}) = T\,In\,A - Q(\sigma)/R$

or $\qquad T(In\,\dot{e} - In\,A) = -Q\dfrac{(\sigma)}{R}$

Noting that $In\,x = 2.3\,Log_{10}x$, the latter result may be expressed in ordinary logarithms as :

$$T(Log_{10}\dot{e} - Log_{10}A) = -0.434Q(\sigma)/R = -f(\sigma)$$

or as expressed by McEvily :

$$Log_{10}\dot{e} = K - \dfrac{1}{T}f(\sigma) \qquad \dots\dots\dots\dots\dots\dots \text{(i)}$$

Evidently a plot of $Log\,\dot{e}$ as ordinates against the reciprocal of temperature $\dfrac{1}{T}$ as abscissae would produce a straight line with negative slope, i.e. $-f(\sigma)$ and with positive intercept on the $Log_{10}\dot{e}$ – axis equal to 'K'; hence 'K'

The Monkman-Grant law may now be used to modify the result expressed as equation (i). If t_R is the time to rupture, then by this law, the extension at rupture is

$$\dot{e}\,t_R = \text{constant, say, } B$$

Equation (i) may now be stated as

$$T\left\{Log_{10}\left(\dfrac{B}{t_R}\right) - K\right\} = f(\sigma)$$

which is equivalent to

$$T(Log_{10} B - Log\ t_R - K) = -f(\sigma)$$

or
$$Log_{10} t_R - Log_{10} B + K = f\frac{(\sigma)}{T}$$

Writing $Log\ B - K = -D$, the latter result may be expressed as

$$Log\ t_R + D = F(\sigma)/T$$

or
$$T(Log\ t_R + D) = f(\sigma)$$

Hence plots of $Log_{10} t_R$ as ordinates against corresponding values of $\frac{1}{T}$ as abscissae for different constant values of stress 'σ' would produce a series of straight lines all with the same negative intercept i.e. $(-D)$ on the $Log\ t_R$ axis. Hence 'D'. See Fig. 31 representing a plot of $Log\ t_R$ vs $1/T$ for constant value of stress for σ_1, σ_2 and σ_3.

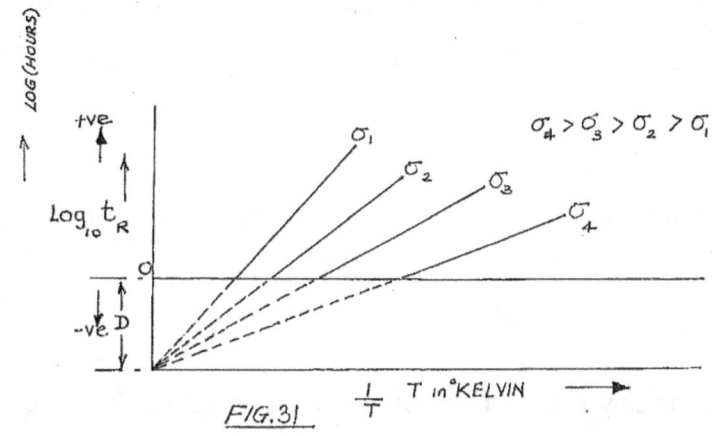

FIG.31

The Larson-Miller parameter is expressed in the form $P_{L,M} = T(Log\ t_R + D)$.

Thus the slopes of the characteristics shown in Fig. 31 as an example are the Larson-Miller parameters for each stress. This is so because if we wrote 'm' as the slope of a typical curve as in Fig. 31 then we may eliminate the equation of any one of the characteristics as

$$Log\ t_R = m\left(\frac{1}{T}\right) - D$$

$\therefore$ $T\ Log\ t_R + TD = m$

~ 225 ~

or $\quad T(Log\ t_R + D) = m = P_{L,M}$

It follows therefore that the results of several creep tests on the same material at, different but constant stress and different but constant temperature levels, can be obtained together with the elapsed time to rupture for each test. For example at a constant stress of say 100 MN/m² the time to rupture (t_R) at a constant temperature of say 200°C may be, say, 25h; at the same stress level of 100 MN/m² but at a different constant temperature of 300°C, t_R may be, say, 20h; at the same stress level of 100oMN/m², t_R at a constant temperature of 500°C may be say, 10h; and so on. Several such results will enable a plot to be done for $Log\ t_R$ vs $\dfrac{1}{T}$ for a constant stress 'σ' of 100 MN/m². Proceeding as outlined in the foregoing, and having determined 'D', values of the Larson-Miller, $P_{L,M}$ may then be evaluated from the slope of the graph. A graph of stress as ordinates versus corresponding values of $P_{L,M}$ as abscissae may then be plotted and used as a master reference. Thus in designing a component which is to operate at an elevated temperature, once the operating stress is predicted, its value can then be entered in this graph and a value of $P_{L,M}$ obtained therefrom. Assuming the value obtained for $P_{L,M}$ is 'u', then

$$u = T(Log\ t_R + D)$$

knowing the absolute temperature 'T', an estimated value of t_R can then be obtained.

Illustrative Example 1

Given that the secondary creep rate of a metal is given by the power law :

$$\dot{e}_s = B\sigma^n$$

in which $\dot{e}_s$ = secondary creep rate, σ = stress and B = constant, calculate the value of B when $\dot{e}_s = 0.3 \times 10^{-10}$ hour and $\sigma = 25 MN/m^2$. Take $n = 4$.

Ans : $B = 7.68 \times 10^{-39}$

Illustrative Example 2

Given that the Larson-Miller parameter for a specimen of iron-nickel-chromium-cobalt alloy is 7000 at a static stress of 400 MN/m², estimate the expected service life of a component at this level of stress at a temperature of

750°C. Take D in a plot of $Log\ t_R$ vs $1/T$ equal to 3.5. Remembering that 'D' was the negative intercept on the $Log\ t_r$ axis, it means that $D = -3.5$, i.e. $-Log\ hours = 3.5$.

Accordingly

$$Log\ t_R = \frac{P_{L,M}}{T} - 3.5$$

Substituting $P_{L,M} = 7000$; $T = 750 + 273 = 1023^\circ K$

$$Log_{10}\ t_R = \frac{7000}{1023} - 3.5$$

$$= 6.8 - 3.5 = 3.3$$

so that $t_R = 10^{3.3}\ hours$

i.e. $t_R = 1995h$, equivalent to 83 days

STRESS RELAXATION AND VISCOELASTICITY

Whereas Creep is the time-dependent response of strain to constant stress at a constant temperature, the time-dependent response of stress to constant strain also at constant temperature is called Stress Relaxation. There are many phenomena that distinguish between the two materials but stress relaxation is the most important characteristically distinguishing feature between metals and plastics.

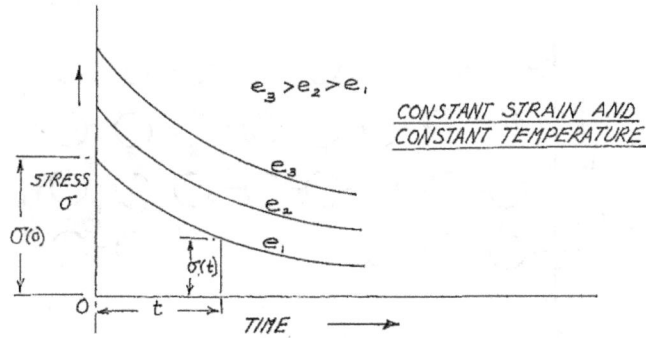

FIG.32: VARIATION OF STRESS 'σ' WITH TIME 't' FOR 3 DIFFERENT VALUES OF CONSTANT STRAIN 'e', IN STRESS RELAXATION

The bolted flanged-joints of steam pipes and heat exchangers which are commonplace on petroleum refining and electricity generating plants are typical examples of installations with mechanical assemblies where the phenomenon occurs. The bolts are initially tightened to specific stress levels using torque wrenches. But with the passage of time the stress in the bolts falls due to stress relaxation.

Fig. 32 shows the variation of stress σ, with time 't' for various values of strain, e.

The theory of stress relaxation relies on the property called viscoelasticity which in essence is a combination of the characteristics of elasticity and viscosity. From what we know of the theory of elasticity so far:

$$E = \frac{\sigma}{e}$$

and from Fluid Mechanics it is known that Newtonian viscosity, η is defined as shearing stress (τ) per unit velocity gradient

i.e.

$$\eta = \frac{\tau}{velocity\ gradient}$$

Picture as in Fig. 33 a set of polymer chains in the rows of an element of any plastic

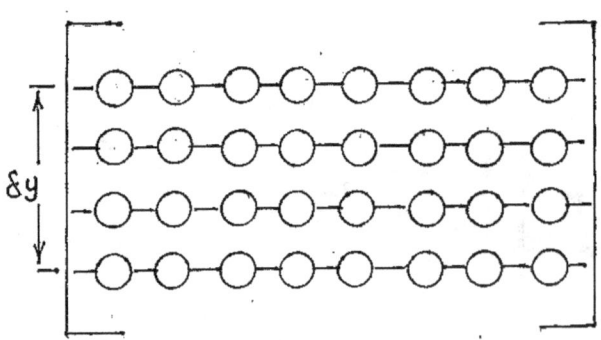

FIG.33: CHAINS OF MOLECULES AT REST

When a shearing stress τ is applied as in Fig. 34, the chains slide as depicted, the difference in velocity over the distance δy being δV. Thus

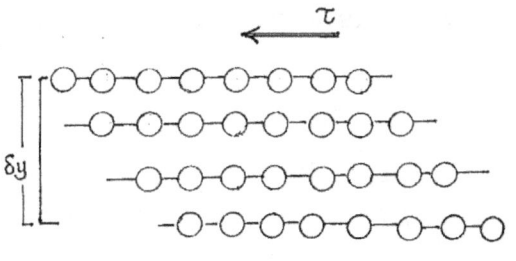

FIG.34: SHEARING STRESS 'τ' CAUSING CHAINS OF MOLECULES TO SLIDE OVER EACH OTHER, IF 'v' IS THE VELOCITY OF SLIDING, THEN VELOCITY GRADIENT FOR ELEMENT OF THICKNESS = δv/δy AND NEWTONIAN VISCOSITY, η = τ / δv/δy

[AFTER ASKELAND]

the velocity gradient is $\dfrac{\delta V}{\delta y}$ which is the limit enables us to write

$$\eta = \frac{\tau}{\left(\dfrac{dV}{dy}\right)}$$

or $\qquad \tau = \eta\left(\dfrac{dV}{dy}\right)$

The dimensions of $\dfrac{dV}{dy} = \dfrac{L}{T} \cdot \dfrac{1}{T} = T^{-1}$ which is equivalent to a strain rate.

Hence,

$$\tau = \eta \text{ (Viscous Strain Rate)}$$

Therefore for the condition of viscoelasticity to exist, two stresses have to be considered as was stated in the foregoing.

Accordingly, the state of stress and strain for linear viscoelasticity may be represented by

Stress = E × (elastic strain) + η × (viscous or inelastic rate or creep strain rate.

$$\sigma = Ee + \eta e_{inelastic}$$

in which $\qquad E$ = Young's modulus ; $e_{inelastic}$ = elastic strain

$\qquad$ = coefficient of viscosity ; and

~ 229 ~

$$e_{inelastic} = \text{viscous or inelastic or plastic strain rate}$$

Checking dimensions:

$$MLT^2 = \left(\frac{ML}{72} \cdot \frac{1}{L^2} \right)(1) + \frac{M}{LT} \cdot \frac{1}{T} = \frac{M}{LT^2} + \frac{M}{LT^2}$$

Stress Relaxation occurs at constant strain and constant temperature. Therefore,

Total Strain = Elastic Strain + Plastic Deformation Creep Strain.

<u>Because this total strain is fixed in time</u>, and total strain in what is required:

$$e_{Total} = e_{elastic} + e_{inelastic}$$

Differentiating w.r.t time 't'

$$\frac{d}{dt}(Total\ Strain) = \frac{d}{dt}(e) + \frac{d}{dt}(e_{inelastic})$$

that is to say $\frac{d}{dt}(e) + \frac{d}{dt}(e_{inelastic}) = 0$ since total strain is fixed in time, i.e.

$$\frac{d}{dt}(total\ strain) = 0$$

or

$$\frac{de}{dt} = -\left(\frac{de}{dt} \right)_{inelastic}$$

From

$$e = \frac{\sigma}{E}$$

$$\frac{de}{dt} = \frac{-1}{E} \frac{d\sigma}{dt}$$

and from the power law for creep strain rate

$$\left(\frac{de}{dt} \right)_{inelastic} = B\sigma^n$$

$$\therefore \quad \frac{1}{E} \frac{d\sigma}{dt} = -B\sigma^n$$

or $\quad \dfrac{d\sigma}{\sigma^n} = -BEdt$

or $\quad \int \sigma^{-n} d\sigma = -BE \int dt$

The limits of integration are for σ : $\sigma(0) =$ stress at time zero to $\sigma(t)$ = stress after time t ; and for time : $t = 0$ to elapse time, say, t

$$\therefore \quad \int_{\sigma(0)}^{\sigma(t)} \sigma^{-n} d\sigma = -BE \int_{0}^{T} dt$$

Integrating,

$$\left[\dfrac{\sigma^{-(n+1)}}{-(n+1)} \right]_{\sigma(0)}^{\sigma(y)} = -BEt$$

$$\left[\dfrac{\sigma^{-(n-1)}}{-(n-1)} \right]_{\sigma(0)}^{\sigma(y)} = -BEt$$

$$\therefore \quad \left[\dfrac{1}{-(n-1) \ \{\sigma(t)\}^{n-1}} \right] - \left[\dfrac{1}{(n-1) \ \{\sigma(0)\}^{n-1}} \right] = -BEt$$

or $\quad \dfrac{1}{\{\sigma(t)\}^{n-1}} - \dfrac{1}{\{\sigma(t)\}^{n-1}} = (n-1)BEt$

Thus with a knowledge of 'B' and 'n' from the power law for secondary creep and the initial value of stress $\sigma(0)$, the value of stress after a period of time 't', $\sigma(t)$ may be determined for a given material of known Young's modulus.

RECOVERY

If a material is undergoing the process of stress relaxation while under the influence of a constant stress, say, σ and this stress is reduced to zero at a time $t = t_r$, then there is an instantaneous reduction of a certain amount of strain. Let us designate it e_e. It is an elastic strain and it is not time dependent. Refer to Fig. 35. For $t > t_r$ a slow reduction of strain continues which, with the passage of time results, in a strain free state. This behaviour which is quite common in plastics is called Recovery.

$\sim 231 \sim$

Strain is normally associated with stress. In recovery as Fig. 36 shows, for $t > t_r$ there is strain but no stress. Clearly therefore one cannot rely on our normal stress vs strain relationship to account for and to analyse the behaviour of a viscoelastic material in its recovery phase. Readers interested in the subject may wish to refer to the following : "Applied Stress Analysis of Plastics : A Mechanical Engineering Approach" by S.I. Krishnamachari, Van Nostrand Reinhold, New York.

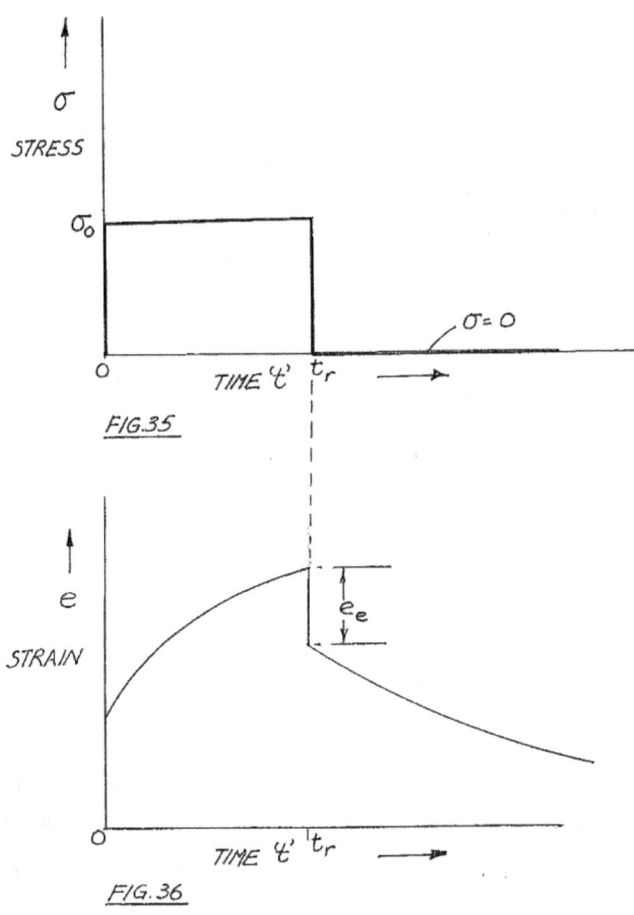

FIG.35

FIG.36

SOLVED AND OTHER PROBLEMS

Q1. The data tabulated below were obtained in a tensile test on a specimen of Duralumin. The gauge length of the test piece was 50mm and its cross-sectional area uniform for this length 8.84(mm)^2. The final extension of the specimen was 10.2mm and the load causing fracture, 3.65kN.

Plot the load extension graph and determine:

(i)   Stiffness
(ii)   Young's modulus;
(iii)   Ultimate tensile strength;
(iv)   Percent elongation on 50mm gauge length; and
(v)   Proof stress corresponding to 0.2 elongation.

LOAD vs EXTENSION

FOR DURALUMIN SPECIMEN

Load kN	Extension mm $\times$ 10^{-2}
.5	4.6
1.3	11.4
1.5	12.2
1.9	16.8
2.0	18.3
2.1	22.9
2.3	28.9
2.4	40.4

The Load vs Extension Graph is labelled Fig. Q1

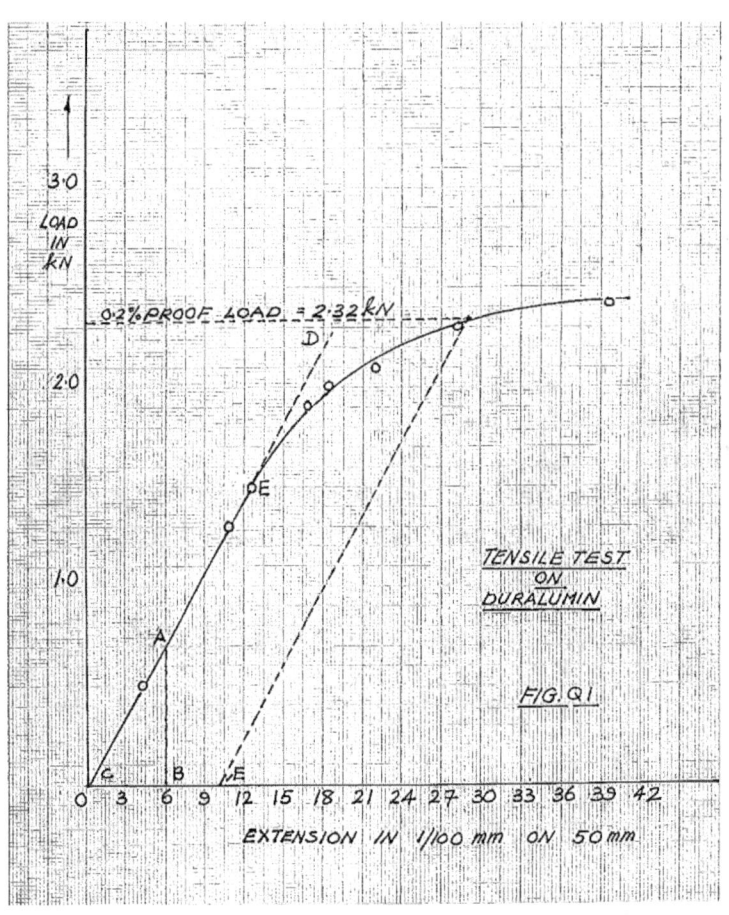

FIG. Q1

(i)   <u>Stiffness (S)</u>

S = Load/Extension (for the straight-line portion AE of load-extension graph)

$$= 0.7kN / 6 \times 10^{-2} mm$$

i.e. $S = 11.7 MN / m$

(ii)   <u>Young Modulus, E</u>

$$E = \text{Stress/Strain} = \frac{load}{area} \times \frac{length}{extension}$$

From the straight line portion of the graph:

$$\frac{load}{extension} = \frac{AB}{CB} = \frac{0.7kN}{(6 \times 10^{-2})} = \frac{700}{6 \times 10^{-2} / 10^{3}}$$

$$\therefore \qquad E = \frac{7 \times 10^7}{6} \times \frac{50}{1000} \quad \frac{10^6}{8.84}$$

i.e. $\qquad E = \frac{7 \times 10^{13}}{6} \times \frac{50}{1000} \times \frac{1}{8.84}$

$$= \frac{70 \times 10^9 \times 50}{6 \times 8.84}$$

$$E = 65.9 GN / m^2$$

(iii)   Ultimate Tensile Strength (UTS):

Load at fracture = $3.65 kN$

Original c.s.a.   = $8.84 (mm)^2$

$\therefore \qquad$ UTS $\quad = \quad \dfrac{3.65 \times 10^3}{8.84 \times 10^{-6}} N / m^2$

i.e.   UTS   = $412 MN/m^2$

(iv)   % elongation on 50mm gauge length:

Total extension on 50mm gauge length = 10.2mm

$$\therefore \quad \% \text{ Elongation} \quad = \quad \frac{10.2}{50} \times 100$$

$$= \ 20.4$$

(v)   0.2% Proof Stress:

This stress is that which is just sufficient to produce under load a non-proportional extension equal to a 0.2% of the gauge length.

Referring to Fig. Q1, it is seen that the line drawn parallel to *AD* and passing through point *E* at extension $10/100 mm \equiv 0.1 mm$ which is 0.2% strain on the 50mm gauge length, cuts the load vs extension graph at 2.32 kN. Accordingly

0.2% Proof Stress $= \quad \dfrac{2.32 kN}{8.84 \times 10^{-6}}$

$$= \frac{2.32 \times 10^9}{8.84}$$

$$= 0.262 \times 10^9 \, N/m^2$$

$$= 262 \, MN/m^2$$

Q2. The results of a tensile test to destruction of an unknown alloy are given in the table below. The specimen was initially 50mm long and diameter 15mm. The maximum load was 60.1kN, the strain at that stage being 5%.

Load	Stress	Strain
kN	MN/m²	%
5.5	29.5	0.03
10	53.6	0.055
20	107.2	0.11
30	161	0.165
37.5	200	0.375
42	225	0.75
45	241	1.125
47	252	1.69
49	263	2.5
50.3	270	3.5%
60.1	Max Load	10%
64	Fracture	30%

The specimen eventually fractured at a load of 64kN, between the marked-off gauge length on the specimen which had increased to 60.15mm. The diameter at fracture was 11.1mm. Plot the stress vs strain curve using the tabulated data and determine (i) Young's modulus of elasticity; (ii) the 0.2% proof stress; (iii) ultimate tensile strength; (iv) % reduction in area; (v) % elongation; and true stress at fracture.

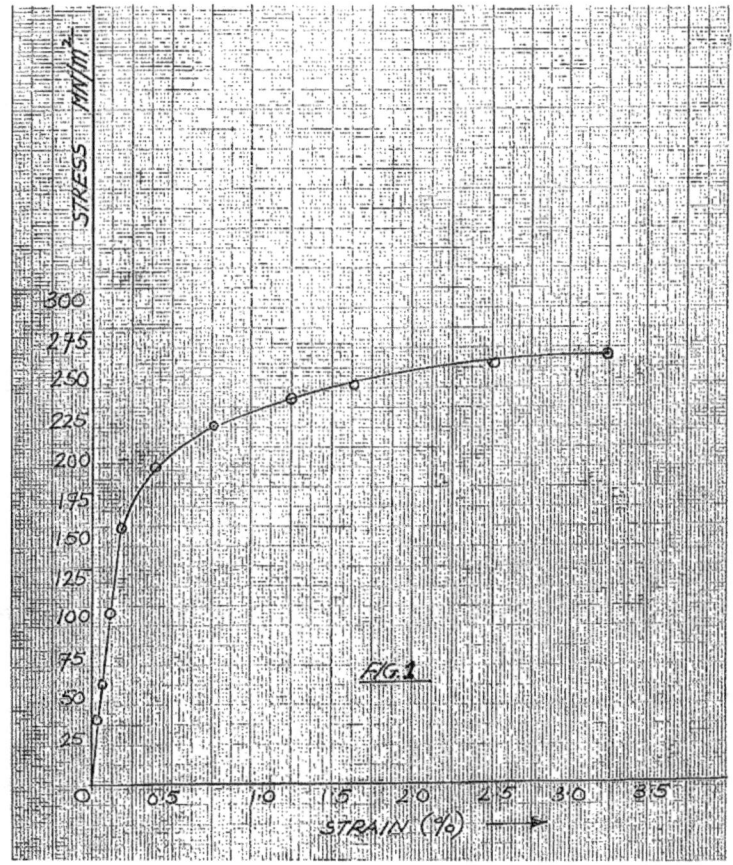

FIG.1

Q3. A three-sheave travelling block is to support a drilling string of weight equivalent to 150kN. Determine the minimum diameter of each cable in the block assuming that no permanent deformation is to occur. The cables are made of high strength steel having a yield stress of $400MN/m^2$. Neglect buoyancy effect due to circulating mud with the string standing in the hole.

During service the yield stress of the steel must not be exceeded, because if there is any kind of permanent deformation of any cable then the cable must be regarded as having failed. Consequently the cable must be taken out of service.

~ 237 ~

For a 3-sheave block there are 6 cables supporting the load. Therefore cross-sectional area of cable under stress = $6(\pi d^2/4)$.

From Stress = Force/area

$$\therefore \quad 400\times10^6 = \frac{1500,000\times4}{6\pi d^2} \qquad \dots\dots\dots\dots\dots\dots \quad (i)$$

where d = diameter of cable

From (i)

$$d^2 = \frac{1500,000\times4}{6\times400\times10^6\times\pi}(m)^2$$

$$= \frac{6.0}{24\times10^3}\times\frac{10^6}{\pi}(mm)^2$$

i.e. $\quad d^2 = 795.9$

or $\quad d = 28.2mm$

Q4. An ingot of a certain metal is to be cold rolled in a rolling mill. The ingot is rectangular in cross-section; its thickness is 'd_0'(mm); and when it passes through the mill its thickness is expected to be reduced to 4.5(mm). The stress exerted on the ingot as it passes through the rolls is 400MN/m². Assume Young Modulus for the metal to be 100GN/m²; also that the ingot does not spread sideways and the rollers do not deflect. Find original thickness 'd_o'

Let $\qquad d_f$ = final thickness of ingot

$$\text{Strain} \ = \ \frac{d_f - d_0}{d_0}$$

Note that strain here is negative because $d_0 > d_f$.

Remember, $\quad e = \dfrac{Final\ thickness - Original\ thickness}{Original\ thickness}$

From $\qquad e = \dfrac{Stress}{Young's\ modulus}$

$$e = \frac{400 \times 10^6}{100 \times 10^9}$$

i.e. $\quad e = \frac{4}{10^3}$

Accordingly, $\qquad -\frac{4}{10^3} = \frac{d_f - d_0}{d_0}$

$$-0.004 d_0 = 4.5 - d_0$$

i.e. $\quad 0.996 d_0 = 4.5 \quad$ or $\quad 4.518 mm.$

$\therefore \qquad d_0 = 4.518 mm$

Q5. A load of 7kg is applied to one end of an aluminium alloy (DTD 683) bar 200mm long rotating at 360 cycles per minute. The bar is 10mm in diameter. Estimate the time to failure. Take $g = 9.81 m/s^2$.

In order to solve this problem it is necessary to know the value of the maximum stress to which the bar is subjected under the load of 5kg as it rotates. According to the statement of the problem, the bar is in effect a rotating cantilever; subjected to a bending stress as it does so.

To obtain this stress, say, σ you will have to take the following relationship on trust at this stage. Its full derivation is given in the chapter on stress due to bending.

$$\frac{\sigma}{y} = \frac{M}{I}$$

Here $y = d/2$ in which d = diameter of the bar

$\qquad M$ = maximum bending moment = $W\ell$ W being the

$\qquad\qquad$ load of 7kg

$\qquad I$ = Moment of inertia of bar about its neutral axis = $\pi d^4 / 64$

Accordingly,

$$\frac{\sigma}{d/2} = \frac{W\ell}{\pi d^4 / 64}$$

from which

$$\sigma = \frac{32}{\pi} \frac{W\ell}{d} = \frac{10.18W\ell}{d^3}$$

Substituting $W = 5 \times 9.81N$; $\ell = 200mm$; $d = 10mm$ and working in metres

$$\sigma = 10.18 \times 9.81 \times \frac{200}{1000} \times \frac{1}{\left(\frac{10}{1000}\right)^3}$$

$$= 10.18 \times 1.4 \times 9.81 \times \frac{1}{5} \times \frac{10^9}{10^3}$$

$$\sigma = 139.8 MN/m^2$$

Referring to the curve for duralumin in Fig. 23 it is seen that for this value of stress, the fatigue life is approximately 10^7 cycles. Because the bar rotates at 360 cycles/minute, the time 't' to failure $= 10^7/360 \equiv 27,800$ minutes or roughly $27,800/(24 \times 60) = 19.3$ days.

Q6. Define Fatigue Limit. A hot rolled mild steel specimen is 300mm long and 25mm in diameter. As a rotating cantilever the specimen is expected to have a fatigue life of 10^7 cycles. What is the maximum allowable load at the end of the bar?

I leave the first part of the question to the student.

For the other part reference must be made to the relevant curve in Fig. 23 where it is seen that for a fatigue life of 10^7 cycles the applied stress is $254MN/m^2$.

From $\qquad \sigma = \dfrac{10.18W\ell}{d^3}$: (see Q5 for derivation)

$$254 \times 10^6 = \frac{10.18W}{\left(\frac{25}{1000}\right)^3} \cdot \left(\frac{300}{1000}\right)$$

$$\therefore \qquad 300(10.18)W = 254 \times 10^6 \times \left(\frac{25}{1000}\right)^3 \times 1000$$

from
$$W = \frac{254 \times 10^6 \times (25)^3 \times 1000}{300 \times 10.18 \times 10^9}$$

$$= \frac{254 \times 15625}{300 \times 10.18}$$

or
$$W \cong 132kg$$

Q7. A 5mm thick, 80mm wide steel rod, 600mm in length, has a circular hole 30mm in diameter located on its longitudinal centre-line as shown in Fig. 1. Determine the maximum tensile force which may be applied along its longitudinal axis as indicated in the diagram without exceeding an allowable stress of 225MN/m². Use Curve labelled 'A' in Fig. 26.

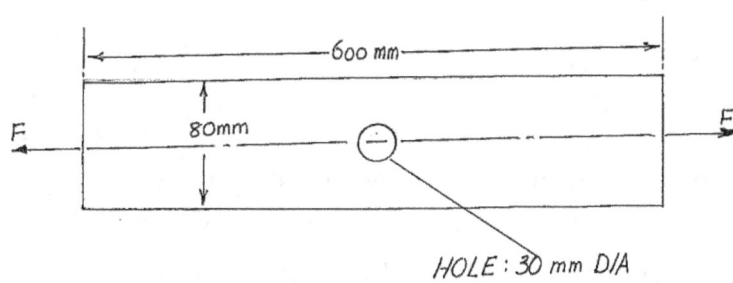

FIG. Q7.

Let   F = maximum tensile force

" σ_n = nominal stress in rod.

By definition,

$$K = \frac{Maximum\ allowable\ stress}{No\min al\ stress\ calculated\ on\ \min imum\ csa}$$

i.e. to say

$$K = \frac{225 \times 10^6}{\sigma_n}$$

Now, $\sigma_n = \dfrac{F}{nett \; csa} = \dfrac{F}{\dfrac{(80-30)}{1000} \times \dfrac{5}{1000}} = \dfrac{F(10^6)}{250} N/m^2$

$\therefore \quad K = \dfrac{225 \times 10^{6 \times 250}}{F(10^6)}$

$\dfrac{r}{b} = \dfrac{15}{80} \approx 0.19$

From Fig. 26 it is estimated $K \approx 2.55$

$\therefore \quad F = \dfrac{225 \times 250}{2.55} N$

$\approx 21,650 N$

i.e. $\quad F \approx 21.7 kN$

Q8. A flat steel bar of rectangular cross-section : thickness 't' and width 'b', has a hole of diameter 'd' drilled through its centre-line as shown in figure below. If the allowable static tensile stress of the steel is σ_A, then what is the maximum permissible static tensile force F_{max} that can be imposed on the bar? Assuming elastic conditions, if the stress concentration factor 'K_T' for uni-axial stress for a rectangular bar is given approximately by $K = 3b/(b+d)$, what then would be the condition to achieve this maximum impost F_{max}?

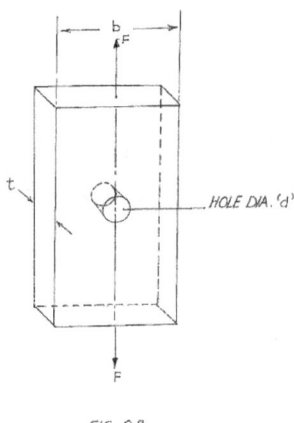

FIG. Q8

The diagram of the bar is at Fig. Q8.

By definition, stress concentration factor K_T is given by:

$$K_T = \frac{Maximum\ value\ of\ actual\ stress}{Nominal\ stress\ calculated\ on\ minimum\ c.s.a.}$$

Let $\qquad \sigma_n \quad = \quad$ nominal stress under maximum load

$\qquad\qquad F_{max} \quad = \quad$ maximum load

Accordingly, when the maximum load F_{max} is acting

$$\sigma_n = \frac{F_{max}}{t(b-d)}, \quad t(b-d) \text{ being the minimum c.s.a.}$$

Now, the allowable tensile stress is σ_A. This in fact is the maximum value of actual stress, that the material can sustain.

$$\therefore \qquad K_T = \frac{\sigma_A}{\sigma_n}$$

becomes,

$$K_T = \frac{\sigma_A}{\dfrac{F_{max}}{t(b-d)}}$$

from which $F_{max} = \dfrac{\sigma_A\, t(b-d)}{K_T}$

The result may be rewritten as

$$F_{max} = \frac{\sigma_A\, tb}{K_T}\left\{1 - \frac{d}{b}\right\}$$

We are also given an approximation for K_T, viz

$$K_T = \frac{3b}{(B=D)}$$

If now we substitute this expression for K_T in the result obtained for F_{max}, then

$$F_{max} = \frac{\dfrac{\sigma_A tb}{3b}\left\{1 - \dfrac{d}{b}\right\}}{(b+d)}$$

$$= \frac{\sigma_A t}{3}(b+d)\left(1 - \frac{d}{b}\right)$$

$$= \frac{\sigma_A t}{3}b\left(1 + \frac{d}{b}\right)\left(1 - \frac{d}{b}\right)$$

$$F_{max} = \frac{\sigma_A tb}{3}\left(1 - \frac{d^2}{b^2}\right)$$

Clearly, when d is small in comparison with 'b', $\left(\dfrac{d^2}{b^2}\right)$ would vanish.

Therefore to satisfy the condition for maximum 'F', the hole diameter on the centre-line should be of the minutest size practicable as in a pin hole, with a ratio r/b of magnitude of not more than, say, $\dfrac{1}{10}$. Then we may write $F_{max} = \sigma_A tb/3$, approximately.

Q9. A slot 50cm with radii of 1.25cm at its ends is milled out of a rectangular low-carbon steel bar 2.5cm thick, 10cm wide and 5m long. See Fig. 1. If a tensile pull $F = 100kN$ is applied to the bar, then determine the maximum stress acting in it. Assume the curve labelled 'A' in Fig. applies. Also, neglecting the effect due to stress raising on the bar's axial deformation, determine the total elongation of the bar. Take $E = 208GN/m^2$.

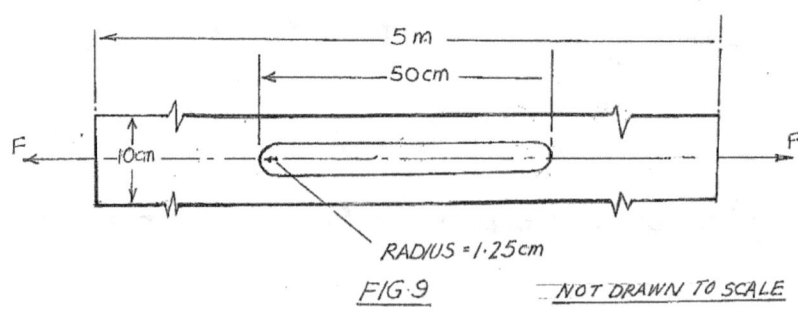

RADIUS = 1.25cm

FIG.9   NOT DRAWN TO SCALE

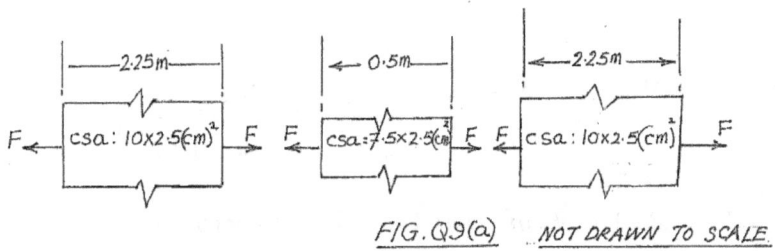

FIG.Q9(a) NOT DRAWN TO SCALE

Minimum cross-sectional area = $2.5\{10-2(1.25)\}(cm)^2 = 18.75(cm)^2$

Radius of ends of slot = 1.25; d = 7.5cm

$$\therefore \quad \text{Ratio} \left[r/w = \frac{1.25}{10} = 0.125 \right]$$

From Fig. 26, $K = 2.5$

By definition :

$$2.5 = \frac{\sigma_{max}}{100kN/18.75(cm)^2}$$

Working in square metres, therefore

$$2.5 = \frac{\sigma_{max}}{100 \times 10^3 / \dfrac{18.75}{10^4}}$$

$$\therefore \quad \sigma_{max} = \frac{2.5 \times 100 \times 10^7}{18.75}$$

$$\sigma_{max} = 133 \times 10^6 \, N/m^2 \quad \text{or} \quad 133MN/m^2$$

Note that this maximum stress is well below the proportional limit for low carbon steel so we would not expect yielding.

As concerns the second part of the question, if the effects of stress concentration are to be neglected, then the problem reduces to one where a constant pull $F=100kN$ is acting on a bar with two end sections 2.25m long and a middle section of length 50cm. Let $\Delta\ell'$ be the elongation of one of the end sections; $\Delta\ell''$ the elongation of the middle section. Considering the bar in its entirety:

~ 245 ~

Total elongation, $\Delta = 2\Delta\ell' + \Delta\ell''$

Now $\Delta\ell'$ is obtained from $\quad : \quad E = \dfrac{F}{\dfrac{(10 \times 2.5)}{100 \times 100} \cdot \Delta\ell'}$

Taking $E = 208 \times 10^9 \, N/m^2$, and working in metres

$$208 \times 10^9 = \frac{100 \times 10^3}{10 \times 2.5} \times \frac{2.25}{\Delta\ell'} \cdot 100 \times 100$$

$$\therefore \quad \Delta\ell' = \frac{2.25 \times 10^9}{208 \times 10^9 \times 2.5 \times 10} m$$

$$= \frac{2.25 \times 10^3}{208 \times 2.5 \times 10}$$

$$\Delta\ell' = 0.433 mm$$

Continuing, we deal with the middle section of length 50cm; $csa = (7.5 \times 2.5)cm^2$.

$$208 \times 10^9 = \frac{100 \times 10^3}{\dfrac{7.5 \times 2.5}{100 \times 100}} \times \frac{50}{100} \cdot \frac{1}{\Delta\ell''}$$

from which $\Delta\ell'' = \dfrac{100 \times 10^3 \times 50 \times 10^4}{208 \times 10^9 \times 7.5 \times 2.5 \times 100}$

$$= \frac{5 \times 10^{10} 10 \times 10^3}{208 \times 10^9 \times 7.5 \times 2.5 \times 100}$$

$$= 0.128 mm$$

so that, total elongation, $\Delta = 2\Delta\ell' + \Delta\ell''$

i.e. $\quad\quad\quad\quad \Delta = 2(0.433) + 0.128 = 0.994 mm$

$$= 0.866 + 0.0013$$

$$\therefore \quad\quad\quad \Delta = 0.994 mm$$

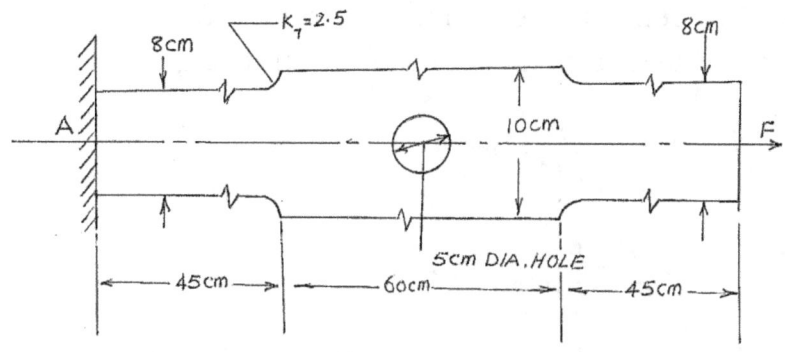

FIG. Q10.

Q10. The steel bar shown in Fig. 1 is 2.5cm thick. The central circular hole on its centre-line is of radius 2.5cm. The stress concentration factors for the fillet radii and hole may be taken as 2.5 and 2.0 respectively. The bar is fixed to a rigid support at 'A'. When a force 'F' is applied as indicated the total elongation of the bar = 0.5mm. Determine the maximum stress in the bar caused by 'F' assuming that stress concentration has no effect on axial deformation. Take E = 200GN/m².

Solution

Because stress concentration has no effect on axial deformation, this means that whatever elongation occurs is due entirely to force 'F' acting on 3 sections of the bar : two of length 45cm and c.s.a. 20(cm)²; and one of 60cm in length and c.s.a. = 25(cm)².

Accordingly for the 2 sections

$$200 \times 10^9 = \frac{F}{20/10^4} \cdot \frac{2(45)}{\Delta \ell' \ \ 100}$$

from which

$$\Delta \ell' = \frac{90F \cdot 10^4}{20 \times 200 \times 10^9 \times} \cdot \frac{1}{100}$$

Similarly for the single section 60cm long

$$\Delta \ell'' = \frac{F \cdot 60 \times 10^4}{25 \times 200 \times 10^9} \times \frac{1}{100}$$

$\Delta\ell'$ and $\Delta\ell''$ being the extension of both 45cm long sections and $\Delta\ell''$ the extension of 60cm long section.   Simplifying

$$\Delta\ell' = \frac{9F}{4\times10^0} \, metres$$

$$\Delta\ell'' = \frac{6F}{5\times10^0} \, metres$$

We are given that the total elongation of the bar = 0.5mm

i.e. $\therefore$ $F\left(\dfrac{9}{4\times10^9}+\dfrac{6}{5\times10^9}\right)=\dfrac{0.5}{1000}$

$$F(2.25+12)=\frac{0.5\times10^9}{1000}=500000$$

from which $F=144.927.5N$

$$F=144.9kN$$

This force gives a nominal stress = F/csa.

Remembering the definition of stress concentration factor 'K' which is given by

$$K=\frac{Maximum \ value \ of \ actual \ stress}{No\min al \ stress \ calculated \ on \ \min imum \ csa}$$

We have to test two situations: one for 'K' due to the fillet radii, the other due to the circular hole.

(i)   For the fillet radii

$$2.5=\frac{Max \ Stress}{0.05F\times10^4}$$

$\therefore$ $Max \ Stress = 181.2MN/m^2$

from which

$$(Max \ Stress)_1 = 181.2MN/m^2$$

For the circular hole

$$2.0 = \frac{(Max\ Stress)_2}{\dfrac{F}{12.5 \times 10^{-4}}}$$

Note that for the section with the circular hole, the minimum c.s.a. = (10-5)2.5cm² =12.5cm².

$$\therefore \quad (Max\ Stress)_2 = \frac{2 \times 144{,}227.5}{0.00125}$$

$$= 231884000 N / m^2$$

$$= 231.9 N / m^2$$

Evidently the maximum stress occurs in the section with the circular hole. Its value = 231.9MN/m².

Q11. In a hypothetical creep test on a material the results are as tabulated below.   Note that the material had a strain of 1.5% before it was subjected to a constant temperature (the value of which need not concern us here) when the test began.   Using the data given, plot the creep curve and calculate the steady state creep rate.

Original length of specimen = 100mm

Length of specimen at start of test = 101.5mm

Temperature $t^{\circ}K$ constant during test.

Length of specimen Mm	Extension Mm	Strain %	Elapsed Time h
101.5	1.5	1.5	0
102.98	2.98	2.98	1200
104.5	4.5	4.5	2700
105.5	5.5	5.5	4200
106.3	6.3	6.3	5400
107.2	7.2	7.2	6600
108.7	8.7	8.7	7800
109.9	9.9	9.9	8400
112	12	12	9000

The creep curve is shown in the accompanying diagram. The slope in the secondary or steady-state phase = CB/AB = (6.3-3.75)/5400-1200 = 9.07×10^4% per hour.

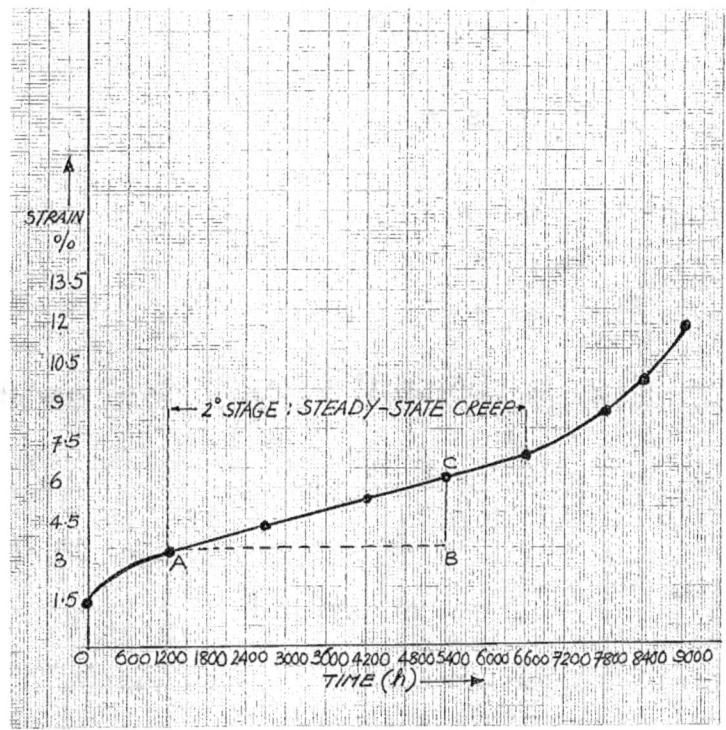

Q12. In a statistical correlation of Brinell Hardness Numerical *(BHN)* with ultimate tensile strength for an alloy steel the following relationship was obtained;

$$\sigma = 0.35(HBN)$$

in which σ =ultimate tensile strength in MN/m². HBN is also expressed in MN/m². [Remember HBN is somewhat of a misnomer, because it does have the dimensions of Stress i.e. Load/Area]; normally one thinks of a number as dimensionless.

The load employed in a Brinell Hardness is 550kg, the indenter diameter being 10mm. The ultimate tensile strength of the test specimen is 820 MN/m². Estimate (i) the BHN of the specimen and also the diameter *'d'* of the impression.

(i)   From $\sigma = 0.35BHN$

$$BHN = \frac{820}{0.35} = 2343 MN/m^2$$

$$= 2343 \times 10^6 \, N/(1000mm)^2$$

$$= 2343N/N(mm)^2$$

But $\quad 1kg = 9.81N \quad [g \approx 9.81m/s^2]$

$\therefore \quad BHN = \dfrac{2343}{9.81}kg/(mm)^2$

$$BHN = 239kg/(mm)^2$$

Recall that by definition BHN is expressed in kg/(mm)².

$$BHN = \dfrac{P}{\dfrac{\pi D}{2}\left\{D - \sqrt{(D^2 - d^2)}\right.}$$

where $\quad P$ = diameter load in kg

$\quad\quad\quad\quad D$ = diameter of indenter ball in mm

$\quad\quad\quad\quad d$ = diameter of indentation

Accordingly

$$239 = \dfrac{550}{3.143(5)\left\{10\sqrt{10^2 - d^2}\right.}$$

i.e. $\quad \left(10 - \sqrt{10^2 - d^2}\right) = \dfrac{550}{239(15.71)} = 0.145$

$\therefore \quad 9.854 = \sqrt{100 - d^2}$

$\quad\quad 97.1 = 100 - d^2$

$\quad\quad d^2 = 2.9$

or $\quad\quad d = 1.7mm$

Q13. (i) Show that the Vickers Diamond Pyramid Hardness Number (VPN) maybe expressed in the form:

$$VPN = \dfrac{1.854P}{d^2}$$

in which P = load in kilograms and d, the diameter measured on the diagonal of the indentation.

(ii)   Discuss the advantages and particular merits of the Vickers Hardness Test over its Brinell counterpart.

Q14. An iron-nickel-chromium-cobalt bar is to operate at a stress of 60MN/m² for 10000h without failing. What is the maximum allowable constant temperature in its operating environment? Take 'D' in the Larson-Miller equation = 5.5 (Log hours). Also, assume the value of $P_{L,M}$= approx. 9000 at $\sigma = 600MN/m^2$

From

$$Log\ t_R = \frac{P_{L,M}}{T} - 5.5$$

$$Log\ 10000 = \frac{9000}{T} - 5.5$$

i.e. $$4 = \frac{9000}{T} - 5.5$$

$\therefore$ $$9.5T = 9000$$

$\therefore$ $$T = 9.47^\circ K$$

$$T = (947 - 273)^\circ K$$

or $$T = 674^\circ C$$

Q15. A heat resistant steel alloy was tested under a certain constant stress and constant temperature of 600°C and ruptured after 1000h. Another specimen of the same material when tested at 677°C under another constant stress it failed after only 100h. Find the value of the constant 'D' in the Larson-Miller parameter equation

Ans : D = 9.33

Q16. Explain how you would proceed to obtain the master curve for the Larson-Miller parameter by using the result obtained for 'D' in Question No. 15. State the other data necessary and the tests necessary to obtain the required result.

Constant Temperature °K	Constant Stress MN/m²	Time to Rupture h
873	σ_1	1000
950	σ_2	100
t1	σ_3	h3
t2	σ_4	h4
t3	σ_5	h5
.	.	.
.	.	.
.	.	.
tn	σ_n	hn

Q17. A ductile circular cylindrical bar 20mm in diameter is subjected to a tensile pull of 12.5kN. Given that the Larson-Miller parameter (LM$_p$) is given by $\dfrac{T}{1000}(36+0.78Log_e t)$ where T = absolute temperature (°K); t = time in hours, determine the elapse of time before the bar fails (a) at 420°C; (b) at 650°C. Use the graph shown at Fig. Q16

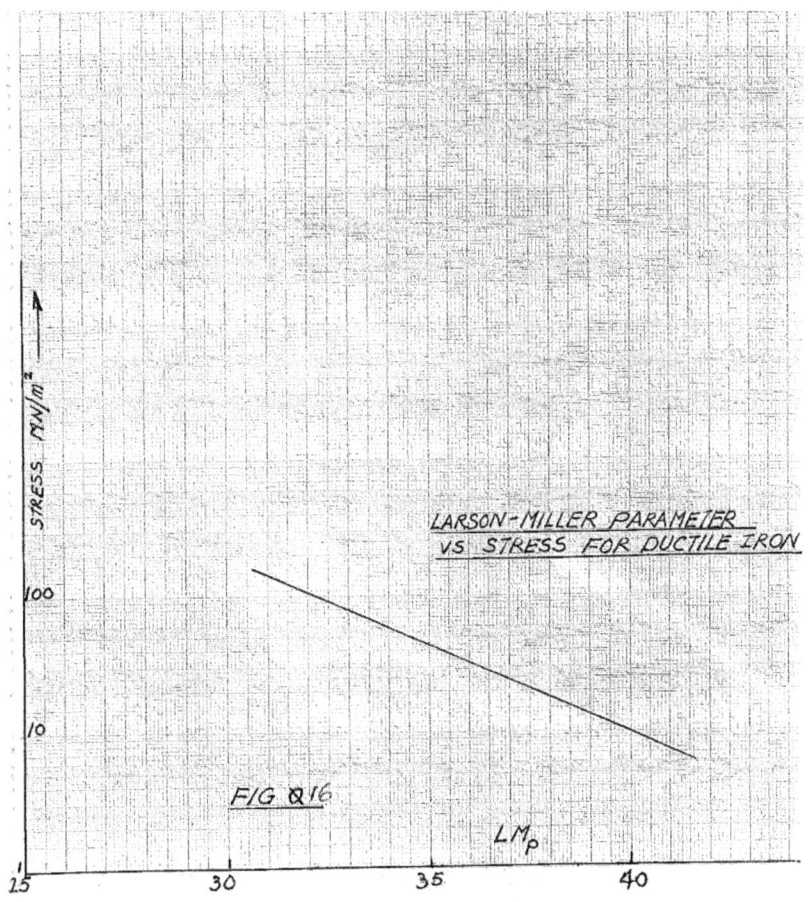

FIG α16

LARSON-MILLER PARAMETER v.s STRESS FOR DUCTILE IRON

Solution

Stress in bar $= \dfrac{12.5 \times 10^3}{\dfrac{\pi}{4}\left(\dfrac{20}{1000}\right)^2}$

$= \dfrac{12.5 \times 10^3 \times 10^6 \times 4}{\pi \times 4 \times 10^2}$

$= 39.78 MN/m^2$, say $39.8 MN/m^2$

From Fig. α, $LMp = 35.4$

(a)　At 420°C　$\therefore$　$35.4 = \dfrac{T}{1000}(36 + 0.78 Log_e t)$

$T = 420 + 273)$ degrees Kelvin

$$\therefore \quad 35.4 = \frac{693}{1000}(36 + 0.78\ Log_e\ t)$$

i.e. $\qquad 51 = 36 + 0.78 Log_e\ t$

or $\ Log_e\ t = 19.23$

$$t = e^{19.23}$$

$\therefore \qquad t = 2.246 \times 10^8\ h$

$$= \frac{2246 \times 10^5}{365 \times 24}$$

$$= 2.56 \times 10^4\ years$$

(b)　　At 650°C

$$35.4 = \left(\frac{650 + 273}{1000}\right)(36 + 78 Log_e\ t)$$

$$= \frac{923}{1000}(36 + 0.78 Log_e t)$$

$$38.35 = 36 + 0.78 Log_e\ t$$

$\therefore \qquad Log_e\ t = 3.0$

i.e. $\quad t = e^3$

or $\quad t = 20\ hours$

Notice the marked reduction in time-to-rupture with temperature increase underlining the sensitivity of creep strain to temperature for ductile iron.

Q18. A cylindrical heat exchanger on a petroleum refinery distillation plant operates at a temperature of 130°C. The ends of the exchanger are closed by circular flanges carrying a set of steel bolts distributed around a common pitch circle. Assuming that the constant 'B' in the power law for secondary creep in a specimen of the steel is 4×10^{-35} with $n = 3$, estimate the magnitude of the stress in each bolt after 10,000h. Each bolt on the rigid flanges was tightened initially at a stress of 50MN/m². Take E = 210GN/m².

The relevant stress-relaxation relationship is

$$\frac{1}{\{\sigma(t)\}^{n-1}} - \frac{1}{\{\sigma(0)\}^{n-1}} = (n-1)BE\Delta t$$

Given, $\qquad B = 4 \times 10^{-35}; \quad n = 3 \quad \sigma(0) = 50MN/m^2$

$$E = 210 \times 10^9 \, N/m^2.$$

$$\therefore \qquad \frac{1}{\{\sigma(t)\}^2} - \frac{1}{(50 \times 10^6)^2} = 2 \times 4 \times 10^{-35} \times 210 \times 10^9 \Delta t$$

Here $\Delta t = 10,000h$, so that the stress in each bolt after 10,000h, viz $\sigma(10,000)$, is obtained from:

$$\frac{1}{\{\sigma(10,000)\}^2} = \frac{1}{0.25 \times 10^{16}} + 168 \times 10^{-25} \times 10^4$$

$$= \frac{4 \times 10^5}{10^{21}} + \frac{168}{10^{21}}$$

$$= \frac{400168}{10^{21}}$$

from which $\{\sigma(10,000)\}^2 \approx \dfrac{10^{21}}{400168} = \dfrac{10^{16}}{4.00168} = \dfrac{10000 \times 10^{12}}{4.00168}$

$$\approx 249.9 \times 10^{12}$$

i.e. $\qquad \sigma(10,000) \approx 15.8 \times 10^6$

or $\qquad \sigma(10,000) \approx 15.8 MN/m^2$

Thus, the stress after 10,000h is 15.8 MN/m².

Q19. If at a constant stress of 50MN/m² and constant temperature of 500°C the secondary creep rate of a steel specimen was found to be $0.5 \times 10^{-10}/h$, determine the constant B in the power law $\in = B(\sigma)^n (hour^{-1})$ for a value of n=3.5.

Ans: $= 5.65 \times 10^{-38} (hour^{-1})$

Q20. The bolts of a steel flange joint of a medium pressure steam line in a crude-oil distillation plant are tightened to an initial stress of 4.5 MN/m². Determine the time taken for this stress in each bolt to relax to one-half its original value given that the secondary or inelastic creep rate '$\in$' is given by $\in = 5.2 \times 10^{-31}(\sigma)^3 \, hour^{-1}$, in which '$\sigma$' is the applied stress in MN/m². Prove any formula you use. Take $E = 200 GN/m²$.

Solution

Starting with the basic relationship for elastic and secondary or inelastic strain rates, i.e.

$$\dot{e}_{elastic} = -\frac{1}{E}\frac{d\sigma}{dt} \qquad \dots\dots\dots\dots\dots\dots\dots \quad (i)$$

$$\dot{e}_{inelastic} = 5.2 \times 10^{-3}(\sigma)^2 \qquad \dots\dots\dots\dots\dots\dots \quad (ii)$$

and assuming that total strain in stress relaxation being constant, implies:

$$\dot{e}_{elastic} = -\ \dot{e}_{inelastic} \qquad \dots\dots\dots\dots\dots\dots\dots\dots \quad (iii)$$

This is so because total strain, say, e_T which is given by

$$e_T = e_{elastic} + e_{inelastic}$$

is fixed in time which means

$$de_T / dt = \dot{e}_{elastic} + \dot{e}_{inelastic}$$

hence (iii)

Expressing (ii) in the form $C\sigma^n$, C being a constant, leads to:

$$C\sigma^n = -\frac{1}{E}\frac{d\sigma}{dt}$$

i.e.
$$\frac{d\sigma}{\sigma^n} = -CEdt \qquad \dots\dots\dots\dots\dots\dots\dots\dots\dots \quad (iv)$$

The limits of integration are for stress $\sigma(0)$ for $t = 0$ and $\sigma(t)$ after lapse of time, say, Δt

Integrating (iv) gives

$$\left[\frac{\sigma^{1-n}}{1-n}\right]_{\sigma(0)}^{\sigma(t)} = -CE\Delta t \qquad \dots\dots\dots\dots\dots\dots \text{(v)}$$

which may be expanded as follows:

$$\{\sigma(t)\}^{1-n} - \{\sigma(0)\}^{1-n} = (n-1)CE\Delta t \qquad \dots\dots\dots\dots\dots \text{(vi)}$$

Since experimental determinations reveal that $n \succ 1$, (and in this problem $n = 3$) it is better to write equation (vi) in the form:

$$\frac{1}{\{\sigma(t)\}^{n-1}} - \frac{1}{\{\sigma(0)\}^{n-1}} = (n-1)CE\Delta t \qquad \dots\dots\dots\dots\dots \text{(vii)}$$

Now, to some substitutions. We are given : $C = 5.2 \times 10^{-31} (h)^{-1}$;
$E = 200 \times 10^9 \, N/m^2$; $n = 3$; $\sigma(0) = 4.5 MN/m^2$; and $\sigma(t) = \dfrac{4.5}{2} = 2.25 MN/m^2$

$$\therefore \qquad \frac{1}{(2.25 \times 10^6)^2} - \frac{1}{4.5 \times 10^6} = 2 \times 200 \times 10^9 \times 5.2 \times 10^{-31} \Delta t$$

i.e. $\qquad \dfrac{10^{-12}}{(2.25)} - \dfrac{10^{-12}}{(4.5)} = 20.8 \times 10^{-20} \Delta t$

or $\qquad 10^{-12}(0.1975 - 0.494) = 20.8 \times 10^{-20}$

from which $\Delta t = 7.12 \times 10^4 \, h$ $\qquad$ i.e. $\qquad \Delta t = 8.12y$

The following unsolved problem is taken from Professor Mc Evily's book on "Metal Failures" published by John Wiley & Sons, Inc., 2002.

Q21. The power-law creep-relation for a Ni-Cr-Mo Steel at 454°C is given as:

$\dot{e} = 10 \times 10^{-20} \sigma^3 (day^{-1})$ where σ is in pounds per square inch.

(a) Show that this relation can be written as $\dot{e} = 1.3 \times 10^{14} \sigma^3 (hour^{-1})$

in which σ is now in megapascal (MPa). Use the conversion *1 MPa = 1MN/m² = 145lb/sq.in*

(b) It is desired that a 2m-long tension member support a load of $4.5 \times 10^4 N$ for 10 years at 454°C without exceeding 2.5mm of creep deformation. What is the maximum stress that can be applied?

(c) If after three years it is decided to lower the total deformation allowed to 1.5mm, what then is the maximum allowable stress?

(d) What cross-section is required?

Given:

$$\dot{e} = 10 \times 10^{-20} \sigma^3 = 10^{-19} \sigma^3 / day$$

$$\therefore \quad Strain\ rate/day(\in) = \frac{1}{24} \in / day$$

i.e. $$(\in)' = \frac{1}{24} \times 10 \times 10^{-20} \sigma^3 / hour$$

or $$(\in)' = \frac{1}{24} \times 10^{-19} \sigma^3 \qquad \dots \dots \dots \dots \dots \dots \dots \text{(i)}$$

By the conversion: $1MPa = 1MN/m^2 = 145lb/sq.in.$

Clearly if 'σ' is in MPa then in order to employ (i) in which 'σ' is in lb/sq. in it is necessary to multiply 'σ' in equation (i) by 145.

$$(\in)' = \frac{1}{24} \times 10^{19} (145\sigma)^3 \ (hour^{-1})$$

$$= \frac{1}{24} \times 10^{19} (1.45 \times 10^3)^3 \ (hour^{-1})$$

$$= \frac{1}{24} \times 10^{13} \times 3.045\sigma^3 \ (hour^{-1})$$

$$= \frac{1}{24} \times 10^{13} \times 30.45 \times 10^{-14} \ (hour^{-1})$$

$$\therefore \quad (\in)' = 1.27 \times 10^{-14} \sigma^3 \ (hour^{-1})$$

that is $$(\in)' \approx 1.3 \times 10^{-4} \sigma^3 (hour^{-1}) \quad QED$$

(b) Strain = $\dfrac{2.5mm}{2000mm} = 1.25 \times 10^{-3}$

~ 260 ~

Based on a 365-day year and 24 hr/day, strain rate

$$= \left(\frac{1.25 \times 10^{-3}}{10 \times 365 \times 24} \right) (hour)$$

Using the result derived for 'σ' in MPa

$$\frac{1.25 \times 10^{-3}}{10 \times 365 \times 24} = 1.3 \times 10^{-14} \sigma^3$$

$$\therefore \quad \sigma_3 = \frac{1.25 \times 10^{-3}}{10 \times 365 \times 24} \times \frac{1}{1.3 \times 10^{-14}}$$

$$= \frac{1.25 \times 10^{-4}}{365 \times 24 \times 1.3} \times 10^{-14}$$

$$= \frac{1250 \times 10^{-7}}{365 \times 32.88} \times 10^{-14}$$

$$= \frac{3.42}{32.86} \times 10^7$$

$$\sigma^3 = \frac{34.2}{32.88} \times 10^6 = 1.04 \times 10^6$$

$$= 1.04 \times 10^6$$

$$\therefore \quad \sigma = 1.01 \times 10^2 = 101 MPA \qquad or \quad 101 MN/m^2$$

(c)   New $\in (hour)^{-1} = \frac{1.5 \times 1}{2000 \times 7 \times 365 \times 24} = \frac{0.75 \times 10^{-3}}{7 \times 365 \times 24}$

$$= \frac{750 \times 10^{-6}}{7 \times 365 \times 24} = \frac{2.05 \times 10^{-6}}{168} = \frac{205 \times 10^{-8}}{168}$$

$$1.22 \times 10^{-8} = 1.3 \times 10^{-14} \sigma^3$$

or $\sigma^3 = \frac{1.22}{1.3} \times \frac{10^{-8}}{10^{-14}} = 0.89 \times 10^6$

$$\sigma = 0.96 \times 10^2$$

i.e. $\sigma = 96 MPa \quad or \quad 96 MN/m^2$

(d)   Cross-section (csa) required:

Load $\neq 5 \times 10^5 N$; Maximum stress $= 101 MN / m^2$

$$\therefore \quad 101 MN / m^2 = \frac{4.5 \times 10^4}{csa}$$

$$\therefore \quad csa = \frac{4.5 \times 10^5}{101 \times 10^6} (m)^2$$

$$= \frac{4.5 \times 10^4 \times 10^6}{101 \times 10^6} (mm)^2$$

$$= \frac{450}{101} \times 10^2 (mm)^2$$

from which $csa = 4.45 \times 10^2 (mm)^2$

$$\therefore \quad csa = 445 (mm)^2$$

CHAPTER 9

SHEAR FLOW AND SHEARING STRESS IN BEAMS OF CONSTANT CROSS-SECTION

So far we have dealt with stresses due to two of the three reactive elements necessary for equilibrium of any transverse section of a beam, viz (i) axial stress attributable to any longitudinal pull or thrust and (ii) bending stress caused by bending moment. In this chapter, shearing stress, the third reactive element, and the one which derives from shear force designated 'V' in Chapter 6 on shearing force and bending moment, is considered.

In pure bending there is, by definition no shearing force. Pure bending is an idealized condition but nevertheless easily attainable experimentally in the engineering laboratory. In practice, bending moment varies from section to section along a loaded beam. Fortunately however, the flexure formula which was derived on the basis of this idealized condition may, as the mathematical theory of elasticity informs us, be applied in cases where bending moment is varying, provided the cross-section of the loaded member is constant. Such cross-section does not necessarily have to be of rectangular shape; it could be of any shape whatsoever, but it must be uniform throughout.

Perhaps it is easier to visualize the distribution of bending stresses in a loaded beam than it is to do the same for the shearing stresses caused by the shearing forces acting on the same beam. Knowing that the shearing force 'V' is at any cross-section perpendicular to the axis of the beam having a cross-sectional area of say 'a' it is perfectly correct to deduce that the average shearing stress actingacross that cross-section is 'V/a', i.e. 'V' divided by 'a'. But can you visualize how that shearing stress is distributed point to point over the section? The fact is that shearing stress is not distributed uniformly over the cross-section at all as I shall proceed to demonstrate shortly.

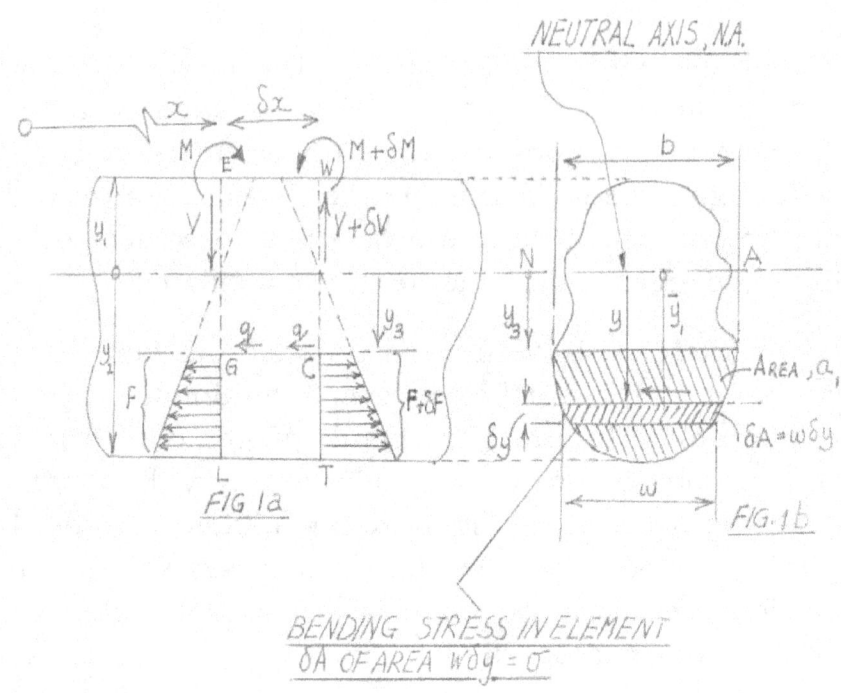

FIG 1a

FIG. 1b

NEUTRAL AXIS, N.A.

BENDING STRESS IN ELEMENT
δA OF AREA $w\delta y = \sigma$

SHEAR FLOW

However, I would like first of all to introduce the concept of shear flow in a loaded beam. Consider the cross-section at 'EL' of a beam where the sagging moment is 'M' and the shearing force is 'V' as in Fig. 1 a. Let 'WT' be the trace of an adjoining parallel cross- section at distance 'δx' From 'EL', where the sagging moment is now M+ δM, and the corresponding shearing force V+ δV. Consider next the forces due to bending stresses acting on the isolated section 'LTCG', in the same diagram. On the surface with the trace 'GL', there are the bending stresses acting normal on it due to bending moment 'M'; on 'CT' the corresponding normal bending stresses due to M+ δM. To make things easier we shall designate the forces due to the bending stresses on 'GL' by 'F' and on 'CT' by 'F + δF'. Looking at Fig. 1a again, it should be clear that some other force must act on the isolated element; otherwise with F+ δF being greater than 'F', the whole element would move to the right. We are talking equilibrium of forces in a free- body situation. Therefore, we designate this "some other force" "acting over the length of beam 'δx" by q. δx' where, note carefully, 'q' is a shearing force per unit length. Thus, for $\Sigma F_x = 0$

$$F + \delta F - F - q\delta x = 0$$

From which

$$q = \frac{\delta F}{\delta x}$$

or, in the limit with δx approaching zero

$$q = \frac{dF}{\delta x}$$

Assuming ' σ ' is the normal stress due to 'M' on the element of thickness δy which is distance 'y' from the neutral axis, then the total force on the infinitesimal element = $\sigma.w\delta y$, 'w' being the width of the element.

See Fig. 1b.

$\therefore$

$$F = \int_{y_2}^{y^1} \sigma w.\delta y [\text{Force} = \text{Stress x Area}]$$

But from the flexure formula: $\frac{\sigma}{y} = \frac{M}{I}$

$$\therefore F = \frac{M}{I} \int_{y_3}^{y^1} wy\, dy \, or \frac{M}{I} \int_{y_3}^{y^1} y dA [wdy = dA]$$

Noting that,

$$\int_{y_3}^{y^1} wy\, dy = a_1\bar{y}_1 \quad [\text{Remember } \bar{y}_1 = \int \frac{ay_1}{a_1}]$$

We may now write,

$$F = \frac{M}{I} a_1 y_1 \, i.e. \, F = \int_{y_3}^{y_1} y dA$$

'a_1' being the area on 'GL' and $\ddot{y}$' the distance of its centre of mass from the neutral axis.

By a similar token, we could write for 'CT'

$$F + \delta F = \frac{M + \delta M}{I} a_1\bar{y}_1 \, or \, \frac{M + \delta M}{I} \int_{y_3}^{y^1} y dA$$

Now,

$$\delta F = F + \delta F - F$$

$$= \left(\frac{M + \delta M}{I}\right) a_1\breve{y}_1 - \frac{M}{I} a_1\breve{y}_1$$

or, in the limit

$$dF = \left(\frac{M + dM}{I}\right) \int_{y_3}^{y_1} y dA - \frac{M}{I} \int_{y_3}^{y_{y_1}} y dA$$

$$dF = \frac{dM}{1} \int_{y_3}^{y_1} y dA$$

~ 265 ~

Early on it was deduced that: $q = \dfrac{dF}{dx}$

But dividing the last result for dF by dx we find

$$\frac{dF}{dx} = \frac{dM}{dx} \cdot \frac{1}{1}\int_{y_3}^{y_1} y\,dA = q$$

And because in the limit $\qquad \dfrac{dM}{dx} = V$

$$q = \frac{V}{1}\int_{y_3}^{y_1} y\,dA$$

$$q = \frac{V}{1}a_1\breve{y}$$

The quantity 'q' is referred to as the Shear Flow. In some texts this latter result is written

$$q = \frac{V}{1}Q$$

where $Q = \int_{y_3}^{y_1} y\,dA$

Remember that shear flow is given in terms of force per unit of the beam's length.

A good way of visualizing shear flow is to imagine two long identical planks resting freely lengthwise on each other, the whole assembly freely supported at the ends and subjected to downward vertical loading, say , at the middle. See in your mind's eye the planks bending convex downwards, their contact surfaces sliding perhaps imperceptibly but nonetheless sliding relative to each other as a result of the vertical load. Just imagine looking at the end of the beams or the mark you placed on each beam and you would observe that horizontal sliding took place due to the vertical loading. Now go out to the garage and do the demonstration. Liken the force causing the sliding per unit length of the plank to 'shear flow'. Think of it as a vertical force causing a horizontal shear flow. Evidently, if we wish to treat both planks as a single beam we shall have either to nail, bolt, glue or otherwise use another fastening agent or device to make them so. Inter- connection of such components by whatever means may be a matter of experience for the untutored but seasoned and experienced builder working by "rule of thumb", but the standard practice has a basis in fundamental engineering theory with which the modern technologist would be required to be familiar; the spacing of nails or the specific fastening agent shall be specified in the working drawings. Here is a problem to solve using the theory of shear flow.

Illustrative Example 1

A long rectangular wooden plank of cross-section 20 cm x 5 cm is to be inter-connected lengthwise to a circular cylindrical wooden rod of the same length and of 20 cm diameter. The two are to be nailed together to form a single cantilever beam subjected to a force of 1kN at its end. Calculate the shear flow at the junction of the composite structure and determine the spacing of the nails required to make the plank and the rod act as a single beam. Assume allowable shear force for each nail to be 40N. See Fig. 2a.

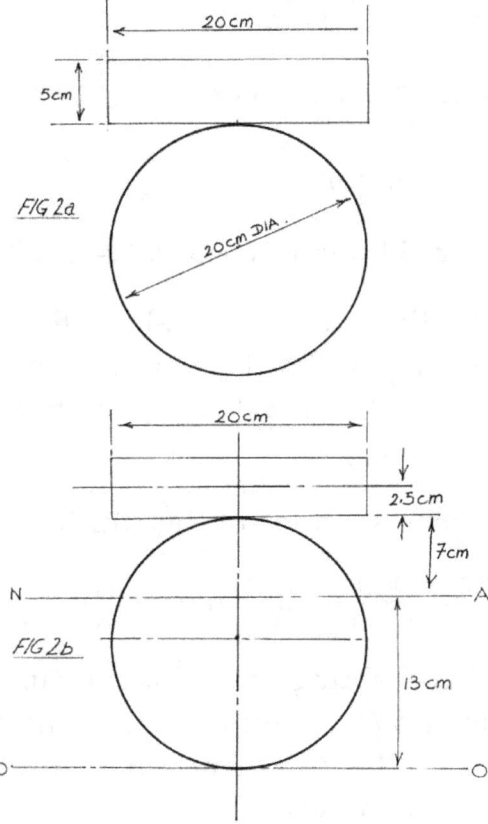

The first thing to do is locate the neutral axis of the composite beam. Let its distance from 'OO' be 'u'.

$$\therefore u = \frac{5 \times 20(22.5) + \pi(10)^2(10)}{100 + (10)^2 \pi}$$

i.e. $u = 13\ cm$

Next we determine the second moment of inertia 'I' for the composite beam about the neutral axis (13 cm from base line '0-0') or 12 cm from top of plank). Employing the Parallel Axis Theorem:

(i)   For the plank, $I = \frac{bd^3}{3} +$ Area $(9.5)^2$

$$= \frac{5(20)^3}{3} + 9025 = 22358 \ cm^2$$

(ii)   For the circular rod:

$$= \frac{\pi r^4}{4} + \frac{\pi d^2}{4}(3^2)$$

$$= \frac{10000 P_1 \pi}{4} + \frac{\pi}{4}(400)9$$

$$= 10933 \ cm^2$$

$\therefore$ total $I_{NA} = 22358 \ cm^2 + 10933 \ cm^2 = 33291 cm^2$

Because what is required is the value of the shear flow at the junction of plank and rod, i.e. at level 'O-O' in Fig. 2b, $\int ydA = A\ddot{y} =$ moment of statical area of 5cm x 20cm plank about the neutral axis of the entire combination, i.e. $= (20 \ x \ 5)(2.5 + 7)$ i.e. $\int ydA = 950 \ cm^3 = Q$.

The shearing force on all cross-sections of the cantilever = 1kN, i.e. V = 1000N

$\therefore$ $q = \frac{1000.950}{33291} = 28.5 \ N/cm$

That is, for the plank and rod acting as a single beam under a constant shear of *1000N*, the shear flow is *28.5 N/cm*. And because the allowable shear force for a nail is *180N*, one nail should be hammered in at intervals of *180N/28.5N/cm = 6.3 cm* along the length of the cantilever.

Based on these calculations the wood technologist would order a spacing of say 6cm between nails.

Referring again to Fig.1b and denoting 't' as the shear stress acting along GC, and 'b' the width of the beam, we may write

$$\tau (b \delta x) = q \delta x$$

or $\tau = \frac{q}{b}$

i.e. $\tau = \frac{V}{Ib} \int ydA$

or alternatively as
$$\tau = \frac{VQ}{Ib}$$

or   Shearing stress $= \frac{V}{Ib} A\bar{y}$

Let us proceed immediately to application of the latter by considering a practical example.

Illustrative Example 2

Show that the intensities of shearing stress, respectively τ_t and τ_m at (i) the top of the web or bottom of the top flange, and (ii) at the middle of the web, of a loaded I- beam of cross – section shown in Fig. 3 are given by:

$$\tau_t = \frac{VW}{8Ib}(H^2 - h^2)$$

and,

$$\tau_m = \frac{V}{8Ib}\{W(H^e - h^e) + bh^2\}$$

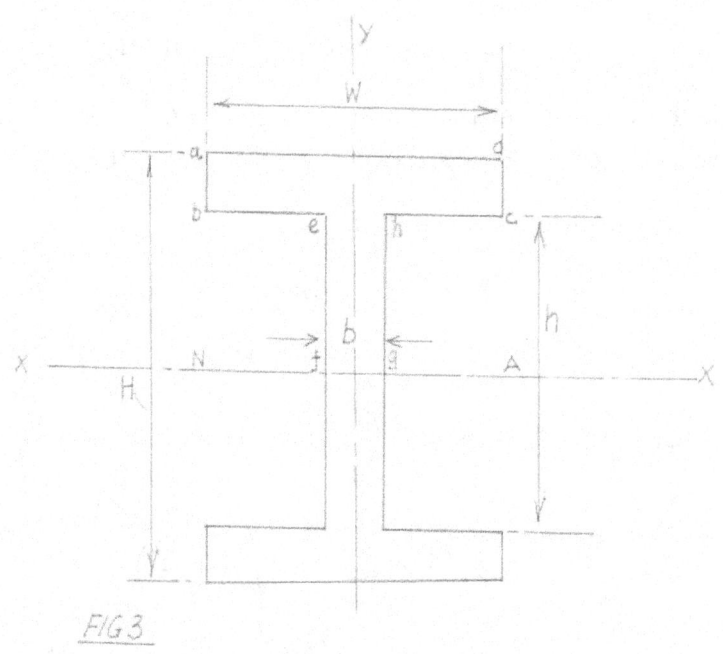

FIG 3

First of all we assume the relationship:

$$\tau = \frac{V}{Ib}\int y\,dA$$

which we write as, $\tau = \frac{V}{Ib} A\breve{y}$

Let us find 'τ_t' at the bottom of the flange 'abcd' in Fig. 3. Note that the bottom of the flange is the top of the web. For any of the flanges assumed identical, 'A' in the above equation for τ_t is the cross- sectional area of 'abcd' and $\breve{y}$ is the distance between the centroid of said area and the neutral axis of the <u>entire</u> section about the neutral axis 'NA' which coincides with 'XX' an axis of symmetry. Accordingly,

$$A_t = W \left(\frac{H-h}{2} \right)$$

and

$$\breve{y}_1 = \tfrac{1}{2}(H - h) + \frac{h}{2} = \tfrac{1}{4}(H + h)$$

so that

$$A_t \breve{y} = W \left\{ \frac{\{H-h\}}{2} \right\} \frac{(H+h)}{4}$$

$$= \frac{W}{8}(H^2 - h^2)$$

therefore

$$\tau^t = \frac{VW}{8lb}(H^2 - h^2), Q.E.D.$$

Consider next the shearing stress τ_m at the middle of the web.

'A' is now the area of a half section, i.e. 'abcd' + 'efgh'; and $\breve{y}$ is the distance of each of these cross- sectional areas from the neutral axis.

Take $A\breve{y}$, in 2 parts as follows:

(i)   For the flange, $A\breve{y} = W \frac{(H-h)}{2} \frac{(H+h)}{4}$

(ii)   For the web, $A\breve{y} = \frac{bh}{2} \left(\frac{1}{2} \cdot \frac{h}{2} \right) = \frac{bh^2}{8}$

∴ Total $A\breve{y}_1 = W \left(\frac{(H^2 - h^2)}{8} \right) + \left(\frac{bh^2}{8} \right)$

so that $T_m = \frac{V}{8lb} \{ W((H^2 - h^2) + bh^2) \}$

Shearing stress in a beam of rectangular cross- section

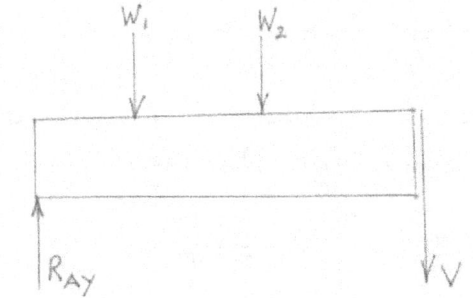

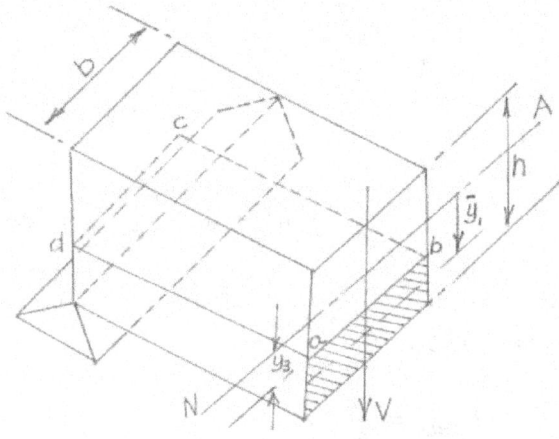

Two views of parts of a beam of rectangular cross- section, width *'b'* and depth *'h'* are shown in Fig. 4a and 4b. The part of the beam is under the influence of a positive shearing force *'V'*. as shown in Fig. 4a.

Employing the same symbology used in the development of the theory of shearing stress in straight beams, let us consider the horizontal plane *'abcd'*, distance *'y₃'* from the neutral axis. See Fig. 4b. Shear flow *'q'* on this plane is given by:

$$Q = \frac{V}{1}a_1y_1$$

With reference to Fig. 4b, $a_1 = b\left(\frac{h}{2} - y_3\right)$

and, $$\breve{y} = \frac{1}{2}\left(\frac{h}{2} + y_3\right) + y_3$$

i.e. $$\breve{y} = \frac{1}{2}\left(\frac{h}{2} + y_3\right)$$

It is known that the second moment of area 'I' of a rectangle of width 'b' and depth 'h' about its neutral axis is : $I = bh^3/12$. Therefore,

$$Q = V\left[b\left(\frac{h}{2} - y_3\right)\left\{\frac{1}{2}\left(\frac{h}{2} + y_3\right)\right\}\right]\frac{12}{bh^3}$$

The student would realize that 'y3' is a variable; hence shear flow varies parabolically with distance from the neutral axis. At the neutral axis where $y_3=0$,

$$q = \frac{3V}{2h}$$

and at the outer surfaces where $y_3 = \pm h/2$,

$$q = \frac{3V}{2h^3}\left(h^2 - 4.\frac{h^2}{4}\right) = 0 \text{ in each case.}$$

Evidently, the maximum value of shear flow occurs when $h^2 - 4y_3^2$ is a maximum, that is when $y_3 = 0$. Hence,

$$q_{max} = \frac{3V}{2h}$$

The intensity of shearing stress 'τ' as before

$$= \frac{q}{b}$$

So we may also conclude that shearing stress 'τ' varies parabolically with depth 'h'. Again the maximum value of 'τ' occurs at the neutral axis so that we may write,

$$\tau_{max} = \frac{3}{2}\frac{V}{bh}$$

The parabolic distribution of shearing stress with depth for a beam of rectangular cross- section is shown in Fig. 5. The shape of the distribution of shear flow is similar, except that the ordinates of the shearing stress curve are multiplied by the width of the beam: the former being τ = q/b; and the latter τ (b) = q.

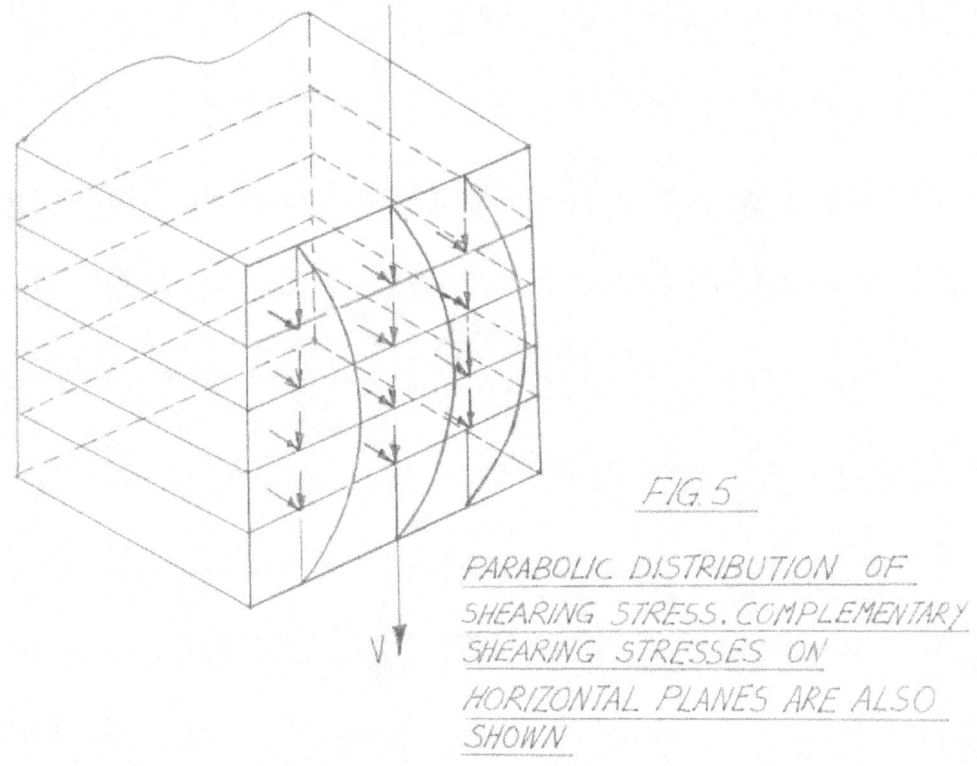

FIG. 5

PARABOLIC DISTRIBUTION OF
SHEARING STRESS. COMPLEMENTARY
SHEARING STRESSES ON
HORIZONTAL PLANES ARE ALSO
SHOWN

Also shown in the same diagram are the matching complementary shearing stresses which act with equal magnitude on horizontal planes. Thus, a wooden beam may fail in a longitudinal plane if the loading is such that the complementary longitudinal or horizontal shearing stress exceed the allowable shearing strength of the wood.

According to Popov (1952) the formula $\tau = \frac{V}{Ib} \int y \, dA$ was derived by

D.I. Jouravsky in 1855. Jouravsky had observed horizontal cracks in the wooden ties of several railway bridges between Moscow and St. Petersburg.

Illustrative Example 3

Derive an expression for the shearing stress on a plane perpendicular to its axis of a circular cylindrical rod subjected to a shearing force 'V' acting in the vertical plane. What is the maximum value of this shearing stress? Refer to Fig. 6a.

~ 273 ~

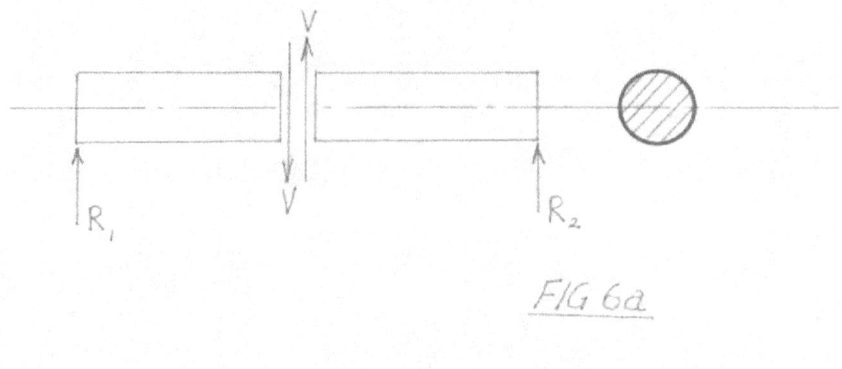

FIG 6a

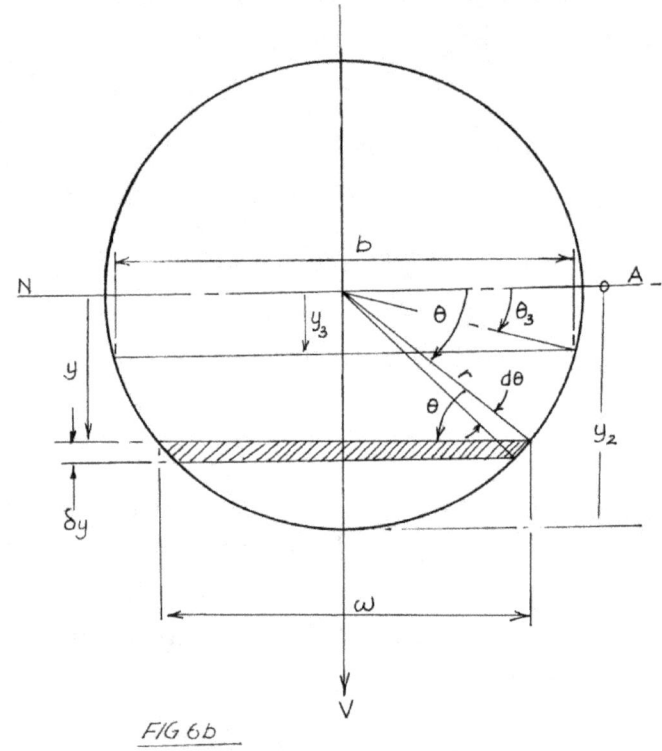

FIG 6b

Consider in Fig. 6b I have shown a typical cross- section through the rod, an elemental area of width 'w' and thickness 'δ'.

From trigonometrical considerations:

$w = 2r\,Cos\theta$

$y = r\,Sin\theta$

Accordingly $\quad dy = r\,Cos\theta d\theta$

Now when $y = y_3, \theta = \theta_3$

Employing our earlier result $q = \frac{V}{I} a_1 \breve{y}_1$, we may write,

$$q = \frac{V}{I} \int_{y_3}^{y_2} wdy.\, y$$

and because I, for the entire section above the neutral axis is $\frac{\pi r^4}{4}$, the latter expression may be restated as

$$q = \frac{4V}{\pi r^4} \int_{y_3}^{y_2} 2r\, Cos\theta.\, r\, Cos\theta\, d\theta.\, rSin\theta$$

Noting that when $y = y_2 = r$, $Sin\,\theta = 1$, i.e. $\theta = \frac{\pi}{2}$

also, when $y = y_3, \theta = \theta_3$, 'q' becomes:

$$q = \frac{8V}{\pi r} \int_{\theta_3}^{\frac{\pi}{2}} Cos^2\,\theta\, Sin\theta\, d\theta$$

$$= \frac{8V}{\pi r}\left[\left(-\frac{Cos_3}{3}\theta\right)\right]_{\theta_3}^{\frac{\pi}{2}}$$

i.e. $\quad q = \frac{8V}{3\pi r} Cos^3\theta_3$

This is the formula for shear flow; remember it has units of force per unit length; in this case per unit length of the circular bar, and because we know $b_3 = 2r\, Cos\,\theta_3$, in orderto convert our 'q' here to shearing stress we have to divide the last expression by 'b_3'. Hence,

$$\tau = \frac{q}{b} = \frac{8V\, Cos^3\theta}{3\pi\, r.2r\, Cos\,\theta_3}$$

$$= \frac{4V.}{3\pi r^2} Cos^2\theta_3$$

Again by trigonometry $\quad \left(\frac{b^2}{2}\right) = r^2 - y_3^2$

and because $\quad r^2 Cos^2\theta_3 = \left(\frac{b^2}{2}\right) = r^2 - y_3^2$

$$Cos^2\theta_3 = (r^2 - y_3^2)/r^2$$

from which $\quad \tau = \frac{4V(r^2 - y^2)}{3\pi r^4}$

which denotes a parabolic distribution of shearing stress with respect to 'y', the distance from N.A.

Observe that when y = 0, the shearing stress is maximum and of value

$$\tau_{max} = \frac{4V}{3\pi\, r^2}$$

The average shearing stress across the cross-section is$\frac{V}{\pi r^2}$, so it may be concluded that the maximum shearing stress is $\frac{4}{3}$ times the average.

Illustrative Example 4

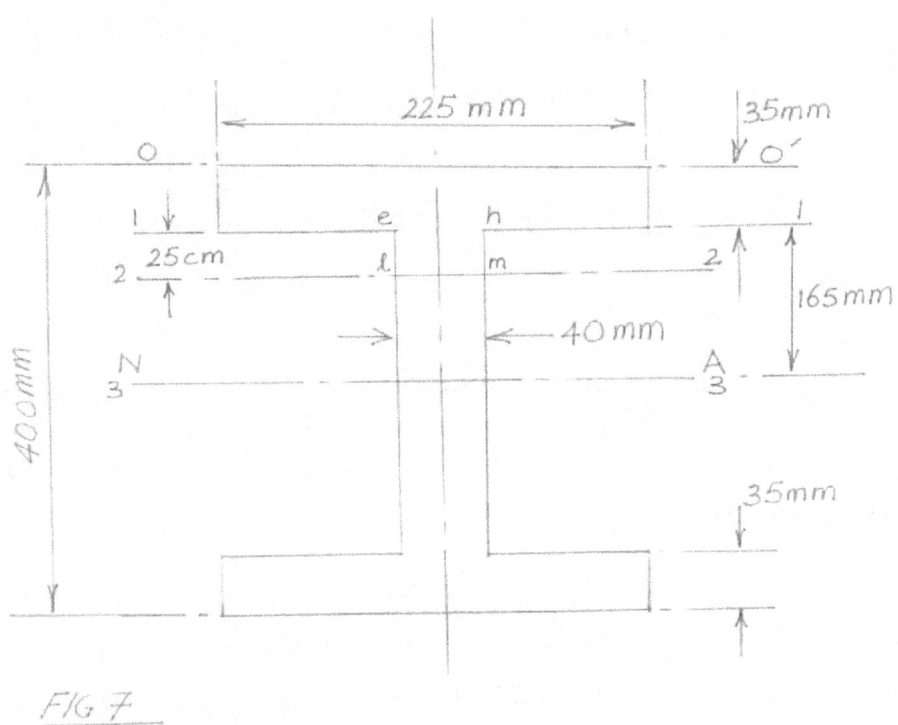

FIG 7

A simply supported I–beam of cross-section shown in Fig. 7, which is not drawn to scale, is subjected to a constant vertical shear V. What error is made by taking the shearing stress in the web at the neutral axis marked 'NA' in the diagram, as equal to V divided by the cross-sectional area of the beam instead of that of the maximum value obtained by using the shearing stress relation $\tau = \frac{V}{1b}\int ydA$ in which the symbols and terms have their usual meaning, 'b' being the thickness of the web = 40 mm.

I have shortened much of the working in what follows; the student should check my results by filling in the details.

Average shearing stress, say, $\tau_A = \dfrac{V}{csa}$

Here csa $= 2\left(\dfrac{225}{1000} \times \dfrac{35}{100}\right) + \left(\dfrac{330}{1000} \times \dfrac{40}{1000}\right)$

$= 28950 \times 10^{-6}\, m^2$

$= \dfrac{V}{28950} x10^{-6} = 34.54\, V/m^2$

Now, the maximum shearing stress τ_{max} occurs in the layer at the neutralaxis and remembering that $\int ydA$ is the sum of the statical moments of area about the neutral axis, it should be clear that we have to do this in 2 parts, thus =

$\int ydA$ = (Area of 225mm x 35mm flange x distance of its centroid of this area from the neutral axis) + (Area of one- half of the web x distance of the centroid of this area from the neutral axis).

Working in metres,

$$\int ydA = \left(\dfrac{225}{1000} \times \dfrac{35}{100}\right) + \left(\dfrac{200-17.5}{1000}\right) + \left(\dfrac{200-35}{1000}\right) \times \left(\dfrac{165}{2(1000)}\right)$$

$= 1981687.5 \times 10^{-9} m^3$

'I' for the whole section about the neutral axis,

$$= \left\{\dfrac{225}{1000} \times \left(\dfrac{400^3}{1000}\right) \times \dfrac{1}{1}\right\}\left\{\dfrac{185}{1000} \times \left(\dfrac{330^3}{1000}\right) \times \dfrac{1}{12}\right\}$$

$= 645971.25 \times 10^{-9} m^4$

Now, b = width (or thickness of the surface where 'τ' is being evaluated

i.e. $\qquad b = 40mm = \dfrac{40}{1000} m$

$$\tau_{max} = \dfrac{V(1981687.5 \times 10^{-9})}{645971.25 \times 10^{-9} \times \dfrac{40}{1000}}$$

$$= V(307) \times \dfrac{1000}{40}$$

i.e $\qquad \tau_{max} = 76.75V, say\ 77V$

Average shearing stress $\tau_A = 34.5\ V$

Maximum shearing stress $\tau_{max} = 77V$

$$\text{Error} = \frac{77 - 34.5}{77} = 0.55$$

Accordingly, the shearing stress is underestimated by 55%; a big difference!

Illustrative Example 5

Determine the distribution of shearing stress for the I – beam with cross section shown in Fig. 7 .The beam carries a load of 500 kN at mid-point as shown inFig. 8a.

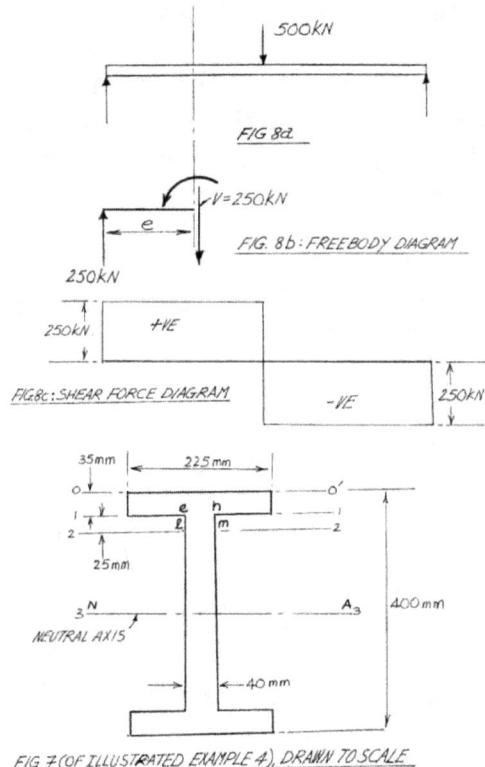

The free-body diagram of that portion of the beam of length 'a' from the left-hand support is at Fig. 8b and the shear force diagram for the whole beam is at Fig. 8c.

Evidently the shearing force 'V' is constant being of magnitude + 250 kN to the left of mid- section and – 250 kN on the right; in our convention that is!

We shall take the value of 'I' for the entire section about the neutral axis as 645971.25 x 10⁻⁹ m⁴. This value was determined in Illustrative Example 4. Let us convert this value of I to (mm)⁴.

Now $(1m)^4 = (10^{12})(mm)^4$. Therefore I = 646 x 10⁶ (mm)⁴.

Since shearing stress, $\tau = \frac{V}{Ib}A\breve{y}$, let usevaluate

$$\frac{V}{I} = \frac{250 \times 1000N}{646 \times 10^6} \, N/(mm)^4$$

$$= 3.86 \times 10^{-4} \, N/(mm)^4$$

and draw up a tabulation to include: Statical Area 'A',(ȳ)and 'b' at each level wetake through the section of the I – beam. Referring to Fig. 7, for level O – Ó: A = o; $\breve{y}$ = o; A$\bar{y}$= 0; b =225 mm. For level 1-1: A = 225 x 35 (mm) ², $\breve{y}$= 182.5mm; b = 225mm. For level 2-2: A has 2 components, viz. 225 x 35 (mm) ² and 25 x 40 (mm) ² and a $\breve{y}$ for each component of area: $\breve{y}$ = 182.5mm for the flange and $\breve{y}$ = 165 – 12.5 = 152.5 mm for the part of the web 'elmh'. Finally for level 3-3 which in fact is the neutral axis, the two components of area A are: 225 x 35 (mm) ² and 165 x 40 (mm) ², the respective $\breve{y}$s being 182.5 mm and 82.5 mm. The relevant data and calculations are shown in the following tabulation.

Level	Statical Area, A (mm)²	$\bar{y}$ mm	$A\bar{y}$ (mm)³	b mm	$\tau = \dfrac{VA\bar{y}}{IB} = \dfrac{3.86}{10^4}\cdot\dfrac{A\bar{y}}{b}$ N/(mm)²
0-0'	0	0	0	225	0
1-1	225 x 35 = 7875	182.5	1437187.5	225	2.46
2-2	225 x 35 = 7875 25 x 40 = 1000	182.5 152.5	1437187.5 + 152500 = 1589687.5	40	15.3
3-3	225 x 35 = 7875	182.5	1437187.5 + 544500 = 1981687.5	40	19.1

The I-beam being symmetrical about its centre- line and neutral axis, the values of shearing stress at the same levels below the neutral axis are the same. The variation of shear stress was plotted over the section. This diagram is shown in Fig. 9. Also shown in this diagram is the direction of shear stresses and complementary shear stresses on an element at the neutral axis of the section. Note that the shear stress on the right-hand face has the same direction as shear force 'V' in the free-body diagram Fig. 8b.

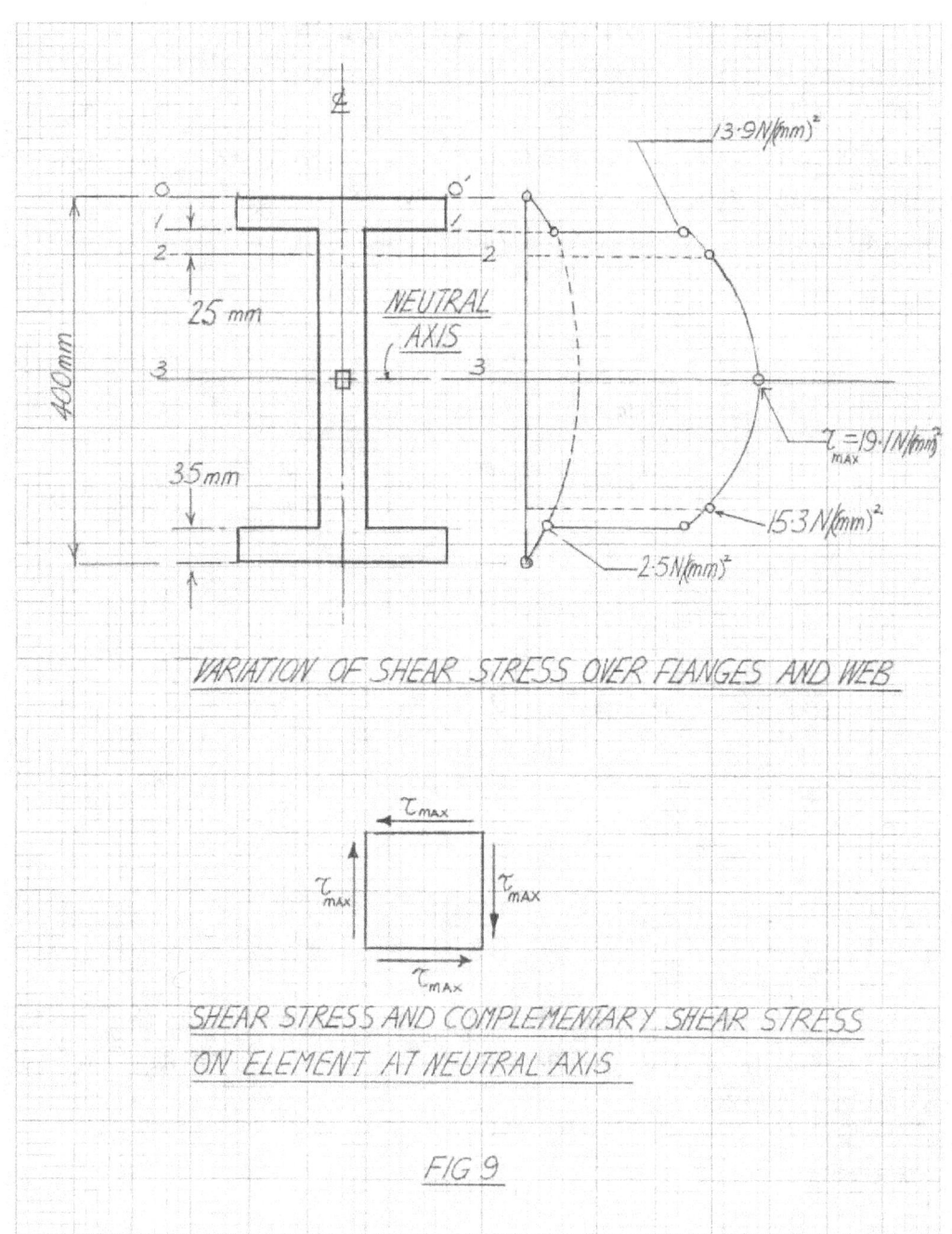

VARIATION OF SHEAR STRESS OVER FLANGES AND WEB

SHEAR STRESS AND COMPLEMENTARY SHEAR STRESS
ON ELEMENT AT NEUTRAL AXIS

FIG 9

SOLVED AND OTHER PROBLEMS

Q1. A revision Problem

State the Parallel Axis Theorem and use it to determine the second moment of area of the figure shown in Fig.1, given that its second moment of area about axis U–U is $8155(cm)^4$.

Ans: $8210(cm)^4$.

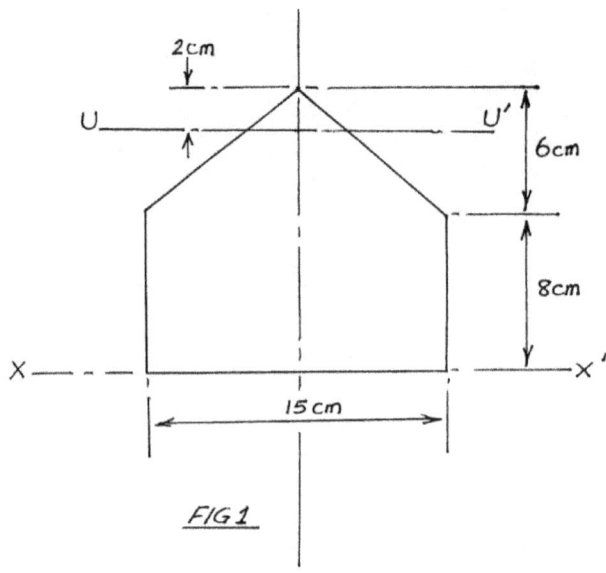

FIG 1

Q2. It was shown earlier in Illustrative Example #3 that the maximum shearing stress τ_{max} due to a shearing force 'V' acting on circular cylindrical bar is given by

$$\tau_{max} = \frac{4V}{3\pi r^2}$$

Using this result and assuming the rod is used as a beam with concentrated load

'W' at mid- span, express the length of the beam in terms of the radius 'R" for which the maximum stress caused by bending is $\frac{1}{2}\tau_{max}$.

Ans. $= \frac{4R}{3}$

Q3. Two pieces of wood each of width 10cm but having thicknesses respectively 3cm and 5cm are glued together to form a single plank. If this plank is subjected to a constant shearing force of 300 N, what is the shear flow between the two pieces of wood? Determine also the shearing stress at the glued surface. Do manufacturers specify the shear strength of their glues?

Assignment:

Do some research on the assumptions made in the derivation of the theory regarding the distribution of shearing stress in a straight beam. Criticise the assumptions made in the derivation.

Q4. A beam has the cross-section shown in Fig. 1 below. Assuming that this beam has to withstand a vertical shear of 750kN throughout, deduce an expression for shear stress $'\tau'$ and plot the distribution across the cross-section of the beam.

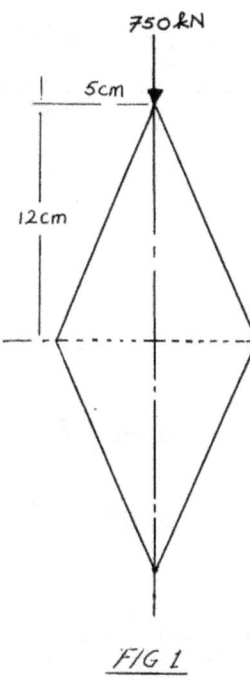

FIG 1

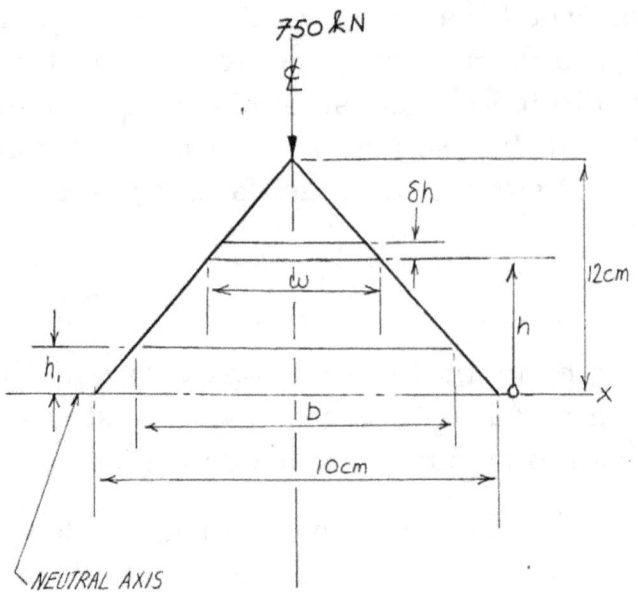

FIG. 2. HALF CROSS-SECTION OF BEAM
(NOT DRAWN TO SCALE)

In Fig. 2, I have shown one-half of the cross-section enlarged, not drawn to scale. I chose dimensions so as to make calculations easier. Shearing stress τ is assumed to be given by

$$\tau = \frac{V}{Ib} \int h \, dA$$

Referring to Fig. 2,

$I = I_{xx}$ $\qquad$ $\delta A = w\delta h$

So that, $\qquad\qquad$ $\tau = \frac{V}{I_{xx}b} \int w \, dh . h$

Observing that b = width of section at h = h, we may now write in our limits of integration thus

$$\tau = \frac{V}{I_{xx}b} \int_{h_1}^{12} wh \, dh$$

By similar triangles $\qquad$ $\frac{b}{10} = \frac{12 - h_1}{12}$

or $\qquad b = \frac{5}{6}(12 - h_1)$

Similarly $\qquad \frac{w}{10} = \frac{12 - h}{12}$

or $\qquad w = \frac{5}{6}(12 - h_1)$

Substituting these results in our expression for shear stress produces

$$\tau = \frac{V}{I_{xx}} \frac{6}{5(12 - h_i)} \int_{h_1}^{12} \frac{5}{6}(12 - h)h\,dh$$

i.e. $\qquad \tau = \frac{V}{I_{xx}} \frac{I}{(12 - h_1)} \int_{h_1}^{12}(12h - h^2)\,dh$

Recalling that 'I' for a triangle about an axis through its centroid is: base x (height)³ / 36 and that its centroid is one- third height from base, we apply the Parallel Axis Theory to find

$$I_{base} = \frac{bh^3}{36} + \frac{1}{2}bh \frac{(h)^2}{3}$$

$$I_{base} = \frac{bh^3}{12}$$

which in this case works out to be, for the entire section:

i.e $\qquad I_{xx} = 2\left\{10\frac{12^3}{12}\right\}(cm)^4$

i.e $\qquad I_{xx} = 2880\ (cm)^4$

Substitution of 'V = 750kN; I_{xx} = 1440 $(cm)^4$ and integrating the expression for τ, we get

$$\tau = \frac{750 \times 1000}{2880\,(12 - h_1)}\left[6h^2 - \frac{h^3}{h_1}\right]12$$

i.e. $\qquad \tau = \frac{290.5}{(12 - h_1)}\left[864 - 576 - 6h_1^2 + \frac{h^3}{3_1}\right]$

i.e. $\qquad \tau = \frac{290.5}{(12 - h_1)}\left[288 - 6h_1^2 + \frac{h^3}{3_1}\right]$

This latter expression gives the distribution, and gives τ in N/ $(cm)^2$. To plot it we assign the following values to h_1, viz. 0, 3cm, 6cm, 9cm and 12cm.

For $h_1 = 0$

$$\tau = \frac{290.5}{12}(288) = 6972 \ N/(cm^2)$$

For $h_1 = 3$ cm

$$\tau = \frac{290.5}{12}(288 - 54 + 9) = 7844 \ N/(cm)^2$$

For $h_1 = 6cm$

$$\tau = \frac{290.5}{6}(288 - 216 + 72) = 6004 \ N/(cm)^2$$

For $h_1 = 9$ cm

$$\tau = \frac{290.5}{12}(288 - 486 + 243) = 4358 \ N/(cm)^2$$

For $h_1 = 12cm$

$$\tau = 0$$

Refer to Fig.3 for shear stress distribution.

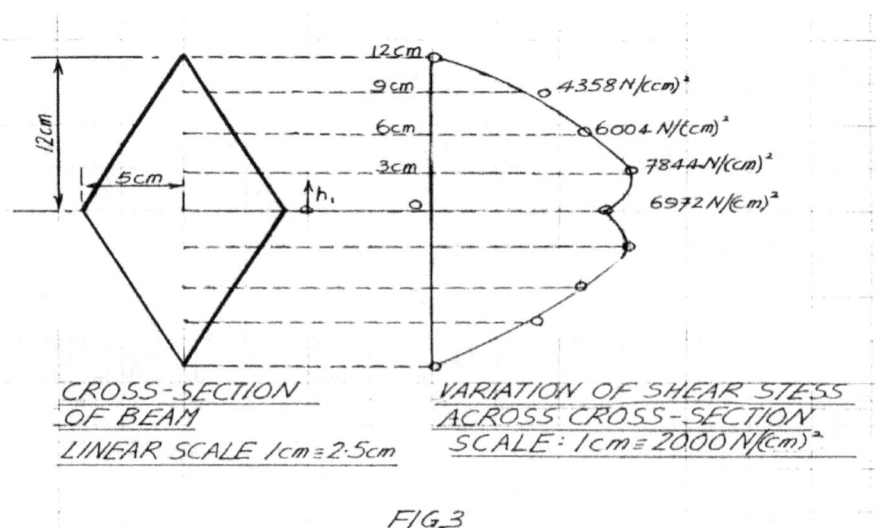

CROSS-SECTION OF BEAM
LINEAR SCALE 1cm ≡ 2.5cm

VARIATION OF SHEAR STESS ACROSS CROSS-SECTION
SCALE: 1cm ≡ 2000 N/(cm)²

FIG.3

Q5.

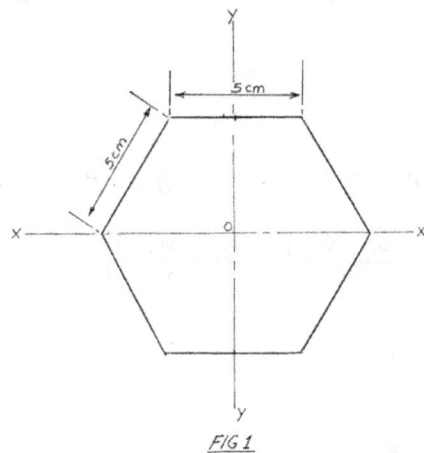

FIG 1

In Fig. 1 above the vertical cross- section of a hexagonal bar of side 5 cm is shown. A load on the bar which is simply supported subjects it to a shear of 900 kN. Deduce an expression for the variation of shearing stress across the section. What is the value of shearing stress at the neutral axis?

This problem is similar to Illustrated Example 3 but a little more complicated. The procedure followed was however the same. Fig. 2 was drawn sufficiently large for you to follow the steps.

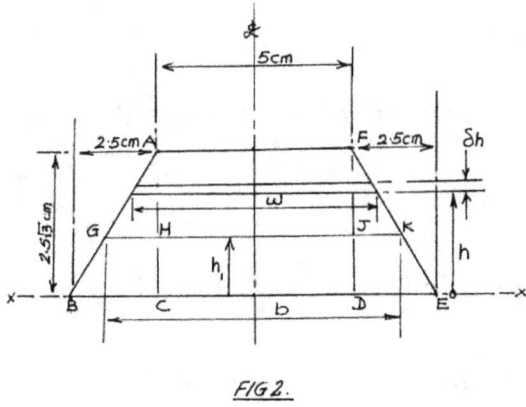

FIG 2.

As before we assume the relationship for shearing stress, viz.

$$\tau = \frac{V}{1b} Ah$$

The cross- section is symmetrical about its centre- line and 'XX'. Axis 'XX' is the neutral axis and 'I' in the above relationship is I_{xx} for the entire cross- section.

Evidently

I_{xx}= 2 (I_{xx} Rectangle ACDF + I_{xx} Triangles ABC, DEF)

$$= 2 \left\{ 5 \frac{(2.5\sqrt{3}^3)}{3} + \frac{2.5(2.5\sqrt{3}^3)}{12} + \frac{2.5(2.5\sqrt{3}^3)}{12} \right\}$$

$$= 2(135.3 + 33.8)(cm^2)$$

$$= 338.2 \ (cm)^4$$

Let width of cross- section 'h_1' from 'XX' = b

$$b = 5 + 2(GH)$$

but

$$= \frac{GH}{2.5} = \frac{2.5\sqrt{3} - h_1}{2.5\sqrt{3}}$$

so that

$$b = 5 + 2 \frac{(2.5\sqrt{3} - h_1)}{\sqrt{3}}$$

i.e. $\quad b = \frac{2(5\sqrt{3} - h_1)}{\sqrt{3}} \ cm$

By a similar token, width 'w' of infinitesimal area 'h' from neutral axis is:

$$w = \frac{2(5\sqrt{} - h_1)}{\sqrt{3}} \ cm$$

Now $\tau = \frac{V}{1b} Ah$ in specific terms here, is in fact

$$\tau = \frac{V}{1_{xx} b} \int_{h_1}^{2.5\sqrt{3}} whdh$$

Putting shear force V = 950 kN = 950 x 1000 N

$$\tau = \frac{950 \times 1000}{338.2} \times \frac{\sqrt{3}}{2(5\sqrt{3} - h_1)} \int_{h_1}^{2.5\sqrt{3}} \frac{2}{\sqrt{3}} \left(5\sqrt{3} - h\right) dh$$

which expression, after some mathematical weariness of the flesh reduces to

$$\tau = \frac{2809}{(5\sqrt{3} - h_1)} \left\{ 28.125 - 5\sqrt{3}(h_1) - \frac{h_1^2}{2} \right\}$$

At the neutral axis $h_1 = 0$, so that

$$\tau_{xx} = \frac{2809}{5\sqrt{3}} (28.125) \; N/(cm)^2$$

i.e. $\tau_{xx} = 9123 N/(cm)^2$

Assignment

Use the result obtained for the variation of shearing stress across the intervals of 2.5cm on both sides of the neutral axis and plot a graph of your results using appropriate linear and stress scales. Compare the shape of the graph with that obtained for Q3.

Q6. Two 20cm x 8 cm steel channels are placed back to back with their top flanges connected by a plate 3cm thick and 24cm wide. This compound section forms the structure of a 1.5 long beam freely supported at its ends and carrying a uniformly distributed load of 50 kN/m run. If the flange is connected to the plate by 2cm diameter rivets, calculate the shear flow between the plate and the channels. Take I_{max} for each channel = 7200(cm)^4 and the area of each channel = 45(cm)^2.

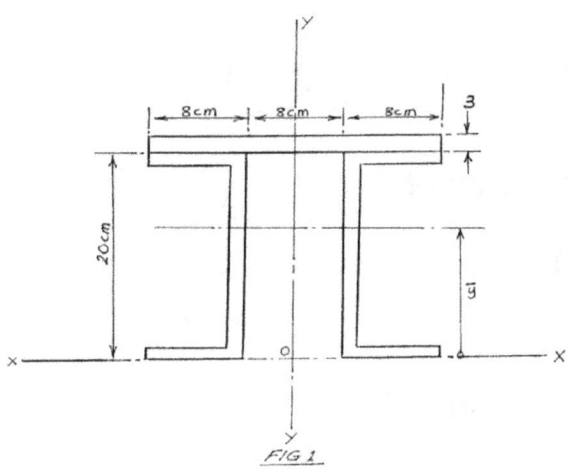

FIG 1

The structural combination is shown in Fig. 1. The first task is to locate the neutral axis (N·A). Let N·A be $\breve{y}$ from the X– X axis, the latter being at the bottom of the channels. See diagram.

Cross- sectional area (c.s.a), say, A for the whole combination

= c s a for plate +c s a for 2 channels = 24(3) + 2(45) = 162(cm)²

Accordingly,

$$\breve{y} = \frac{24(3)(20-1.5)+2(45)(10)}{162}$$

$$= \frac{1332 = 900}{162}$$

$$\breve{y} = 13.8 \ cm$$

In words, the neutral axis is distant 13.8 cm from X-X or (20 + 3 – 13.8) cm or 9.2 cm from the top of the plate.

The next step is to determine the moment of inertia for the entire figure about the neutral axis.

We are given that I about an axis through its axis of symmetry parallel to

X-X = 7200 (cm)⁴. Therefore for the two channels, employing the given data and the Parallel Axis Theorem, we have for the channels

$I_{NA} = 2 \{7200 + 45 (13.8 – 10)^2\}$

= 2 (7200 + 649.8)

= 15700 (cm)⁴

and for the plate

$$I_{NA} = \frac{24 \times (3)3}{12} + 24(3) \ (9.25 – 1.5)^2$$

= (54 + 4263)(cm)⁴

= 4323(cm)⁴

Total I = (15700 + 4323)(cm)⁴

= 20023(cm)⁴

Given udl = 50 kN/metre run. Therefore total load on beam = 75kN. Accordingly each support reaction = 37.5kN, so that the shearing force = 37.5kN.

We need to work out the value of shear flow 'q'.

It was shown that

$$q = \frac{V}{1}\acute{A}h \text{ where}$$

A = area of plate above junction with channels. Here, V = 37.5kN; I= 20023 (cm)^4 or 20023 x 10^{-8} (m)^4 A = 24 x 3 = 72 (cm)^2 = 72 x 10^{-4} m^2, and ħ is the distance from the axis through the centroid of the plate to the neutral axis of the entire structural combination, i.e. (9.22 − 1.5) cm = 7.72 cm.

Working in kN and metres

$$q = \frac{37.5}{20023} \; x \; 10^6 \; x \; \frac{72}{10^4} \; x \; \frac{7.72}{100}$$

i.e. $\qquad q = 104 \; kN/m$

Thus we can determine the shear stress in the rivets.

Q7. A steel beam has a T- section as shown in Fig.1 . Given that I through the centroid of the section = 5660(cm)^4, find the shearing stress 'τ' at the levels indicated in the diagram, assuming the beam is subjected to a vertical shear force of 200kN throughout. Note one of the levels is the Neutral Axis (N.A). Plot the variation of 'τ' across the section.

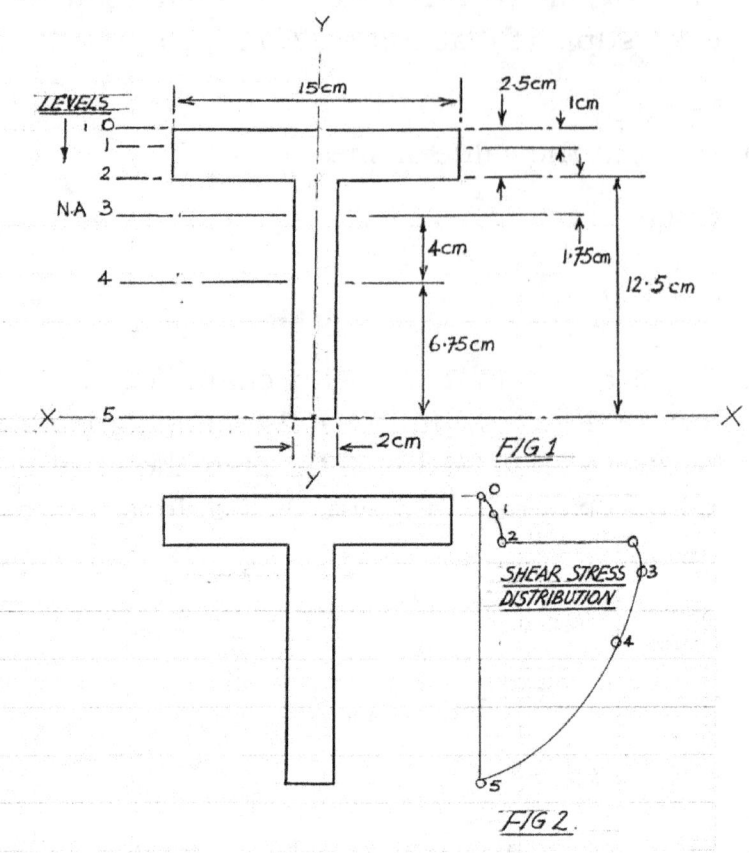

FIG 1

SHEAR STRESS DISTRIBUTION

FIG 2.

Let $\breve{y}$ = distance of centroid from x – x axis

$$\breve{y} = \frac{15(2.5)(12.5+1.25)+12.5(2)(6.25)}{15(2.5)+12.5(2)}$$

$$= \frac{515.625+156.25}{62.5}$$

$\breve{y}$ = 10.75cm from x –x axis or 4.25cm from top surface of flange

Now $\tau = \frac{V}{1b}\,A\bar{h}$

$\frac{V}{I}$ is a constant so let us evaluate it separately

Here V = 200 x 1000 N ; I = 5660(cm)^4

$$\frac{V}{I} = \frac{200 \ x \ 1000}{5660} = 35.3 \ N/(cm)^4$$

The data to be used for the plot are best compiled in a tabulation such as I have shown below. The distribution is shown plotted in Fig. 2.

Level	Statical Area (cm)^2	$\bar{y}$ cm	A$\bar{y}$ (cm)^3	b cm	$\tau = 35.3 \dfrac{A\bar{y}}{b}$ N/(cm)^2
0-0	0	4.25	0	15	0
1-1	15 x 1 = 15	3,75	56,25	15	133.4
2-2	15 x 2.5 = 37.5	3	112.5	15 2	266.3 1985.6
3-3 (Neutral Axis)	15 x 2.5 = 37.5 2 x 1.75 = 3.5	3 0.875	112.5 + 3.0625 115.5625	2	2039.6
Levels 4 and 5 are below N.A.					
4-4	6.75 x 2 = 13.5	7.375	99.5625	2	1757
3-3	4 x 2 = 8	16	99.5625 + 16 115.5625	2	2039.6

Q8. By considering an infinitesimal element on the cross-section of a solidrect-angular beam shown in Fig.1 show that the expression for shearing stress distribution across it is given by $\tau = \dfrac{6V}{bd^3}\left(\dfrac{d^2}{4} - h_1^2\right)$ in which V = vertical shearing force on the beam and h_1 is the distance from the neutral axis of the cross- section.

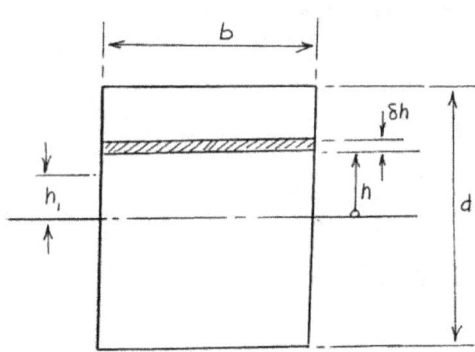

FIG 1

Q9. A solid beam of square cross-section is subjected to a vertical shear 'V'. Derive an expression for the shear stress distribution over a cross-section of the beam. See Fig. 1

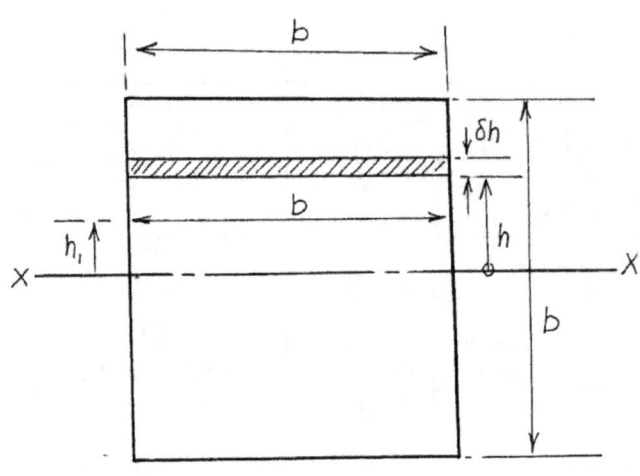

<u>FIG 1</u>

Considering the infinitesimal area $b\delta h$

$$\tau = \frac{V}{Ib} \int_{h_1}^{b/2} bdh.h$$

$$= \frac{V}{Ib} \int_{h_1}^{b/2} hbh$$

Putting $I_{xx} = \frac{b^4}{12}$ and integrating

$$\tau = \frac{12V}{b^4} \left[\frac{h^2}{2}\right]$$

$$i.e. \, \tau = \frac{6V}{b^4} \left[\frac{b^2}{4} - h_1^2\right]$$

This is the requested relationship.

Writing $A = b^2$

At $h_1 = 0, \tau = \frac{3V}{2A}$

At $h_1 = \frac{1}{8}b, \tau = \frac{45V}{64A}$

At $\quad h_1 = \frac{1}{4}b, \tau = \frac{9V}{8A}$

At $\quad h_1 = \frac{1}{2}b, \tau = 0$

Q10. The internal diameter of a thin- walled circular cylindrical steel tube of density 'ρ' tapers uniformly from radius 'R' at one end to R/2 at the other as in Fig. 1. The tube is freely supported at its ends over a length 'L'. Given that $\tau_{\max} = k\tau_{\text{average}}$ calculate the maximum shearing stress in the tube due to its self weight.

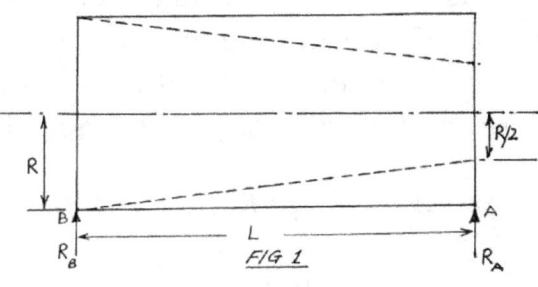

FIG 1

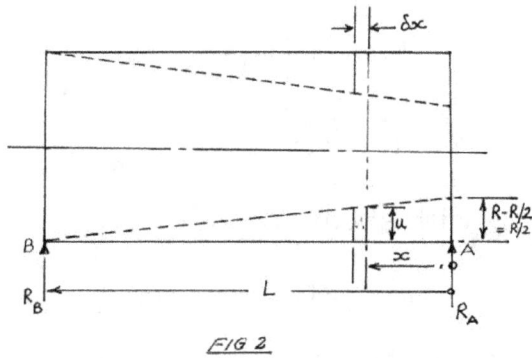

FIG 2

[Note: In Figs. 1 and 2 the wall thickness was exaggerated for the purpose of illustration. But remember wall thickness 't' is described as thin.]

First of all let us determine the radius of the tube at distance 'x' from the larger end.

Referring to Fig.2

$$\frac{u}{R-\frac{R}{2}} = \frac{L-x}{L}$$

$$U = (L-x)\frac{R}{2}, \text{ from which } R_x = R - U = \frac{R(L+x)}{2L}$$

so that

$$R_x = \frac{R}{2} - u$$

$$= R\left(\frac{L+x}{2L}\right)$$

Now let us calculate the self weight of the tube. Let t = thickness of the tube.

Therefore at section distance 'x' from larger end, an elemental volume = $2\pi R_x.t.\delta x$, δx being the infinitesimal length of the element. Accordingly we may write $\delta V = 2\pi R_x t \delta x$ and self weight of this volume = $2\rho\pi R_x t \delta x$, where ρ = density. Therefore total weight, say, W of the tube is:

$$W = \int_0^L 2\rho\pi R_x\, t dx$$

$$= \int_0^L 2\pi\, t\, R\left(\frac{L+x}{2L}\right)$$

$$= 2\rho\pi\, \frac{tR}{2L}\left[Lx + \frac{x^2}{2}\right]^L$$

i.e.

$$W = \rho\, \frac{\pi tR}{L}\left(\frac{3L^2}{2}\right)$$

or

$$\pi\rho tLR = \frac{2W}{3}$$

Let us start dealing with shear. We are given

$$\tau_{max} = k\tau_{average} = k\tau_{av}$$

But at end A for average shear

$$\tau_{av} = \frac{R_A}{2\pi Rt}$$

and

$$\tau_{av} = \frac{R_B}{2\pi\frac{R}{2}}t = \frac{R_B}{\pi Rt} \quad \text{at end B}$$

Similarly, for maximum shear

$$\tau_{max} = \frac{kR_A}{2\pi Rt} \quad \text{at end A}$$

~ 296 ~

$$\tau_{max} = \frac{kR_B}{\pi Rt} \qquad \text{at end B}$$

Referring to Fig.1 and taking moments about B

$$R_A L = \int_0^L 2\rho\pi R_x . t dx (L - x)$$

where $\int 2p\pi R_x \, t \, dx \, (L - x)$ is the moment of the elemental self- weight portion of the tube, such portion being $(L - x)$ from B.

Accordingly,

$$R_A L = 2\rho\pi \, t \int_0^L \frac{R}{2}\left(\frac{L+x}{L}\right)(L-x)dx$$

$$= \frac{\rho\pi \, t \, R}{L} \int_0^L (L^2 - x^2)dx$$

$$= \rho\pi \, t \, R \left(\frac{2L^2}{3}\right)$$

so that $\qquad R_A = \frac{2L}{3} . \rho\pi \, tR$

but $\qquad \rho\pi \, tRL = \frac{2}{3}W$

so that, $\qquad R_A = \frac{2}{3} . \frac{2W}{3}$

i.e $\qquad R_A = \frac{5}{9}W$

so that τ_{max} at A $\qquad = k \, \tau_{av} = k.\frac{R_A}{2\pi\frac{Rt}{2}} = \frac{4W}{9\pi Rt}$

It is evident that at B, $R_B = \frac{5W}{9}$

therefore $\qquad \tau_{max}$ at B $\quad = k.\frac{5W}{9\,(2\pi\,Rt)}$

$$= \frac{5\,kW}{18\pi\,Rt}$$

Therefore the maximum shear stress occurs at end A.

Q11. A beam has a cross- sectional area in the form of an isosceles triangle for which thebase'b' is equal to one- half its height 'h'. Show that the maximum shearing stresscaused by a vertical shear occurs at a distance one- half of the height above the base. (Popov)

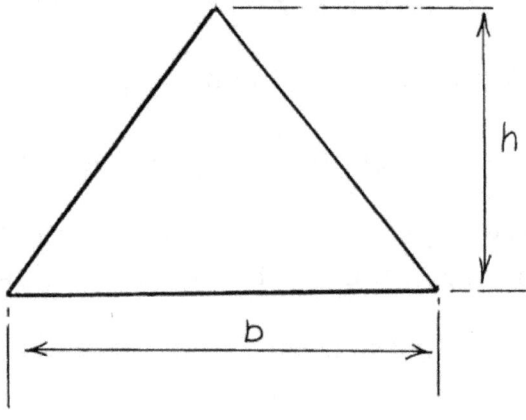

<u>FIG1</u>

Comment:

Follow the procedure demonstrated in Problems 4 and 5 and" show them what you have got".

Q12. Deduce the relationship giving the shearing stress distribution over the cross- section of a thin- walled aluminum tube of uniform section with radius 'R'. The tube is used as a freely-supported beam and carries a concentrated load 'W' at mid- span. Neglect self- weight of beam. What is the maximum shearing stress and where does it occur?

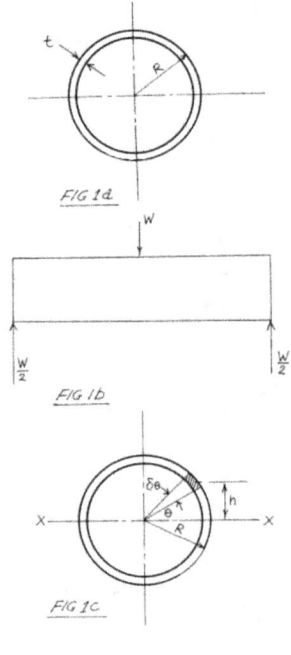

FIG 1d

FIG 1b

FIG 1c

Employing the relationship $\tau = \frac{V}{1b} A\bar{h}$

and noting that $V = \frac{W}{2}$; and I_{xx} for a thin- walled tube $= \pi R^3 t$

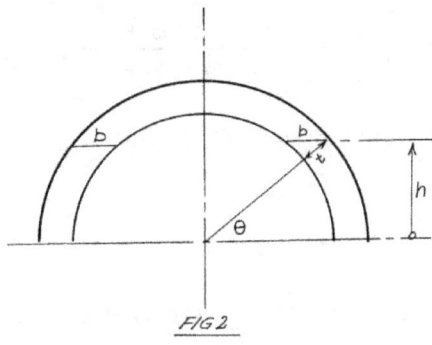

FIG 2

Referring to Fig. 2, the width to be considered 'h' above xx may be approximated using $\frac{t}{b} = Cos\ \theta$ or $b = \frac{t}{Cos\ \theta}$

therefore $\qquad$ Total $\quad b = \frac{2t}{Cos\ \theta}$

For the elemental area shown shaded in Fig.1c we may write

$$\tau = \frac{V}{1b} \int dAh$$

$$\frac{h}{R} = Sin\ \theta, and$$

$$\delta A = R\delta\theta.t$$

therefore $\qquad \delta Ah = R\delta\theta.t.RSin\ \theta$

$$\therefore\ \delta Ah = R^2 t\ Sin\theta d\theta$$

so that for the whole tube above X-X

$$A\bar{h} = 2 \int_\theta^{\frac{\pi}{2}} R^2\ tSin\ \theta d\theta$$

$$= 2R^2 T\left[-Cos\theta_\theta^{\pi/2}\right]$$

$$= 2R^2\ tCos\ \theta$$

therefore
$$\frac{Va\bar{h}}{1b} = \frac{V}{\pi R^3 2t} Cos\,\theta.2R^2t\,Cos\,\theta$$

$$= \frac{V\,Cos^2\theta}{\pi\,Rt}$$

But the cross-sectional area of the thin tube = $2\pi\,Rt$

therefore
$$\tau = \frac{2\,V\,Cos^2\theta}{csa\,of\,tube = A_c}$$

say,
$$\tau = \frac{2V}{A_c}Cos^2\theta$$

from which
$$\tau_{max} = \frac{2V}{A_c}$$

Here is a nice one for you to try your hand at, if you would pardon the prepositional ending.

Q13. The cross-section of a box beam is shown in Fig. 1. The beam is freely supported on knife edges and is subjected to a vertical shear force of 100kN. What is the maximum shearing stress induced in the beam?

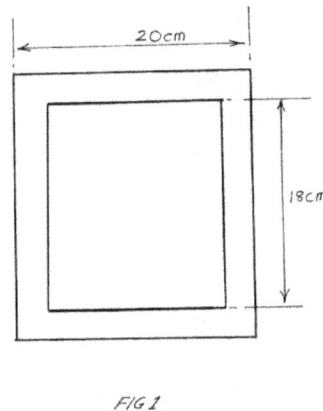

20cm

18cm

FIG 1

Q14. Assuming that the allowable shear stress for purple heart wood (Sapater) is 6650kN/m², determine the capacity of a beam having a solid cross-section 12cm x 30cm to resist a vertical shear when placed with (i) the 30cm side vertical; and, (ii) the 12cm side vertical.

In Q.8 you would have shown that for a rectangular beam of depth d,

$$\tau = \frac{V}{2I}\left\{\left(\frac{d}{2}\right)^2 - h_1^2\right\}$$

Evidently τ_{max} occurs where $h_1 = 0$

therefore $\qquad\qquad\qquad\qquad \tau_{max} = \dfrac{Vs^2}{8l}$

When the 30cm side is placed vertical $I_{xx} = \dfrac{12 \times (30)^2}{12}$ (cm)⁴ where I_{xx} = moment of inertia with reference to the neutral axis. Accordingly I_{xx} = 27000(cm)⁴

$$\tau_{max} = \frac{V.(30)^2}{8(27000)}$$

$$\tau_{max} = \frac{V \ (cm)^2}{240 \ (cm)^4}$$

If $\tau_{max} = 6650\text{kN/m}^2$, then let's see what kind of shearing force the beam can withstand.

$$\tau_{max} = \frac{6650}{10000} \ kN/cm^2$$

therefore $\qquad \dfrac{0.665 \ kN}{(cm)^2} = \dfrac{V}{240} \dfrac{1}{cm^2}$

i.e. $\qquad\qquad\qquad\qquad V = 159.6 \text{ kN}$

Accordingly a concentrated load of 319.2kN has to be placed at the mid-point of the simply supported beam to induce a maximum shearing stress of 6650 kN/m². With the 12cm side vertical,

$$I_{xx} = \frac{30 \ (12)^3}{12} \ (cm)^4$$

i.e. $\qquad\qquad\qquad I_{xx} = 4320 \ (cm)^4$

The corresponding τ_{max} is given by

$$\tau_{max} = \frac{V.(12)^2}{8(4320)}$$

$$= \frac{V}{240} \frac{1}{(cm)^2}$$

As before $\qquad\qquad\qquad 0.665 = \dfrac{V}{240}$

from which $\qquad\qquad\qquad V = 159.6 \text{ kN}$

In other words it makes no difference how the beam is oriented. Why is this so?Because:

$$\tau_{max} = \frac{Vd^2}{8l}$$

$$= \frac{Vd^2}{8.bd^3} \cdot 12$$

$$= \frac{3V}{2\,bd}$$

$$\tau_{max} = \frac{3}{2A}$$

Clearly because cross-sectional area remains the same no matter how you orient the beam τ_{max} will always be the same.

Q15 Prove that the direction of the shearing stresses at a section through a beam is the same as that of the shearing force at that section.

CHAPTER 10

ELASTIC STRESS ANALYSIS OF THIN-WALLED CYLINDERS AND SPHERICAL SHELLS UNDER INTERNAL PRESSURE

The terms 'shells' and 'thin-walled' in the title of this Chapter refer to vessels which are designed, invariably for the containment of fluid under internal pressure as opposed to external pressure, and in respect of which the weight of the fluid contained is itself disregarded, such vessels being manufactured from steel plates or curved sheet steel. Three notable examples of these engineering structures which are, and continue to be of great practical and economic importance in industry are:

(i)   air tanks served by the air compressors found in most modern engineering workshops;
(ii)   boiler drums; and
(iii)   pressure vessels and piping used in chemical processing and petroleum-refining installations.

Another example is the well-head separator used in oil-well production. This is a vessel designed to withstand very high internal pressures. Typical separators are 1.5m diameter in the main section and anything like 20m-25m in length overall; wall thickness – still small in comparison with vessel diameter – could be anywhere in the range of 30mm-35mm; and working pressures in the vicinity of 50 bars. The design of these vessels is based on design codes laid down by such professional bodies as Lloyd's Register of Shipping, American Society of Mechanical Engineers, American Petroleum Institute and the British Standards Institution. The basic formulae derived in what follows here are simply just that: basic. There are many coefficients which have to be incorporated in these formulae to allow for such things as corrosion wastage or erosion, operating temperature, type of dished end- plates, type of welds and a host of other conditions. If you choose to make a career as a designer or inspector of pressure vessels then you would need to consult the relevant design codes.

Thin-walled, means that skin-or wall-thickness is small in comparison with diameter, the implication being that the wall of the vessel may be treated as a membrane. Thus, in these vessels the stresses in their walls are assumed to be constant across wall thickness. By contrast, thick-walled vessels including some pipes of which a gun barrel is a good example, have such a pronounced variation of stress across wall thickness that such variation cannot be disregarded.

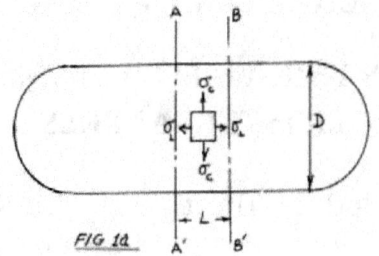

FIG 1a

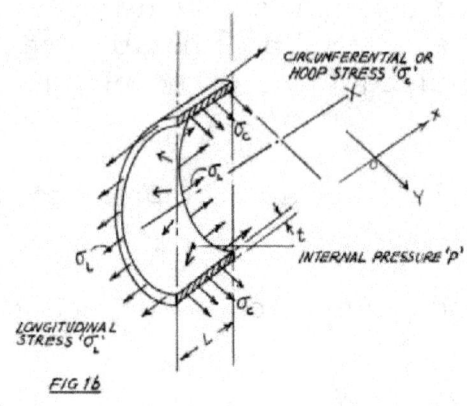

CIRCUMFERENTIAL OR
HOOP STRESS 'σ_c'

$σ_c$

$σ_c$

t

INTERNAL PRESSURE 'p'

$σ_L$

LONGITUDINAL
STRESS 'σ_L'

$σ_c$

FIG 1b

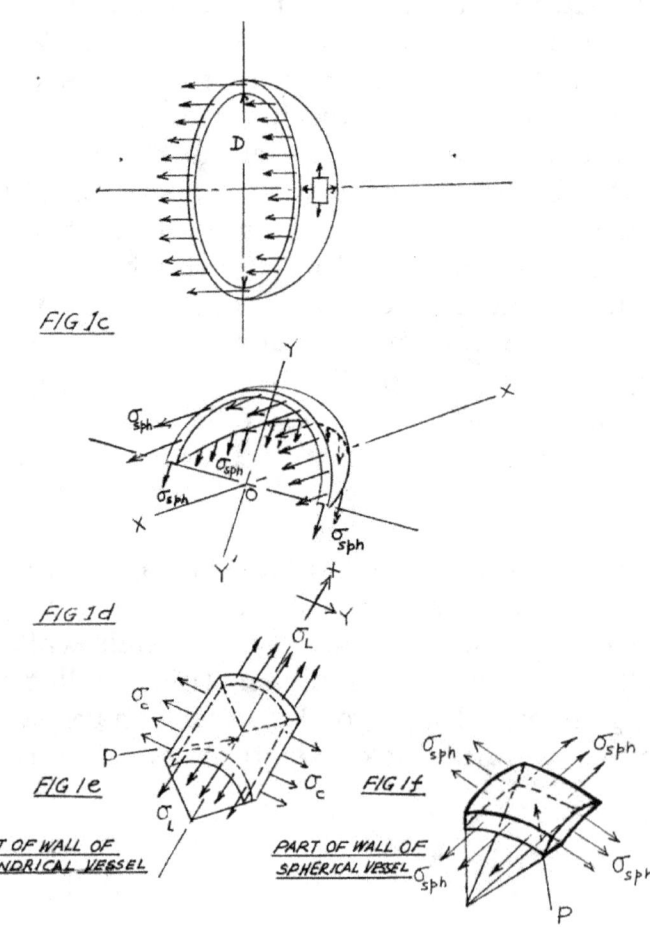

FIG 1c

FIG 1d

FIG 1e

PART OF WALL OF
CYLINDRICAL VESSEL

FIG 1f

PART OF WALL OF
SPHERICAL VESSEL

~ 304 ~

WALL STRESSES

Fig. 1a represents the outline of a horizontal thin-walled circular cylindrical vessel of internal diameter $'D'$ subjected to an internal pressure $'p'$. The major assumptions undergirding the theoretical analysis of stresses in such vessels including thin-walled spheres are:

(i)   no variation of stress across wall thickness;
(ii)   weight of the fluid contained in the vessel is negligible and therefore disregarded; and
(iii)   any stresses due to incompatibility of strain at ends (as in cylindrical vessels) are ignored.

Fig. 1b shows a section of the vessel shown in Fig. 1a. This section is delimited by the planes A-A' and B-B' perpendicular to the axis of the vessel.

Let us determine quickly these stresses by considering a length of $'L'$ of one-half of a circular cylindrical vessel of diameter $'D'$ and wall thickness $'t'$. See Fig. 1b.

Assuming a gauge pressure of $'p'$ within the cylinder, it is seen that for horizontal equilibrium, i.e. for $\Sigma F_x = 0$,

[Force in wall of one-half of cylinder   =   Force due to internal (gauge)
due to longitudinal stress, $'\sigma_L'$ acting   pressure on projected circular
on area, $\dfrac{\pi Dt}{2}$ area, in x-x direction $\dfrac{1}{2} \dfrac{\pi D^2}{4}$]

i.e. $\dfrac{\sigma_L Dt}{2} = \dfrac{p\pi D^2}{8}$

because we considered only one-half of a complete cross-section.

Accordingly,

i.e. $\sigma_L(2lT) = \dfrac{pD}{4t}$

Referring to the same diagram and this time dealing with circumferential or hoop stress σ_c, we have for $\Sigma F_y = 0$.

[Force in wall of cylinder due to   =   Force due to internal gauge
due to circumferential or hoop stress,   pressure $'p'$ on projected area:
σc acting on area: $Lt + Lt = 2Lt$ $'LD'$]

i.e. $\sigma_C(2lT) = pLD$

from which
$$\sigma_C = \frac{pD}{2t}$$

Thus, circumferential or hoop stress 'σ_c' often times designated 'σ_h' in this text is twice the longitudinal stress σ_L. It is easy to extend the foregoing analysis to the case of a spherical pressure vessel. Refer to Fig. 1c. Just simply consider any section through the sphere separated by a plane passing through the centre of the sphere. For a wall stress of say 'σ_{hsp}, sphere diameter 'D' and wall-thickness 't', we have for equilibrium in a direction perpendicular to any such plane:

$$'\sigma_{hsp} \pi Dt = \frac{p\pi D^2}{4}$$

from which
$$\sigma_{hsp} \pi = \frac{pD}{4t}$$

which is the value of the stress in a thin-walled spherical vessel: $\sigma_{hsp} = \sigma_{spn}$ in Fig. 1e. It should be noted that in deriving the equations for longitudinal stresses and hoop stresses for thin cylindrical vessels, the condition of symmetry of loading on the sections considered implied that no shearing stresses were acting in the walls of the vessel; only the stresses σ_c and σ_l. Hence σ_c and σ_L are called principal stresses. However, because the cylindrical portion of a cylindrical pressure vessel expands a different amount from that of the closed ends in the region of the junction between the two, use of flat ends would only exacerbate local bending and shearing stresses. In order therefore to mitigate these effects, the ends of cylindrical pressure vessels are always curved. I used the word 'mitigate' advisedly because even with hemispherical ends, shearing forces and bending moments still act at the joint of cylinder and hemi-sphere to produce what are called 'discontinuity stresses'. If you continue with your studies in this field you will meet discontinuity stresses later on when you consider the theory of plates and shells.

When I was a regional technical trainee with that great company UBOT Limited, Point Fortin prior to proceeding on scholarship to the United Kingdom, first to study at the famous Loughborough College of Advanced Technology (now Loughborough University) and later at the Imperial College of Science and Technology (now Imperial College of Science Technology and Medicine), it was not uncommon to come across during my intensive training in the refinery and exploitation engineering departments, pressure vessels seamed by treble-riveted double-strapp butt joints, i.e. two plates or straps of steel - one on the inside of the vessel the other on the outside – evenly spaced on both sides of the butt joint formed from the plates of the pressure vessel, "butting" against each other; both straps taking 3 rows of rivets on each side of the joint.

It is however very unlikely that a modern-day thin-walled vessel would be manufactured with riveted joints. If a vessel is seamless, assuming 'σ_w' is the safe working stress, then the permissible working pressure 'p' is obtained by a straightforward consideration of the following obtained from our earlier derivations:

$$\sigma_w = \frac{pD}{2t}$$

i.e. $\quad p = \frac{2t\sigma_w}{D}$

$$\sigma_w = \frac{pD}{4t}$$

i.e. $\quad p = \frac{4t\sigma_w}{D}$

the specific value of 'p' to be chosen being the lesser of the two. If however the vessel has welded seams or welded joints where the cylinder meets the hemispherical ends, then efficiency factors are generally used in which case the safe working pressure is the smaller of the following two expressions.

$$p = \frac{4t\sigma_w\eta_c}{D}$$

$$p = \frac{2t\sigma_w\eta_L}{D}$$

in which η_c = efficiency of the circumferential seam weld; and η_L = efficiency of the longitudinal seam weld.

UNIT OF PRESSURE

The base unit of pressure in SI is the pascal abbreviated *Pa; 1Pa = 1N/m²*. Because this unit is extremely small, the International Standard Organisation (ISO) decided to allow the 'bar' to be used as a permitted unit, it being approximately 1 atmosphere; 1 bar = *10⁵N/m² = 10⁵Pa*.

STATE OF STRESS

As shown in Figs. 1d and 1e, a state of triaxial stress actually exists in the walls of the cylinder and sphere. However, because the internal pressure 'p' which is in fact a compressive stress, is negligible in comparison with σ_1, σ_2 (cylinder) and $\sigma_1 = \sigma_2 = \sigma_{spn}$ (sphere), it is disregarded:

$\frac{(D)}{T}$ in the relation between σ_1, σ_2 is large, therefore $\sigma_1, \sigma_2 >> p$.

Thus, for most practical purposes it is generally taken that the state of stress on the inside surface of these vessels is biaxial, the stresses σ_1 and σ_2 being principal stresses: at right angles to each other with no shear stress on either plane on which σ_1 and σ_2 act; σ_1 is the maximum stress and σ_2 the minimum stress. As was said before 'p' is disregarded which means that in considering the criterion of failure on the basis of maximum shear stress we must be careful to remember that the stress condition $\sigma_1 > \sigma_2 >> (p=0)$ implies that the maximum shear stress is not $(\sigma_1 - \sigma_2)/2$ as one might suppose, but rather $(\sigma_1 - 0)/2$ or $\sigma_1/2$ in the $(\sigma_1 - p)$ – plane:

CIRCUMFERENTIAL, DIAMETRAL AND LONGITUDINAL STRAIN

Let $\delta D =$ increase in diameter (D) of cylinder or sphere under internal pressure

$\therefore \quad \dfrac{\delta D}{D} =$ diametral strain

New diameter $\quad\quad\quad\quad\quad = D + \delta D$
New circumference $\quad\quad\quad = \pi(D + \delta D)$
$\therefore$ increase in circumference $\quad = \pi(D + \delta D) - \pi D = \pi\delta D$

and so circumferential strain $\quad = \dfrac{\pi\delta D}{\pi D} = \dfrac{\delta D}{D}$

which means that circumferential strain = diametrical strain.
We could also arrive at the same conclusion by writing circumference, say,
$C = \pi \, x$ diameter (D) i.e. $C = \pi D$

Taking logarithm of both sides
$Log\,C = Log\,\pi + Log\,D$

And differentiating to obtain
$$\frac{dC}{C} = \frac{dD}{d}$$
For the assumed biaxial state of stress in a thin-cylinder wall containing pressure 'p', Hooke's Law reduces to:

$$e_c = \frac{1}{E}(\sigma_c - v\,\sigma_2)$$

$$e_L = \frac{1}{E}(\sigma_L - v\,\sigma_1)$$

In which $\quad\quad\quad \sigma_c =$ circumferential stress
$\quad\quad\quad\quad\quad\quad e_c =$ circumferential strain
$\quad\quad\quad\quad\quad\quad \sigma_L =$ longitudinal stress
$\quad\quad\quad\quad\quad\quad e_L =$ longitudinal strain
$\quad\quad\quad\quad\quad\quad v =$ Poisson's ratio

substituting $\quad\quad\quad \sigma_c = \dfrac{pD}{2t}$

$$\sigma_L = \frac{pD}{4t}$$

$$e_c = \frac{pD}{4t}(2 - v)$$

$$e_L = \frac{pD}{4t}(1 - 2v)$$

So that
$$\frac{e_c}{e_L} = \frac{(2-v)}{(1-v)}$$

Consider now the case of a thin-walled circular cylindrical vessel with hemispherical ends.

At the junction of cylinder and hemisphere, the circumferential strain on the cylinder side is

$$e_c = \frac{1}{E}(\sigma_c - v\sigma_2)$$

In which σ_c, σ_L are the principal stresses and 't' the wall thickness of the cylinder

$$\therefore \qquad e_c = \frac{pD}{4tE}(2 - v)$$

Designating the thickness of the hemisphere's wall to be t_{hsp}, then the circumferential strain on the hemisphere side is

$$e_{sph} = \frac{1}{E}(\sigma_{sph} - v\sigma_{hsp})$$

i.e
$$= \frac{pD}{4t_{hsp}E}(1 - v)$$

Therefore
$$\frac{e_c}{E_{hsp}} = \frac{(2-v)}{t}\frac{(t_{hsp})}{1-v} = \frac{(2-v)}{1-v}\frac{t_{hsp}}{t}$$

Evidently when
$$t = t_{hsp}, \frac{(e_c)}{e_{hsp}} = \frac{(2-v)}{1-v}$$

Therefore with the same wall thickness in cylinder and hemisphere there will be unequal deformation at the junction, resulting in local bending and shearing stresses in that region.

In order that there be equality of strains at the junction of cylinder and hemisphere i.e. for $\dfrac{e_c}{e_{hsp}} = 1$

i.e
$$\frac{(2-v)t_{hsp}}{1-v} \text{ must be equal 1}$$

so that
$$\frac{(t)}{t_{sph}} = \frac{(2-v)}{1-v}$$

Because Poisson's ratio is always less than 1, this means that to achieve equality of strains the thickness of the cylinder wall must be thicker than that of the wall of the hemisphere by the ratio $\frac{(2-v)}{1-v}$.

This means a reduction in overall design working-pressure to that capable of being contained safely by the hemispherical ends.

VOLUMETRIC STRAIN

The conduct of hydraulic-pressure testing of boilers and other pressure vessels is generally speaking a statutory requirement internationally. In such tests the vessels are filled with water at atmospheric pressure and extra water is pumped in. The increase in volume of the vessel beyond a specific limit at a particular pressure is interpreted as evidence of a safe or of a defective vessel. Two things are of interest here. First, is that under the influence of internal pressure the vessel itself undergoes an increase in its physical dimensions. Second, is the contraction in the volume of the water produced by the hydraulic pressure intensity of the test.

Thus, the total volume of water pumped into a test vessel comprises an amount equal to the increase in volume of the vessel plus the amount necessary to make up the volume by which the water itself would have shrunk, as it were, due to the imposed hydraulic pressure on it.

Determination of the latter volume involves a knowledge of the Bulk Modulus of the water. This modulus often designated by the letter 'K' is defined as follows:

$$K = \frac{Hydraulic\ Pressure}{Volmeteric\ Strain}$$

Therefore if a certain volume of water 'V_w' shrinks by an amount dV_w when a hydraulic pressure 'P' is applied, then

$$K = \frac{P}{dV_w/V_w}$$

i.e.
$$dV_w = \frac{PV_w}{K}$$

What is the volumetric strain of a thin- walled cylinder under internal pressure? If its internal diameter 'D' increases by 'δD' and its length 'L' by δL, then,

increase in volume $= \frac{\pi}{4}(D + \delta D)^2(L + \delta L) - \frac{\pi D^2}{4}L$

$$= \frac{\pi}{4}\{D^2 + 2D(\delta D) + (\delta D)\}^2 (L + \delta L) - \frac{\pi D^2}{4} L$$

$$= \frac{\pi}{4}\{D^2L + D^2(\delta L) + 2D (\delta D)L + 2D (\delta D)(\delta L)\}$$

$$= + (\delta D)^2L + (\delta D)^2\delta L - \frac{\pi D^2 L}{4}$$

Neglecting second- order small quantities, the increase in internal volume = $\frac{\pi}{4} \{d^2\delta L + 2D (\delta D)L\}$ therefore Volumetric strain = $\frac{increase\ in\ volume}{original\ volume}$

or
$$\frac{dV}{V} = \frac{\pi}{4} \left\{ \frac{D^2\delta L + 2D(\delta D)L}{\frac{\pi}{4}D^2L} \right\}$$

therefore
$$\frac{dV}{V} = \frac{\delta L}{L} + \frac{2\delta D}{D}$$

and so
$$dv = V \left\{ \frac{\delta L}{L} + \frac{2\delta D}{D} \right\}$$

or, in words dV= Original volume (longitudinal strain + twice diametral strain)

Note again that the result obtained in the foregoing for volumetric strain could have been obtained quite easily by writing

$$V = \frac{\pi D^2}{4} L$$

from which
$$Log\ V = Log\ \frac{\pi}{4} + 2Log\ D + Log\ L$$

so that
$$\frac{dV}{V} = 2\frac{dD}{D} + \frac{dL}{L}$$

The student is left to show that the volumetric strain of a sphere is given by:

$$\frac{dV}{V} = 3\frac{dD}{D}$$

Illustrative Example 1

An air vessel is fabricated from a piece of seamless steel pipe 250mm internal diameter, 10mm thick. The working stress is limited to 40MN /m², and the efficiency of the circumferential seam welds used to fit 'caps' to seal the cylinder is taken as 50%. What is the maximum working pressure?

There is no longitudinal welding, the original stock being seamless. Accordingly,in considering circumferential stress we have to regard the material as being at full strength i.e. η_L = 100%

Therefore $\qquad\qquad\qquad\qquad \sigma_c \dfrac{pD}{4t}$

or $\qquad\qquad\qquad\qquad p = \dfrac{4t\sigma_c}{D}$

Putting σ_c as the working stress = 40 MN/m²

$$p = 4 \, x \, \frac{10}{100} \, x \, \frac{40 \, x \, 10^6}{\left(\frac{250}{1000}\right)}$$

$$= \frac{40 \, x \, 40 \, x \, 10^6}{250}$$

$$= 6.4 \, x \, 10^6 \ NM/m^2$$

$$\therefore p = 64 \ bars$$

When we consider longitudinal stress, the circumferential welding is regarded as only 50% strong. Putting σ_L = 40 MN/m², so that on the basis of

$$p = \frac{2t\sigma_L\eta_L}{D}$$

$$p = 2 \, x \, \frac{10}{1000} \, x \, \frac{40 \, X \, 10^6}{\frac{250}{1000}} \, x \, 0.5$$

$$= 1.6 \, x \, 10^6 \ N/m$$

$$= 16 \ bars$$

Therefore Maximum Working Pressure = 16 bars.

Illustrative Example 2

A spherical steel vessel 0.75m internal diameter and 10mm thick is being pressure tested hydraulically. The vessel is filled with water at atmospheric pressure and extra water is then pumped in until the gauge pressure is 25 bars. Determine

(i)   the maximum stress in the wall of the sphere;
(ii)   the increase in volume of the sphere; and
(iii)   the extra water that had to be pumped in to achieve the pressure of 25 bars.

Take E for steel = 208 GN/m²; Poisson's ratio ν = 0.25 and Bulk Modulus of water = 3 GN/m².

Maximum stress σ_{sph} in wall of sphere of diameter 'D' and wall thickness 't' at gauge pressure 'p' of 25 bars = pD/4t

therefore

$$\sigma_{sph} = \frac{25 \times 10^5 \times 0.75}{4 \times \left(\frac{10}{1000}\right)}$$

$$= 46.8 \times 10^6 \text{ N/m}^2$$

i.e.

$$\sigma_{sph} = 46.8 \text{ MN/m}^2$$

Increase in volume of the sphere is obtained from : Volumetric strain = 3 (Diametral Strain), since volume 'V' of the sphere may be written in the form

$$V = kD^3.$$

i.e.

$$\frac{dV}{V} = 3\left(\frac{dD}{D}\right)$$

or

$$dV = 3\left(\frac{dD}{V}\right)V \qquad \qquad \dots\dots\dots\dots\dots(i)$$

But $\frac{dD}{D}$ = diametral strain also equal circumferential strain $\frac{dC}{C}$

$$e_c = \frac{1}{E}\left(\sigma_{sph} - v\sigma_{sph}\right)$$

$$= \frac{1}{E}\,\sigma_{sph}(1-v)$$

or

$$e_c = \frac{pD}{4tE}\,(1-v)$$

Putting Poisson's ratio, v = 0.25

$$\frac{dc}{c} = \frac{dD}{D} = \frac{25 \times 10^5 \times 0.75\,(1-0.25)}{4 \times \left(\frac{10}{1000}\right)208 \times 10^9}$$

$$= \frac{25 \times 10^5 \times 0.75 \times 0.75 \times 1000}{4 \times 10 \times 208 \times 10^9}$$

$$\frac{dD}{D} = 16.9 \times 10^{-6}$$

Accordingly, by (i),

$$dV = 3\,(16.0 \times 10^{-6}) \times \frac{\pi D^3}{6}$$

$$= \left\{3 \times 16.9 \times 10^{-6} \times \frac{\pi}{6}.(0.75)^3\right\}m^3$$

$$= 3 \times 16.9 \times 10^{-6} \times \pi \times \frac{(0.75)^3}{6} \times 10^6 \; cc$$

$$= 21.4 \; cc$$

Increase in volume of sphere $= 21.4 \; cc$

Extra water to be pumped in is obtained by considering the 'bulk strain' sustained by the water at 25 bars.

Bulk modulus or volume modulus 'K' is defined as:

$$K = \frac{p}{bulk\ strain} = \frac{p}{dV/V}$$

Here k = 3 GN/m²

therefore

$$\frac{dV}{V} = \frac{25 \times 10^5}{3 \times 10^9}$$

$$= \frac{25}{3} \times 10^{-4}$$

Volume of water 'V' was initially $= \frac{4}{3}\pi r^3 = \frac{\pi D^3}{6}$

$$= \pi \frac{(0.75)^3}{6} \; m^3$$

so that decrease in volume of water, dV

$$= \frac{25}{3} \times 10^{-4} \times V$$

$$= \frac{25}{3} \times 10^{-4} \times \frac{\pi}{6} V \,(0.75)^3 \; m^2$$

$$= \frac{25}{3} \times \frac{1}{10^{-4}} \times \frac{\pi}{6} \times (0.75)^3 \times 10^6 \; cc$$

$$= 184 \; cc$$

Therefore, total extra water pumped in after the spherical vessel was filled with water at atmospheric pressure = Vol. of water to fill the increase in volume of the sphere itself and Vol. of water to make up for its shrinkage under pressure = 21.4 cc + 184 cc , say, 206 cc approximately.

SOLVED AND OTHER PROBLEMS

Q1. A steel water pipe 600mm in diameter has to withstand hydraulic pressure due to a head of 140m of water. What should the thickness of the line be if the working intensity of pressure in the metal of the line is restricted to 75MN /m^2 after the line has lost 1/10 mm of its thickness due to corrosion? Take density of water = 1000kg /m^3 and g ≈ 10m /sec^2 Pressure 'p' due to 140 head of water is obtained from

$$p = \rho gH \qquad \dots\dots\dots\dots\dots\dots\dots(i)$$

Where   p = pressure in N/m^2
                 ρ = density in kg/m^3
                 g = gravitational acceleration ≈10m/sec^2
                 H = Pressure head in m

Substituting values in (i)

$$p = 1000 \times 10 \times 140$$
$$= 1.4MN /m^2$$

Now equating circumferential or hoop stress to the maximum stress intensity of 75 MN/m^2

$$\frac{pr}{t} = 75$$

where   r = pipe radius, mm

            t = pipe thickness, mm

from which $1.4 \, x \, \frac{300}{t} = 75$

or   = 5.6 mm

Because this is the thickness after the pipe has lost 0.1mm due to corrosion, the original thickness should be 5.6 + 0.1 = 5.7 mm.

Ans: 5.7mm

Q2. A cylindrical pressure vessel having hemispherical ends is 1 metre in length overall. Its outside diameter is 200mm and wall thickness is 15mm. See Fig.1. Determine the change in internal volume of the vessel when subjected to internal pressure of 130 bars. Assume E = 210 x 10^9 N/m^2 and Poisson's ratio, v = 0.3. See Fig. 1.

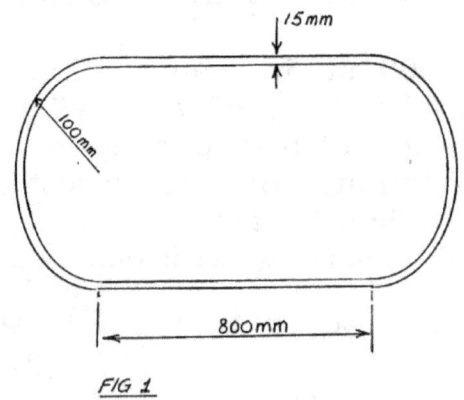

FIG 1

Circumferential strain (e_c) = Diametral strain (e_d).

Now $$e_c = \frac{1}{E}\left(\frac{pt}{t} - v\frac{pr}{2t}\right)$$

or $$e_c = \frac{pr}{tE}\left(\frac{1-v}{2}\right) \qquad \dots\dots\dots\dots\dots(i)$$

Longitudinal strain $$e_L = \frac{1}{E}\left(\frac{pr}{2t} - v\frac{pr}{t}\right)$$

or $$e_L = \frac{pr}{tE}\left(\frac{1-v}{2}\right) \qquad \dots\dots\dots\dots\dots(ii)$$

Equations (i) and (ii) relate to the cylindrical portion of the vessel. Note,

p = internal pressure, N/m²
r = internal radius of cylinder, mm
t = thickness, mm
E = Young's Modulus, N/m²

Still considering the cylindrical portion of the vessel

$$\frac{dV_c}{V_c} = 2e_d + e_L;\ e_d = \text{diametral strain};\ e_L = \text{longitudinal strain}$$

$$= \frac{2pr}{tE}\left(1 - \frac{v}{2}\right) + \frac{pr}{tE}\left(1 - \frac{v}{2}\right)$$

$$= \frac{pr}{tE}\left(2 - v + \frac{1}{2} - v\right)$$

$$= \frac{pr}{tE}\left(\frac{5}{2} - 2v\right)$$

therefore $$dV_c = \frac{pr}{tE}\left(\frac{5}{2} - 2v\right)V_c$$

~ 316 ~

where $\qquad$ V_c = volume of cylinder
$$= \pi r^2 \cdot L$$

$$dV_c = \frac{\pi p r^3 L}{tE} \left(\frac{5}{2} - 2v\right) \qquad \ldots\ldots\ldots\ldots\ldots\ldots(iii)$$

Turning now to the two hemispherical ends, we may combine them into a single sphere and then determine its volumetric strain thus :

$$\frac{dV_{sp}}{V_{sp}} = 3 \text{ (diametral strain of sphere)}$$

Now diametral strain of sphere

$$= 3 \left\{\frac{1}{E}\left(\frac{pr}{2t} - v\,\frac{pr}{2t}\right)\right\}$$

$$= \frac{3pr}{2tE}\,(1 - v)$$

or $\qquad$ $dV_{sph} = \frac{3pr}{2tE}\,(1-v)\,V_{sph}$

$$= \frac{3pr}{2tE}\,(1-v)\cdot\frac{4}{3}\,\pi r^3$$

i.e. $\qquad$ $dV_{sph} = \frac{2p\,\pi\,r^4}{tE}\,(1-v) \qquad \ldots\ldots\ldots\ldots\ldots(iv)$

Total increase in volume = $dV_c + dV_{sph}$

therefore total increase in volume

$$= \frac{\pi p r^2 L\,(5-2v)}{tE} + \frac{2p\pi r^4}{tE}\,(1-v)$$

or Total increase in volume

$$= \frac{\pi r^3}{tE}\left\{L\left(\frac{5}{2} - 2v\right) + 2r(1-v)\right\}$$

Substituting numerical values in the last equation, we get

Total increase in volume $= \dfrac{\pi.130 \times 10^5.(85)^3}{15 \times 210 \times 10^9}\,\{800\,(1.9) + 170(.7)\}$

$$= \frac{\pi.130 \times 10^5 (85)^3\,(1639)}{15 \times 210 \times 10^9}$$

Change in internal volume = 13051 cubic millimeters
Ans = 13051 (mm)^3

Q3. A thin circular hoop 5 mm thick, 50 mm wide and 1 metre diameter is subjected to an internal loading of 30 Newtons per millimeter of inside circumference. Find the circumferential stress and the change in diameter of the hoop. Take E = 220GN /m²

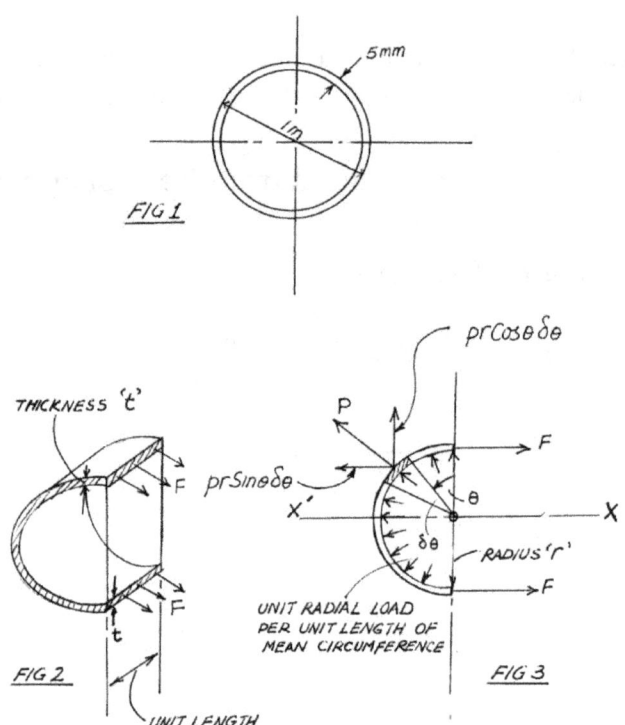

FIG 1

THICKNESS 't'

FIG 2

UNIT LENGTH

prCosθδθ

prSinθδθ

P

F

θ

δθ

RADIUS 'r'

F

F

X'

X

UNIT RADIAL LOAD PER UNIT LENGTH OF MEAN CIRCUMFERENCE

FIG 3

To determine the force 'p' in the ring, which in this case is tensile – the force that is - consider a section of the ring of unit length perpendicular to the plane of the paper, cut by a horizontal diametral plane. See Fig.2. A cross- section of such a section in the plane of the paper is shown in Fig. 3 which is the free- body diagram showing the hoop tension 'F' newtons balancing the internal loading p newtons per metre, let us say. Note that there is no longitudinal tension in the ring because the loading is purely circumferential; it is a ring not a tube with closed ends.

Now consider an element of the ring subtending angle dθ at the centre. Evidently the resultant force on this element is p r d θ, r being the radius of the ring. The component of this force parallel to the X-X axis is p r d θ (sinθ). For equilibrium of the section therefore

$$2F = 2 \int_0^{\pi/2} pr Sin\theta d\theta$$

$$\therefore \quad F = pr \left[-Cos\theta \right]_0^{\pi/2}$$

i.e. $\quad F = pr$

That is to say the radial tension is equal to the uniform radial load per unit length of mean circumference multiplied by the mean radius, which we take as 'r' because the ring is thin anyway. It should be noted that the sum of com-ponents of force $\int_0^{\pi/2} pr\, Cos\theta\, d\theta$ completely balance out the components $\int_0^{\pi/2} prd\, Cos\,(\theta)$ and therefore we need not consider them in the analysis.

From the hoop tension F = pr, the hoop stress σ_h is simply

$$\sigma_h = \frac{F}{cross-sectional\ area}$$

Employing the data given, loading is 30 N/mm. i.e 3×10^4 N/m; radius of ring $= \frac{1}{2}$ metre.

$$\therefore \qquad F = 3 \times 10^4 \times \frac{1}{2}$$

i.e $\qquad\qquad F = 1.5 \times 10^4$ Newtons.

and circumferential stress:

$$\sigma_h = \frac{F}{csa}$$

Cross- sectional area (csa) $= \left(\frac{5}{100} \times \frac{50}{1000}\right)$

$$= 250 \times 10^{-6}\ m^2$$

Therefore $\qquad\qquad \sigma_h = \frac{1.5 \times 10^4}{250 \times 10^6}$

$$= 60 \times 10^6\ N/m^2$$

Circumference of Hoop, $\quad C = \pi D$

Therefore $\qquad\qquad Log\ C = Log\ \pi + Log\ D$

i.e. $\qquad\qquad \frac{dC}{C} = \frac{dD}{D}$

Therefore circumferential strain = Diametral strain

$$\frac{dC}{C} = \frac{60 \times 10^6}{220 \times 10^9} = \frac{dD}{D}$$

$$dD = D \times \frac{3}{11} \times 10^{-3}$$

$$D = 1 \text{ metre}$$

$$dD = 1 \times \frac{3}{11} \times 10^{-3} \text{ metre}$$

$$= \frac{3}{11} \text{ mm}$$

Change in diameter $\qquad = \frac{3}{11} \text{ mm}$

Q4. A fully loaded petroleum product tanker transporting liquid of density 850 kg /m³ is travelling on a roadway at a speed of 60km /h. The driver brings the vehicle to an abrupt stop in 2 seconds.

Calculate the longitudinal force per metre length of perimeter of the tank and also the lateral or circumferential force per metre length of the tank.

The dimensions of the tank are: diameter = 2m; length = 6.5m. Neglect static hydrostatic pressure.

Ans: Longitudinal Force $\qquad$ = 24.4 kN/m;
Lateral (circumferential) Force = 48.8 kN/m.

Q5. A mechanical engineer employed as a designer desires to eliminate the weakness inherent in the use of longitudinal seams in a thin cylindrical vessel he is designing. He employs in the design of a cylinder 3000mm diameter and 15mm thick which is to operate at a pressure of 15 bars, butt- welded seams inclined at 30° to the cylindrical surface. Calculate the stresses normal and tangential to the weld and the intensity of the resultant stress. Derive any formulae you use for the normal and tangential stresses (William Embleton).

Using our customary symbology, the principal stresses are the hoop stress σ_h (or σ_c) and the longitudinal stress σ_L. Designating,

p = internal pressure
r = internal radius
t = thickness of cylinder wall

it is known that $\sigma_h = \frac{pr}{t}$ and $\sigma_L = \frac{pr}{2t}$

Employing the data given:

$$\sigma_h \qquad = 15 \times 10^5 \times 3 \ / \ 2 \ (15/1000)$$
i.e. $\qquad \sigma_h \qquad = 15 \times 10^7 \text{ N/m}^2 = 150 \text{ MN/m}^2$
so that $\qquad \sigma_L \qquad = 7.5 \times 10^7 \text{ N/m}^2 = 75 \text{ MN/m}^2$

In order to derive expressions for the normal and tangential stresses at butt- welded seams at $\theta°$ to the cylindrical surface consider the element shown in Fig.1

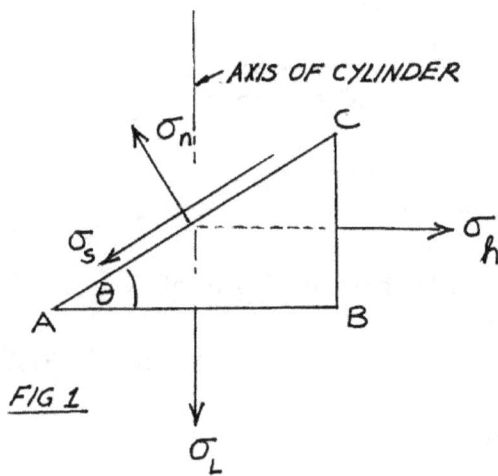

FIG 1

Let σ_N = normal stress on surface inclined at $\theta°$ as shown in Fig. 1
Let σ_s = tangential stress along surface AC

Assuming the triangular element to be of unit thickness and remembering that stresses are tensors :

$$\sigma_N \ (AC \times 1) \ Sin \ \theta = \sigma_h \ (BC \times 1)$$

and

$$\sigma_N \ (AC \times 1) \ Cos \ \theta = \sigma_L \ (AB \times 1)$$

therefore

$$\sigma_N AC \ Sin\theta = \sigma_h \ BC \qquad \dots\dots\dots\dots\dots\dots\dots\dots\dots\dots\dots(i)$$

$$\sigma_N AC \ Cos\theta = \sigma_L \ AB \qquad \dots\dots\dots\dots\dots\dots\dots\dots\dots\dots\dots(ii)$$

Squaring (i) and (ii) and adding the results

$$\sigma_N \ \{ \ (AC)^2 \ Sin^2\theta + (AC)^2 \ Cos^2\theta) \ \} = \sigma_h^2 \ (BC)^2 + \sigma_L^2 \ (AB)^2$$

from which

$$\sigma_N^2 = \sigma_h^2 \left(\frac{BC^2}{AC}\right) + \sigma_L^2 \left(\frac{AB^2}{AC}\right)$$

i.e.

$$\sigma_N^2 = \sigma_h^2 \ Sin^2\theta + \sigma_L^2 \ Cos^2\theta \qquad \dots\dots\dots\dots\dots(iii)$$

Resolving forces parallel to σ_h we get

$$\sigma_s \ (AC \times 1) \ Cos\theta = \sigma_h \ (BC \times 1)$$

and parallel to σ_L

$$\sigma_s \ (AC \times 1) \ Sin\theta = -\sigma_L \ (AB \times 1)$$

Therefore $\qquad \sigma_s \ Cos\theta = \sigma_s Cos\theta = \sigma_h \left(\dfrac{BC}{AC}\right)$

$$\sigma_s \ Sin\theta \ = -\sigma_L \left(\dfrac{AB}{AC}\right)$$

or $\qquad \sigma_s \ Cos\theta = \sigma_h \ Sin\theta \qquad$ ....(iv)

and $\qquad \sigma_s \ Sin\theta = -\sigma_L \ Cos\theta \qquad$ ....(v)

Multiplying (iv) by $Cos\theta$ and (v) by $Sin\theta$ and adding the results, we arrive at

$$\sigma_s = (\sigma_h - \sigma_L) \ Sin\theta \ Cos\theta \qquad ....(vi)$$

Employing these results:

$$\sigma_N{}^2 = (150 \times 10^6)^2 \ (Sin 30°)^2 + (75 \times 10^6)^2 \ (Cos 30°)^2$$

therefore normal stress, $\quad \sigma_N \approx 99 \ MN/m^2$

and $\qquad\qquad\qquad \sigma_s \ = (\sigma_h - \sigma_L) \ Sin\theta \ Cos\theta$

$$= (150 - 75)10^6 \left(\dfrac{1}{2}\right)\left(\dfrac{\sqrt{3}}{2}\right)$$

therefore Tangential stress, $\sigma_s = 32.4 \ MN/m^2$

Designating 'σ_R' as resultant stress on surface AC

$$\sigma_R = \sqrt{(\sigma_N{}^2 + \sigma_s{}^2)} = 104.2 \ MN/m^2$$

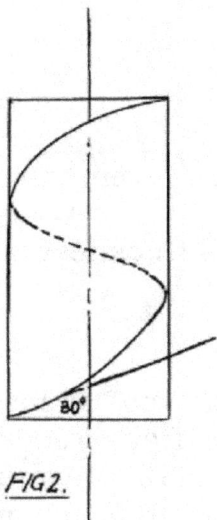

FIG2.

The helical spiral showing the butt-welded joint is sketched in Fig.2. Using the helical seam, as the calculation shows, reduces the resulting stress that would arise in the case of a longitudinal seam. In the case of a longitudinal seam the resultant stress would be $\sqrt{(150)^2 + (75)^2} = \sqrt{28125} = 167.7 \ MN/m^2$.

Q6. A cylindrical tank 15m diameter and 6m high is to be designed for the storage of gas oil of density 860kg/m³. What thickness of bottom plate would you specify if the maximum stress in tension for the steel from which the bottom plates are to be made is 140 MN /m². Assume that the plates are to be butt- welded and take joint efficiency of welds as 75%. Because corrosion from water and acid is inevitable, increase the thickness estimated by 100%. Neglect stresses at junction of bottom plates and tank bottom. Take g = 10m/s²

Maximum hydrostatic pressure on bottom plates = pgh

$$= \frac{860 \ kg}{m^3} . 6m . \frac{10m}{sec^2}$$

$$= 51600 \ \frac{kg.m}{sec^2} . \frac{m}{m^3}$$

$$= 51600 \ N/m^2$$

Axial stress in wall of vessel due to this pressure is limited to 140 MN/m²

$$140 \ x \ 10^6 = \frac{51600 \ x \ 7.5 \ x \ 1000}{t \ x \ 0.75} \qquad \text{.................(i)}$$

in which   t = thickness of bottom plate in mm

From (i)   t = 3.68 mm

I would specify 3¾ mm or even 4mm.

Adding 100% calculated thickness

Ans: t = 8 mm

Q7. A common sight on the roof of many buildings in New York City is a water tank constructed from wooden staves. See Fig. 1. A bus ride or stroll along, say, 9th or 10th Avenue will confirm this. You would see many such tanks, some housed in rather elegant looking structures, architecturally embellished to suit the buildings they serve. A typical tank is 3 m inside diameter and 4 m in height. If on such a tank the steel hoops used are of rectangular cross- section, say, 30mm by 6mm, determine the spacing between them assuming that the maximum allowable stress in the steel hoops is 84 MN /m². Take density of water = 1000kg /m³ and g ≈ 10m /sec²

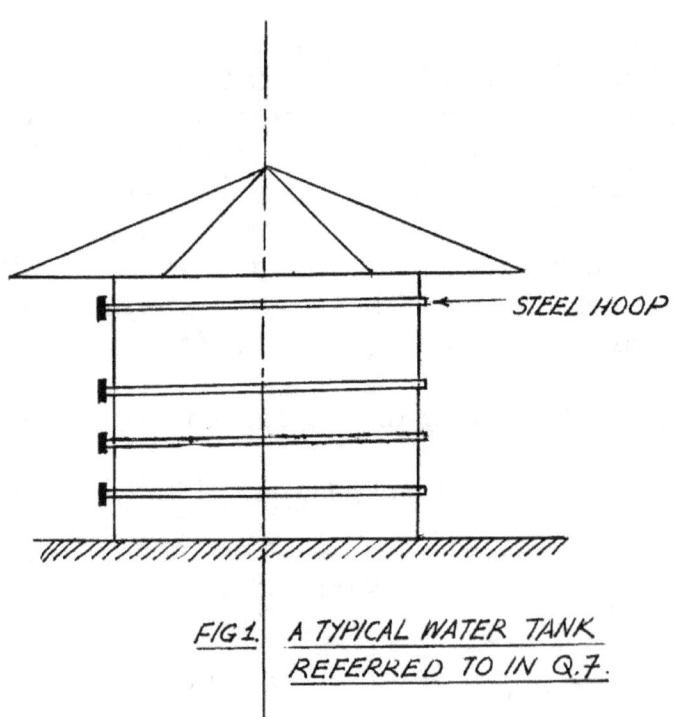

FIG 1. A TYPICAL WATER TANK REFERRED TO IN Q.7.

The first thing we have to note is that because the pressure and hence the thrust against the sides of the tank due to hydrostatic head varies with this head, the spacing of the hoops will vary; spacing will increase as we move from the bottom of the tank where the pressure is greatest to the top where it is least.

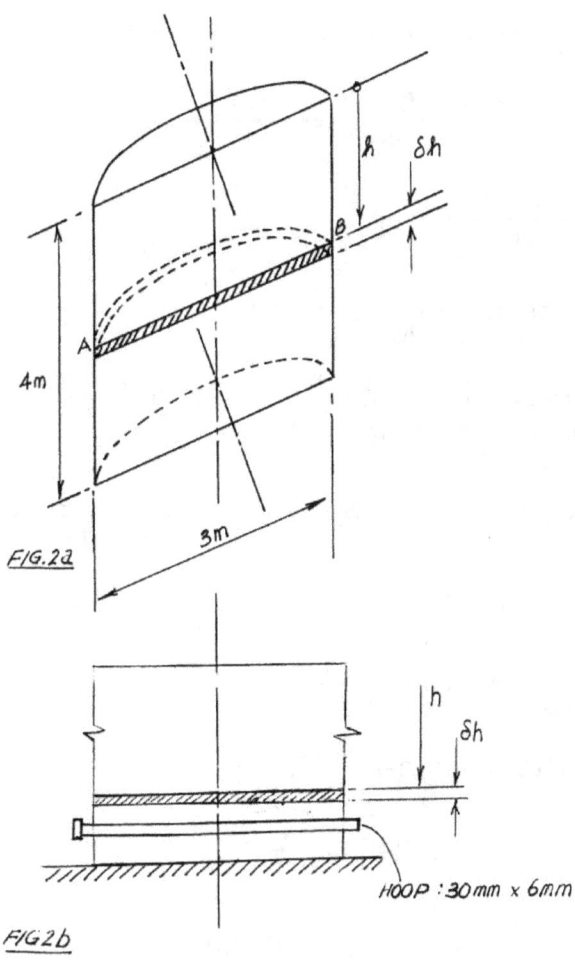

FIG.2a

FIG 2b

In Figs 2a and 2b, AB is the width of any strip of the cylinder the hydrostatic thrust against which is resisted by the hoop. The internal force tending to separate the tank in this strip is:

$$3000 \text{(millimeters)} \times \delta h \text{(millimeters)} \times p_h$$

In which p_h = hydrostatic pressure 'h' mm below surface of water in tank.

Now $\quad p = \rho g h$

$$= \frac{1000\,kg}{(1000\,mm)^3} \; x \; \frac{10m}{sec^2} \; x\; h \; (in\, mm)$$

~ 325 ~

Remembering that $\frac{kg.m}{sec^2}$ are the dimensions of Newton, then

$$p = \frac{1000}{1000^3\ (mm)^3} \times Newton \times 10.h\ (mm)$$

$$= \frac{10\ Newton\ (h)}{(1000)^2\ (mm)^2}\ (N/mm^2)$$

Therefore the internal force tending to separate the tank is:

$$= \frac{3000\ mm \times dh(mm) \times 10\ Newton \times h}{(1000)^2\ mm^2}$$

$$= \frac{30\ (Newton)hdh}{1000}\ Newton$$

or better expressed $\frac{3\ hdh}{100}\ Newton$

We must be careful to note also that this hydrostatic thrust is resisted by 2 cross-sectional areas of the hoop.

therefore Resisting Hoop Force $= 2 \times \dfrac{30}{1000} \times \dfrac{6}{1000} \times 84 \times 10^6$

As expressed, this resisting force has the dimensions of Newton.

therefore 30240 Newton $= \dfrac{3}{100}\ hdh.$

Considering the bottom-most strip.

$$30240 = \frac{3}{100} \int_{h1}^{4000} hdh$$

i.e. $\qquad \dfrac{4000^2 - h_1^2}{2} = 1008000$

or $\qquad\qquad h_1^2 = 4000^2 - 2016000$

from which $\qquad\quad h_1 = (13984000)^{1/2}$

i.e. $\qquad\qquad\quad h_1 = 3739.5\ mm$

(say) $\qquad\qquad\quad h_1 = 3739\ mm,$

as in Fig. 3

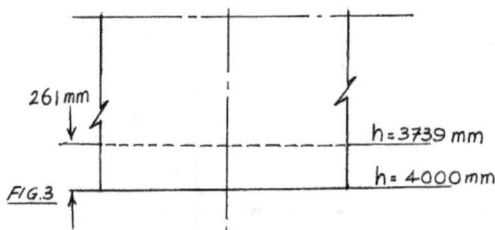

FIG.3

Now the centre-line of the hoop must be in the plane of the centre of pressure of the cylindrical strip 'ABCD' shown in Fig. 3.

In order to determine the centre of pressure, consider the moment about the water surface of the element $(\rho g h\, dh).h$, so that the total moment for 'ABCD'

$$= \rho g \int_{3739}^{4000} h^2\, dh$$

Therefore if $\bar{h}_1$ is the depth of the centre of pressure of 'ABCD' below the surface of the water, then

$$\bar{\bar{h}}_1 = \frac{\rho g \int_{3739}^{4000} h^2\, dh}{\rho g \int_{3739}^{4000} h\, dh}$$

i.e.

$$\bar{h} = \frac{2}{3} \left[\frac{4000^3 - 3739^3}{4000^2 - 3739^2}\right]$$

or

$$\bar{h} = \frac{2}{3} \left(\frac{11728328000}{2019879}\right)$$

$$\bar{h} = 3871\ mm$$

The bottom hoop is therefore placed as shown in Fig.4 symmetrically about the plane of the centre of pressure.

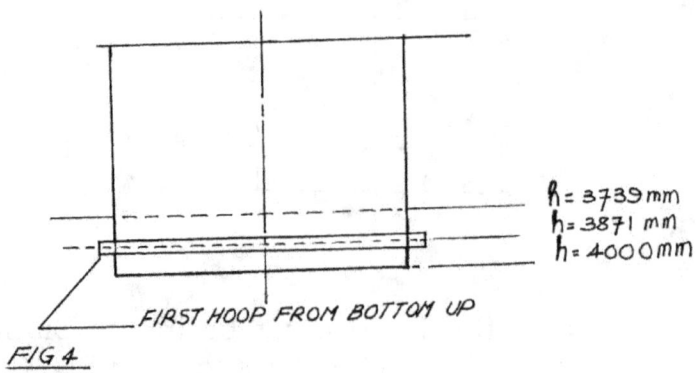

$R = 3739\,mm$
$h = 3871\,mm$
$h = 4000\,mm$

FIRST HOOP FROM BOTTOM UP

FIG 4

Similarly, $\qquad h_2 = (3739)^2 - 2016000$

or $\qquad h_2 = \sqrt{11964121}$

i.e $\qquad h_2 = 3459$ mm

and $\qquad \bar{h}_2 = \dfrac{2}{3}\left[\dfrac{(3739)^3 - (3459)^3}{(3739)^2 - (3459)^2}\right]$

i.e. $\qquad \bar{h}_2 = 3600\ mm$

again $\qquad h_3^2 = (3459)^2 - 2016000$

or $\qquad h_3 = 3154$ mm

and $\qquad \bar{h}_3 = \dfrac{2}{3}\left[\dfrac{(3459)^3 - (3154)^3}{(3459)^2 - (3154)^2}\right]$

i.e. $\qquad \bar{h}_3 = \left(\dfrac{1001073500}{2016965}\right)$

therefore $\qquad \bar{h}_3 = 3309\ mm$

The spacing of the bottom two hoops is shown in Fig. 5.

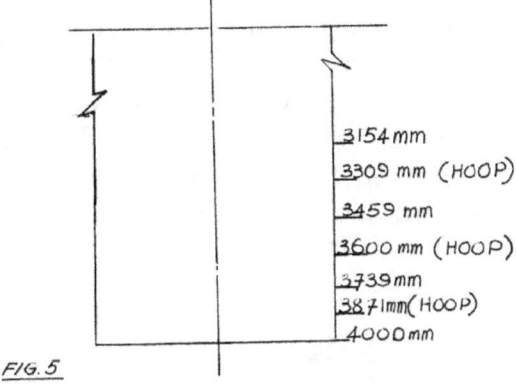

3154 mm
3309 mm (HOOP)
3459 mm
3600 mm (HOOP)
3739 mm
3871 mm (HOOP)
4000 mm

FIG. 5

Not drawn to scale.

The remaining calculations along the lines illustrated in the foregoing are left to the designer.

Q8.   A boiler tube is to be designed to withstand internal pressure of 45 bars without hoop or circumferential stress exceeding 75 MN/m². The internal diameter of the tube is 60mm and the material from which the tube is manufactured is 8000kg /m³. What should be the mass per metre run of the tube?

Equating the maximum circumferential or hoop stress to the value given, we have: $\frac{pr}{t} = 75 \ x \ 10^6 \ N/m^2$

where: p = internal pressure in N/m²
               r = Internal radius of tube, mm
               t = thickness of tube, mm

therefore $\frac{45 \ x \ 10^5 \ x \ 30}{t} = 75 \ x \ 10^6$

or $t = 1.8 \ mm$

Cross- sectional area of tube is approximately = $\pi \ x \ \frac{60}{1000} \ x \ \frac{1.8 \ mm}{1000} \ sq. \ metres$

For a 1 metre length of tube, the volume = $\frac{\pi \ x \ 60 \ x \ 1.8 \ x \ 1}{10^6} \ cubic \ metres$

Now mass of 1 cubic metre of material   = 8000 kg

therefore mass of 1 metre length of tube = $\frac{\pi \ x \ 60 \ x \ 1.8 \ x \ 1 \ x \ 8000 \ kg}{10^6 \ x \ 1}$

Mass of tube = 2.7 kg/metre

Q9. The thin- walled cylindrical pressure vessel in Fig.1 is to be constructed with an elliptical inspection manhole in its walls. Show by means of a sketch how you would orient the manhole with particular reference to its major and minor axes in relation to the length (or height) and circumference of the vessel. State your reasons including why you would not use a circular manhole.

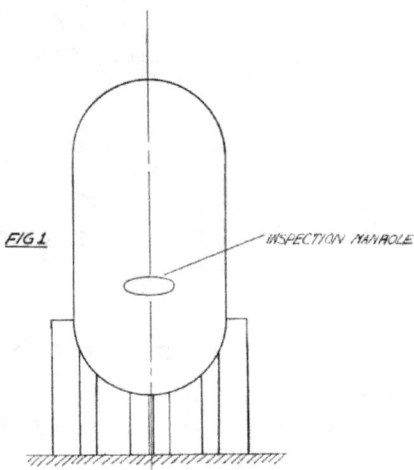

FIG 1   INSPECTION MANHOLE

With 'p' as working pressure; 'r' as inside radius; and 't' as wall thickness:

Circumferential stress $= \dfrac{pr}{t}$

Axial stress $\qquad = \dfrac{pr}{2t}$

Evidently the circumferential or hoop stress being the greater stress, the smaller of the projected area of material to be removed is the product of the minor axis and the thickness 't' of the plate from which the vessel is made. Referring to Fig.2, 2bt < 2at. Hence the greater of the two stresses namely the hoop stress should be associated with minor axis.

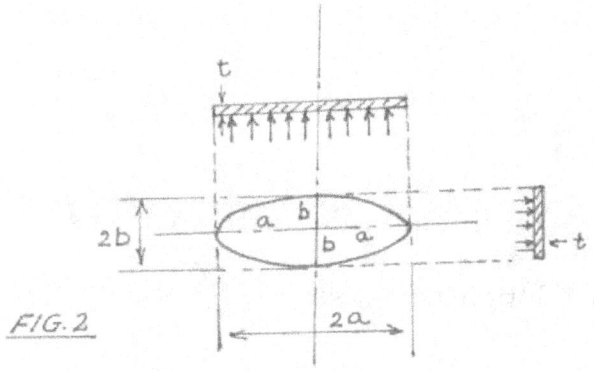

FIG.2

Accordingly, the minor axis must be parallel with the vertical axis of the cylinder as in Fig. 2. If the vessel were lying in a horizontal position then the minor axis of the elliptical manhole would have to be parallel with the horizontal axis of the vessel.

Clearly, for a circular manhole the corresponding projected areas are equal being 'Dt" where D is the diameter of the hole. There is no design advantage. The student should note however that the presence of a hole elliptical or otherwise in the wall of the vessel gives rise to stress concentrations around the hole.

Q10. A thin cylindrical vessel has hemispherical ends as shown in Fig.1. What must be the ratio of the thickness of the cylindrical portion to the thickness of the hemispherical section so that extra stresses are not to be developed in the junction between cylindrical and spherical portions?

Take Poisson's ratio, $v = 0.25$

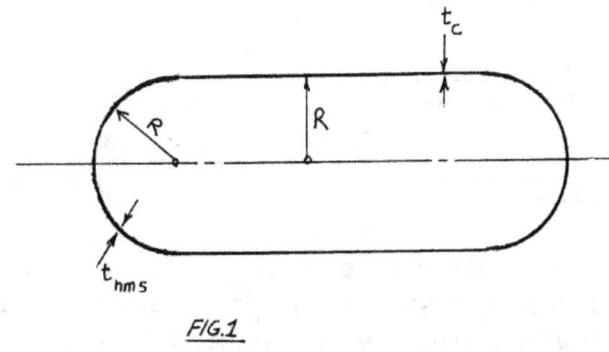

FIG.1

R = radius of cylinder and hemispheres

t_c = thickness of cylindrical portion of vessel

t_{hms} = thickness of hemispherical ends

The principal stresses in the cylinder are

$$\sigma_{hoop} = \frac{pR}{t_c}; \ \sigma_{long} = \frac{pR}{2t^c}$$

where σ_{hoop} = circumferential or hoop stress;

σ_{long} = longitudinal stress

Now, circumferential or hoop strain, e_{hoop}, in cylindrical portion of vessel is given by

$$e_{hoop} = \frac{1}{E}\left(\sigma_{hoop} - v\sigma_{long}\right) \qquad \dots\dots\dots\dots\dots\dots\dots\dots\dots\dots(i)$$

where   E = Young's Modulus

v = Poisson's ratio

$$e_{hoop} = \frac{1}{E}\left(\frac{pR}{t_c} - \frac{vpR}{2t_c}\right)$$

$$= \frac{pR}{t_c E}\left(1 - \frac{v}{2}\right)$$

i.e $$e_{hoop} = \frac{pR}{t_c E}\frac{(2-v)}{2} \qquad \dots\dots\dots\dots\dots\dots\dots\dots\dots\dots(ii)$$

Again if the ends are free of restraint which was assumed in the case of the circumference of the cylinder, the circumferential strain, e_{hms}, in the hemispherical ends is given by

$$e_{hms} = \frac{1}{E}\left(\frac{pR}{2t_{hms}} - v\frac{pR}{2t_{hms}}\right) \qquad \dots\dots\dots\dots\dots\dots\dots\dots\dots\dots(iii)$$

i.e $$e_{hms} = \frac{pR}{Et_{hms}}\left(\frac{1}{2} - \frac{v}{2}\right)$$

$$e_{hms} = \frac{pR}{Et_{hms}}\frac{(1-v)}{2} \qquad \dots\dots\dots\dots\dots\dots\dots\dots\dots\dots(iv)$$

To obtain the actual extensions in the case of equations (ii) and (iv) we need only to multiply the strains by $2\pi R$. But it is obvious that because $_{hoop} \neq e_{hms}$ there is discontinuity at the ends. In order to avoid the stresses that would otherwise be the case, both hoop strains should be the same.

Putting $e_{hoop} = e_{hms}$ we get $\dfrac{t_c}{t_{hms}} = \dfrac{2-v}{1-v}$

For $\qquad v = 0.25$

therefore $\qquad \frac{t_c}{t} = 2 - 0.25 / 1 - 0.25$

Ans: $\qquad t_{hms} = \frac{3}{7} t_c$

By virtue of this relationship it means that the thickness of the material of the hemispherical cap is roughly 40% of that of the cylindrical portion; the weakest part of the vessel is now its ends.

Q11. A cylindrical pressure vessel 300 cm inside diameter and 10 m high is part of a chemical process plant. The wall thickness of the vessel is 40 mm. Determine the circumference and diametral elongation of the vessel under an internal pressure of 5 bars. Assume E= 200 GN/ m² and Poisson's ratio 'μ' = 0.25.

Let $\qquad \delta_h$ = hoop stress or circumferential stress

$\qquad \delta_L$ = longitudinal stress

$\qquad e_c$ = hoop strain or circumferential strain

Now, $e_c = \frac{1}{E} (\sigma_h - v\sigma_L)$ $\qquad\qquad$ ....(i)

and $\quad \sigma_h = \frac{pr}{t} N/m^2$ $\qquad\qquad$ ....(ii)

where $\qquad$ p = internal pressure

$\qquad$ t = wall thickness of vessel

$\qquad$ r = internal radius of vessel

$\therefore \qquad \sigma_h = \frac{15 \times 10^5 \, N/m^2}{\frac{40}{1000}} \times \frac{150}{100}$

i.e. $\qquad \sigma_h = \frac{15}{16} \times 10^7 \, N/m^2$

so that, $\quad \sigma_L = \frac{15}{16} \times 10^7 \, N/m^2$

Substituting in (i) we get:

$$e_c = \frac{1}{210} \times 10^9 \left(\frac{15}{8} \times 10^7 - 0.25 \times \frac{15}{16} \times 10^7 \right)$$

i.e. $\qquad e_c = \dfrac{1}{210} \times 10^9 \dfrac{15}{8} \times 10^7 \left(\dfrac{1-0.25}{2}\right)$

$$= \dfrac{1}{210} \times 64$$

i.e. $\qquad \dfrac{dC}{C} = 0.000078$

from which $dC = C(0.000078)$

$\therefore \qquad dC = \pi \times 300 \times 0.000078$ cm

$$= \pi \times 300 \times 10 \times 0.000078 \text{ mm}$$

i.e. $\qquad dC = 0.736$mm

because $\qquad C = \pi D$ ....(ii)

where $\qquad C$ = circumference of vessel

and $\qquad D$ = diameter of vessel

taking logs of each side of (iii)

Log C $\qquad = $ Log π + Log D

therefore $\qquad \dfrac{dC}{C} = \dfrac{dD}{D}$

that is circumferential strain = diametral strain

$\therefore \qquad d/D = D \times 0.000078$cm

$$= 300 \times 10 \times 0.000078 \text{mm}$$

or $\qquad dD = 0.234$mm

Q12. In an experiment to determine Poisson's ratio a thin steel cylindrical tube is capped at its ends and instrumented to measure axial elongation caused by (i) internal hydraulic pressure and (ii) by direct axial force. When an axial pull of 50 kN is applied to the caps at the ends of the tube the elongation on a 200 mm gauge length marked on the tube is 0.024 mm. When this axial pull is removed and internal pressure of 15 bar applied, the elongation of the same gauge length is 0.018 mm. The tube is of internal diameter 250 mm and wall thickness 2.5 mm. Calculate the value of Poisson's ratio for the material of the tube.

Axial stress from axial pull acting alone $= \dfrac{50 \times 10^3 N}{\pi d T}$

where   D   =   internal diameter of tube, m

t   =   thickness of tube, m

Note the use of πDt because of small thickness of tube instead of $\pi(D^2 - D^2)/4$

$\therefore$ Axial stress $= \dfrac{50 \times 10^3}{\pi \times \frac{250}{1000} \times \frac{2.5}{1000}}$

Now, because

$$E = \frac{Axial\ stress}{Axial\ strain} = \frac{50 \times 10^3 \times 10^6 \times 200}{\pi \times 625 \times 0.024}$$

i.e. E = 212×10^9 N/m²

now longitudinal strain, e_L $= \dfrac{1}{E}(\sigma_L - v\sigma_h)$

$\therefore$ longitudinal extension $= \dfrac{1}{E}(\sigma_L - v\sigma_h) \times Gauge\ length$

$$= \frac{1}{212} \times 10^9 \left\{\frac{pr}{2t} - \frac{vpr}{t}\right\} \times 200$$

$$= \frac{pr}{t} \times \frac{1}{212} \times 10^9 \left(\frac{1}{2} - v\right) \times 200$$

Substituting relevant values

$$0.018 = \frac{15 \times 10^5}{212 \times 10^9} \times \frac{125}{2.5}\left(\frac{1}{2} - v\right) \times 200$$

or $\left(\dfrac{1}{2} - v\right)$ $= \dfrac{0.018 \times 212 \times 10^9 \times 2.5}{15 \times 10^5 \times 125}$

from which $\left(\dfrac{1}{2} - v\right)$ = 0.25

so that, Poisson's ratio $v = 0.25$

Q13. In the experimental arrangement shown in Fig.1(Popov) for obtaining controlled ratios of principal stresses (Popov), the ends of a piece of thin – walled steel tubing are closed by caps and the resulting closed cylinder pressurized. The capped tubing is also subjected to an axial tensile load 'F' in a testing machine. In one such test, a tube of diameter 250 mm and 2.5 mm wall thickness is subjected to internal pressure of 15 bars, the axial tensile load being 50 kN. What is the magnitude of the longitudinal extension in a gauge length of 200 mm of the tube under these conditions? Take E = 210 GN/m² and Poisson's ratio, ν = 0.25.

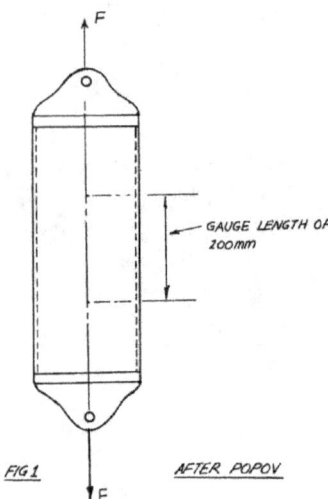

GAUGE LENGTH OF 200mm

FIG 1   AFTER POPOV

Let σ_h =   circumferential or hoop stress due to internal pressure

        σ_L =   longitudinal stress due to internal pressure

∴ σ_h = $\dfrac{15 \times 10^5 \; N/m^2 \times 12.5}{2.5}$

i.e. σ_h =   75 x 10⁶ N/m²

and σ_L =   37.5 x 10⁶ N/m²

Now, because of the axial tensile load F, the additional longitudinal stress is σ'_L where,

        σ'_L = $\dfrac{50 \times 10^3 \, N}{\pi \times \frac{250}{1000} \times \frac{2.5}{1000}}$

in which   D = internal diameter of pipe

∴ σ'_L = $\dfrac{50 \times 10^3 \times 10^6}{\pi \times 250 \times 2.5}$

∴ σ'_L = $\dfrac{80 \times 10^6}{\pi} \; N/m^2$

∴   Total longitudinal stress σ_T = $\sigma_L + \sigma'_L$

                                                = $\left(37.5 + \dfrac{80}{\pi}\right) 10^6 \; N/m^2$

~ 336 ~

Axial strain, e_{AX} $= \frac{e_L}{200}$; the gauge length = 200 mm

and, e_L = axial extension

But $e_{AX} = \frac{1}{E}(\sigma_T - v\sigma_h)$ .... (i)

Substituting in (i) we get

$$\frac{e_L}{200} = \frac{1}{210 \times 10^9}\left\{\left(37.5 + \frac{80}{\pi}\right)10^6 - 0.25(75 \times 10^6)\right\}$$

$$= \frac{1}{210 \times 10^9}10^6\left(37.5 + \frac{80}{\pi} - \frac{75}{4}\right)$$

$$\frac{210 \times 10^6 \times e_L}{200} = 18.75 + \frac{80}{\pi}$$

i.e. $1050\, e_L = 44.21$

from which $e_L = 0.042$ mm

Ans: longitudinal extension on 200 mm length = 0.042 mm

Q14. A cylindrical pipe 600 mm diameter and 7500 mm long is made of steel plate 12 mm thick. The ends are closed by plates bolted to flanges on the pipe. Show that if the pipe be then subjected to hydraulic pressure of 30 bars, it will stretch longitudinally by about 0.036 mm. Take E = 210 GN /m² and Poisson's ratio = 0.3.

(Case)

When the pipe is subjected to hydraulic pressure of 30 bars, the force on the ends is $\frac{\pi}{4}\left(\frac{600^2}{1000}\right) \times 30 \times 10^5$ Newtons.

If σ_L is the longitudinal stress developed in the pipe wall as a result of this force, then, working in Newton and metres,

$$\sigma_L \times \pi\left(\frac{600^2}{1000}\right) \times \frac{t}{1000} = \frac{\pi}{4}\left(\frac{600^2}{1000}\right) \times 30 \times 10^5$$

Note that σ_L is in N /m² and t = thickness of pipe in mm.

therefore $\sigma_L\, 600.12 = 600 \times \frac{600}{4} \times 30 \times 10^5$

or $\sigma_L = 37.5$ MN/m²

Now, the hoop stress, σ_L , developed in the cylinder as a result of the applied hydraulic pressure $= 30 \times 10^5 \times \frac{300}{12}$

i.e. $\qquad \sigma_L = 75 \times 10^6$ MN/m²

so that the longitudinal strain in the cylinder (e_L) must be given by

$$e_L = \frac{1}{E} \{\sigma_L - v\,(\sigma_h - p)\}$$

where $\qquad\qquad v$ = Poisson's ratio

and $\qquad\qquad\quad p$ = hydraulic pressure

therefore $\qquad e_L = \frac{1}{210 \times 10^9} \{37.5 - 0.3\,(75 - 3)10^6\}$

Note $\qquad\qquad p = 30 \times 10^5$ N/m² = 3 MN/m²

therefore Strain (e_L) $= \frac{1}{210 \times 10^9} (15.9) \times 10^6$

so that $\quad$ Extension $= \frac{1}{210 \times 10^9} \times 7500$

$$= \frac{75}{2100}\ mm$$

$$= 0.036\ \text{mm} \qquad\qquad \text{Q.E.D}$$

Q15. A welder constructed a closed cylindrical vessel having flat ends, the internal measurements being: length = 600 mm, diameter = h 250 mm. The thickness of the cylinder is 4 mm. Calculate hoop and longitudinal stresses when the vessel's internal pressure is 25 bars. Ignoring the extra stresses at the ends of the cylinder, determine also the increase in length, diameter and volume due to this pressure. Take E = 210 GN / m² and Poisson's ratio = 0.3.

Circumferential or hoop stress (σ_h) $= 25 \times 10^5 \times \frac{125}{4}$

i.e. $\qquad\qquad\qquad \sigma_h = 781.25 \times 10^5$ N/m²

$$= 78.1\ \text{MN/m}^2$$

so that

Longitudinal stress $\qquad = 39$ MN /m²

Longitudinal strain $(e_L) = \frac{1}{E} \{39 - 0.3 (78.1)\}$

$$= \frac{1}{210 \times 10^9} (39 - 23.4)10^6$$

$$= \frac{1}{210 \times 10^9} (15.6)10^6$$

$$= \frac{15.6}{210 \times 10^3} = 0.74 \times 10^6$$

increase in length $\quad = \frac{15.6}{210 \times 10^3} \times 600 \; mm$

$$= 0.045 \text{ mm}$$

Circumferential strain $(e_c) = \frac{1}{E} \{78.1 - 0.3 (39)\}$

$$= \frac{1}{210 \times 10^9} (66.4)10^6$$

$$e_c = \frac{66.4}{210 \times 10^3}$$

$$= 0.32 \times 10^{-4}$$

Circumference (C) $\quad = \pi \text{ (Diameter(D))}$

$$C = \pi D$$

therefore $\quad \text{Log } C = \text{Log } \pi + \text{Log } D$

$$\frac{dc}{C} = \frac{dD}{D}$$

or Circumferential strain = Diametral strain

therefore $\quad \frac{dD}{D} = 0.32 \times 10^4$

or $\quad dD = 0.32 \times 10^{-4} \times 250$

i.e $\quad dD = 0.079 \text{ mm}$

Volume of cylinder (V) $\quad = \pi D^2.L$

Therefore $\quad \text{Log } V = \text{Log } \pi + 2 \text{ Log } D + \text{Log } L$

or $\quad \frac{dV}{V} = 2 \frac{dD}{D} + \frac{dL}{L}$

or, Volumetric Strain = 2 (Diametral Strain) + Longitudinal Strain
therefore Volumetric Strain = 2 $(0.32 \times 10^{-4}) + (0.074 \times 10^{-4})$

$$= 0.714 \times 10^{-4}$$

so that increase in volume $= 0.714 \times 10^{-4} \times V$

where
$$V = \pi \times \frac{250^2}{4} \times 600 = 29456250 \ (mm)^3$$

$$= 2103 \ (mm)^3$$

Ans: 78.1 MN/m² ; 39 MN/m²; 0.045 mm; ; 0.079 mm ; 2103 mm³.

Q16. A steel tube 30 mm internal diameter and 1.5 mm thick is 500 mm long. Its ends are closed by 2 plugs through one of which a hole is made for a pressure connection. What is the internal pressure in the tube when its capacity has increased by 15 (mm)³.

Take E= 210 GN/m² and Poisson's ratio = 0. 25.

Volumetric Strain = 2 (Diametral Strain) + Longitudinal Strain .... (i)

Also, Circumferential Strain = Diametral Strain.

therefore, Equation (i) may be rewritten as

Circumferential Stress $= p \times \dfrac{15}{1.5}$

$$= p \times 10^5 \times \frac{15}{1.5} \ N/m^2 = p \times 10^6 \ N/m^2$$

where $\quad$ p = internal pressure

Longitudinal stress $= \dfrac{P \times 10^6}{2} \ N/m^2$

Circumferential strain $(e_c) = \dfrac{1}{E} \left\{ p \times 10^6 - \dfrac{1}{4} \left(\dfrac{p \times 10^6}{2} \right) \right\}$

$$= \frac{7}{8} \times p \times 10^6$$

Now, Longitudinal Strain $= \dfrac{1}{E} \left\{ p \times 10^6 - \dfrac{1}{4} \left(\dfrac{p \times 10^6}{2} \right) \right\}$

$$= \frac{1}{4E} \times p \times 10^6$$

∴ Volumetric strain $= \dfrac{14}{8E} p \times 10^6 + \dfrac{1}{4E} p \times 10^6$

$$= \frac{2}{E} p \times 10^6$$

Now, volume of tube

$$= \frac{1}{2} \frac{(30)^2}{4} \ x \ 500$$

$$= \frac{1}{2} \ x \ \frac{900}{4} \ x \ 500$$

$$= \frac{45}{8} \ x \ 10^4 \ (mm)^2$$

∴ increase in capacity, dV = V times Volumetric Strain

i.e. Volumetric Strain

$$= V \left\{ 2 \left(\frac{7p \ x \ 10^6}{8E} \right) + \left(\frac{P \ x \ 10^6}{4E} \right) \right\}$$

$$= \left(\frac{45 + 10^4}{8} \right) + \left(\frac{2p \ x \ 10^6}{E} \right)$$

therefore

$$15 = \frac{2p \ x \ 10^6 \ x \ 45 \ x \ 10^4}{210 \ x \ 10^9 \ x \ 8}$$

so that

$$p = \frac{15 \ x \ 8 \ x \ 210 \ x \ 10^9}{2 \ x \ 10^6 \ x \ 45 \ x \ 10^4}$$

$$= \frac{15 \ x \ 8 \ x \ 21}{2 \ x \ 45}$$

i.e.

p = 28 bars

Q17. A long steel tube 90mm inside diameter and 2.25mm wall thickness is closed by a flange at end end. Under internal pressure the maximum wall stress is 120 MN/m². determine the per cent increase in the volume of the tube. Take E = 210 GN/m² and Poisson's ratio = 0.25.

Ans: 0.143%

Q18. A long horizontal cylinder has an internal diameter of 200 mm and wall thickness 5 mm. The material from which the cylinder was constructed has a yield stress of 315 MN /m². Water pressure is applied to the inside of the cylinder until it yields. What are the magnitudes of this pressure assuming: (1) maximum shear stress and (2) maximum strain are the criteria of yielding? Take E = 210 GN / m² and Poisson's ratio = 0.25.

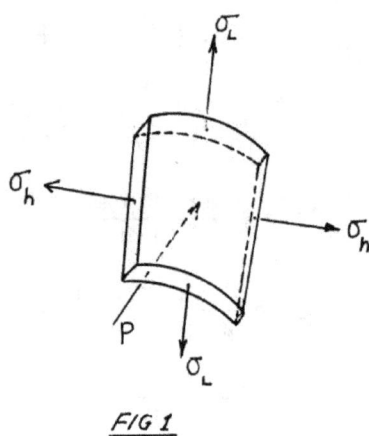

FIG 1

A portion of the cylinder's shell is shown in **Fig.1.**

Maximum shearing stress in plane of σ_h and σ_L

$$= \frac{\sigma_h - \sigma_L}{2} \qquad \qquad \dots\dots\dots\dots\dots\dots(i)$$

Maximum shearing stress in plane of σ_L and p

$$= \frac{\sigma_L - 0}{2}$$

$$= \frac{\sigma_L}{2} \qquad \qquad \dots\dots\dots\dots\dots(ii)$$

because 'p' is negligible in comparison with σ_h and σ_L.

Similarly, maximum shearing stress in the $\sigma_h - p$ plane is

$$= \frac{\sigma_h - 0}{2} \qquad \qquad \dots\dots\dots\dots\dots(iii)$$

$$= \frac{\sigma_h}{2}$$

Now, $\sigma_h = \dfrac{100p}{5}$

If p is in the bars, then

$$\sigma_h = \frac{p \; x \; 10^5 100}{5}$$

$$= 20 \; x \; 10^5 p$$

Using equation (iii), which gives the highest of the 3 maxima

$$= \frac{20 \; x \; 10^5 p}{2} = 315 \; x \; 10^6$$

from which p = 315 bars.

Therefore, using the maximum shear theory, the pressure causing yield is 315 bars.

For case (ii), the maximum strain at yield is: $\frac{315 \; x \; 10^6}{210 \; x \; 10^9}$

But maximum strain $= \frac{1}{E} \; (20 \; x \; 10^5 p - 0.25 \; x \; 10 \; x \; 10^5 p)$

Therefore, $\frac{315 \; x \; 10^6}{210 \; x \; 10^9} = \frac{p}{210 \; x \; 10^9} \; (20 \; x \; 10^5 - 2.5 \; x \; 10^5)$

or $\quad\quad\quad$ p = 180 bars

Q19. A locomotive-type boiler is constructed from 15 mm thick steel plate. It is 1 m internal diameter and 3 metres in length. By how much will the thickness of the plate be reduced when the internal pressure is 40 bars? Assume E = 210 GN/m² and Poisson's ratio = 0.25.

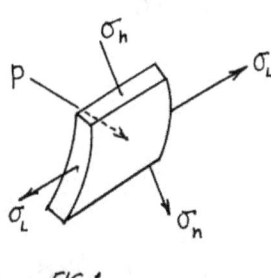

FIG 1

Let $\quad$ p = compressive stress due to internal pressure

$\quad\quad\quad$ σ_h = circumferential or hoop stress

$\quad\quad\quad$ σ_L = longitudinal stress

Now $\sigma_h = 40 \times 10^5 \times \dfrac{500}{15} = \dfrac{2 \times 10^9}{15}$ N/m^2

$$\sigma_L = \dfrac{10^9}{15} \ N/m^2$$

$$p = 40 \times 10^5 \ N/m^2$$

If dT = change in thickness of plate

and T = original thickness of plate, then

$$\dfrac{dT}{T} = \dfrac{1}{E}\{-p - v(\sigma_h + \sigma_L)\} \qquad\qquad\ldots\ldots\ldots\ldots\ldots\ldots\ldots\ldots(i)$$

Note that the radial pressure gives rise to a compressive stress and therefore p is negative. In thin shells p may normally be neglected and the wall- thickness strain obtained from

$$\dfrac{dT}{T} = \dfrac{1}{E}\{-v - v(\sigma_h + \sigma_L)\} \qquad\qquad\ldots\ldots\ldots\ldots\ldots\ldots\ldots\ldots(ii)$$

Here we shall use both equations and compare our results.

In the case of (i)

$$\dfrac{dT}{T} = \dfrac{1}{210 \times 10^9}\left\{-40 \times 10^5 - 0.25\left(\dfrac{2 \times 10^9}{15} + \dfrac{1 \times 10^9}{15}\right)\right\}$$

$$= \dfrac{1}{210 \times 10^9}\left(-40 \times 10^5 - 0.25 \times \dfrac{3 \times 10^9}{15}\right)$$

$$= \dfrac{1}{210 \times 10^9}\left(-40 \times 10^5 - 5 \times 10^7\right)$$

$$= \dfrac{-1}{210 \times 10^9} \times 54 \times 10^6$$

i.e. $\dfrac{DT}{T} = \dfrac{-54}{210000}$

$\therefore \quad dT = \dfrac{-54 \times 15}{210000} = \dfrac{810}{210000}$ or dT = -0.00386 mm

Using (ii)

$$\dfrac{dT}{T} = \dfrac{1}{210 \times 10^9}\left\{0.25\left(3x \ \dfrac{10^9}{15}\right)\right\}$$

$$= \dfrac{1}{210 \times 10^9} \times 5 \times 10^7$$

therefore dT = -0.00357

The difference based on the result from equation (i) is roughly 7.5%.

Q20. The pressure vessel of an air compressor is manufactured from steel plate 20 mm thick. The vessel is 6 m long, 80 cm internal diameter and is fitted with a Bourdon gauge calibrated in tenths of a bar. If it is found during a hydraulic test on the vessel, that a marked off gauge length of 500 mm on the vessel, in a longitudinal direction , increases by 0.0244 mm in this same direction, then what should be the reading on the Bourdon gauge. Take E = 210 GN/m² and Poisson's ratio = 0.25.

Let e_L = longitudinal extension on marked- off gauge length of 500mm

Let p = gauge reading in bars

Let r = internal radius of vessel

Let t = wall thickness of vessel

Let v = Poisson's ratio

Then e_L 　　$= \dfrac{1}{E}\left\{\dfrac{pr}{2t} - v\,\dfrac{pr}{t}\right\}$

　　　　　　$= \dfrac{pr}{Et}\left(\dfrac{1}{2} - \dfrac{1}{4}\right)$

　　　　　　$= \dfrac{1}{4}\dfrac{pr}{ET}$

so that $\dfrac{0.0244}{500} = \dfrac{p \times 10^5 \times 400}{4 \times 210 \times 10^9 \times 20}$, p in bars from which

　　p 　$= \dfrac{0.0244 \times 4 \times 210 \times 10^9 \times 20}{500 \times 10^5 \times 400}$

　　p 　$=$ 　20.496, say p = 20.5 bars.

Q21. A mild steel cylindrical tube 5 mm thick and internal diameter 100 mm is plugged at its ends. When subjected to internal hydraulic pressure the maximum direct stress in the tube is 135 MN/m². What is the per cent increase in the capacity of the tube? Take E = 200 GN/m² and Poisson's ratio = 0.25.

Maximum stress is the hoop or circumferential stress, σ_c; Longitudinal stress, $\sigma_L = \sigma_c /2$.

As shown in previous examples, volumetric strain of cylinder $\frac{dV}{v}$ is given by:

$$\frac{dV}{v} = 2e_d + e_L \qquad \dots\dots\dots\dots\dots\dots\dots\dots\text{(i)}$$

where e_d (= e_c) is the diametral strain and e_L = longitudinal strain, e_c = hoop strain

$$e_d = \frac{1}{E}(\sigma_c - v\sigma_L) \qquad \dots\dots\dots\dots\dots\dots\dots\dots\text{(ii)}$$

$$\text{and} \quad e_L = \frac{1}{E}(\sigma_L - v\sigma_c) \qquad \dots\dots\dots\dots\dots\dots\dots\dots\text{(iii)}$$

$$\text{therefore} \quad \frac{dV}{V} = \frac{1}{E}\{\sigma_c(2-v) + \sigma_L(1-2v)\} \qquad \dots\dots\dots\dots\dots\dots\dots\text{(iv)}$$

$$= \frac{\sigma_c}{E}\left\{(2-v) + \frac{1}{2}(1-2v)\right\} \quad \left[Note: \sigma_L = \frac{\sigma_c}{2}\right]$$

$$\text{i.e.} \quad \frac{dV}{V} = \frac{\sigma_c(5-4v)}{2E}$$

Substituting 135 MN/m² for σ_c ; v = 0.25

and E = 200 GN/m²

$$\frac{dV}{V} = \frac{135 \times 10^6}{200 \times 10^5}\left\{5 - 4\left(\frac{1}{4}\right)\right\}$$

$$= \frac{135 \times 4}{200 \times 10^5}$$

therefore $\frac{dV}{V}$ as percent increase in capacity

$$\frac{dV}{V}(100) = 0.27\%$$

Ans: = 0.27%

Q22. LPG is to be stored in a steel spherical vessel 75 cm in diameter. The two halves of the vessel are to be welded together. The maximum design pressure is 25 bars. Assuming an efficiency of 80% for the welded joint and an allowable stress of 70 MN/m², calculate the wall thickness of the vessel.

Ans: 8.4 mm.

Q23. Prove that for two principal stresses 'p$_{xx}$' and 'p$_{yy}$' acting at a point in a material for which Poisson's ratio is 'v' and Young's modulus 'E' strain energy per unit volume is $\frac{1}{2E}\left(p_{xx}^2 + p_{yy}^2 - 2vp_{xx}p_{yy}\right)$.

A mechanical engineer is asked to design a thin- walled cylindrical steel vessel to withstand an internal pressure of 20 bars. The vessel's internal diameter is fixed at 500mm. If he uses maximum strain energy as the criterion of failure then what wall thickness will he estimate if a factor of safety equal 3 is taken? Samples of the construction steel from which the vessel is to be fabricated were sent to a Materials' Testing laboratory for testing and it was found that the average yield stress in simple tension was 300 MN/m². Assume Poisson's ratio = 0.25.

Let e$_{xx}$ and e$_{yy}$ be the strains in the direction of p$_{xx}$ and p$_{yy}$ respectively.

By definition

Strain Energy per

Unit volume, say, $U = \frac{1}{E}p_{xx}e_{xx} + \frac{1}{2}p_{yy}e_{yy}$ ....(i)

$$e_{xx} = \frac{1}{E}(p_{xx} - vp_{yy})$$

and $$e_{yy} = \frac{1}{E}(p_{yy} - vp_{xx})$$

therefore $$U = \frac{p_{xx}}{2E}(p_{xx} - vp_{yy}) + \frac{p_{yy}}{2E}(p_{yy} - vp_{xx})$$

i.e. $$U = \frac{p_{xx}^2}{2E} - \frac{vp_{xx}p_{yy}}{2E} + \frac{p_{yy}^2}{2E} - \frac{vp_{xx}p_{yy}}{2E}$$

therefore $$U = \frac{1}{2E}\{p_{xx}^2 + p_{yy}^2 - 2vp_{xx}p_{yy}\} \; Q.E.D.$$

Cylinder Design:

Let σ_h = hoop stress

σ_L = longitudinal stress

p = internal pressure

t = thickness of cylinder wall

r = inside radius of cylinder

Now $\qquad \sigma_h = \dfrac{pr}{t}$

and $\qquad \sigma_L = \dfrac{pr}{2t}$

Units:

With p in N/m²; r in metres and t in metres, σ_h must be in N/m². Note that since the ratio $\dfrac{r}{t}$ is dimensionless they must carry the same units.

To avoid confusion, work with metres so as to match the unit N/m²; t in metres also.

$$\sigma_h = \frac{20 \times 100000}{\left(\frac{t}{1000}\right)} \; x \; \frac{500}{1000}$$

i.e. $\qquad \sigma_h = \dfrac{20 \times 100000 \times 500}{t}$

$$= \frac{10^9}{t} \; N/m^2$$

$$\sigma_L = \frac{5 \times 10^8}{t} \; N/m^2$$

Using the expression derived in the first part of the question, and putting $\sigma_h = p_{xx}$ and $\sigma_L = p_{yy}$

$$U = \frac{1}{2E} \left\{ \frac{(10^9 \times 10^9)+}{t^2} + \frac{(5.10^8 \times 5.10^8)}{t^2} - 2\,(0.25)\frac{10^9}{t} \; x \; \frac{5.10^8}{t} \right\}$$

$$= \frac{1}{2E} \left\{ \frac{10^{18}}{t^2} + \frac{25.10^{16}}{t^2} - \frac{2.5 \times 10^{17}}{t^2} \right\}$$

$$= \frac{10^{16}}{2E} \left\{ \frac{100}{t^2} + \frac{25}{t^2} - \frac{25}{t^2} \right\}$$

$$U = \frac{10^{18}}{2Et^2}$$

Now the maximum strain energy under uniaxial direct yield stress σ_{yy} is $\dfrac{\sigma_{yy}^2}{2E}$. With a factor of safety = 3, σ_{yy} = 100 MN/m² (i.e. 300 MN/m²/3).

$$\frac{100 \times 10^6 \times 100 \times 10^6}{2E} = \frac{10^{18}}{2Et^2}$$

i.e $\qquad 10^{16} = \dfrac{10^{18}}{t^2}$

or $\qquad t^2 = 10^2 \qquad\qquad \therefore t = 10 \; mm$

Q24. A copper tube having an elliptical cross-section is plugged at its ends. The major and minor axes of the tube's cross- section are 10mm and 6mm respectively. Hydraulic pressure is applied to the inside of the tube by way of an opening in one of the plugs. What should be the thickness of the tube in order that the hoop stress may not exceed 35 MN/m² under internal hydraulic pressure of 70 bars? Also, what would be the approximate longitudinal stress in the tube?

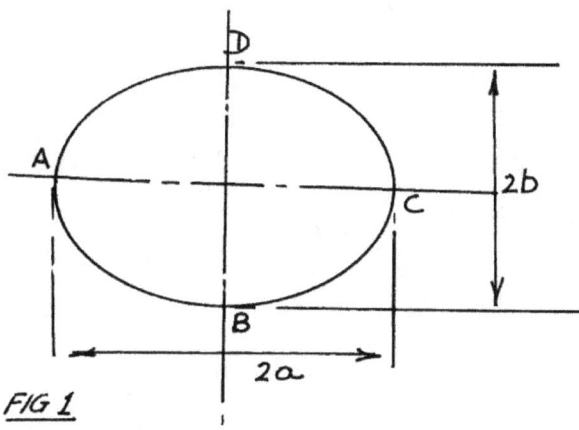

FIG 1

The cross- section of the tube is shown in Fig.1. The intensity of hoop stress varies from point to point along the periphery of the ellipse as I shall now demonstrate.

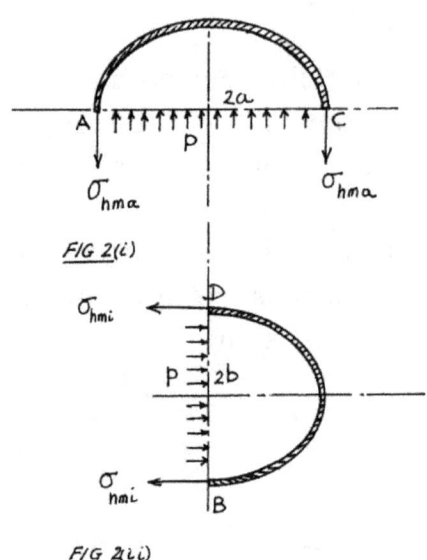

FIG 2(i)

FIG 2(ii)

Considering equilibrium of section AC shown in Fig. 2(i) of length L.

$$2\sigma_{hma} \times L \times t = p \times 2a \times L$$

or
$$\sigma_{hma} = \frac{pa}{t}$$

i.e.
$$\sigma_{hma} = \frac{pa}{t} \times \left(\frac{Major\ Axis}{2}\right) \qquad \dots\dots\dots\dots\dots(i)$$

where t = thickness of tube.

Similarly, considering equilibrium of section BD of length L.

$$2 \times \sigma_{hmi} \times t \times L = p \times 2b \times L$$

i.e.
$$\sigma_{hmi} = \frac{pb}{t}$$

$$\frac{p}{t} \times \left(\frac{Minor\ Axis}{2}\right) \qquad \dots\dots\dots\dots\dots(ii)$$

Therefore the hoop stress varies in value from $\frac{p}{t} \times \left(\frac{Major\ Axis}{2}\right)$ at A (and C)

to $\frac{p}{t} \times \left(\frac{Minor\ Axis}{2}\right)$ at B (and D).

If σ_{Le} is the longitudinal stress, then for the equilibrium of section perpendicular to the axis of the tube, shown in Fig.2(ii).

$$\sigma_{Le} \times \text{perimeter of tube} \times t = p \text{ (c.s.a. of tube)}$$

Now
$$\text{csa of tube} = \pi \left(\frac{Major\ Axis}{2}\right)\left(\frac{Minor\ Axis}{2}\right)$$

~ 350 ~

$$= \pi ab$$

Therefore $\quad\quad\quad\quad\quad\quad \sigma_{Le} = \dfrac{p\pi ab}{T\ (perimetre\ of\ tube)} \quad\quad\quad\quad \ldots\ldots\ldots\ldots(iii)$

The perimeter of an ellipse can only be approximated because of the algebraic expression for its absolute value is the sum of an infinite series. There are however a number of approximations, the most commonly used being: Perimeter of Ellipse $= \pi\ (a+b)$.

I shall use this.

Using equations (i) and (ii) and substituting the numerical values given

$$\sigma_{hma} = \dfrac{p}{t}\ x\ \left(\dfrac{Major\ Axis}{2}\right) \quad\quad\quad\quad \ldots\ldots\ldots\ldots(i)$$

$$35\ x\ 10^6 = \dfrac{70\ x\ 10^5}{t}\ x\ \dfrac{10}{2}$$

t, in millimeters: i.e. $\quad\quad\quad$ t = 1mm

Using (ii)

$$35\ x\ 10^6 = \dfrac{70}{t}\ x\ 10^5\ x\ \dfrac{6}{2}, \quad\quad \text{i.e.}\ \ t = 0.6mm$$

Because the thickness from (ii) will fail in (i), <u>the thickness should be 1mm.</u>

Now $\quad\quad\quad\quad\quad\quad \sigma_{LE} = \dfrac{p\pi ab}{T\ (perimetre\ of\ tube)} \quad\quad\quad\quad \ldots\ldots\ldots\ldots(iii)$

where units are:

σ_{Le} : N /m^2

p   : N /m^2

a,b,t,perimeter : mm

Therefore $\quad\quad\quad \sigma_{LE} = \dfrac{70}{1\ x\ \pi\ x\ (5+3)}\ x\ 10^5\ x\ \pi\ x\ \dfrac{10}{2}\ x\ \dfrac{6}{2}$

$$= \dfrac{70}{8\ x\ 4}\ x\ 10^5\ x\ 10^5\ x\ 6^3$$

$$= 13.1\ x\ 10^6\ N/m^2$$

i.e. $\quad\quad\quad\quad\quad\quad \sigma_{LE} = 13.1\ MN/m^2$

Q25. A mechanical engineer is engaged in the design of a cylindrical pressure vessel for an agricultural–product processing factory. As a first approach she considers her design on the basis of the maximum shearing distortion theory, for which the value of the yield point stress in simple tension is 210 MN /m². If the internal diameter of the vessel is fixed at 1 m, and its working pressure is 15 bars, what thickness will the designer estimate?

According to the maximum shear distortion theory or the Huber-Hencky-von Mises Theory, it is assumed that the failure of the material is caused only by the energy of shearing deformation. Symbolically,

$$2y_{yp}^2 = (\sigma_h - \sigma_L)^2 + \qquad (\sigma_L - \sigma_3)^2 + (\sigma_3 - \sigma_h)^2 \qquad \dots\dots\dots\dots\dots\dots(i)$$

where y_{yp} = yield point stress of material in simple tension.

$\sigma_h > \sigma_L > \sigma_3$ are the principal stresses on three perpendicular axes.

If we neglect σ_3, the local compressive stress equal to the internal pressure, then

$$y_{yp}^2 = \sigma_h^2 - \sigma_h\sigma_L + \sigma_L^2 \qquad \dots\dots\dots\dots\dots\dots(ii)$$

Now for the pressure vessel being designed

$$\sigma_h = 30 \ x \ 10^5 \ x \ \frac{500 \ mm}{t}$$

where t = wall thickness in millimeters

i.e. $\qquad \sigma_h = \frac{15 \ x \ 10^8}{t} \ N/m^2$

Also $\qquad \sigma_L = \frac{7.5 \ x \ 10^8}{t} \ N/m^2$

Substituting in (ii) we get

$$(210 \ x \ 10^6)^2 = (15 \ x \ \frac{10^8}{t})^2 - 15 \ x \ 7.5 \ x \ \frac{10^{16}}{t^2} + (7.5 \ x \ 10^7)^2$$

$$= 225 \ x \ \frac{10^{16}}{t^2} - \frac{112.5}{t^2} \ x \ 10^{16} + \frac{56.25}{t^2} \ x \ 10^{14}$$

$$44100.10^{12} = \frac{168.75 \ x \ 10^{16}}{t^2}$$

~ 352 ~

$$44100 = 10^4 \left(\frac{168.75}{t^2}\right)$$

from which
$$t^2 = \frac{1687500}{44100}$$

$$t^2 = 38.3$$

therefore
$$t = 6.2 \, mm, say, 6 \, ^1/_4 \, mm$$

Q26. An elliptical opening was made in the wall of a cylindrical air vessel for the purpose of routine inspection. A short inspection pipe of elliptical cross-section was welded to the edges of this opening. A vertical cut was made and an elliptical flange welded to the end of the pipe. This opening was then closed by an elliptical blank flange carrying 15 bolts, each 20 mm in diameter. When the vessel is pressurized to 25 bars, what will be the stress in each bolt, assuming a stress concentration factor of 3 at the root of the thread of each bolt? See Fig. 1.

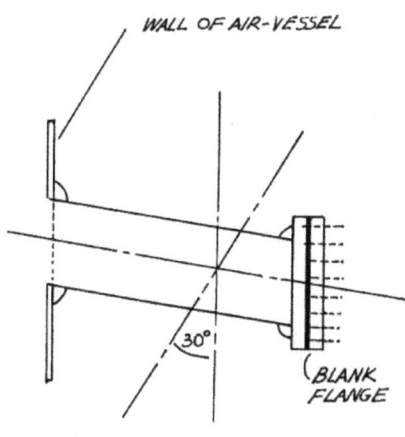

WALL OF AIR-VESSEL

30°

(BLANK FLANGE

PLANE OF ELLIPTICAL CROSS-SECTION:
MAJOR AXIS = 20 cm; MINOR AXIS = 10 cm

FIG 1

Ans: 0.433 MN/m²

Q28. Explain the stress discontinuity at the junction between the ends of a thin-walled pressure vessel and its spherical end caps. Illustrate your answer with neat sketches and explain the discontinuity.

This page is intentionally left blank.

CHAPTER 11

ELASTIC STRESS ANALYSIS OF CIRCULAR BARS IN TORSION

From an engineering standpoint torque means twisting moment or turning moment, and the condition or state of being twisted, engineering-wise that is, is described as torsion. Thus, common around-the-house mechanical components such as the drive shaft of the family car in the garage and the crankshaft of the lawn mower in the workshop are said to be in a state of torsion when transmitting power.

By definition, work done is equal to the product of a force, say, 'F' and the distance moved by the force in its line of action, say 'x'. Hence work done $(W.D) = F(x)$. Now if the same force 'F' is moving in its line of action at a uniform rate of say 'x' units of distance during 't' units of time, i.e. at, say, a velocity $v = \frac{x}{t}$ then the work done per unit of time is the product of the force 'F' and the velocity 'v' in its line of action. The engineering term for rate of doing work is Power. Hence we may write:

$$\text{Power} = \text{Force} \times \text{Velocity} = Fv$$

When this same force 'F" is moving in say an arc of constant radius 'r', velocity 'v' may be expressed as:

$$v = r\omega$$

where ω = angular velocity

$$\therefore \text{Power} = Fv = Fr\omega$$

Now twisting moment or torque, $T = F \times r$ by definition, so that for the force 'F' acting with radius arm 'r' and at angular velocity 'ω', we may write, Power = Torque (angular velocity): $Fr\omega$.

The units for (i) power, (ii) torque, and (iii) angular velocity in SI are respectively (i) watt, abbreviated 'W', with dimensions kgm^2/s^3; (ii) newton metre abbreviated N m, with dimensions kgm^2/s^2; and (iii) radian per second, expressed as rad/s or rads^{-1}, with dimensions t^1.

When circular shafts, either solid or hollow, are in torsion, shearing stresses are induced in them. In this section we shall concern ourselves with such stresses within the elastic range of the material, that is up to the proportional limit.

The formula derived in the textbooks relating torque (T), shearing stress (τ), angle of twist (θ) length (ℓ), modulus of rigidity (C), radius (r) and polar moment of inertia (J) is given by:

$$\frac{T}{J} = \frac{\tau}{r} = \frac{C\theta}{\ell}$$ . . . .   (i)

It is straightway to be noted however that while there are shafts and other components of a non wholly-circular cross-sectional area that resist torque in power transmission, as for example the 'kelly' or 'grief stem' of an oil–well drilling-string, the torsion formula, as the relationship at (i) is commonly called, is not to be applied in such cases; the formula applies only to solid and hollow circular shafts of uniform or uniformly varying circular cross-section throughout their length.

The assumptions underlying the derivation of the torsion formula are:

(i)   material is isotropic;

(ii)   all strains and stresses associated therewith are within the limit of proportionality;

(iii) shearing strain is directly proportional to the distance from the central axis of the shaft which implies that shearing stress is also directly proportional to radial distance from this same axis; and,

(iv)   plane sections perpendicular to axis of a circular member before torsion remain plane after application of torque, that is to say there is no warping of such sections or no deformation other than the shear strain acting in planes perpendicular to the axis. This assumption as we shall see is a wholly idealized one.

In order to derive the Torsion Formula consider a solid circular cylindrical shaft of length 'L' and of uniform radius 'R' subjected to a turning moment or torque, say, T at the shaft's free end; the other end of the shaft is fixed to a rigid support. The axis of 'T' is the same as the axis of the shaft. The shaft is in equilibrium because 'T' at the free end is balanced by 'T' in the opposite sense at the support or fixed end. The shaft is therefore said to be in a state of 'pure torsion' and each cross-section is accordingly in a state of pure shear.

Now refer to Fig. 1.

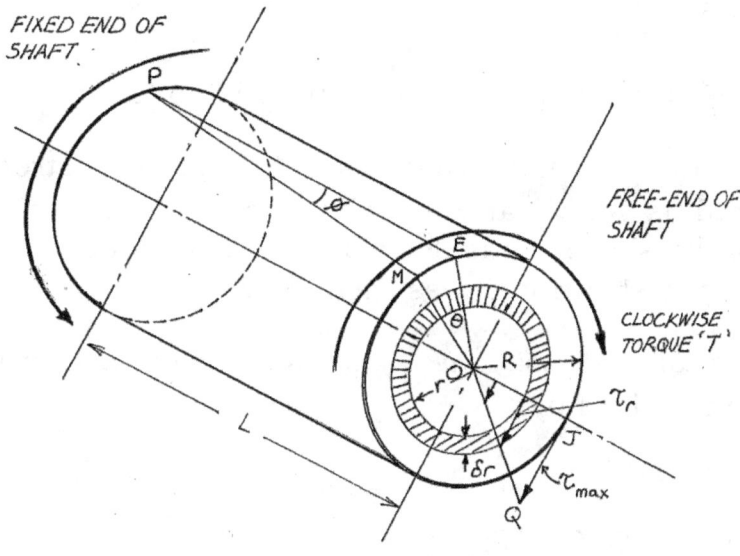

ANTI-CLOCKWISE TORQUE 'T'

FIXED END OF SHAFT

FREE-END OF SHAFT

CLOCKWISE TORQUE 'T'

FIG. 1

It is not difficult to visualize the effect of 'T' on each cross-section of the shaft: causing displacement of each cross-section relative to each other, starting with zero displacement at the support, to EM at the free end; a line such as PE parallel to the axis of shaft originally displaced to PM. In the shaft itself angular displacement is 'θ'. On the surface of the shaft angular displacement of PE to PM = 'ϕ'. L = constant length of the shaft.

Now ϕ being small we may write

$$EM = L\phi = R\theta, \text{ i.e. } \phi = R\theta/L$$

By Hooke's law for shearing stress 'τ'

$$C = \frac{\tau}{\phi} \quad \text{or} \quad \tau = C\phi$$

'C' being the modulus of rigidity of the material of the shaft, assumed constant.

Employing these results:

From $\qquad \tau = C\phi \qquad$ and $\qquad \phi = R\theta/L$

$$\tau = \frac{CR\theta}{L} \quad \text{or} \quad \frac{\tau}{R} = \frac{C\theta}{L}$$

~ 357 ~

Thus, the latter may be written as

$$\tau = KR$$

where $K = \dfrac{C\theta}{L}$, C being itself a constant; and so too 'L'.

This means for any constant value of 'θ', shearing stress 'τ' is directly proportional to radius 'R'. Clearly therefore the maximum value of 'τ' occurs at the outer skin of the shaft. Consider next a ring at radius 'r' and infinitesimal thickness δr. Let the shearing stress there be τ_r.

Accordingly the total force, say, δF_r on this elemental ring is obtained from: stress × area, and is given by:

$$\delta F_r = \tau_r (2\pi r \delta r)$$

By similar triangles in OJQ of Fig. 1

$$\frac{\tau_r}{\tau_{max}} = \frac{r}{R}$$

$$\therefore \quad \tau_r = \tau_{max} \frac{r}{R}$$

so that the force δF_r acting on the infinitesimal element may now be written as:

$$\delta F_r = \tau_{max} \cdot \frac{r}{R} \cdot 2\pi \, r \delta r$$

$$= \frac{\tau_{max}}{R} 2\pi \, r^2 \delta r$$

and the moment of δF_r about the centre of the shaft is

$$\delta F_r (r) = \frac{\tau_{max} \, 2\pi \, r^3 \delta r}{R} = \delta T_r$$

To obtain the total resisting moment we have to integrate this latter expression. Hence total torque is:

$$\int dT_r = T = \frac{2\pi \tau_{max}}{R} \int_0^R r^3 \, dr$$

$$\therefore \quad T = \frac{\tau_{max} \, \pi \, R^3}{2}$$

This result may be re-expressed as

$$T = \frac{\tau_{max}}{R} \cdot \frac{\pi R^4}{2}$$

You would at once recognize $\frac{\pi R^4}{2}$ as the polar moment of inertia of the shaft about an axis through its centre and perpendicular to its cross-section. Thus, denoting the polar moment of inertia by 'J', we may express the final result as

$$T = \frac{\tau_{max}}{R} \cdot J$$

$$\therefore \quad \frac{T}{J} = \frac{\tau_{max}}{R}$$

Evidently $\tau_{max} / R = \frac{\tau}{r}$, so that employing the relationship: $\frac{\tau}{R} = \frac{C\theta}{L}$ we may write,

$$\frac{T}{J} = \frac{\tau}{r} = \frac{C\theta}{L}$$

This is the Torsion Equation. Frame it and put it up there with the others over your bedhead and rehearse its development in your mind before you say your prayers.

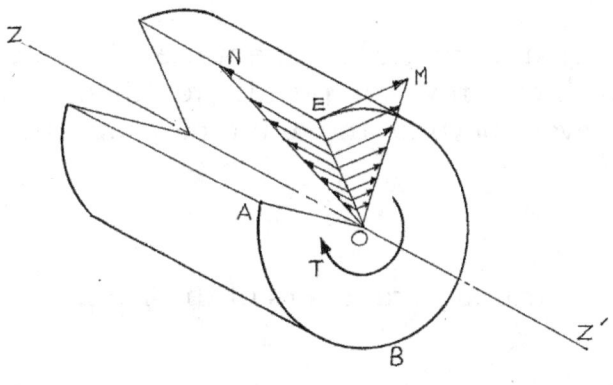

FIG 2.

In Fig. 2 the distribution of shearing stress on a section of the shaft whose face 'ABE' is perpendicular to shaft-axis ZZ' is shown. See triangle OME. The complementary shear stress distribution which is at 90° to ABE and parallel to the axis of the shaft is also shown. See triangle OEN. These complementary stresses act across planes perpendicular to the axis of the shaft and cause some distortion in the direction of their line of action resulting in some warping. Thus, the assumption (iv) referred to at the outset, on which the Torsion formula is based about plane sections remaining plane could be criticized on this ground. Further, the complementary shearing stresses must cause a square such as 'ABCD' scribed on the surface of a circular cylindrical shaft shown in Fig. 3, which would on the basis of the assumption about shearing strain move to $A'B'C'D'$, take up instead the position $A''B''C''D''$. Hence the assumption about strains varying linearly across sections starting from the axis of the shaft is also a point of contention.

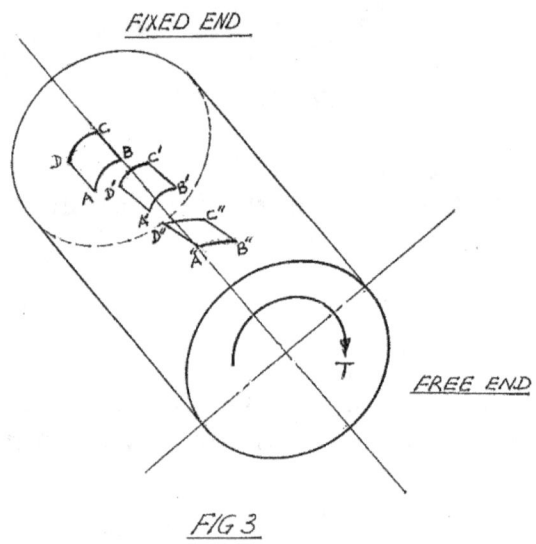

FIG 3

These criticisms notwithstanding, experimental evidence proves however, that the assumptions are not only completely justified, but that the Torsion Formula works well even beyond the limit of proportionality of a material. (Popov).

UNITS

The units in which the various parameters of the Torsion Formula are generally expressed are:

Torque (T)	:	newton metre : N m
Polar moment of Inertia (J)	:	$(\text{metre})^4$: m^4
Shear stress (τ)	:	newton per $(\text{metre})^2$: N/m^2

Radius (r)	:	metre : m		
Modulus of Rigidity (C or G)	:	newton per (metre)^2 : N/m^2		
Angle of Twist (θ)	:	radian : rad		
Length (ℓ)	:	metre : m		

The equation $\dfrac{T}{J} = \dfrac{\tau}{r}$

is sometimes written in the form:

$$T = \tau_{max} Z_o \qquad\qquad \dots\dots\dots\dots\dots\dots \text{(ii)}$$

where τ_{max} = maximum shear stress, which as you know occurs in the outer fibres of the shaft; and

Z_o = Polar modulus of the cross-section of the shaft $= \dfrac{J}{r}$

Remember that the value of 'T' in equation (ii) to cause the maximum shear stress, τ_{max}, must be the maximum value of the torque on the shaft. When you are given that a shaft is transmitting say, 'W' watts at N rev/min, the value of torque obtained from

$$W = \text{Torque} \times \frac{2\pi N}{60}$$

is an average value = T_{avg}. In the absence of any further information this would be the value of 'T' to be used in (ii). However, it is sometimes given that the maximum torque is, say, 'j' times the average torque where $j \succ 1$; in which case the value of 'T' to generate τ_{max} is 'jT_{avg}' and

$$j \times T_{avg} = \tau_{max} Z_o$$

or $\qquad T_{max} = \tau_{max} \cdot Z_o$

Illustrative Example 1

A hollow shaft 250mm external diameter and 180mm internal diameter is used in a mechanical power-transmission system. The maximum torque on the shaft is 1.25 times the mean torque. What is the power transmitted at 75 rpm assuming that the maximum allowable working shearing stress in the shaft is not to exceed 65MN/m²? Determine also the value of the maximum torque.

Recall

$$T_{max} = \tau_{max} Z_o$$

$$Z_o = \frac{J}{r}$$

i.e.
$$Z_o = \frac{\pi}{32}\left\{\left(\frac{250}{1000}\right)^4 - \left(\frac{180}{1000}\right)^4\right\} \cdot \frac{1}{\frac{250}{2(1000)}}$$

$$= \frac{\pi}{32} \cdot \frac{1}{10^8}\left\{(25)^4 - (18)^4\right\} \cdot \frac{2000}{250}$$

$$= \frac{\pi}{32} \cdot \frac{1}{10^8}(390625 - 104976)$$

$$= \frac{\pi}{4.10^8}(285649)$$

$$= 2.244 \times 10^{-3} m^3$$

$$\therefore \quad T_{max} = 65 \times 10^6 \times 2.244 \times 10^{-3} Nm$$

$$= 145.86 \times 10^3 Nm$$

Given $T_{max} = 1.25 T_{avg}$

T_{avg} being the average torque

$$\therefore \quad T_{avg} = \frac{T_{max}}{1.25}$$

$$= \frac{145.86}{1.25} \times 10^3$$

$$T_{avg} = 116.5 \times 10^3 Nm$$

Power transmitted at 75 rpm?

~ 362 ~

Angular velocity at 75 rpm $= \dfrac{2\pi}{60}(75)$

$\therefore$ Power transmitted $= T_{avg}\,w = 7.68$ rad s^{-1}

$= 116.7 \times 10^3 \times 7.86$

$= 917.26 \times 10^3$ watts, say 917.3 kW

A favourite examination question deals with determination of shaft diameter to satisfy, simultaneously, criteria such that (i) a certain value of maximum shear stress and (ii) a specified angle of twist in a given length, respectively, must not be exceeded. A typical question is that represented by the following.

Illustrative Example 2

The shearing stress in a hollow circular cylindrical shaft transmitting 400kW at 150 revs/min with a uniform torque is not to exceed 40MN/m². The ratio of outside to inside diameter is $\dfrac{5}{4}$ and the angle of twist is not to exceed 1.5° in 3 metres. What are the minimum external and internal diameters of the shaft?

Take C = 80 GN/m²

$$\text{Power} = T \times w$$

$\therefore \quad 400 \times 10^3 = T \times 2\pi \cdot \dfrac{150}{60}$

$\therefore \quad T = 25.5 \times 10^3\,Nm.$

Recalling

$$\frac{T}{J} = \frac{\tau_{max}}{D_o/2}$$

Let us work in metres but with D_o, D_i expressed in millimetres

DIAMETER FOR STRESS CRITERION

$$\frac{T}{\frac{\pi}{32}\left\{\left(\frac{D_o}{1000}\right)^4 - \left(\frac{D_i}{1000}\right)^4\right\}} = \frac{2\tau_{max}}{\left(\frac{D_o}{1000}\right)}$$

$$\frac{T}{\frac{\tau}{32}\cdot\frac{1}{10^{12}}(D_o^4 - D_i^4)} = 2\frac{\tau_{max}}{D_o}\cdot 1000$$

$$T\cdot\frac{10\cdot 32}{\pi}\cdot\frac{1}{D_o^4\left\{1-\left(\frac{D_i}{D_o}\right)^4\right\}} = \frac{2\pi_{max}10^3}{D_o}$$

$$\frac{D_i}{D_o} = \frac{4}{5}$$

$$\frac{T\cdot 16\cdot 10^9}{\pi D_o^3\left\{1-\left(\frac{4}{5}\right)^4\right\}} = \tau_{max}$$

$$\therefore\quad \frac{25.5\times 10^3\times 16\times 10^9}{\pi D_o^3(0.5904)} = 40\times 10^6$$

$$\therefore\quad D_o^3 = \frac{25.5\times 10^6\times 16}{40\times\pi\times 0.5904}$$

$$= 5.4985432\times 10\times 10^6$$

$$\therefore\quad D_o = 1.764\times 10^2$$

$$= 176mm.$$

DIAMETER FOR ANGLE OF TWIST CRITERION

Remember that 'θ' in the Torsion Formula is in radians.

Because $\quad 360° = 2\pi\,$rad

$$1.5° = \frac{0.5}{360}\times 2\pi\ \text{rad.}$$

$$\therefore \quad \frac{T}{\frac{\pi}{32}\left\{\left(\frac{D_o}{1000}\right)^4 - \left(\frac{D_i}{1000}\right)^4\right\}} = \frac{C}{3} \cdot \frac{1.5}{360} \times 2\pi$$

i.e.

$$\frac{25.5 \times 10^3 \times 32 \times 10^{12}}{D_o^4\left\{1 - \left(\frac{D_i}{D_o}\right)^4\right\}} = \frac{80 \times 10 \times 2\pi}{3 \times 360} \times \frac{3}{2}$$

$$\therefore \quad D_o^4(0.5904) = \frac{25.5 \times 10^6 \times 32 \times 6 \times 360}{80 \times 2\pi \times 3}$$

$$D_o^4 = \frac{25.5 \times 10^6 \times 32 \times 6 \times 360}{80 \times 2\pi \times 0.5904 \times 3}$$

$$= 1979.47 \times 10^6$$

$$\therefore \quad D_o = 211mm.$$

Evidently this value of D_o will satisfy both criteria because if $D_o < 211mm$, say equal to 176mm the angle of twist criterion will not be the satisfied; $D_i = \frac{4}{5}D_o$ so that $D_i = 211mm$, the inside diameter $D_i = 168.8mm$.

FLANGED SHAFT COUPLING

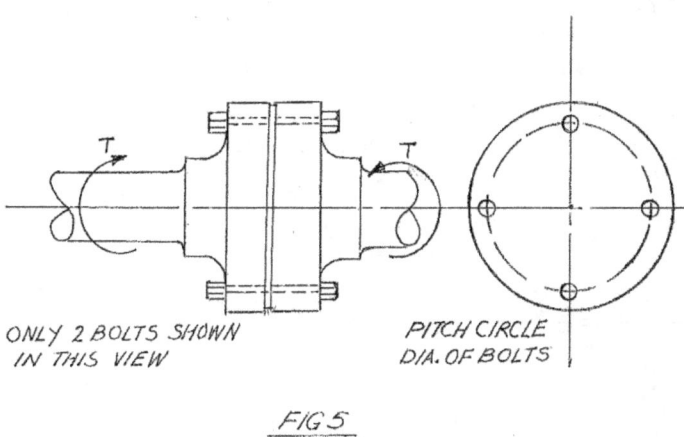

ONLY 2 BOLTS SHOWN
IN THIS VIEW

PITCH CIRCLE
DIA. OF BOLTS

FIG 5

In Fig. 5 a typical flanged coupling in equilibrium joining 2 shafts transmitting torque 'T' is shown.

Assuming 'd' as diameter of 1 bolt, 's' as maximum allowable shear stress in a bolt and 'pcd' as the pitch circle diameter around which the bolts are fitted.

Bolts in such flanged shaft coupling arrangements are designed on the basis that the sum total of their moment of resistance to shear about the shaft axis is equal to the torque transmitted. Therefore if there are n bolts, then:

$$T = n \times s \times \frac{\pi d^2}{4} \times \frac{(pcd)}{2}$$

SHAFTS OF EQUAL STRENGTH

Shafts are of equal strength when they transmit the same torque at the same maximum shear stress. I reiterate: because shear stress is taken as being directly proportional to distance from shaft centre, it should be clear that the maximum value most of the stresses induced in resisting a torque are located in the outermost skin of the shaft. This is of great practical significance because savings of weight would be effected by employing say, a hollow shaft to replace a solid one as for example in aircraft engine design. The material properties of such hollow shafts are of immense importance. Consider a solid circular cylindrical shaft of diameter 'D' designed to transmit a certain power at a specified maximum allowable shear stress concerning which decisions related to minimizing mass are of paramount importance. Suppose this shaft is to be replaced by a hollow one of outside diameter 'd$_o$' and inside diameter 'd$_i$' the material of construction being the same as that of the solid shaft and ratio $\frac{d_i}{d_o} = n.$ We have for the solid shaft

$$\frac{T}{\frac{\pi D^4}{32}} = \frac{\tau_{max}}{\frac{D}{2}}$$

i.e. $\qquad T = \frac{\pi D_s^3 \tau_{max}}{16}$ $\qquad$ . . . .

and for hollow shaft

$$\frac{T}{\frac{\pi}{32}(d_o^4 - d_i^4)} = \frac{T_{max}}{\frac{d_o}{2}}$$

from which $\qquad \dfrac{32T}{\pi d_o^4 \left\{ 1 - \left(\dfrac{d_i}{d_o} \right)^4 \right\}} = \dfrac{2\tau_{max}}{d_o}$

Putting $d_i/d_o = n$, the latter results becomes

$$\frac{16T}{\pi \, d_o^3(1-n^4)} = \tau_{max}$$

For the same strength of shafts, both T and τ_{max} are the same

$$\therefore \qquad \frac{T}{\tau_{max}} = \frac{\pi \, D_s^3}{16} = \frac{\pi \, d_o^3 \, (1-n^4)}{16}$$

or $$D_s^3 = d_o^3(1-n^4)$$

Suppose the diameter of a solid circular shaft is 150mm and it is to be replaced by a hollow counterpart, inside diameter being $\frac{3}{4}$ outside diameter, i.e. $n = 3/4$, then,

$$(150)^3 = d_o^3\left\{1-\left(\frac{3}{4}\right)^4\right\}$$

$$= d_o^3(1-0.3164)$$

$$= 0.6836d_o^3$$

$$\therefore \qquad 337500 = 0.6836d_o^3$$

$$d_o^3 = 493098$$

$$\therefore \qquad d_o = 169mm$$

and $$d_i = 127mm$$

If ρ = density of material of shaft

ℓ = length of shaft

then, mass of solid shaft, $$M_s = \pi\left(\frac{150}{4}\right)^2 \times \ell \times \rho,$$

and " " hollow ", $$M_H = \frac{\pi}{4}(169^2-127^2)\times \ell \times \rho$$

$$\therefore \qquad M_s = 17674\ell\rho$$

and
$$M_H = \frac{\pi}{4}(12432)\ell\rho$$

$$M_H = 9765\ell\rho$$

$\therefore$ Percent lightness of hollow shaft compared with solid shaft

$$100\left(\frac{17674\ell\rho - 9765\ell\rho}{17674\ell\rho}\right) = 44.75\%$$

TORSION OF THIN-WALLED TUBES : BREDT'S FORMULA

It was not my intention originally to treat with the topic of torsion of non-circular sections. However, because of the importance of shearing stress determination in hollow sections of arbitrary shape and with walls of variable thickness -- small in comparison with their other dimensions -- it was thought appropriate to introduce the subject of torsion in these sections. Such stress determinations play an important part in the design of aircraft structures for example. The derivation is after Popov.

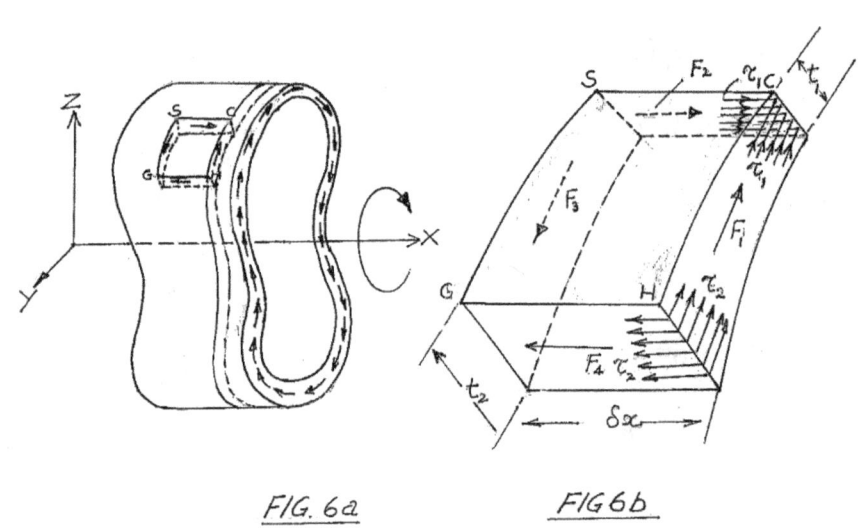

FIG. 6a   FIG 6b

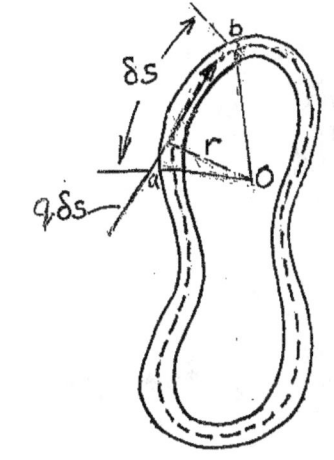

Consider therefore the infinitesimal element of the hollow closed tube shown in Fig. 6a; the surface of the element is marked SGHC. The element is shown magnified in Fig. 6b which is its free-body diagram. Its thickness is constant in the longitudinal direction, but of magnitude t_1 at side SC and t_2 at GH as shown in the diagram 6b. For equilibrium in the direction of the X-axis $F_2 = F_4$; τ_1 and τ_2 are complementary shearing stresses. Therefore $\tau_1\, t_1\, \delta x = r_2\, t_2\, \delta x$ or $\tau_1\, \tau_1 = \tau_2\, t_2$. So that if we were to choose arbitrarily other elements in planes normal to the axis of the tube it could fairly be deduced that $\tau_1\, t_1 = \tau_2\, t_2 = \tau_3\, t_3 = \ldots \tau_n\, t_n$. That is to say the product $\tau\, t =$ constant. This constant is referred as the shear flow and is generally denoted by the symbol 'q'; and, expressed as newton/metre i.e.force per unit length. Refer to Chapter 9.

In Fig. 6c an infinitesimal length 'δs' of the tube measured along the median line of the perimeter is shown with the tangential force $q\delta s$ acting on a radius 'r' measured from an arbitrary point 'O'. Consequently we may express the infinitesimal torque δT on the element 'δs' as follows : $\delta T = q\delta s(r)$. Therefore for the entire median line of length 'L$_m$' enclosing the tube

$$\int \delta T = q \int_0^{L_m} r\,ds$$

Referring again to Fig. 6c, the area of triangle OAB which for the sake of convenience is designated A$_1$ may be approximated to $\frac{1}{2}r\delta s$, i.e. $\frac{1}{2}(height)(base)$.

Thus, $A_1 = \frac{1}{2}r\delta s$ or $2A_1 = r\delta s$.

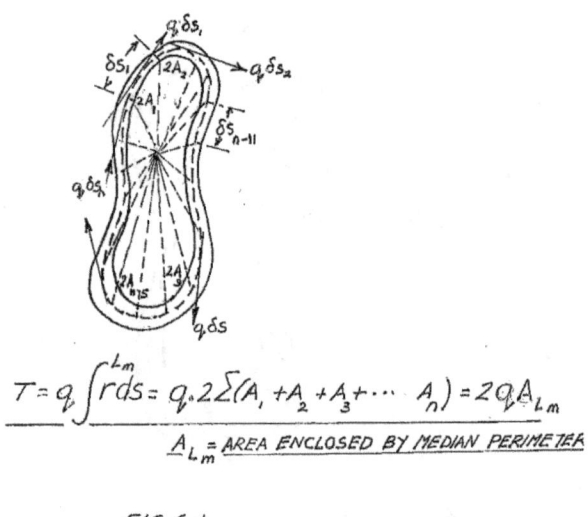

$$T = q_v \int^{L_m} r\,ds = q_v . 2\Sigma(A_1 + A_2 + A_3 + \cdots A_n) = 2q_v A_{L_m}$$

A_{L_m} = AREA ENCLOSED BY MEDIAN PERIMETER

FIG 6d

In Fig. 6d I have shown the entire area enclosed by the median line as made up of triangles $A_1, A_2, A_3, ... A_n$. Accordingly:

$$\int_o^{L_m} r\,ds = 2\sum A_1 + A_2 + A_3 + ...A_n$$

$$= 2A$$

$$\therefore \quad T = 2Aq \quad \text{or} \quad q = T/A_m$$

where 'A_m' is the total area enclosed by the median line L_m. And because shearing stress 'τ' is expressed as q/t

$$\tau = \frac{T}{2A_m t}$$

This formula is known in the engineering literature after Rudolph Bredt a German engineer who studied in Karlsruhe and in Zurich, Switzerland. He also worked in England and later became a manufacturer of cranes in which connection he developed his torsion theory.

It is interesting to note that if one were to apply the Torsion formula which was developed earlier in this chapter to a thin-walled circular tube of outside and inside radii, R_o and R_i respectively then on the basis of

$$\frac{T}{J} = \frac{\tau}{R_o}$$

we obtain

$$\frac{I}{\frac{\pi}{2}\left(R_o^2 - R_i^2\right)} = \frac{\tau}{R_o}$$

which may be rewritten in the form

$$\frac{2T}{\pi\left(R_o - R_i\right)\left(R_o + R_i\right)\left(R_o^{2+} R_i^2\right)} = \frac{\tau}{R_o}$$

But thickness of the tube $'t' = R_o - R_i$ and because t is small $R_o \approx R_i \approx R_m$

$$\therefore \qquad \frac{2T}{\pi(t)} \frac{1}{2R_o} \cdot \frac{1}{2R_o^2} = \frac{\tau}{R_o}$$

i.e. $\qquad \dfrac{T}{2\pi R_o^2\, t} = \tau$

which may be approximated to

$$\tau = \frac{T}{2\tau R_m^2\, t}$$

The beauty of Bredt's formula is that it is applicable to thin closed-walled tubes of any shape. The approximate solution derived for a thin circular tube using the Torsion Formula $\dfrac{T}{J} = \dfrac{C\theta}{\ell} = \dfrac{\tau}{r}$ simply would not work for, say, a hollow tube of square or triangular cross-section for example.

Illustrative Example 4

A torque of 120Nm is applied at A to a handwheel attached to a gear wheel having n_2 teeth, this wheel fitted to a thin-walled brass tube of length 2m, mean diameter 40mm and thickness 5mm. If the angle of twist in the tube is 1°, what is the gear ratio n_1/n_2. Take C for brass = 39GN/m². See Fig. 7 and 7a. Use Bredt's formula to determine shear stress in the brass tube.

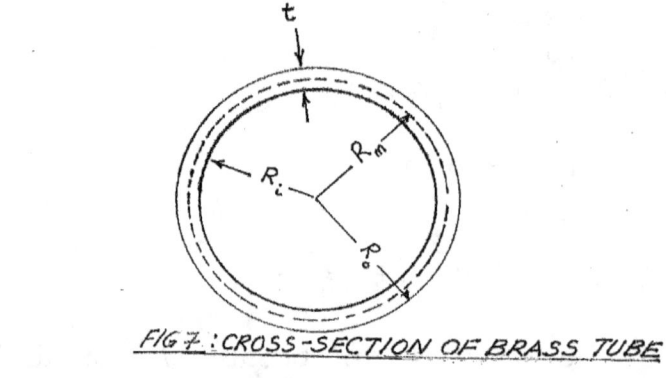

FIG 7 : CROSS-SECTION OF BRASS TUBE

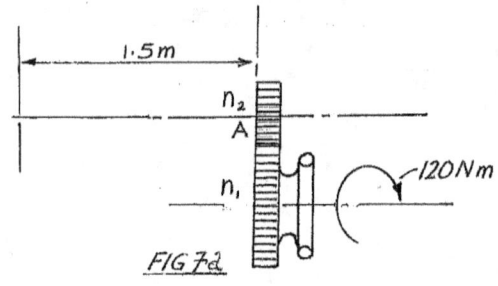

FIG 7a

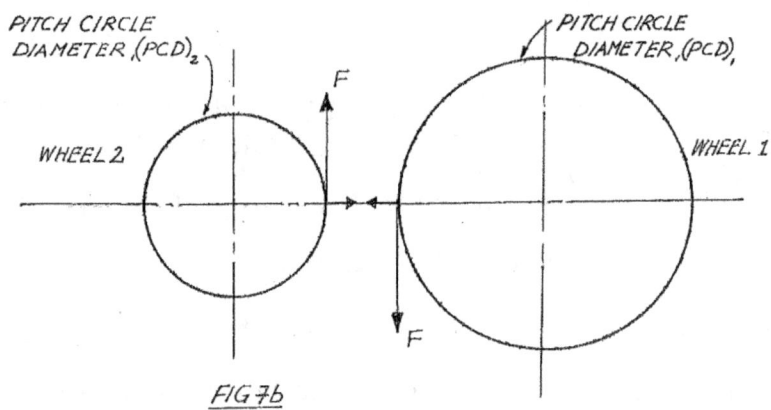

FIG 7b

Torque on wheel 1 = 120Nm

Designating F as force at (PCD), to produce this torque, then referring to Fig. 7b

$$F \times (PCD) = 120Nm$$

$$\therefore \qquad F = \frac{120}{(PCD)/2}$$

Therefore torque on wheel $\quad 2 \quad = \quad \dfrac{120}{\dfrac{(PCD_1)}{2}} \times \dfrac{(PCD)_2}{2}$

$\therefore \quad$ Torque on wheel 2 $\quad = \quad \dfrac{120(PCD)_2}{(PCD)_1}$

With n_1, n_2 as the number of teeth on gear wheels 1 and 2 respectively, we have

$$\frac{n_2}{n_1} = \frac{(PCD)_2}{(PCD)_1}$$

Torque on wheel 2 $\quad = \quad 120\left(\dfrac{n_2}{n_1}\right) kN$

Applying Bredt's formula, the shear stress, τ, in the thin-walled tube affixed to wheel 2 is given by

$$\tau = \frac{120\left(\dfrac{n_2}{n_1}\right)}{2 \times \pi \left(\dfrac{20}{1000}\right)^2 \times \left(\dfrac{5}{1000}\right)}$$

the mean radius and thickness of the tube being 20mm and 5mm respectively

$$\therefore \quad \tau = 120 \left(\frac{n_2}{n_1}\right) \cdot \frac{10^6 \times 1000}{2 \times \pi \cdot (20)^2 \, 5}$$

Writing the gear ratio $n_1/n_2 = k$

$$\tau = \frac{120}{k} \times \frac{10^9}{2 \times \pi \times 400 \times 5}$$

$$\tau = \frac{120 \times 10^6}{4\pi k}$$

Further $\quad \dfrac{\tau}{r_m} = \dfrac{C\theta}{\ell}$

$$\therefore \quad \frac{120 \times 10^6}{4\pi \, k \left(\dfrac{20}{1000}\right)} = 39 \times \frac{10^9}{2} \times \theta$$

Angle of $1° = \dfrac{1}{360} \times 2\pi$ radian

$\therefore \quad \dfrac{120 \times 10^3}{4\pi \; k \times 20} = \dfrac{39 \times 10^3 \times 1}{360 \times 2} \times 2\pi$

$4\pi \; k \times 20 \times 39 \times 1 \times 2\pi = 120 \times 360 \times 2$

$\therefore \quad k = \dfrac{120 \times 360 \times 2}{4\pi \times 20 \times 39 \times 2\pi} = 1.4$

With a gear ratio of 1.4, possible configurations are n_1 = 35 teeth; n_2 = 25 teeth or n_1 = 70 teeth; n_2 = 50 teeth and son on.

ANGLE OF TWIST $\emptyset$ FOR A THIN-WALLED TUBE UNDER TORSION 'T'

In order to develop an expression for angle of twist ϕ for a thin-walled tube of arbitrary cross-section subjected to torsion 'T', the work done by 'T' in twisting the tube through angle 'ϕ' is equated to the total strain energy in the tube due to the shearing stress induced by the torsion 'T'. Working from first principles, consider the shear strain energy in an element such as that in Fig. 6e.

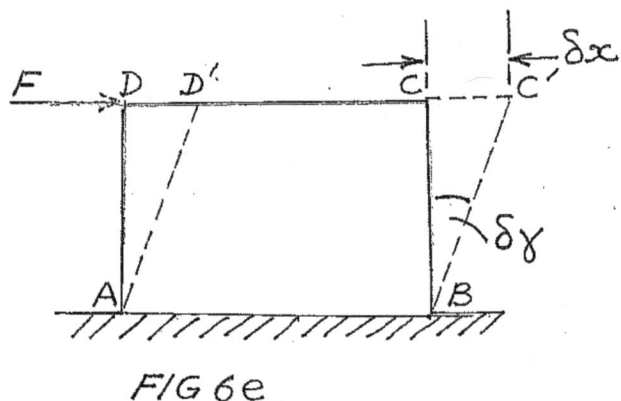

FIG 6e

on which the gradually, applied force 'F' produces a movement 'δx' in its direction. Evidently the work done 'δU' is the area under the "F" vs 'x' characteristic, i.e. $F\delta x/2$. Remember "F" starts from zero and achieves a value of 'F' when the distance travelled in its line of action is 'δx' force being proportional to distance travelled. By contrast if a force is already at value 'F' and moves a distance, say, 'x' at that magnitude, then the work done is F(x).

Accordingly, $\qquad U = \dfrac{F\delta x}{2} = \dfrac{1}{2} F \cdot \ell \cdot \delta \gamma$

If shear stress is designated τ and 'a' is the area over which τ acts, then
$F = \tau a$

$$U = \frac{F\delta x}{2} = \frac{1}{2} \cdot \tau a \cdot t \cdot \delta \gamma$$

But by definition modulus of rigidity $G = \dfrac{\tau}{\delta \gamma}$.

Here $\delta \alpha = \dfrac{\tau}{G}$, so that

$$U = \frac{F\delta x}{2} = \frac{1}{2} \tau a \cdot \ell \cdot \frac{\tau}{G}$$

$$U = \frac{1}{2} \frac{\tau}{G}(a\ell)$$

Thus, the strain energy or resilience of the tube is $\dfrac{1}{2}\tau^2$ (Volume).

Applying this relationship to the element shown in Fig. 6, the infinitesimal strain energy may be written in the form

$$\delta u = \frac{1}{2} \frac{\tau^2}{2G}(t\ \delta s\ \delta x)$$

Substituting $\tau = \dfrac{q}{t}$ i.e. shear flow/t in the latter produces

$$\delta u = \frac{1}{2G} \frac{q^2}{t^2} \cdot t \cdot \delta s \cdot \delta x$$

$$= \frac{1}{2G} \frac{q2}{t} \delta s \cdot \delta x$$

$$\therefore \quad \int dU = \frac{1}{2G}q^2 \cdot \int_o^L \frac{ds}{t} \int_o^L dx$$

The limits of integration being : For dx : from 0 to L = length of the tube and for $\dfrac{ds}{t}$, the distance along the median perimeter with 't' taking appropriate values as 's' changes

Accordingly
$$U = \frac{q^2 L}{2G} \int_o^{L_m} \frac{ds}{t}$$

This result may be expressed in terms of the applied torque 'T'. Thus from

$$q = \frac{T}{2A_m}$$

we may now write

$$U = \frac{T^2 L}{8G\, A_m^2} \int_o^{L_m} \frac{ds}{t}$$

In tables of the properties of cross-sections you would come across values for equal and un-equal angles, rolled hollow-squares and so on, of a property called torsional constant. For a thin-walled tube it is denoted by the letter J and is defined as

$$J = \frac{4A_m^2}{\int_o^{L_m} \frac{ds}{t}}$$

This 'J' is not to be confused with the symbol for polar moment of inertia. I prefer to designate it J_{TC}. The resilience of the tube may therefore be re-expressed as:

$$U = \frac{T^2 L}{2GJ}$$

and if we equate this with the work done by T in twisting the tube through angle ϕ, then from $\frac{T\phi}{2} = \frac{T^2 L}{2GJ}$ we find that $\phi = TL/GJ_{TC}$.

Illustrative Example 6

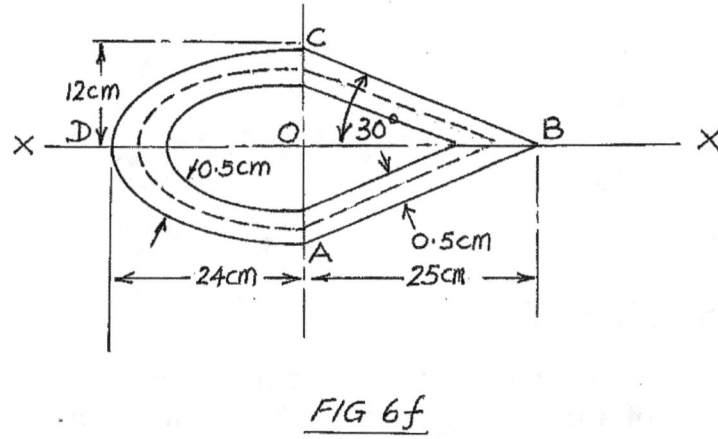

FIG 6f

A thin tube is of uniform cross-section as shown in Fig. 6f throughout its length. ACD is a semi-ellipse of major axis 'a' = 48cm and minor axis 'b' = 24cm; the other portion of the tube is of triangular cross-section. OB = 25cm; thickness of the tube = 0.5cm throughout. If the applied torque is of magnitude 80Nm, then what is the value of shear stress in the wall of the tube.

Let A_m = total area enclosed by the perimeter of the median line around the tube

A_m is divided into 2 parts: one for the semi-elliptical area, and the other for the triangular portion;

Semi-Elliptical area $= \pi ab/2$

$$= \frac{\pi(12-0.25)(24-0.25)m^2}{2\times(100)^2}$$

$$= \frac{\pi(11.75)(23.75)}{2\times10^4}$$

$$= 438.3\times10^{-4}\,m^2$$

Triangular area $= \dfrac{1}{2}\dfrac{(24-0.5)(25-0.5)}{(100)^2}m^2$

$$= \frac{1}{2}\frac{(23.5)(24.5)}{10^4}$$

$$= 287.8\times10^{-4}\,m^2$$

Accordingly, $A_m = (438.3 + 287.8) = 726.1\times10^{-4}\,m^2$

~ 377 ~

But shear stress $\tau = \dfrac{T}{2A_m t}$

Given that T = 50Nm,

$$\tau = \dfrac{80 \times 10^4 \times 100}{2 \times 726.1 \times 0.5}$$

$$= 0.11 \times 10^6 \, N/m^2 \qquad\qquad \text{i.e.} \qquad \tau = 110 kN/m^2$$

Illustrative Example 7

The thin-walled tube of uniform cross-section as shown in Fig... throughout its length is made up of a triangular portion ABC and a semi-circular portion. The outer radius of the semi-circular portion is 12cm and its wall thickness 1cm. OB = 24cm and the thickness of the triangular portion 0.5cm. What is the angle of twist per unit length assuming a torque of 80Nm is applied to the tube? Take modulus of rigidity = 25MN/m². Take length of each of the median lines along sides AB and BC of the triangular portion of the tube = 33.6cm. Neglect the effects of stress concentration at the junction of the two cross-sections.

Fig. 6g

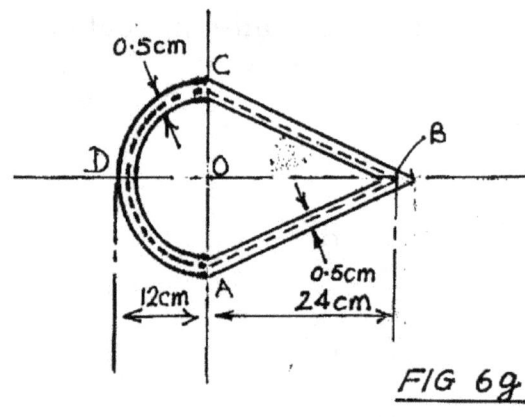

FIG 6g

Let us first evaluate the torsional constant 'J' for the entire tube.

Writing

$$J = \dfrac{4A_m^2}{\displaystyle\int_0^{L_m} \dfrac{ds}{t}}$$

we have for the semi-circular portion of the tube

$$A_m = \frac{\pi r_m^2}{2} = \frac{\pi (11.5)^2}{2 \times 10^4} = 207.7 \times 10^{-4} m^2, \text{ and}$$

$$\int_o^{L_m} \frac{ds}{t} = \frac{\pi r_m}{1}. \text{ For } r_m = 11.5 cm \text{ and } t = 1cm$$

$$\therefore \quad \int_o^{L_m} \frac{ds}{t} = \pi \frac{(11.5)}{1} = 36.1$$

$$\therefore \quad \text{J (semi-circular portion)} = \frac{4(207.7)^2 \times 10^{-4}}{36.1} = 4780 \times 10^{-8} m^4$$

Next, we consider the triangular portion of the tube. With reference to Fig. 6g its A_m = (24 x 23)/2 = 276 x 10^{-4} m^2.

$A_m = 287.8 \times 10^{-4} m^2$ from the determination made in Illustrative Example 6.

We have to obtain the length of each of the median lines along AB and BC. With OB = 24cm, we have $(11.5)^2 + (24)^2$ = (Median Line along CB)^2

$$\therefore \quad \text{Median Line along CB = 26.6 cm}$$

$$\therefore \quad 2\int_o^{33.6} \frac{ds}{t} = \frac{2(33.6)}{0.5} = 134.4$$

So that $J = \dfrac{(287.8)^2 \times 10^{-4}}{134.4} = 2465 \times 10^{-8} m^4$

Accordingly total 'J' for the section = $7245 \times 10^{-8} m^4$ We are given T = 80Nm, so that angle of twist per length of tube $\phi / L = \dfrac{T}{GJ}$

i.e.
$$\phi / L = \frac{80 \times 10^8}{25 \times 10^6 \cdot 7245} = \frac{8000}{25 \times 7245}$$

$$\phi / L = 0.044 \text{ rad.}$$

Freebody Diagrams for Torsion

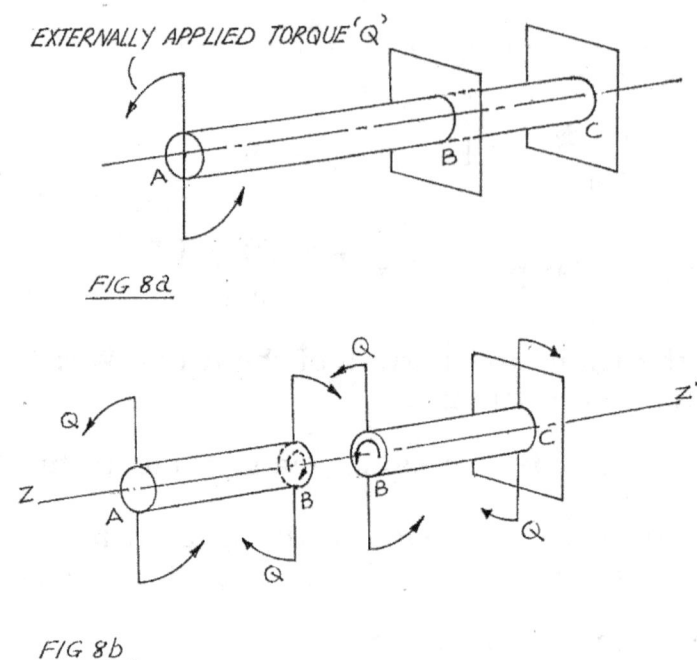

FIG 8a

FIG 8b

When we considered forces in frames and took sections through them, the internal forces 'exposed' had to be equated on the basis of application of the laws of statical equilibrium, to any externally exposed loads and any external reactions in order that static equilibrium be maintained. By a similar token we saw in the case of a beam that, for the most general loading, three elements of resistance viz a horizontal force (horizontal pull or compression), a vertical force (shear force); and, moment were necessary in order to maintain equilibrium at any section taken perpendicular to the axis of the beam.

Torsion in shafts is no exception. At any section of a shaft perpendicular to its axis, any externally applied torque must be balanced by an internal resisting torque. In Fig. 8(a) a shaft AC is shown with an externally applied anti-clockwise torque at end A. At B a section is taken perpendicular to the axis of the shaft, that is, we cut the shaft at B, said cut being perpendicular to the axis of the shaft!

In Fig. 8(b), the front portion of the shaft 'AB' is brought forward and at the face 'B' where the section was taken, an internal resisting clockwise torque 'Q' equivalent to the action of the shearing stresses gives rise to what Duncan refers to as a moment of resistance to torsion, and this clockwise torque balances the external torque 'Q' at 'A'. Note that the arrows of the internal shearing stresses are in the same direction as that of the internal resisting torque.

~ 380 ~

Evidently for equilibrium, when the two faces of the shaft exposed by the parting plane at B are brought together, the directions of the internal resisting torque and shearing stresses in the part of the shaft 'BC' must be opposite their counterparts in AB. And clearly at 'C' the reactive torque there must be in the direction shown not only to balance that at B of shaft 'BC' but 'Q' at A of AC when the 2 parts of the shaft are brought together again, i.e. when there is no cut.

For statical equilibrium in statically determinate members in torsion, the governing equation at any section of the shaft is $\sum M_z = 0$, ZZ' being the axis of rotation of the torques.

Let us apply this to a particular case. First, consider the shaft 'AB' shown in Fig. 9(a) where a torque T_c is applied at a section 'C' between A and B

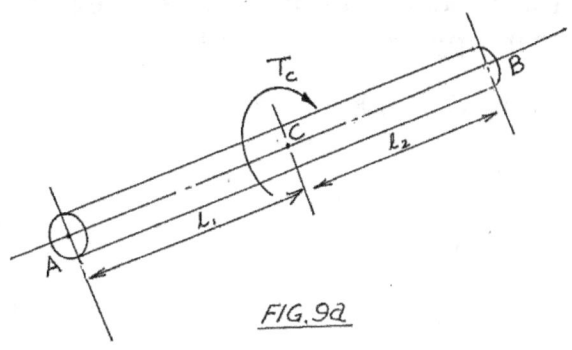

FIG.9a

If T_1 is the torque in AC and T_2 in BC as in Fig. 9)b)

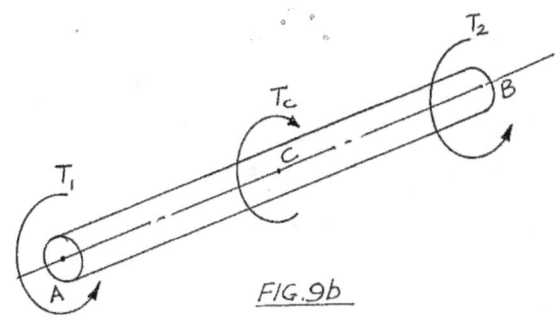

FIG.9b

applying $\sum M_z = 0$

~ 381 ~

$$T_1 - T_c + T_2 = 0$$

$$T_2 = (T_c - T_1)$$

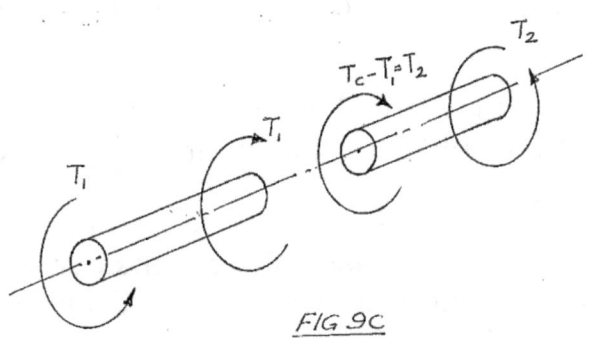

FIG 9C

It is to be noted that there is no convention there with respect to direction of torque in the sense that a convention was adopted for shear force and bending moment. A clockwise torque could be positive or negative in any equation; but one has to be consistent in any specific situation.

For that portion of the shaft CB in Fig. 9(c)

$$(T_c - T_1) - T_2 = 0$$

i.e. $$T_c - T_1 - T_2 = 0$$

or $$T_2 = T_c - T_1$$

as before

Twisting moments T_1 and T_2 may be expressed in terms of T_c by resorting to the Torsion Formula

$$\frac{T_1}{J_1} = \frac{C\theta_1}{\ell_1}$$

and $$\frac{T_2}{J_2} = \frac{C\theta_2}{\ell_2}$$

For $J_1 = J_2$ and identical C

$$\theta_1 = \theta_2$$

because the angle of twist must be the same for each section in this case.

$$\therefore \quad T_1 \ell_1 = T_2 \ell_2$$

Substituting in $T_2 = T_c - T_1$

$$T_2\left(\frac{\ell_2}{\ell_1} + 1\right) = T_c$$

$$T_2 = \frac{T_c \ell_1}{L}$$

where $L = \ell_1 + \ell_2$

and $T_1 = \frac{T_c \ell_2}{L}$

Notice the inverse relationship between T_1 and T_2 and the lengths of the section under their respective influence.

TORSION OF SHAFTS IN GEARED SYSTEMS

Consider first of all the torque required to accelerate a train of gears as for example the 2-gear speed reducer shown in Fig. 10a.

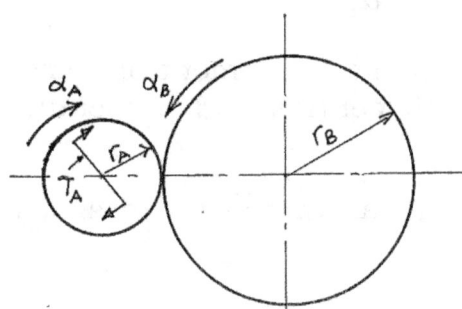

FIG 10a: 2-GEAR SPEED REDUCER

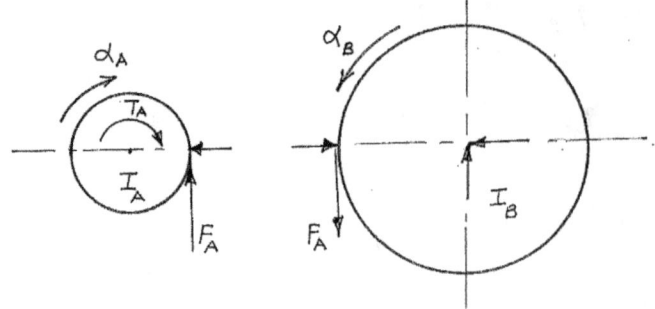

FIG 10b : FREE-BODY DIAGRAM OF
SYSTEM SHOWN IN FIG.10a

Figure 2 is the free-body diagram of the system. Torque or couple T_A on the driver 'A' has to accelerate both gear 'A; and gear 'B': Gear 'A' at α_A radians/s² and 'B' at α_B radians/s².

The Newtonian equation of motions are:

$$\text{For 'A' :} \qquad T_A - F_A r_A = I_A \alpha_A \qquad \ldots\ldots\ldots\ldots\ldots\ldots \text{(i)}$$

and $\qquad$ For 'B' : $\qquad F_A r_B = I_B \alpha_B \qquad \ldots\ldots\ldots\ldots\ldots\ldots$ (ii)

where I_A and I_B are respectively the polar moment of inertia of the gear wheels with respect to axes through their centres and perpendicular to the plane of each gear.

From the geometry of the motion, assuming no loss of motion between the gears:

$$\alpha_A r_A = \alpha_B r_B$$

From (ii) $\qquad F_A = \dfrac{I_B \alpha_B}{r_B}$

and when this result for F_A is substituted in (i) we obtain

$$T_A - I_B \cdot \frac{r_A}{r_B} \cdot \alpha_B = I_A \alpha_A$$

Now putting $\alpha_B = \dfrac{\alpha_A r_A}{r_B}$ in the latter and with some minor arrangement, we obtain

$$T_A = \alpha_A \left\{ I_A + \left(\frac{r_A}{r_B} \right)^2 I_B \right\} \qquad \ldots\ldots\ldots\ldots\ldots\ldots \text{(iii)}$$

It should be noted that $\alpha_B / \alpha_A = r_A / r_B$ is in fact the gear ratio 'G' between the gears 'A; and 'B'. Hence we may write, alternatively

$$T_A = \alpha_A \left(I_A + G^2 I_B \right) \qquad \ldots\ldots\ldots\ldots\ldots\ldots \text{(iv)}$$

Let us now extend the analysis to a system comprising three gear trains as in Fig. 11a.

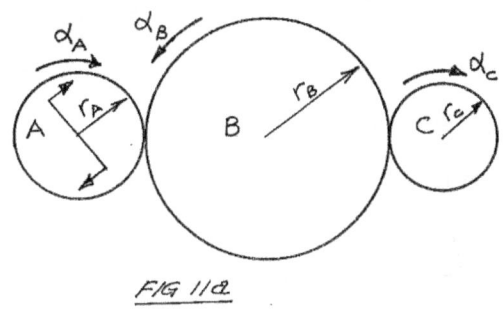

FIG 11a.

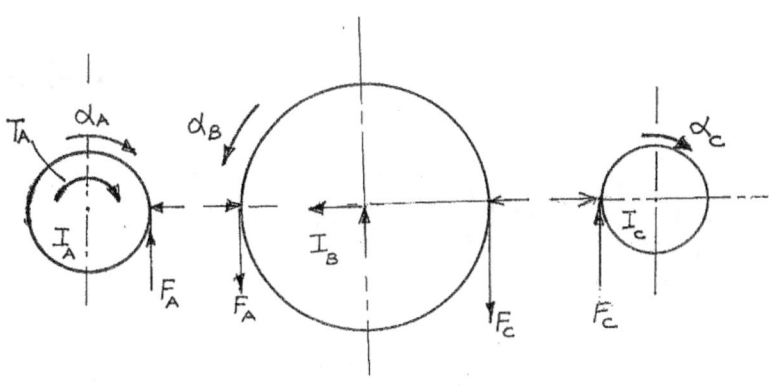

FIG 11b : FREE-BODY DIAGRAM
OF SYSTEM SHOWN IN FIG 11a

The free-body diagram of the system is shown in Fig. 11b.

The equations of motion are:

$$T_A - F_A r_A = I_A \alpha_A \qquad \dots\dots\dots\dots\dots \text{(v)}$$

$$(F_A - F_C) r_B = I_B \alpha_B \qquad \dots\dots\dots\dots\dots \text{(vi)}$$

$$F_C r_C = I_C \alpha_C \qquad \dots\dots\dots\dots\dots \text{(vii)}$$

Substituting from (vii) $F_c = \dfrac{I_c \alpha_c}{r_c}$ in (vi) yields

$$F_A - \frac{I_c \alpha_c}{r_c} = \frac{I_b \alpha_B}{r_B}$$

$$\therefore \qquad F_A = \frac{I_B \alpha_B}{r_B} + \frac{I_c \alpha_c}{r_c}$$

Substituting the latter in (v) produces

$$T_A - r_r \left(\frac{I_B \alpha_B}{r_B} + \frac{I_c \alpha_c}{r_c} \right) = I_A \alpha_A \qquad \dots\dots\dots\dots \text{(viii)}$$

Now,

$$\alpha_A r_A = \alpha_B r_B = \alpha_c r_c$$

If in (viii) the substitutions : $\alpha_B = \dfrac{\alpha_A r_A}{r_B}$ and $\alpha_c = \dfrac{\alpha_A r_A}{r_c}$ are made then

$$T_A - I_B \alpha_A \left(\frac{r_A}{r_B} \right)^2 - I_c \alpha_A \left(\frac{r_A}{r_c} \right)^2 = I_A \alpha_A$$

so that

$$T_A = \alpha_A \left\{ I_A + I_B \left(\frac{r_A}{r_B} \right)^2 + I_c \left(\frac{r_A}{r_c} \right)^2 \right\} \qquad \dots\dots\dots\dots \text{(ix)}$$

Note carefully the r_A / r_B is the gear ratio between the driver and the first driven gear and r_A / r_c is the gear ratio between the driver and the third gear which in this case is the gear labeled 'C'.

Illustrative Example 8

An electric motor drives a gear 'A' fitted to its steel shaft which is 30mm in diameter. 'A' is part of a 2-gear speed reducer See Fig. 10a. The angular acceleration of 'A' is 15 rads/s². Determine the maximum shear stress in the shaft given that the shear modulus of steel = 80GN/m². The pitch-circle diameter of gear 'A' = 50mm and of the driven gear 150mm. The polar moments about the axes of rotation are I_A = 0.075 kg m² and I_{DRIVEN} = 0.25kg m². Calculate also the angle of twist if the shaft is 3cm long.

The first order of business is to find the torque 'T_A' necessary to accelerate both the driver i.e. gear 'A' and the driven gear; let us call it gear 'B'. For this purpose we employ:

$$T_A = \alpha_A \left(I_A + G^2 I_B \right)$$

where $\quad \alpha_A \quad =$ angular acceleration of A

$I_A, I_B =$ respective polar moment of inertia of inertia about

axes of rotation

$G \quad =$ gear ratio $= r_A / r_B = 25/75 = \dfrac{1}{3}$

Accordingly

$$T_A = 15\left\{ 0.075 + \left(\frac{1}{3} \right)^2 0.25 \right\}$$

i.e. $\quad T_A = 1.545 kg/m^2/s^2$

or $\quad T_A = 1.55 Nm$

Observe that in the foregoing the inertia of the shaft which presumably is keyed to the gear has been ignored.

Continuing

$$\frac{T}{J} = \frac{\tau}{r}$$

$\therefore \quad \dfrac{1.55}{\dfrac{\pi}{32}\left(\dfrac{30}{1000} \right)^4} = \dfrac{\tau}{\dfrac{30}{2000}}$

from which

$$\tau = 0.3 Mn/m^2$$

Also
$$\frac{\tau}{\dfrac{30}{2000}} = \frac{C\theta}{L}$$

or
$$0.3 \times \frac{10^6}{30} \times 2000 = 80 \times 10^9 \frac{\theta}{\dfrac{3}{100}}$$

$$\therefore \quad 20 = 80 \times 10^3 \times 100 \times \frac{\theta}{3}$$

i.e.
$$\theta = \frac{60}{80 \times 10^5}$$

or $\quad \theta = 7.5 \ X \ 10^{-6} \ rad.$ rad.

in short "zilch"

SOLVED AND OTHER PROBLEMS

Q1. What is the theoretical diameter of a steel shaft which is to transmit a torque of 7500Nm, assuming the allowable shear stress is 60MN/m²?

Let   d = diameter of shaft, mm

T = torque on shaft, N m

J = polar moment of inertia, m⁴

Q = allowable shear stress, N/m²

Now

$$\frac{q}{\left(\dfrac{d}{2000}\right)} = \frac{T}{J}$$

or $$\frac{60 \times 10^6 \times 2000}{d} = \frac{7500}{\dfrac{\pi}{32}\left(\dfrac{d}{1000}\right)^4}$$

∴ $$\frac{60 \times 2 \times x10^9}{d} = \frac{7500 \times 32 \times 10^{12}}{\pi d^4 d^3}$$

$$d^3 = \frac{7500 \times 32 \times 10^{12}}{\pi \times 60 \times 2 \times 10^9}$$

$$= \frac{2000 \times 10^3}{\pi}$$

$$d^3 = 636.33 \times 10^3$$

∴ $$d = 10\sqrt{636.3} \quad = 85.8mm$$

Theoretical *d*, say 86mm.

Q2. The solid circular cylindrical rod of length 'L' and diameter 'd' shown in Fig. 1 is fixed at both ends. An external torque 'T' is applied to the rod in a plane perpendicular to the axis of the shaft and at a distance ℓ_1 from its left-hand end, where $\ell > (L-\ell)$. If τ_w is the maximum allowable shear stress in the rod, show that diameter 'd' is given by :

$$d = \left(\frac{16T\,\ell_1}{L\pi\,\tau_w} \right)^{\!\frac{1}{3}}$$

(After Timoshenko): The problem is due to Professor Timoshenko, but the solution is mine.

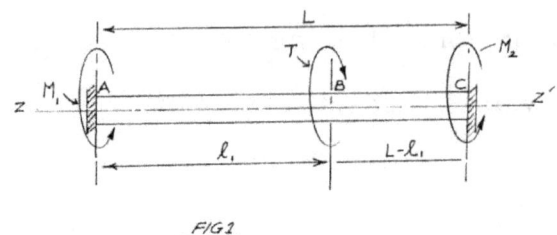

FIG 1

Let M_1 = Torque on AB

" M_2 = Torque on BC

"   T   = Torque at B

For equilibrium about $ZZ' - axis$: $\sum Moments_{ZZ'} = 0$

   $M_1 + M_2 - T = 0$

i.e. $T = M_1 + M_2$ . . . . (i)

For equal angles of twist in AB and BC: In AB, $\theta = M_1\,\ell_1 / CJ$;   In BC, $\theta = M_2(L - \ell_1 / CJ)$. Therefore for equal '$\theta$s'

   $$M_1\,\ell_1 = M_2(L - \ell_1)$$

   $$M_1 = M_2\left(\frac{L - \ell}{\ell_1} \right)$$ . . . . (ii)

By (i) $$T = M_2\left(\frac{L - \ell_1}{\ell_1} \right) + M_2$$

   $$= M_2\,\frac{(L - \ell_1 + \ell_1)}{\ell_1}$$

or $$M_2 = \frac{T\ell}{L}$$

so that
$$M_1\ell_1 = \frac{T\ell_1}{L}(L-\ell_1)$$

$$M_1 = T\frac{(L-\ell_1)}{L}$$

Now, because $\ell_1 > (L-\ell_1)$ it follows that $M_2 > M_1$ and so the shaft must be designed with M_2 as the torque to generate 'τ_w'. Therefore, relying on the Torsion Formula we have for the general expression

$$\frac{Torque}{J} = \frac{\tau}{r}$$

where the symbols have their usual meaning,

$$\frac{M_2}{J} = \frac{\tau_w}{r}$$

Substituting $M_2 = \frac{T\ell_1}{L}$; $J = \pi d^2/32$ and $r = \frac{d}{2}$ we get

$$\frac{Tl_1}{L\left(\dfrac{\pi d^4}{32}\right)} = \frac{\tau_w}{\left(\dfrac{d}{2}\right)}$$

from which $d = \left(\dfrac{T\ell_1}{L\pi\,\tau_w}\right)^{1/3}$

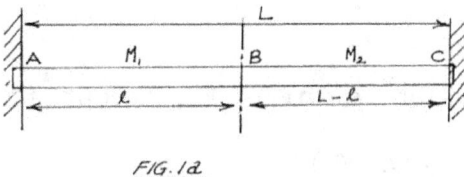

FIG. 1a

The student would note the inverse relationship between the torques of individual sections and their lengths, thus

$$M_1 = T\frac{(L-\ell)}{L} \qquad \text{i.e.} \qquad M_1 = k)L-\ell)$$

$$M_2 = \frac{T\ell}{L}. \qquad \text{i.e.} \qquad M_2 = kL \quad \text{where } k = \frac{I}{L}$$

so that $\dfrac{M_1}{M_2} = \dfrac{L-\ell}{\ell}$. See Fig. 1a.

which leads us to the next problem also due to Timoshenko.

Q3.   The solid circular cylindrical shaft of length 'L' and diameter 'd' shown in Fig. 1 below is fixed at both ends. Separate torques of T_1 and T_2 are applied to the shaft at distances ℓ_1 and $\ell_1 + \ell_2$ respectively from the left hand end of the shaft. Determine the torque acting on each section of the shaft.

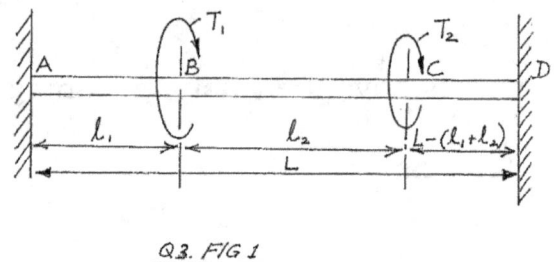

Q3. FIG 1

Using the knowledge of the inverse relationship between torque of individual sections and lengths in Question 1, we could consider the torque distribution between the sections when T_1 and T_2 are each acting separately and add the separate effects

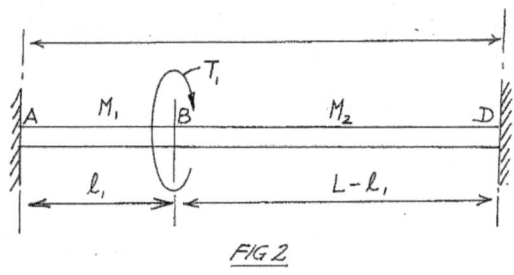

FIG 2

Taking T_1 acting alone as in Fig. 2, we can straightway write

$$M_1 = T_1 \frac{(L-\ell_1)}{L}$$

$$M_2 = \frac{T_1\,\ell_1}{L}$$

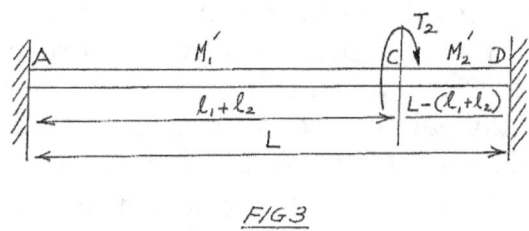

FIG 3

For T_2 acting alone as in Fig. 3

$$M_1' = \frac{T_2\{L-(\ell_1+\ell_2)\}}{L}$$

$$M_2' = \frac{T_2(\ell_1+\ell_2)}{L}$$

Adding the separate effects, the torque on AB -- let us call it T_{AB} -- is given by

$$T_{AB} = M_1 + M_1'$$

i.e.
$$T_{AB} = \frac{T_1(L-\ell_1)}{L} + T_2\frac{\{L-(\ell_1+\ell_2)\}}{L}$$

Similarly, designating T_{CD} as the torque on CD

$$T_{CD} = M_2 + M_2'$$

i.e.
$$T_{CD} = \frac{T_1\ell_1}{L} = \frac{T_2(\ell_1+\ell_2)}{L}$$

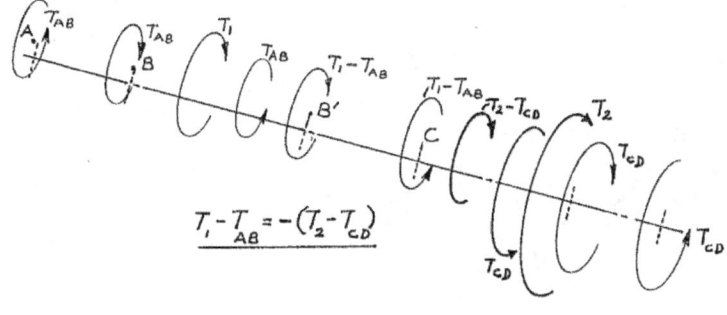

FIG 4

In Fig. 4, I have shown a Freebody diagram of the shaft. Let us refer to it. Starting from the left-hand side, section 'AB' is shown under the effect of torque T_{AB} : counter-clockwise at 'A' and clockwise at B. The shaft having been cut by an imaginary plane at B we consider next the section

B'C. To restore equilibrium we must apply a counter-clockwise torque 'T$_{AB}$' at *B'* to balance the clockwise torque at 'B'; and because a clockwise torque 'T$_1$' is at B, the nett result is a clockwise torque T$_1$ – T$_{AB}$ at *B'*. Hence that part of the shaft *B'C* is under the effect of a torque T$_1$ - T$_{AB}$: clockwise at *B'* and counter-clockwise at C. Lastly for that portion of the shaft *C'D* the torque acting on it is T$_{CD}$. As the shaft is cut by an imaginary plane at C, a counter-clockwise torque T$_{CD}$ has to be applied to restore equilibrium, and because there is an external clockwise torque of T$_2$ at C, the net effect is a clockwise torque T$_{CD}$. has to be applied to restore equilibrium and because there is an external clockwise torque of T2 at C, the net effect is a clockwise torque T$_2$ – T$_{CD}$. But at C we had an anticlockwise torque of T$_1$ – T$_{AB}$. Accordingly

$$T_1 - T_{AB} = -(T_2 - T_{CD}) = T_{CD} - T_2$$

T$_1$ - T$_{AB}$ is the torque on the part of the shaft BC.

Also, based on the above analysis and referring to Fig. 4,

$$T_{BC} = T_1 - T_{AB}$$

i.e. $$T_{BC} = T_1 - \left[T_1 \frac{(L - \ell_1)}{L} + T_2 \{ L - (\ell_1 + \ell_2) \} \right]$$

$$= T_1 - T_1 \frac{L}{L} + \frac{T_1 \ell_1}{L} - T_1 \frac{L}{L} + T_2 \frac{\ell_1}{L} + \frac{T_2 \ell_2}{L}$$

which reduces to

$$T_{BC} = \frac{\ell_1}{L}(T_1 + T_2) - T_2 \left(1 \frac{\ell_2}{L} \right)$$

Just as a check, let us find T$_{BC}$ using

$$T_{BC} = T_{CD} - T_2$$

$$= \frac{T_1 \ell_1}{L} + T_2 \frac{(\ell_1 + \ell_2)}{L} - T_2$$

i.e. $$T_{BC} = \frac{\ell_1}{L}(T_1 + T_2) - T_2 \left(1 - \frac{\ell_2}{L} \right), \text{Check}$$

Summarising,

$$T_{AB} = T_1 \frac{(L - \ell_1)}{L} + T_2 \left\{ \frac{L - (\ell_1 + \ell_2)}{L} \right\}$$

$$T_{BC} = \frac{\ell_1}{L}(T_1 + T_2) - T_2(1 - \frac{\ell_2)}{L}$$

and $$T_{CD} = \frac{T_1 \ell_1}{L} + T_2 \left(\frac{\ell_1 + \ell_2}{L} \right)$$

Q4. A steel tube 2m long, 40mm outside diameter and 0.75mm thick is twisted by a couple of magnitude 75Nm. Find the maximum shear stress, the maximum tensile stress and the angle through which the tube twists. Assume C = 80 x 10⁹ N/m².

$$\frac{T}{J} = \frac{q}{r} = \frac{C\theta}{L}$$

. . . . (i)

Units:

T is in Newton metres   i.e. N m

J "   "   m⁴

q "   "   Newtons per (metre)² i.e. Nm⁻²

r "   "   metres

C "   "   Nm⁻²

θ "   "   radians

L "   "   metres

$$\therefore \qquad \frac{75}{\frac{\pi}{32} \left\{ \left(\frac{40}{1000} \right)^4 - \left(\frac{38.5}{1000} \right)^4 \right\}} = \frac{Q_{max}}{\frac{40}{2000}}$$

from which $$q_{max} = \frac{75 \times 32 \times \frac{40}{2000} \times 10^2}{\pi(40^4 - 38.5^4)}$$

$$= \frac{48000 \times 10^9}{\pi \times 117.75 \times 3082.25}$$

$$\therefore \qquad q_{max} = 0.042 \times 10^9 \, N/m^2$$

or $\qquad q_{max} = 42 \times 10^6 \, N/m^2$

$$= 42 \, MN/m^2$$

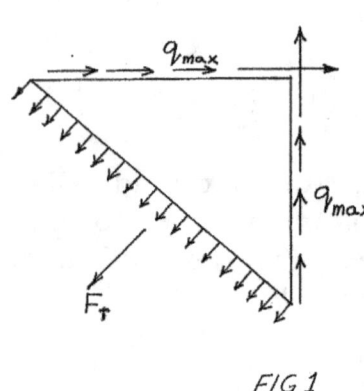

<u>FIG 1</u>

In Fig. 1 ABC represents a small triangular element under the influence of complementary shear stress 'q_{max}'. AB = BC. Let the thickness of the element be unity perpendicular to the figure.

$\therefore \qquad$ Shear force along AB $= q_{max} \cdot (AB \times 1)$

Also $\qquad$ " $\qquad$ " $\qquad$ " $\qquad$ BC $= q_{max} \cdot (BC \times 1)$

Let F_T be the resultant (on plane AC) required to maintain equilibrium.

$$\therefore \qquad F_T \cdot (AC \times 1) = \sqrt{(q_{max} \, AB)^2 + (q_{max} \, BC)^2}$$

$$F_T \cdot AC = q_{max} \sqrt{(AB)^2 + (BC)^2}$$

But $\qquad AC = \sqrt{(AB)^2 + (BC)^2}$

$$\therefore \qquad F_T = q_{max}.$$

Accordingly, the maximum tensile stress is also of magnitude 42MN/m².

From (i)

$$Q = \frac{qL}{Cr}$$

$$= \frac{42 \times 10^6 \times 2}{80 \times 10^9 \times \dfrac{20}{1000}}$$

$$= \frac{42 \times 10^6 \times 2 \times 1000}{80 \times 10^9 \times 20}$$

$$\theta = \frac{42}{800} \text{ radians}$$

$$= \frac{42}{800} \times \frac{360}{2\pi}$$

$$= \frac{198}{20\pi} \text{ or } \theta = 3^\circ$$

Q5. The single composite steel shaft shown in Fig. 1 is made up of two parts: AB is a solid circular cylindrical portion 300mm long and diameter 20mm; BC is 500mm long and of hollow circular cross-section, 25mm outside diameter, d_i mm inside diameter. If equal and opposite torques 'T' are applied to the ends of the shaft, what is the value of 'd_i' when maximum allowable shear stress in the solid and hollow parts of the shaft is the same? Determine also total angle of twist in degrees in terms of 'T' and modulus of rigidity 'C'.

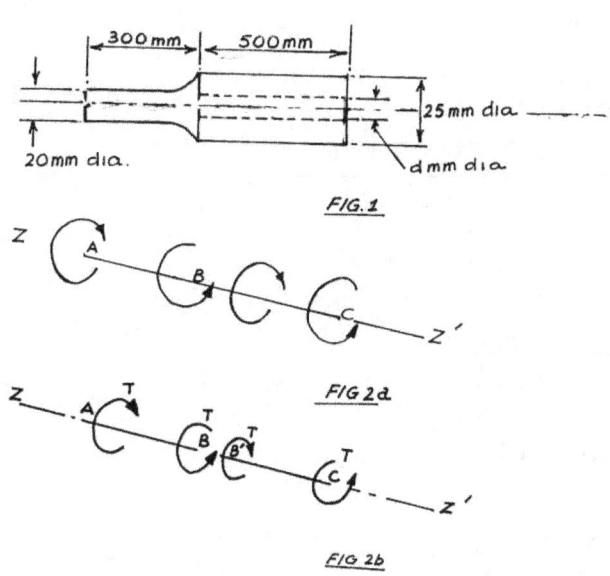

FIG. 1

FIG 2a

FIG 2b

I have shown free body diagrams Figs. 2a and 2b for the loading arrangement. 'B' is at the junction of the solid portion of the shaft and its hollow portion. For a clockwise turning moment at 'A', the balancing torque at 'B' in respect of the solid portion must therefore be anti-clockwise. This means that at the same 'B' but this time in respect of the hollow portion of the shaft, the torques to balance the anti-clockwise turning moment we put there to balance the clockwise 'T' at A, must clearly act in a clockwise manner as indicated. If 'T' in one sense is, say, positive, then 'T' in an opposite sense must be negative.

It should also be clear from Fig. 2(b) that for the separate portions of the shaft AB and BC, $\sum M_z$ also results in +T-T = 0 for each section. Thus, each portion of the shaft is in equilibrium under the action of twisting moment 'T'. Applying the Torsion Formula to each section of the shaft in turn, we have:

For AB:

$$\frac{T}{J_{AB}} = \frac{\tau_{AB}}{d_{AB}/2}$$

Working in metre units

i.e.
$$\frac{T}{\frac{\pi}{32}\left(\frac{20}{1000}\right)} = \frac{\tau_{AB}}{\frac{20}{2(1000)}}$$

$$\frac{T \cdot 32 \times 10^{12}}{\pi(20)^4} = \frac{\tau_{AB} \times 2 \times 1000}{20}$$

$$T = \frac{\tau_{AB} \times 2 \times 1000 \times \pi \times (20)^4}{32 \times 10^{12} \times 20} \qquad \cdots \cdots \cdots \cdots \text{(i)}$$

For BC:

$$\frac{T}{J_{BC}} = \frac{\tau_{BC}}{d_{BC}/2}$$

Working in metre units

$$\frac{I}{\frac{\pi}{32}\left\{\left(\frac{d_o}{1000}\right)^4 - \left(\frac{d_1}{1000}\right)^4\right\}} = \frac{\tau_{BC}}{\frac{d_o}{2(1000)}}$$

$$\frac{T}{\frac{\pi}{32} \cdot \frac{1}{10^{12}} \left(d_o^4 - d_i^4\right)} = \frac{\tau_{BC} \times 2 \times 1000}{25}$$

$$\therefore \qquad \frac{T \cdot 32 \times 10^{12}}{\pi \left\{(25)^4 - d_i^4\right\}} = \frac{\tau_{BC} \times 2 \times 1000}{25}$$

$$\text{or} \qquad T = \frac{\tau_{BC} \times 2 \times 1000 \times \pi \left\{(25)^4 - d_i^4\right\}}{32 \times 10^{12} \times 25} \qquad \cdots\cdots\cdots \text{ (ii)}$$

Equating (i) and (ii)

$$\frac{\tau_{AB} \times 2 \times 1000 \times \pi \times (20)^4}{32 \times 10^{12} \times 20} = \frac{\tau_{BC} \times 2 \times 1000 \times \pi \left\{(25)^4 - d_i^4\right\}}{32 \times 10^{12} \times 25}$$

Allowable working stresses are the same, i.e. $\tau_{AB} = \tau_{BC}$

$$\therefore \qquad (20)^3 = \frac{(25)^4 - d_i^4}{25}$$

$$25(8000) = 290625 - d_i^4$$

$$200000 = 290625 - d_i^4$$

$$\therefore \qquad d_i^4 = 190625$$

$$\text{or} \qquad d_i = 20.89mm, \text{ say}$$

$$d_i \approx 21mm.$$

Further, from

$$\frac{T}{J} = \frac{C\theta}{\ell}$$

where the symbols have their usual meanings; remember 'θ' is in radians.

$$\therefore \qquad \frac{T}{J_{AB}} = \frac{C\theta_{AB}}{\ell_{AB}}$$

and

$$\frac{T}{J_{BC}} = \frac{C\theta_{BC}}{\ell_{BC}}$$

so that

$$\theta_{AB} = \frac{T \cdot \ell_{AB}}{J_{AB} C}$$

and

$$\theta_{BC} = \frac{T \ell_{BC}}{J_{BC} C}$$

Working in metres

$$\therefore \qquad \theta_{AB} = \frac{T \cdot \left(\dfrac{300}{1000}\right)}{\dfrac{\pi}{32}\left(\dfrac{20}{1000}\right)^4 C}$$

$$= \frac{300}{1000}\,T \cdot \frac{10^{12}}{\pi (20)^4} \frac{32}{C}$$

Similarly

$$\theta_{BC} = \frac{T \cdot \left(\dfrac{500}{1000}\right)}{\dfrac{\pi}{32}\left\{\left(\dfrac{25}{1000}\right)^4 - \left(\dfrac{21}{1000}\right)^4\right\}}$$

$$= \frac{500}{1000}\,T \cdot \frac{1}{\pi} \cdot \frac{32 \times 10^{12}}{(25)^4 - (21)^4}$$

$$\theta_{BC} = \frac{500}{1000}\,T \cdot \frac{1}{\pi} \cdot \frac{32 \times 10^{12}}{200000}$$

Total angle of twist $= \theta_{AB} + \theta_{BC}$

$$= \frac{300T}{1000} \cdot \frac{10^{12} \cdot 32}{\pi\, 160000 \cdot C} + \frac{500T}{1000} \cdot \frac{32 \times 10^{12}}{200000} \times \frac{1}{\pi}$$

$$= \frac{T}{\pi C}\left(\frac{6 \times 10^7}{} + \frac{8 \times 10^7}{}\right)$$

$$= \frac{14 \times 10^7 T}{\pi C} \text{ radians}$$

$$\therefore \quad \text{Total angle of twist} = \frac{14 \times 10^7 \times T}{AC} \times \frac{360}{2\pi} \text{ degrees}$$

$$= \frac{510 \cdot 5 \times 10^7 T}{C} \text{ degrees.}$$

Q6. A solid circular cylindrical steel shaft of diameter 20mm and 750mm long has its ends rigidly fixed. A couple of magnitude 70Nm is applied to the shaft 250mm from one end. Determine (i) the couples at the fixed ends of the shaft; (ii) the maximum shearing stress induced in the shaft and (iii) the angle of twist of the section where the couple was applied. Take $C = 84 \times 10^9 N/m^2$. Refer to Fig. 1

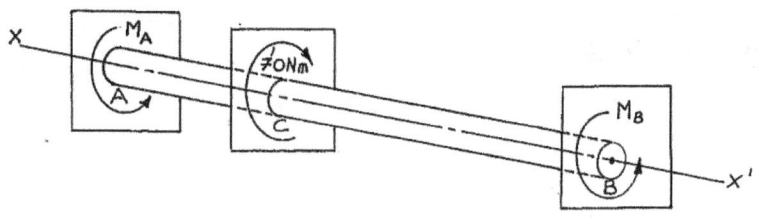

FIG 1

Plane 'C' which is normal to the axis of the shaft passes through the section where the couple of 70Nm is applied.

Let the couples acting in planes 'A' and 'B' at the fixed ends be designated 'M_A' and 'M_B' respectively.

Distance between planes 'A' and 'C' = 250mm

" " " 'C' " 'B' = 500mm

Considering moments about the 'X'-axis

$$\sum M_x = 0 \qquad \qquad \dots \dots \dots \dots \text{(i)}$$

$$\therefore \quad M_A + M_B - 70 = 0 \qquad \dots \dots \dots \dots \text{(ii)}$$

i.e. $\qquad M_A = M_B = 70 \qquad \dots \dots \dots \dots \text{(iii)}$

Considering length of shaft between 'A' and 'C'

$$\frac{M_A}{r_{shaft}} = \frac{C\theta_A}{250} \qquad \dots\dots\dots\dots\dots\dots \text{(iv)}$$

Similarly for the length of shaft between 'C' and 'B'

$$\frac{M_B}{r_{shaft}} = \frac{C\theta_B}{500} \qquad \dots\dots\dots\dots\dots\dots \text{(v)}$$

Because each section of the shaft must rotate through the same angle

$$\theta_A = \theta_B$$

$$\therefore \qquad \frac{250 M_A}{C\, r_{shaft}} = \frac{500 M_B}{C\, r_{shaft}}$$

from which $M_A = 2M_B$ $\qquad \dots\dots\dots\dots\dots\dots$ (vi)

Substitution in (iii) gives

$$M_B = \frac{70}{3}\, Nm$$

and $\qquad M_A = \dfrac{140}{3}\, Nm$

Evidently the maximum shearing stress will occur in the shaft between A and C

$$\therefore \qquad \frac{q_m}{\left(\dfrac{10}{1000}\right)} = \frac{\dfrac{140}{3}}{\dfrac{\pi}{32}\left(\dfrac{20}{1000}\right)^4}$$

$$\therefore \qquad 100 q_m = \frac{140 \cdot 32 \cdot 10^6}{3 \cdot \pi \cdot (.16)}$$

or $\qquad q_m = \dfrac{1.40 \times 32 \times 10^6}{3\pi \cdot 0.16}$

$$29.7 \times 10^6\, N/m^2$$

i.e. $\qquad q_m = 29.7\, MN/m^2$

To determine angle of twist we may consider either length of the shaft. Let us take the section between 'A' and 'C' for which we have already the shearing stress

$$\frac{q_{max}}{r_{shaft}} = \frac{C\theta_A}{L}$$

$$\therefore \quad \frac{29.7 \times 10^6}{\left(\dfrac{10}{1000}\right)} = \frac{84 \times 10^9 \times \theta_A}{\left(\dfrac{250}{1000}\right)}$$

i.e. $\quad \theta_A = \dfrac{29.7 \times 10^6 \times 25}{84 \times 10^9} \;$ rad

But $\quad 2\pi$ rad $= 360°$

$$1 \;\; \text{rad} \;\; = \;\; \frac{360}{2\pi}$$

$$\therefore \quad \theta_A^o = \frac{29.7 \times 10^6 \times 25 \times 360}{84 \times 10^9 \times 2\pi}$$

$$\theta_A = 0.5 \;\; \text{degrees}$$

Q7. Using the data in Question 6 and assuming AB = 60cm long, determine the lengths AC and CB of the shaft.

Q8. A shaft of total length 4000mm is built in to rigid supports at both ends. It is twisted by two couples at two planes perpendicular to its axis, the first of magnitude 4500 Nm a distance of 800mm from the left hand end and the second 9500 Nm at a distance of 1200 mm from the right-hand end. What is the twisting moment in each portion of the shaft?

Refer to Fig. 1.

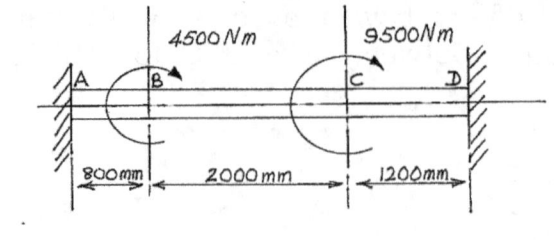

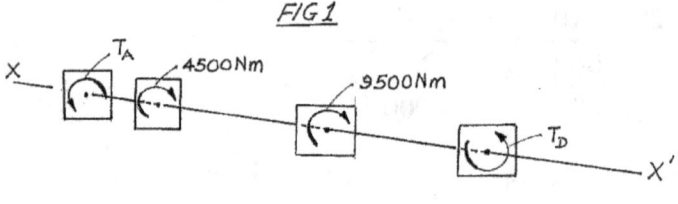

FIG 2

Referring to Fig. 2, it is evident that for equilibrium of moments about X-axis:

$$\sum M_x = 0$$

i.e. $T_A + T_D - 4500 - 9500 = 0$ . . . .   (i)

where T_A, T_D are the twisting moments at A and D respectively to maintain equilibrium.

Let   J   =   polar moment of inertia of the shaft, constant throughout in this case

C   =   torsional modulus

θ =   angle of twist

Considering the portion of the shaft 'AB'

$$\frac{T_A}{J} = \frac{C\theta_A}{\left(\dfrac{800}{1000}\right)}$$ . . . .   (ii)

For portion of the shaft BC

$$\frac{T_{B-C}}{J} = \frac{C\theta_{B-C}}{\left(\dfrac{2000}{1000}\right)}$$ . . . .   (iii)

~ 404 ~

Similarly, for section DC

$$\frac{T_D}{J} = \frac{C\theta_D}{\left(\dfrac{1200}{1000}\right)} \qquad \dots\dots\dots\dots\dots\dots\dots \text{(iv)}$$

Remember that at section 'B' the angle of twist relative to say a fixed marker at 'A' is θ_A. Likewise at section 'C' the angle of twist relative to the same marker is $\theta_A + \theta_{B-C}$. If we placed an identical marker at 'D', angular displacement of the section of the shaft at 'C' relative to that marker as we move along the shaft from 'D' must, in order to satisfy compatibility of displacement, be equal to the displacement of 'C' relative to the marker at A.

i.e. $\qquad \theta_D = \theta_A + \theta_{B-C} \qquad \dots\dots\dots\dots\dots\dots \text{(v)}$

so that $\dfrac{J_D}{JC} \cdot \left(\dfrac{1200}{1000}\right) = \dfrac{T_A}{JC}\left(\dfrac{800}{1000}\right) + \dfrac{T_{B-C}}{JC}\left(\dfrac{2000}{1000}\right)$

from which

$$3T_D = 2T_A + 5T_{B-C}$$

But $\qquad T_{B-C} = T_A - 4500 \qquad \dots\dots\dots\dots\dots\dots \text{(vii)}$

$\therefore \qquad 3T_D = 2T_A + 5T_A - 22500$

or $\qquad 3T_D = 7T_A - 2250 \qquad \dots\dots\dots\dots\dots\dots \text{(vii)}$

from which $T_D = \dfrac{7T_A}{3} - 7500$

Substituting this result in (i) gives

$$T_A + \frac{7}{3}T_A = 14000$$

$$\frac{10T_A}{3} = 14000$$

$\therefore \qquad T_A = 4200 Nm$

$$T_D = 14000 - T_A$$

$\therefore \qquad T_D = 9800 Nm$

Q9. The jackshaft in Fig. 1 is in two sections comprising a single whole as shown in Fig. 1 below. The part to the left is 80mm diameter and 750mm in length and that to the right is 1000mm long with diameter 800mm. The diagram is not drawn to scale. The ends of the shaft are built into rigid supports. If a torque of 750Nm is applied at the shoulder of the shaft, what is the torque in each section of the shaft? Determine also the maximum shearing stress and the angle through which the shoulder section rotates. Take C = 84 x 10^9 N/m². Neglect torsional stress concentration at the fillet of the shoulder.

Ans: T_A = 606Nm; T_B = 155Nm; q_{max} = 6MN/m² and θ = 0.077°.

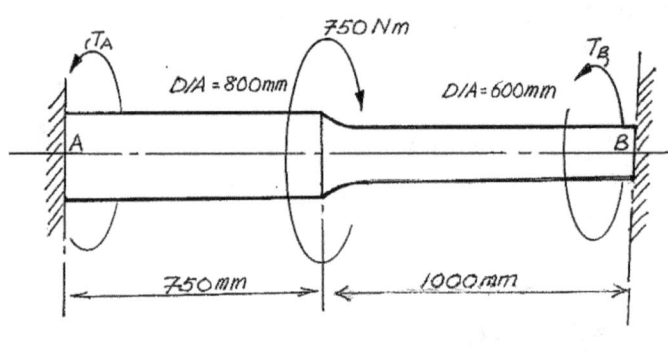

FIG 1

Q10. In developing an algorithm for evaluating diameter sizes for a solid circular cylindrical shaft of diameter 'd' transmitting W kilowatts at a speed of N revs/minute a technician arrives at the following formula for computing 'd':

$$d = 10\tau \left(\sqrt[3]{W/N} \right)$$

in which 'd' is in mm, and τ = maximum allowable value of stress in MN/m².

Determine what the imputed value of maximum allowable shear stress is in the shaft, the value of this stress being in MN/m².

Let T = maximum torque, N m

   w = angular velocity, rad s^{-1}

If power transmitted = W kilowatts, then

   T $\times$ w = W $\times$ 1000

~ 406 ~

Also
$$w = \frac{2\pi N}{60}$$

$$\therefore \quad T \times \frac{2\pi N}{60} = 1000W$$

i.e.
$$T = \frac{60 \times 10^3 W}{2\pi N} \text{ Newton metre}$$

Recalling the Torsion Formula

$$\frac{T}{J} = \frac{\tau}{r}$$

where the symbols have their usual meaning, then if τ is the maximum shear stress in MN/m^2 and 'd' is in mm, then

$$\frac{\dfrac{60 \times 10^3 W}{2\pi N}}{\dfrac{\pi}{32}\left(\dfrac{d}{1000}\right)^4} = \frac{\tau \times 10^9}{\dfrac{d}{2(1000)}}$$

i.e.
$$\frac{60 \times 10^3 \times 10^{12} \times 32W}{2\pi N \times \pi \times d^4} = \frac{2 \times \tau 10^9 \times 10^3}{d}$$

or
$$d^3 = \frac{480 \times 10^3 W}{\tau \pi^2 N}$$

$$= \frac{48.62 \times 10^3}{\tau} \cdot \frac{W}{N}$$

$$\therefore \quad d = 10\left(\frac{48.62}{\tau}\right)^{1/3}\left(\frac{W}{N}\right)^{1/3}$$

from which it follows that the imputed value of τ may be obtained by writing

$$\left(\frac{48.62}{\tau}\right)^{1/3} = 1$$

or
$$\tau = 48.62 \, MN/m^2.$$

i.e.   the imputed value of maximum allowable stress in the technician's algorithm is $\tau = 48.62 MN/m^2$.

Q11.   Case has reported that a good rule of thumb to follow in when specifying torsional stiffness in a solid steel shaft transmitting power is to allow a maximum angle of torsion of 1° for a length equivalent to 20 diameters. Prove that for a shaft for which C = 84 x 10⁹ N/m², this rule translates into a maximum allowable shearing stress of 36.7 MN/m².

Let   q = maximum allowable shearing stress, N/m².

D = diameter of shaft, mm

C = torsional modulus, N/m²

L = length of shaft, mm

θ = angle of twist, radians

$$\therefore \quad \frac{\frac{q}{D}}{2} = \frac{\frac{C_\theta}{\ell}}{1000}$$

i.e. $$\frac{2q}{D} = \frac{C \cdot \theta}{20D}$$

i.e. $$q = \frac{C\theta}{40} \qquad \dots\dots\dots\dots\dots\dots\dots\dots\dots \text{(i)}$$

In equation (i) θ is in radians. The rule of thumb is 1° in 20 diameters.

Because 2π radians = 360

$$1° = \frac{2\pi}{360} \text{ radians}$$

$\therefore$ Equation (i) becomes

$$q = \frac{C}{40} \cdot \frac{2\pi}{360}$$

Substituting C = 84 x 10⁹ N/m²

$$q = \frac{84 \times 10^9 \cdot 2\pi}{40 \cdot 360}$$

$$= 36,668,333 N/m^2$$

say $36.7 \times 10^6 N/m^2$ or

$36.7 MN/m^2$

Q12. A hollow bronze tube is shrunk on to a solid Nickel-Chrome-Molybednum steel tube of the same length thereby producing a composite shaft. A cross-section of the shaft which is uniform throughout is shown in Fig. 1. A torque of 1000 Nm is applied to the shaft. Determine the maximum shear stress in the bronze and Ni-Cr-Mo tubes. Take shear modulii to be: bronze = 45GN/m²; Ni-Cr-Mo steel = 80GN/m².

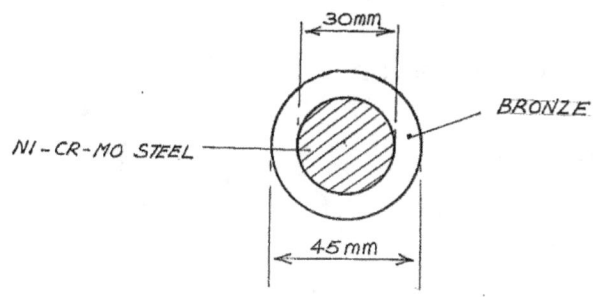

FIG 1: COMPOSITE SHAFT

Note that the maximum stress for reach component of the composite shaft is in the outermost fibres.

Ans: $\tau_{bronze} = 48.5 MN/m^2$; $\tau_{steel} = 28.7 MN/m^2$.

Q13. In a torsion test employing a mild steel specimen 200mm in length and 20mm diameter it was observed that for a torque of 125Nm the angle of twist was 0.020 radians, and also that the limit of proportionality of the specimen in shear was reached at 225Nm. What is the value of the shearing or torsional modulus C? Determine also the stress corresponding to the limit of proportionality.

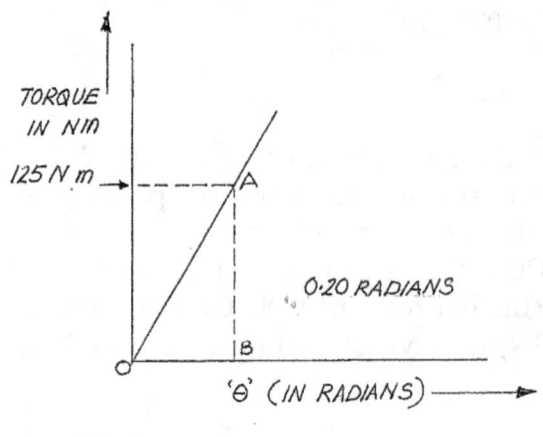

FIG 1

From theoretical considerations

$$\frac{T}{J} = \frac{C\theta}{L}$$

i.e.

$$T = \frac{CJ}{L}\theta$$

Therefore for a constant 'C', 'J' and 'l' a plot of T versus θ as in Fig. 1 should produce a straight line passing through the origin, the slope of which is a constant equal to $\frac{CJ}{L}$

From the plot shown:

$$\text{Slope} = \frac{AB}{OB}$$

i.e. $\text{Slope} = \dfrac{125 Nm}{0.020\ rad}$

$\therefore$ $\dfrac{CJ}{L} = \dfrac{125}{0.02}$

or $C = \dfrac{125}{0.02} \times \dfrac{200}{1000} \times \dfrac{1}{J}$

$$= \frac{125}{0.02} \times \frac{200}{1000} \times \frac{32 \times 10^{12}}{\pi(20)^4}$$

$$= \frac{125}{0.02} \times \frac{200}{1000} \times \frac{32}{\pi} \times \frac{10^{12}}{16 \times 10^4}$$

$$= \frac{12500 \times 2 \times 10^{12}}{10 \times \pi \times 10^4}$$

$$= \frac{250}{\pi} \times 10^9$$

$$\therefore \qquad C = 79.5 \times 10^9 \, N/m^2.$$

Torque at proportional limit = 225 N m. If q_L = shearing stress corresponding to this limit, then

$$\frac{q_L}{10/1000} = \frac{225.32}{\pi(20/1000)^4}$$

$$\therefore \qquad q_L = \frac{225.32 \times 10^{12}}{\pi \times 100 \times 16 \times 10^4}$$

$$= \frac{450}{\pi} \times 10^6$$

$$= 143 \times 10^6 \, N/m^2$$

$$q_L = 143 MN/m^2.$$

Q14. A torque of 5600 Newton metre is applied to a solid shaft of diameter 75mm. What is the magnitude of the shearing stress in the most extreme fibres of the shaft? By a suitable sketch show the direction of the shearing stress in relation to the applied torque.

We may use the relationship:

$$\frac{T}{J} = \frac{q}{r} \qquad \dots\dots\dots\dots\dots\dots\dots\dots\dots\dots\dots\dots\dots \text{(i)}$$

in which,

T = torque in Newton metre;

J = polar moment of inertia in (metre)⁴;

q = shearing stress in Newtons per square metre,

i.e. N/m² and

r = radius in metre

Now T = 5600 N m

$$J = \frac{\pi d^4}{32}$$

$$= \frac{\pi}{32}(0.3164)10^{-4}$$

i.e. $J = 0.0311 \times 10^{-4} m^4$.

$$r = \left(\frac{75}{2 \times 1000}\right) m$$

Substituting in equation (i) which we have rearranged as,

$$q = \frac{T_r}{J}$$

we get

$$q = \frac{5600 \times 75}{0.311 \times 10^{-5} \times 2 \times 10^3}$$

$$= \frac{560 \times 10^6 \times 75^3}{0.311 \times 2 \times 10^3}$$

$$= \frac{7}{0.311} MN/m^2$$

i.e. $\underline{q = 22.5 MN/m^2}$

This is the value of shearing stress in the outermost fibres of the shaft. See Fig. 2 for directions of shearing stress.

<u>Note directions of shearing stress 'q' are the same as directions of internal resisting torques</u>

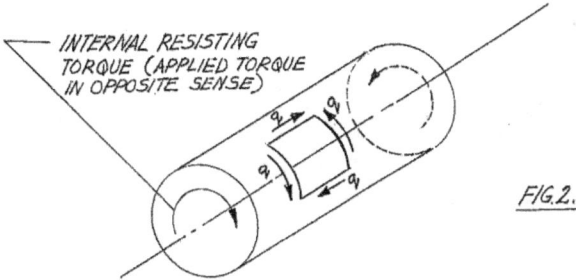

INTERNAL RESISTING
TORQUE (APPLIED TORQUE
IN OPPOSITE SENSE)

FIG. 2.

NOTE DIRECTIONS OF SHEARING STRESS 'q' ARE
THE SAME AS DIRECTIONS OF INTERNAL RESISTING
TORQUE

Q15. A phosphor-bronze tube 0.5m long and 75mm outside diameter is shrunk on to a solid shaft of the same length, 50mm diameter, and made of monel metal.

If the working shearing stresses in the phosphor bronze and monel metal are restricted to 30MN/m² and 20MN/m² respectively, then what is the maximum axial torque the compound shaft can transmit?

Take modulus of rigidity of phosphor bronze = 40GN/m² and that for monel metal = 25GN/m².

From the torsion formula

$$\frac{T}{J} = \frac{C\theta}{\ell} = \frac{q}{r} \qquad \dots\dots\dots\dots\dots\dots\dots\dots\dots \text{(i)}$$

where the symbols have their usual meaning, it is evident when θ/ℓ is a constant, say k, we may write

$$\frac{q}{r} = kc \qquad \dots\dots\dots\dots\dots\dots\dots\dots\dots \text{(ii)}$$

Let q_{pb} = maximum stress in phosphor bronze tube

      q_m = maximum stress in monel metal shaft

      r_{pb} = radius of phosphor-bronze tube where q_{pb} occurs,

         i.e. outside radius of phosphor-bronze tube

Let r_m = radius of monel metal shaft where q_m occurs,

         i.e. outside radius of monel metal shaft

" C_{pb} = modulus of rigidity of phosphor-bronze

" C_m = modulus of rigidity of monel metal.

Accordingly from (ii)

$$\frac{q_{pb}}{r_{pb}} = C_{pb}k$$

and

$$\frac{q_m}{r_m} = c_m k$$

$$\therefore \qquad \frac{q_{pb}}{q_m} \cdot \frac{r_m}{pb} = \frac{C_{pb}}{C_m} \qquad \dots\dots\dots\dots\dots\dots\dots\dots\dots \text{(iii)}$$

Assuming that the maximum stress occurs in the monel metal, then putting $q_m = 20MN/m^2$ in (iii)

$$\frac{q_{pb}}{20} \cdot \frac{50}{2} \cdot \frac{2}{75} = \frac{C_{pb}}{C_m}$$

We are given that $C_{pb} = 40GN/m^2$ and $C_m = 25GN/m^2$

$$\therefore \qquad \frac{q_{pb}}{20} \cdot \frac{50}{75} = \frac{40}{25}$$

from which $q_{pb} = 48MN/m^2$.

This value of q_{pb} exceeds the maximum permissible working stress in the phosphor bronze tube.

Therefore the assumption of a maximum working stress occurring in the monel metal shaft turns out to be incorrect.

Using the same relationship at (iii) let us see what happens in the monel metal shaft when the maximum stress in the phosphor bronze tube is at its maximum permissible working stress of $30MN/m^2$. Therefore we now put $q_{pb} = 30 \ MN/m^2$ in equation (iii) and treat q_m as the unknown.

Accordingly

$$\frac{30}{q_m} \cdot \frac{50}{2} \cdot \frac{2}{75} = \frac{40}{25}$$

i.e. $\qquad q_m = \dfrac{25 \times 20}{40}$

$$\therefore \qquad q_m = 12.5 MN/m^2$$

It is seen therefore that when the phosphor bronze tube is at its maximum working stress of 30MN/m², the working stress in the monel metal shaft is 12.5MN/m².

This condition fulfils the limitation put on working stresses in the separate components of the compound shaft.

Now,

$$\text{Total axial torque} = \text{axial torque in phosphor-bronze}$$
$$+ \text{axial torque in monel metal}$$

From (i) $\quad \dfrac{T}{J} = \dfrac{q}{r}$

$\therefore \quad \dfrac{T_{pb}}{J_{pb}} = \dfrac{q_{pb}}{r_{pb}}$

$\therefore \quad T_{pb} = \dfrac{q_{pb} \cdot J_{pb}}{r_{pb}}$

Similarly

$$T_m = \dfrac{q_m \cdot J_m}{r_m}$$

$\therefore \quad \text{Total axial torque} = \dfrac{q_{pb}J_{pb}}{r_{pb}} + \dfrac{q_m J_m}{r_m}$

in which $q_{pb} = 30 \text{ MN/m}^2$ and $q_m = 12.5 \text{ MN/m}^2$

Now $\quad J_{pb} = \dfrac{\pi}{2}\left(r_o^4 - r_i^4\right) = \dfrac{\tau}{2}\left\{\left(\dfrac{75}{2000}\right)^4 - \left(\dfrac{50}{2000}\right)^4\right\}$

For the monel metal shaft $J_{pb} = \dfrac{1}{2} \cdot \pi\left(\dfrac{50}{1000}\right)^2$

$\therefore \quad \text{Total axial torque} = 30\times10^6 \dfrac{\pi}{32(1000)^4} \dfrac{(5625-2500)\ (5625+2500)}{75/2000}$

$$12.5\times10^6 \cdot \dfrac{1}{2}\pi\left(\dfrac{50}{2000}\right)^4 \dfrac{2000}{50}$$

$$= 10^6 \left\{ \frac{2000 \times 30\pi (3125)(8125)}{32 \times 10^{12} \times 75} + \frac{12.5}{32} \cdot \frac{\pi}{10^{12}} \cdot \frac{50^3 \cdot 2000}{10^9} \right\}$$

$$= 10^6 \left\{ \left(\frac{199443}{10^8} \right) + \left(\frac{306836}{10^9} \right) \right\}$$

$$= 2301.2 Nm \quad \text{or} \quad 2.3kNm.$$

Q16. The allowable shearing stress in a hollow shaft of outside diameter 100mm and inside diameter 75mm which is to transmit a torque, is. 55MN/m². What is the maximum torque the shaft can transmit? Determine also the value of shearing stress at the inner surface of the shaft for that same torque.

The maximum shearing stress occurs in the outmost fibres of shaft, i.e. at r = 50mm. Now the polar moment of inertia of the shaft, say, J is given by:

$$J = \frac{\pi}{32} \left\{ \left(\frac{100}{1000} \right)^4 - \left(\frac{75}{1000} \right)^4 \right\} m^4$$

$$= \frac{\pi}{32} \left\{ \left(\frac{1}{10} \right)^4 - \left(\frac{3}{40} \right)^4 \right\} m^4$$

$$= \frac{\pi}{32} (0.0001 - 0.00003164)$$

$$= \frac{\pi}{32} (0.00006836)$$

$$= 0.000006712$$

$$= 6.712 \times 10^{-6} m^4$$

Using the relationship

$$\frac{q}{r} = \frac{T}{J}$$

where the quantities q, r, T, and J were as previously defined and with units :

$$q \ : \ N \ m^2$$

r : m

T : N m

J : m⁴

Here, $q = 56 \times 10^6 \ N/m^2$

$r = 50mm$

$= \dfrac{50}{1000} m$

$= \dfrac{1}{20} m.$

$J = 6.712 \times 10^{-6} \ m^4$

∴ $T = \dfrac{qJ}{r}$

$= \dfrac{56 \times 10^6}{\left(\dfrac{1}{20}\right)} \times \dfrac{6.712}{10^6}$

$= 20 \times 56 \times 6.712 \, Nm$

$= 7517.44 \, Nm$

$T \approx 7517 \, Nm.$

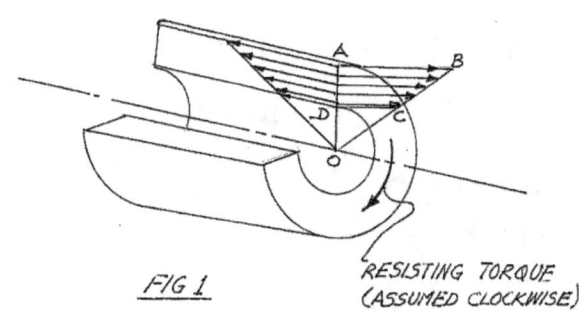

FIG 1

RESISTING TORQUE
(ASSUMED CLOCKWISE)

If the internal resisting torque is as indicated in Fig. 1.., then the direction of the shearing stress in outermost fibres of shaft is $\xrightarrow[AB]{}$.
 $\xrightarrow[DC]{}$ is the direction of the shearing stress in the fibres of the inside

wall and its magnitude q_i is obtained from $\dfrac{q_i}{56MN/m^2} = \dfrac{OD}{OA} = \dfrac{37.5}{50}$, i.e.

$$q_i = 42 MN/m^2$$

Q17. A component of a piece of equipment used on a chemical plant has a copper rod of diameter 25mm fitted tightly inside a tube made of Inconel of outer diameter 50mm so that there can be no relative movement between rod and tube. Determine the torque which when applied to the composite shaft will cause a maximum shear stress of 30MN/m² in the Inconel. Assume a linear distribution of stress in each of the materials and that the torsion modulus of Inconel is roughly twice that of copper. Find also the maximum shear stress in the copper.

Let T_c = torque in copper rod, N m

T_I = " " Inconel tube, N m

∴ Total torque $T_t = T_c + T_I$ .... (i)

Now

$$\frac{T_C}{\dfrac{\pi}{32}\left(\dfrac{25}{1000}\right)^4} = \frac{C_c \theta_c}{\ell} \qquad \dots\dots\dots\dots\dots\dots \text{(ii)}$$

Similarly

$$\frac{T_I}{\dfrac{\pi}{32}\left\{\left(\dfrac{50}{1000}\right)^4 - \left(\dfrac{25}{1000}\right)^4\right\}} = \frac{C_I \theta_I}{\ell} \qquad \dots\dots\dots\dots \text{(iii)}$$

where C_c = torsion modulus of Copper

C_I = " " " Inconel

From (ii) $\qquad \theta_c = \dfrac{T_c \cdot 32 \cdot 10^{12} \cdot \ell}{\pi \times (25)^4 \cdot C_c}$

From (iii) $\qquad \theta_I = T_I \cdot \dfrac{32}{\pi} \cdot \dfrac{10^{12}}{(1875)(3125)}$

Because there is no relative movement

$$\theta_c = \theta_I$$

$$\therefore \quad \frac{T_c 32 \cdot 10^{12} \cdot \ell}{\pi \cdot 625 \times 625 \cdot C_c} = \frac{T_I \cdot 32 \times 10^{12} \cdot \ell}{C_I \times \pi \cdot 1875 \times 3125}$$

Note it is assumed that the rod and tube are of the same length

$$\therefore \quad \frac{15 T_c}{C_c} = \frac{T_I}{C_I}$$

But $\quad C_I = 2C_c$

$$\therefore \quad \frac{15 T_c}{Cc} = \frac{T_I}{2Cc}$$

or $\quad 30 T_c = T_I$

We are given that the maximum shear stress in the Inconel is 30MN/m²

$$\therefore \quad \frac{T_I}{\dfrac{\pi}{32} \times \dfrac{1}{10^{12}}(1875)(3125)} = \frac{30 \times 10^6}{\left(\dfrac{25}{1000}\right)}$$

or $\quad T_I = \dfrac{30 \times 10^6 \times 1000 \times \pi 1875 \times 3125}{32 \times 10^{12} \times 25}$

$$= 6906 Nm.$$

$$\therefore \quad T_t = T_c + T_I$$

$$= \frac{T_I}{30} + T_I$$

$$= \frac{31}{30} T_I$$

$$= \frac{31}{30(690.6)} Nm$$

$$T_t = 714 Nm$$

Maximum shear stress in copper from

$$\frac{T_c}{\dfrac{\pi}{32}\left(\dfrac{25}{1000}\right)^4} = \frac{q_c}{\left(\dfrac{25}{2000}\right)}$$

$$\therefore \quad \frac{690.6}{30 \times \pi} \times \frac{32 \times 10^{12}}{625 \times 625} = q_c \cdot \frac{2000}{25}$$

i.e. $\quad q_c = \dfrac{690.6 \times 32 \times 10^{12} \times 25}{30 \times \pi \times 625 \times 625 \times 2000}$

$$= 14.99 \times 10^6$$

$$\therefore \quad q_c = 15 \times 10^6 \, N/m^2$$

$$q_c = 15 MN/m^2$$

Q18. A solid steel shaft is to be designed to transmit 20kW at 150 rpm. Design considerations limit the angle of twist to no more than 1° in a length of 25 diameters. Determine the diameter of the shaft and estimate the maximum intensity of shearing stress. Take C = 84 x 10⁹ N/m².

Let   d = diameter of shaft in mm

"   w = angular velocity of shaft in radians/s

"   T = shaft torque in Newton metres

$\therefore \quad 20000 = T\,w \qquad \dotfill \qquad$ (i)

But $\quad w = \dfrac{2\pi N}{60}$

where N = speed of shaft in rpm

$$\therefore \quad w = 2\pi \times \frac{150}{60}$$

$$= 15.7 \text{ radians/sec}$$

$\therefore \quad$ from equation (i)

$$T = \frac{20000}{15.7} \, Nm$$

$$= 1274 Nm$$

Now,

$$\frac{T}{J} = \frac{C\theta}{\ell} = \frac{q_m}{r} \qquad \dotfill \qquad \text{(ii)}$$

where

$$J = \text{polar moment of inertia, m}^4$$

$$C = \text{torsional modulus, N/m}^2$$

$$\theta = \text{angle of twist, radians}$$

$$\ell = \text{length of shaft, m}$$

Given that 'θ' is restricted to 1° in a length of 25 diameters, this may be restated as $\dfrac{2\pi}{360}$ radians in a length of $\dfrac{25d}{1000}$, because an angle of 360° is equivalent to 2π radians and diameter 'd' as previously defined, is in millimetres.

Substituting in (ii)

$$\therefore \qquad \frac{1274 \cdot 32}{\pi\left(\dfrac{d}{1000}\right)^4} = \frac{84\times10^9}{25d}\times\frac{2\pi}{360}\times1000$$

i.e.

$$\frac{1274\times32\times10^{12}}{\pi d^4} = \frac{84\times10^9\times2\pi\times1000}{25d\times360}$$

$$d^3 = \frac{1274\times16\times25\times360}{84\times\pi\times\pi}$$

$$d^3 = 221087.3$$

or $\qquad d = 60.4 \quad$ say $\quad 61mm$

Let $\quad q_m = $ maximum intensity of shearing stress in N/m²

$\qquad r = $ radius of shaft in mm $\left(\dfrac{d}{2000}\right) mm$

$$\therefore \qquad \frac{q_m}{\left(\dfrac{d}{2000}\right)} = \frac{C\theta}{L} = \frac{84\times10^9\times2\pi\times1000}{25\times60\cdot4\times360}$$

i.e. $\qquad 2q_m = \dfrac{84\times10^9\times2\pi}{25\times60.4\times360}$

or $\qquad q_m = 0.486\times10^6$, say, $0.49 Mn/m^2$

Q19. The allowable shearing stress parallel to the grain of a wooden shaft, is 700kN/m². What is the maximum torque such a solid shaft of 200mm diameter can transmit with the grain parallel to the axis of the shaft?

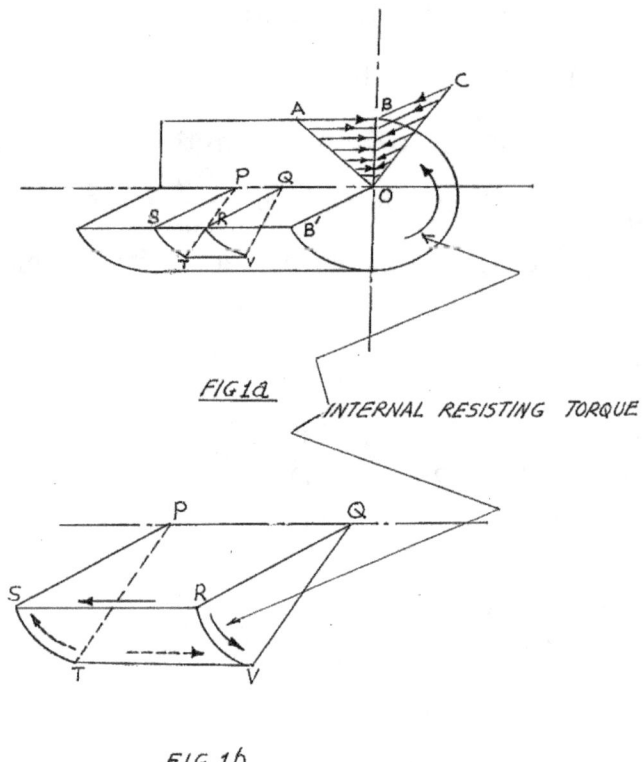

FIG 1a

INTERNAL RESISTING TORQUE

FIG. 1b

In Fig. 1a the assumed direction of the resisting torque is anti-clockwise. Accordingly the direction of the shearing stress caused by this torque, in the plane of the shaft *BB'* is as indicated. This shearing stress distribution is given by triangle 'OBC'. The complementary shearing stress distribution in a plane parallel to the axis of the shaft is OAB and the direction of stress is from A to B.

We have shown these complementary shearing stresses in a wedge of the shaft, in Fig. 1b.

Because the shear strength of wood is of a value which is very much less parallel to the grain than across it, the allowable shearing strength of 700kN/m² is taken as the complementary shearing stress induced in planes normal to the axis of the shaft, resulting from the applied torque

$$J = \frac{\pi}{32}\left(\frac{200}{1000}\right)^4 m^4$$

$$= 1.6 \times 10^{-4} m^4$$

~ 422 ~

Now
$$\frac{q}{r} = \frac{T}{J}$$

$$\therefore \quad \frac{700 \times 10^3}{\left(\dfrac{100}{1000}\right)} = \frac{T}{1.6 \times 10^{-4}}$$

$$\therefore \quad T = \frac{700 \times 10^3 \times 1.6 \times 1000}{100 \times 10^4}$$

$$T = 11.2 \times 100$$

$$T = 1120 Nm$$

If a torque is applied on the basis of a computed value of shearing strength 'across the grain', then the natural consequence would be failure of the shaft by splitting along planes parallel to the grain, because the induced complementary shearing stress parallel to the axis of the shaft will then be greater than the natural shear strength of the wood 'along the grain'.

Q20. A steel jackshaft supports 4 pulleys as shown in Fig. 1 below. Determine (i) the unknown tension 'P' to maintain equilibrium, (ii) the resultant bearing reactions at each end of the shaft, and (iii) the shearing stresses in each of its five sections: A -1; 1-2; 2-3; 3-4; and 4-5. The diameter of the pulleys in millimetres, from 1 to 4 are respectively: 150,200, 100,100.

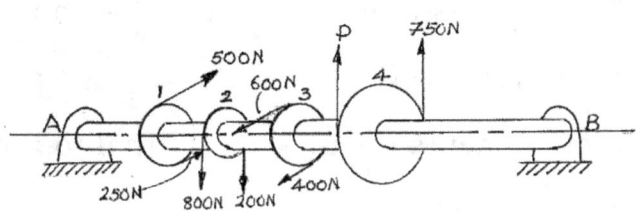

FIG 1

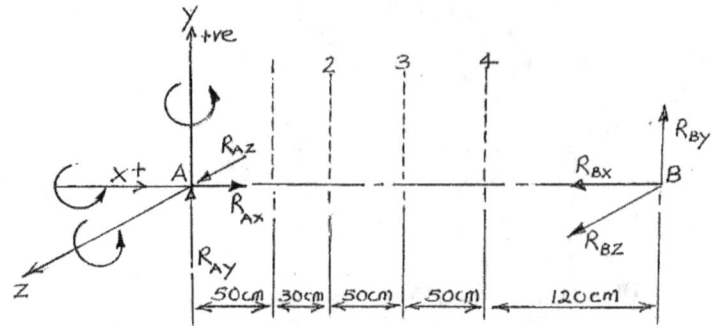

FIG.2 : COORDINATE AXES AND DIRECTION OF
POSITIVE FORCES AND MOMENTS TOGETHER
WITH COMPONENTS OF BEARING REACTIONS

Having arbitrarily fixed the directions of positive forces and moments, it was assumed that the reactions at bearings 'A' and 'B' take the directions indicated.

There are six quantities which form the basis of three-dimensional equilibrium, namely:

$$\sum R_x = 0 \quad ; \quad \sum R_y = 0 \quad ; \quad \sum R_z = 0$$

$$\sum M_x = 0 \quad ; \quad M_y = 0 \quad ; \quad \sum M_z = 0$$

These merely state that the sum of all forces along the three coordinate axes equals zero, and so too the moments.

Inspection of Fig. 1 reveals that there are no X-component of forces

$$\therefore \quad R_{AX} = 0 \quad ; \quad R_{BX} = 0$$

Taking the sum of moments about X-axis equal to zero, i.e. $\sum M_x = 0$, we have

$$-500\left(\frac{75}{1000}\right) + 250\left(\frac{75}{1000}\right) + 800\left(\frac{100}{1000}\right) - 200\left(\frac{100}{1000}\right)$$

$$+600\left(\frac{50}{1000}\right) - 400\left(\frac{50}{1000}\right) - P\left(\frac{50}{1000}\right) + 750\left(\frac{50}{1000}\right) = 0$$

i.e. $\quad -250\left(\frac{75}{1000}\right) + 600\left(\frac{100}{1000}\right) + 200\left(\frac{500}{1000}\right) - P\left(\frac{50}{1000}\right)$

$$+750\left(\frac{50}{1000}\right) = 0 \qquad \cdots\cdots\cdots\cdots\cdots \quad (i)$$

or $\qquad \dfrac{75}{4}+60+10-\dfrac{P}{20}+37.5=0$

$$\therefore \qquad \dfrac{P}{20}=107.5-\dfrac{75}{4}$$

from which $\dfrac{P}{20}=88.75$

or $\qquad$ P = 1775N

Applying $\sum R_y = 0$

$$R_{AY}+800-200+P+750+R_{BY}=0 \qquad \dots \dots \dots \text{(ii)}$$

Also, $\sum M_y = 0$

As before, working in Newtons and metres:

$$500\left(\dfrac{50}{100}\right)+250\left(\dfrac{50}{100}\right)-600\left(\dfrac{130}{100}\right)-400\left(\dfrac{130}{100}\right)-R_{BZ}\left(\dfrac{300}{100}\right)=0$$

i.e. $\qquad 250+125-780-520-3R_{BZ}=0 \qquad \dots \dots \dots \dots \dots \text{(iii)}$

$$\therefore \qquad 3R_{BZ}=-1300+375$$

$$=-925$$

i.e. $\qquad \underline{R_{BZ}=-308.3N}$

The negative sign indicates that R_{BZ} acts in a direction opposite to that shown in Fig. 2.

Considering now $\sum R_z = 0$

$$R_{AZ}-500-250+600+400+R_{BZ}=0$$

i.e. $\qquad R_{AZ}+1000-750+R_{BZ}=0 \qquad \dots \dots \dots \dots \dots \text{(iv)}$

Substituting R_{BZ} = -308.3N

$$R_{AZ}+250-308.3 = 0$$

i.e $\quad \underline{R_{AZ} = 58.3N}$

Finally, $\sum M_Z = 0$

$\therefore \quad -800\left(\dfrac{80}{100}\right) - 200\left(\dfrac{80}{100}\right) + P\left(\dfrac{180}{100}\right) + 750\left(\dfrac{180}{100}\right)$

$$+ R_{BY}\left(\dfrac{300}{100}\right) = 0$$

i.e. $\quad -640 - 160 + 1775(1.8) + 7.5(180) + 3R_{BY} = 0 \qquad \cdots \cdots \cdots$ (v)

or $\quad -800 + 3195 + 1350 + 3R_{BY} = 0$

$\therefore \quad 3R_{BY} = -3745$

or $\quad \underline{R_{BY} = -1248.3N}$

Substituting this value of R$_{BY}$ in (ii) produces

$R_{AY} = -1525 - R_{BY}$

$= -1525 - (-1248.3)$

$\underline{R_{AY} = -276.7N}$

Again, the negative sign tells us that the direction of R_{AZ} assumed in Fig. 2 is opposite what is shown in the diagram. Collecting results we have :

$R_{AX} = 0 \qquad ; \quad R_{AY} = -276.7N;$

$R_{AZ} = 58.3N \quad ; \qquad P = 1775N$

$P_{BZ} = 0 \qquad ; \quad R_{BY} = -1248.3N$

$R_{BZ} = -308.3N$

Resultant bearing reactions

Let resultant bearing reactions be S$_{RA}$ and S$_{RB}$ at 'A' and 'B' respectively.

Accordingly,

$$S_{RA} = \sqrt{R_{AX}^2 + R_{AY}^2 + R_{AZ}^2}$$

$$= \sqrt{(0)^2 + (-276.7)^2 + (58.3)^2}$$

$$= \sqrt{76563 + 3399}$$

$$= \sqrt{79962}$$

$$\therefore \qquad \underline{S_{RA} = 282.7N}$$

Similarly,

$$S_{RB} = \sqrt{R_{BX}^2 + R_{BY}^2 + R_{BZ}^2}$$

$$\therefore \qquad S_{RB} = \sqrt{(0)^2 + (-1248.3)^2 + (-308.3)^2}$$

$$= \sqrt{1558253 + 95049}$$

$$= \sqrt{1653302}$$

i.e. $\qquad \underline{S_{RB} = 1286N}$

<u>Shearing Stresses</u>

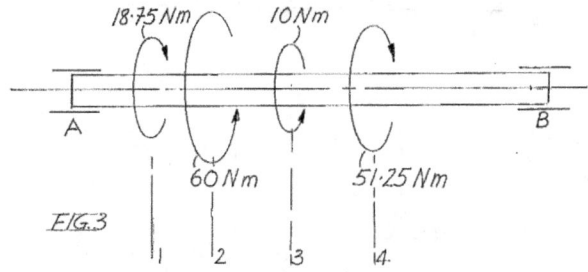

FIG.3

Torque at pulley 1 $\quad = \quad (500 - 250)\dfrac{75}{1000} = 18.75Nm \circlearrowleft$

"   "   " $2 \quad = \quad (800 - 200)\dfrac{100}{1000} = 60Nm \circlearrowright$

"   "   " $3 \quad = \quad (600 - 400)\dfrac{50}{1000} = 10Nm \circlearrowright$

"   "   " $4 \quad = \quad (1775 - 750)\dfrac{50}{1000} = 51.25Nm \circlearrowleft$

Note: $\sum T = 0$, i.e. assuming clockwise torques +ve: $(51.25 - 10 - 60 + 18.75)Nm = 0$. Refer to Fig. 3

Torque between A and pulley 1 = 0

Torque between pulleys 1 and 2 = 18.75N m ↺

"   "   "   2 and 3 = 41.25N m↺

"   "   "   3 and 4 = 51.25N m↺

"   "   "   4 and 5 = 0

The determination of the shearing stresses in the various sections is left as an exercise for the student. To help you along the way I have drawn the FBD, Fig. 4. By referring to it you would see that the section of the shaft between sections 1 and 2 of length 30cm is under the influence of a torque of 18.75Nm; that between sections 2 and 3 of length 50cm under the influence of a torque of 41.25Nm and between 3 and 4 of length 50cm, 51.25Nm. Note that the nett torque at section 2 is 60Nm anti-clockwise and at 3, 10Nm anti-clockwise. Now you do the rest.

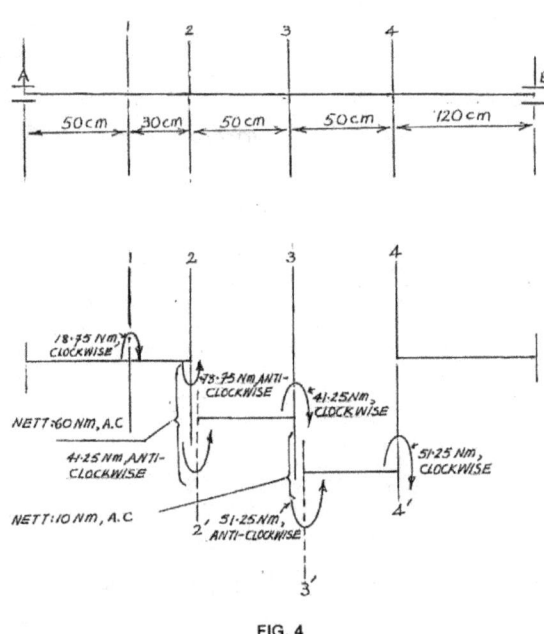

FIG. 4

Q21 The solid circular cylindrical steel shaft carrying the three pulleys shown in Fig. 1 is driven by a 25kW motor at 150 rpm. Determine the torsional stresses in the sections of the shaft A-1; 1-2; 2-3 and 3-4. The shaft is 50mm diameter throughout.

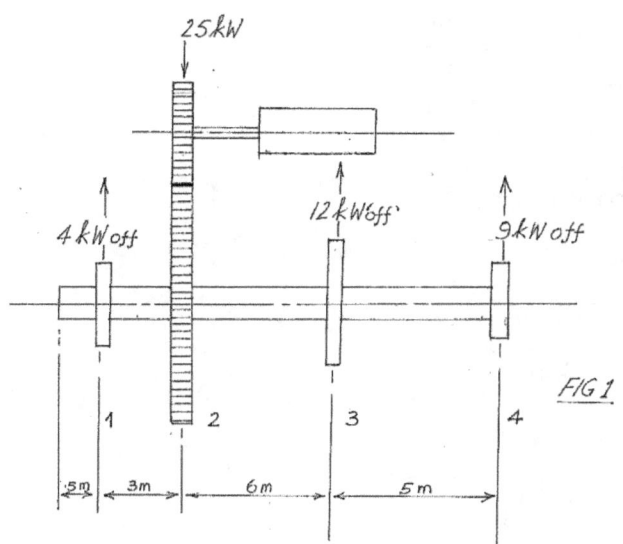

From the basic relationship:

Power (in Watts) = Newton metre per second

= Torque × angular velocity

≡ N m × w

we may determine the torques at pulleys 1, 2, 3 and 4.

Accordingly at 1

4000 = Torque (T in N m) × w

Now
$$w = \frac{2\pi N}{60} \text{ rad/second}$$

$$= \frac{2\pi \cdot 150}{60} \text{ rad/s}$$

$$= 15.7 \text{ rad/s}$$

∴
$$T_1 = \frac{4000}{15.7}$$

$$= 254.7 Nm$$

T_1, say, 255N m

∴
$$T_2 = \frac{25000}{15.7}$$

$$= \frac{25000}{4000} \times 255$$

$$= 1594 Nm$$

Similarly by proportion:

$$T_3 = \frac{12000}{4000} \times 255$$

$$= 765 Nm$$

and

$$T_4 = \frac{9000}{4000} \times 255$$

$$= 574 Nm$$

These torques may, for the sake of convenience, be distributed on the shaft in accordance with the sketch on the following page, Fig. 2

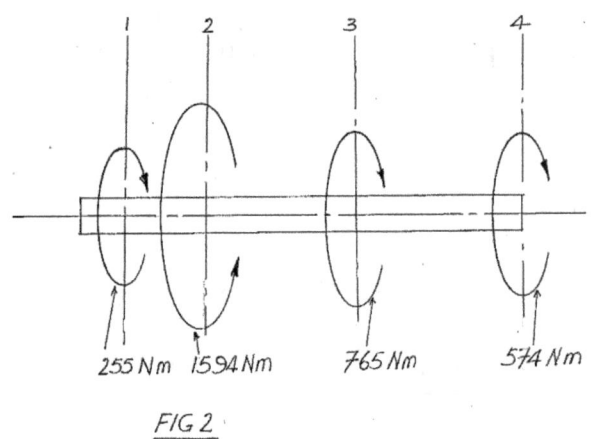

FIG 2

Torque in section 4 − 3 = 574 ↻

" " " 3 − 2 = 1339 ↻

" " " 2 − 1 = 255 ↺

" " " 1 − A = 0(255↺ + 255↻)

where all torques are in Newton metres. Polar moment of inertia (J) for shaft

$$= \frac{\pi}{32} \left(\frac{50}{1000} \right)^4$$

$$= 0.613 \times 10^{-6} \, m^4$$

Now, torsional stress 'q' $= \dfrac{Torque \times radius}{J}$

so that $\quad q_{1-A} = 0$

$$q_{2-1} = \frac{255}{0.613} \times \frac{25}{1000} \times 10^6$$

$$= 10.4 \times 10^6 \, N/m^2$$

$\therefore \quad q_{2-1} = 10.4 \, MN/m^2$

Similarly

$$q_{3-2} = \frac{1339}{0.613} \times \frac{25}{1000} \times 10^6$$

or proportionally,

$$q_{3-2} = \frac{1339}{255} \times 10.4 \, MN/m^2$$

i.e. $\quad q_{3-2} = 54.6 \, MN/m^2$

For section 4-3

$$q_{4-3} = \frac{574}{255} \times 10.4 \, MN/m^2$$

$$= 23.4 \, MN/m^2$$

Collecting our answers we summarise as follows:

$$q_{1-A} = 0;$$

$$q_{2-1} = 10.4 \, MN/m^2;$$

$$q_{3-2} = 54.6 \, MN/m^2; \quad \text{and}$$

$$q_{4-3} = 23.4 \, MN/m^2$$

Q.22 The crankshaft of a lawnmower engine is shown in the accompanying diagram, Fig. 1. With the crank in the position shown the effective force in the connecting rod is 10kN. The torque 'W' developed by the engine is taken off at 'Q', just beyond bearing 'A'. Assuming that the bearings provide no restraint on the shaft, calculate the twisting moment on the crank pin for the position in Fig. 1 of the crank.

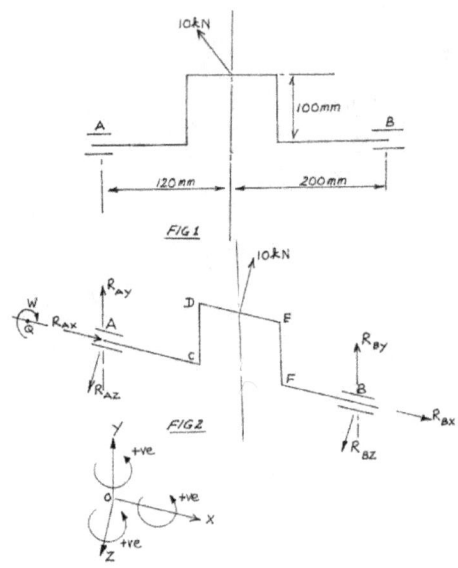

Ans: $R_{AZ} = R_{BX} = 0$; $R_{AY} = R_{BY} = 0$; $R_{AZ} = 6.25$ N; $R_{BZ} = 3.75$.

Twisting Moment on crank-pin = 375 Nm, clock wise about x-axis.

Q23. The single-throw crankshaft shown in Fig. 1 below is supported by two bearings at A and B. The centre of the crankpin is 300mm and 350mm respectively from A and B. The maximum force on the crank pin is 50kN and occurs when the angle between it and the connecting rod of the piston is 90°. Energy is taken off at a point just beyond bearing A. Determine the bearing reactions and also the twisting moment on the crankpin.

Ans: $R_{AX} = R_{AY} = 0$; $R_{AZ} = 27kN$; $R_{BX} = R_{BY} = 0$; $R_{BZ} = 23kN$

Twisting moment on crank pin = 4.1kNm (clockwise).

~ 432 ~

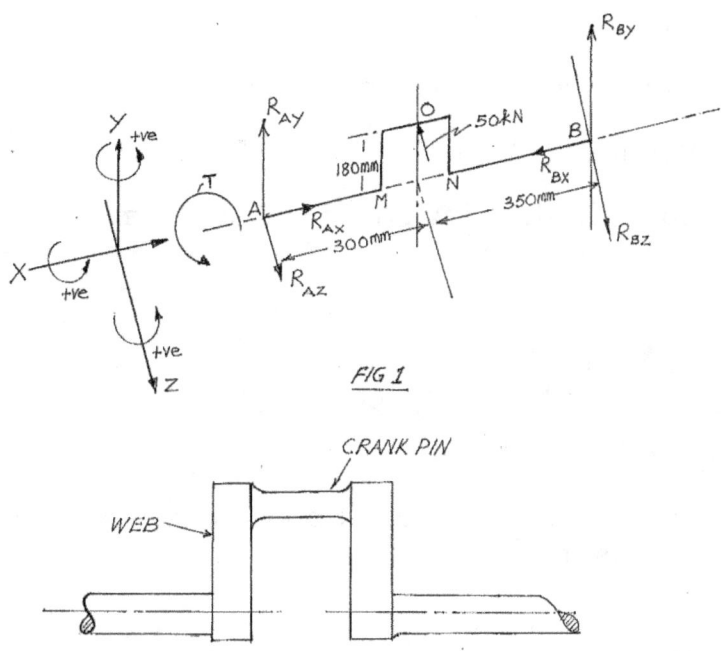

FIG 1

CRANK PIN

WEB

Q24. The flange coupling in Fig. 1 has 6 bolts each having a cross-sectional area of 324 (mm)². in a 200mm diameter bolt circle and 4 bolts each having a cross-sectional area of 400 (mm)² each in a bolt circle of 120mm diameter. If the allowable stress in the bolts is 100 MN/m² what is the torque-carrying capacity of the coupling?

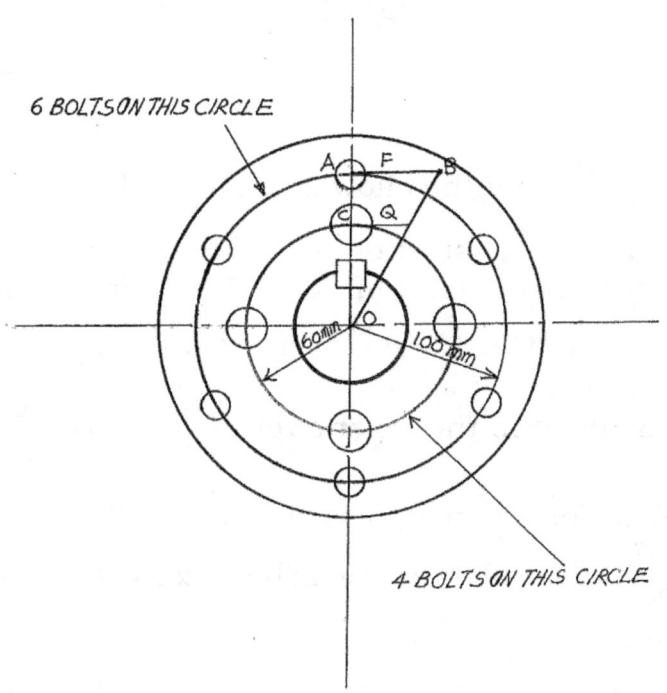

6 BOLTS ON THIS CIRCLE

4 BOLTS ON THIS CIRCLE

Considering the outer bolts:

Let A = csa of 1 bolt in outer bolt circle, m²

" F = force in 1 " " " " , N

" S = allowable shearing stress in bolt, N/m²

∴ F = sA

$$= 100 \times 10^6 \times \frac{324}{10^6}$$

$$= 32400 \text{ N}$$

Considering triangle OAB, let us suppose that the force in the inner bolt circle is Q Newtons, then

$$\frac{Q}{F} = \frac{60}{100}$$

i.e. $Q = \frac{60}{100}(32400) \ N$

$$= 19440N$$

If we divide this force by the csa of a bolt in the inner bolt circle, then the stress in an inner bolt is obtained. This stress is $\dfrac{19440}{\left(\dfrac{400}{10^6}\right)}$ i.e. 48.6 x 10⁶ N/M².

This figure is well within the allowable limit of 100 x 10⁶ N/m².

On the other hand, if we start off by assuming that the bolts in the inner circle are stressed to the limit of 100 x 10⁶ N/m², then the force on each bolt is this circle is $100 \times 10^6 \times \dfrac{400}{10^6} N$ i.e. 40000N. On this basis therefore

the force on each bolt in the outer circle = $40000 \times \dfrac{100}{60} N$ i.e. 66667N.

For this level of force, the stress in each bolt in the outer bolt circle would be $\dfrac{6667}{\left(\dfrac{324}{10^6}\right)} = 205 \times 10^6 \ N/m^2$, which exceeds the allowable stress.

Evidently therefore the torque carrying capacity of the coupling = $F \times OA + Q \times OC$ where F and Q are permissible forces in the bolts and OA and OC are their respective radius arms, F being 32400N for 1 bolt, outer circle; Q = 19440N for 1 bolt, inner circle.

$$\therefore \text{Torque-carrying capacity} = 6 \times 32400 \left(\frac{100}{1000} \right) + 4 \times 19440 \times \frac{60}{1000} \quad \text{of coupling}$$

$$= 19440 + 4665.6$$

$$= 24105.6 \text{ N m}$$

$$= 24.1 \text{ kN m}$$

Q25. A solid steel rod of circular cross-section throughout and varying uniformly from a diameter of 'D' metres at one end to $\frac{'D'}{5}$ metres at the other is used as a horizontal cantilever. See Fig. 1. The larger end is fixed rigidly to a vertical wall and a torque T Newton metres is applied to the free end. What is the torsional strain energy of the rod which is 'k' metres long; and determine also the angle of rotation of the free end. Take 'C' as the torsion modulus.

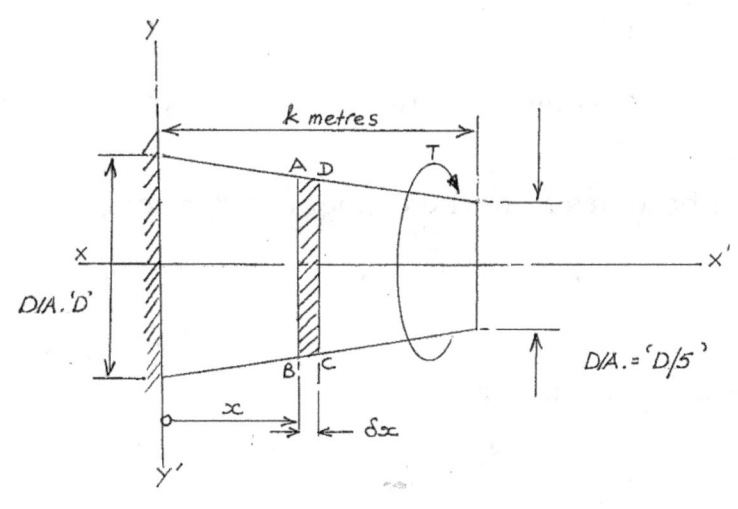

FIG 1

Consider cross section of the rod 'x' metres from YY.

Let the angle of rotation of 'CD' relative to AB be $\delta\theta$ radians, the thickness of the strip ABCD being δx

$$\therefore \qquad \frac{T}{J_x} = C\frac{\delta\theta}{\delta x} \qquad \dots\dots\dots\dots\dots\dots\dots\dots\dots\dots \text{(i)}$$

from which

$$\delta\theta = \frac{T}{CJ_x}\delta x \qquad \dots\dots\dots\dots\dots\dots\dots\dots \text{(ii)}$$

or $\qquad \theta = \int_0^k \frac{Tdx}{CJ_x}$

and because T and C are constant

$$\theta = \frac{T}{C}\int_0^k \frac{dx}{J_x} \qquad \dots\dots\dots\dots\dots\dots\dots\dots\dots\dots \text{(iii)}$$

Now, strain energy of the same element ABCD = dU

$$\therefore \qquad dU = \frac{1}{2}T\cdot d\theta \qquad \dots\dots\dots\dots\dots\dots\dots\dots\dots\dots \text{(iv)}$$

But from (ii)

$$d\theta = \frac{Tdx}{CJ_x}$$

J_x being the polar moment of inertia of the section 'x' from the lefthand support.

Therefore (iv) becomes from the element under consideration

$$dU = \frac{1}{2}\frac{T^2}{CJ_x}dx$$

and for the entire cantilever

$$\int dU = \frac{1}{2}\frac{T^2}{C}\int_o^k \frac{dx}{J_x}$$

i.e. $\qquad U = \frac{1}{2}\frac{T^2}{C}\int_0^k \frac{dx}{J_x} \qquad \dots\dots\dots\dots\dots\dots\dots\dots \text{(v)}$

Diameter of rod decreases from D to $\dfrac{D}{5}$ in k metres, therefore fall per metre is $\dfrac{1}{k}\left(D-\dfrac{D}{5}\right)=\dfrac{4D}{5k}$ / metre

$\therefore$ Diameter of rod at x, say $'D_x'=D-\dfrac{x}{k}\dfrac{4D}{5}$

i.e. $D_x=D\left(1-\dfrac{4x}{5k}\right)$

so that

$$J_x=\dfrac{\pi D_x^4}{32}$$

$$=\dfrac{\pi}{32}\left\{D\left(1-\dfrac{4x}{5k}\right)\right\}^4$$

$$=\dfrac{\pi D^4}{32}\left(1-\dfrac{4x}{5k}\right)^4 \qquad \dots\dots\dots\dots\dots\dots \text{(vi)}$$

Substituting this result in (v) we obtain

$$U=\dfrac{1}{2}\dfrac{T^2}{C}\cdot\dfrac{32}{\pi D^4}\int_0^k\dfrac{dx}{\left(1-\dfrac{4x}{5k}\right)^4} \qquad \dots\dots\dots\dots\dots \text{(vii)}$$

Let $\left(1-\dfrac{4x}{5k}\right)=p$

$\therefore$ $-\dfrac{4}{5k}dx=dp$

or $dx=-\dfrac{5k}{4}dp$

Changing the limits of integration, when x = 0; p = 1,   when x = k ; $p=\dfrac{1}{5}$

Therefore (vii) becomes

$$U = \frac{1}{2}\frac{T^2}{C}\frac{32}{\pi D^4}\int_1^{1/5} - \frac{5k}{4}\frac{dp}{p^4}$$

$$= \frac{1}{2}\frac{T^2}{C}\frac{32}{\pi D^4}\left[+\frac{5k}{12p^3}\right]_1^{1/5}$$

$$= \frac{16T^2k}{\pi CD^4}\left[\frac{5}{12}\left\{\frac{1}{\left(\frac{1}{5}\right)^3} - \frac{1}{1}\right\}\right]$$

i.e. $$U = \frac{16T^2K}{\pi CD^4}\left(\frac{5}{12}\cdot 124\right)$$

$\therefore$ $$U = \frac{2480T^2K}{3\pi CD^4} \qquad\qquad \cdots\cdots\cdots\cdots\cdots \text{(viii)}$$

Using equation (iii)

$$\theta_{end} = \frac{T}{C}\int_0^k \frac{k\,dx}{J_x}$$

$$= \frac{T}{C}\int \frac{32}{\pi D^4}\cdot\frac{dx}{\left(1-\frac{4x}{5k}\right)^4}$$

$$= \frac{32T}{\pi CD^4}\int_0^k \frac{dx}{\left(1-\frac{4x}{5k}\right)^4}$$

Making the substitutions done previously we obtain

$$\theta_{ens} = \frac{32}{\pi}\frac{TK}{CD^4}\cdot\frac{5}{12}(124)$$

$$= \frac{4960\ Tk}{2\pi\ CD^4} \qquad\qquad \cdots\cdots\cdots\cdots\cdots \text{(ix)}$$

$\therefore$ Angle of rotation of free end θ_{end} is $\dfrac{4960\ Tk}{3\pi\ CD^4}$ of radians. Of course

we could simply have equated the result at (viii) to $\dfrac{1}{2}T\theta_{end}$, i.e.

$$\frac{1}{2}T\theta_{end} = \frac{2480\ T^2K}{3\pi\ CD^4}$$

from which

$$\theta_{end} = \frac{4960\ TK}{3\pi\ CD^4} \quad \text{as at (ix)}$$

Q26. A solid circular cylindrical steel shaft of 100mm diameter is made from stock having a limit of proportionality stress of 145MN/m². What torque can be transmitted by this shaft when 10mm of the outermost fibres have yielded? Take C = 78 x 10⁹N/m².

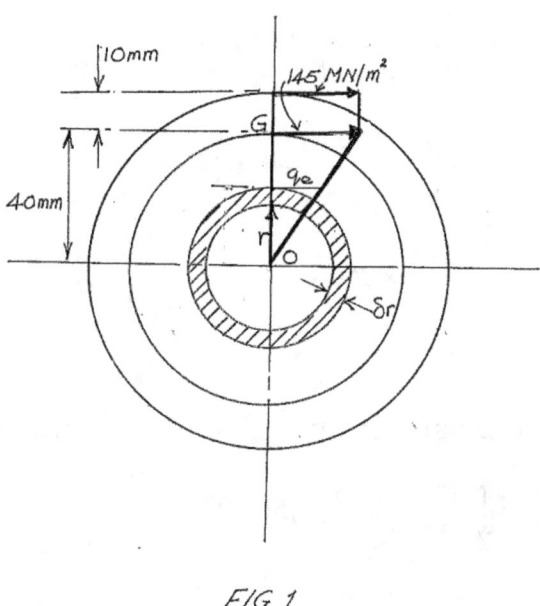

FIG 1

Shearing stress q_e varies linearly with respect to radius in the elastic zone, i.e. up to radius 'OG'. In the plastic region shearing stress is assumed constant at 145MN/m². Considering the elastic region of the shaft, the elastic torque 'Tₑ' is given by

$$T_e = \int_{r=0}^{r=40/1000} 2\pi r^2 q_e\, dr \qquad \ldots\ldots\ldots\ldots\ldots\ldots\ldots\ (i)$$

But

$$\frac{q_e}{145\,MN/m^2} = \frac{\left(\dfrac{r}{1000}\right)}{\left(\dfrac{40}{1000}\right)} = \frac{r}{40}$$

i.e. $\quad q_e = 145 \cdot \dfrac{r}{408}\,MN/m^2$

$$= \frac{29r}{8}\,MN/m^2$$

Substituting in (i)

$$T_e = \frac{29}{8}\int_0^{40/1000} 2\pi r^3 dr \ \ MN \ m$$

$$= \frac{29}{8}\left[\left(\frac{\pi}{2}r^4\right)\right]_o^{30/1000}$$

$$= \frac{29\pi}{16}\left(\frac{40}{1000}\right)^4 \ MN \ m$$

$$= \frac{29\pi}{16}\times\frac{4^4\times10^4}{10^{12}}\times10^6\,N \ m$$

$$= \frac{29\pi}{1600}\times 256$$

$$= 14.6\,Mn.$$

Considering the plastic region of the shaft, the 'plastic' torque is given by

$$T_p = \int_{r=40/1000}^{r=50/1000} 2\pi r^2 (145\times10^6)dr$$

$$= 145\times10^6 \times 2\pi$$

$$= 145\times10^6 \times 2\pi\left[\frac{r^3}{3}\right]_{40/100}^{50/1000}$$

$$= \frac{2\pi}{3}\times145\times10^6\left[\left(\frac{50}{1000}\right)^3 - \left(\frac{40}{1000}\right)^3\right]$$

$$= \frac{2\pi}{3} \times 145 \times 10^6 \times \frac{10^3}{10^9} \left(5^3 - 4^3\right)$$

$$= \frac{2\pi}{3} \times 145 \times 61$$

$$= 18533 Nm$$

∴   Total Torque   = Elastic Torque + Plastic Torque

                                  = 18533 + 15

                                  = 165548 Nm

Q27. Two solid circular cylindrical shafts are connected by gears. See Fig. 1. The left-hand shaft is 50mm diameter and is driven by a torque applied at 'A', which induces a shearing stress of 60MN/m². What should the diameter of the right-hand shaft be if the shearing stress there is also limited to 60MN/m². Determine also the angular rotation of pulley 'A' if pulley 'D' is held fixed. Assume C = 64 x 10⁹ N/m².

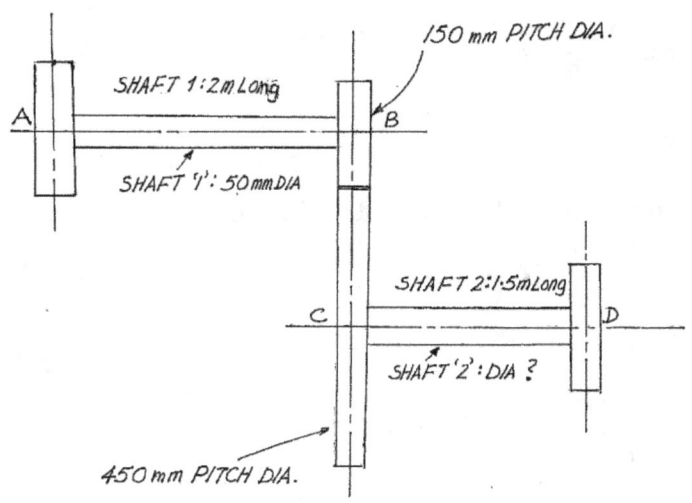

<u>FIG 1</u>

Consider the shaft labeled '1'. The torque applied at 'A;, T_A, may be computed from the induced shearing stress. Thus.

$$\frac{60 \times 10^6}{\left(\dfrac{25}{1000}\right)} = \frac{T_A}{\dfrac{\pi}{32}\left(\dfrac{50}{1000}\right)^4}$$

i.e. $\qquad 40 \times 60 \times 10^6 = \dfrac{32T_A \times 10^6}{\pi \times 6.25}$

$\qquad \therefore \qquad t_a = \dfrac{2400 \times \pi \times 6.25}{32}$

$\qquad\qquad\qquad = 1473 \ Nm.$

Assuming no power loss at gearing we may write

$$T_A w_A = T_D w_D$$

where w_a, w_D are the angular velocity of shafts 1 and 2 respectively. Now the linear velocity (v) at the pitch line circles for each gear must be equal and because $w = \dfrac{v}{r}$, where 'r' is pitch line radius we may write:

$$\frac{T_A}{r_1} = \frac{T_D}{r_2}$$

i.e. $\qquad \dfrac{T_A}{75} = \dfrac{T_D}{225}$

or $\qquad 3T_A = T_D$

$\therefore \qquad$ Torque at pulley D $= 3 \times 1473$ N m

$\qquad\qquad\qquad = 4419$ N m

Let the diameter of shaft '2' be d mm

$\qquad \therefore \qquad \dfrac{4419}{\dfrac{\pi}{32}\left(\dfrac{d}{1000}\right)^4} = \dfrac{60 \times 10^6}{\left(\dfrac{d}{2000}\right)}$

i.e. $\qquad \dfrac{4419 \times 32 \times 10^{12}}{\pi \times d^4} = \dfrac{60 \times 10^6 \times 2000}{6}$

or $\qquad\qquad d^3 = \dfrac{4419 \times 1600}{6 \times \pi}$

$\qquad\qquad d^3 = 374928$

$\therefore \qquad\qquad d = 72mm$

With pulley 'D' fixed, the angle of twist at the 450mm pitch diameter gear is obtained from

$$\frac{4419}{\frac{\pi}{32}\left(\frac{72}{1000}\right)^4} = \frac{C\theta_2}{1.5}$$

i.e. $\quad \dfrac{4419 \times 32 \times 10^6}{\pi \times 26.873} = \dfrac{84 \times 10^9 \cdot \theta_2}{1.5}$

from which $\quad \theta_2 = \dfrac{1.5 \times 4419 \times 32 \times 10^6 \times 360}{\pi \times 26.873 \times 84 \times 10^9 \times 2\pi}$

$$= 1.7^o$$

This angular twist translates to $3 \times 1.7^o = 5.1^o$ at the pinion gear on shaft 1. But relative to this pinion pulley A moves through an angle θ_1, which may be obtained from

$$\frac{1473}{\frac{\pi}{32}\left(\frac{50}{1000}\right)^4} = \frac{C\theta_1}{2}$$

$\therefore \quad \theta_1 = \dfrac{2 \times 1473 \times 32 \times 10^6 \times 360}{\pi \times 6.25 \times 84 \times 10^9 \times 2\pi}$

$$= 3.27^o, \text{ say } 3.3^o$$

$\therefore$ Relative to pulley 'D', pulley 'A' has an angular rotation =

$5.1^o + 3.3^o = 8.4^o$.

Q28. An arrangement of a 2-gear speed reducer is shown schematically in the Figure 1 below. The gears may be treated as solid homogeneous cylinders. The maximum torque at driver 'X' is required to accelerate 'Y' at an angular acceleration of 10 radians per sec². The mass of 'X' and 'Y' are respectively 20 kg and 50 kg. Determine, the maximum torque acting on driver 'X', working from first principles.

Angular accl'n of X = α_x Angular accl'n of Y=10 rad/s

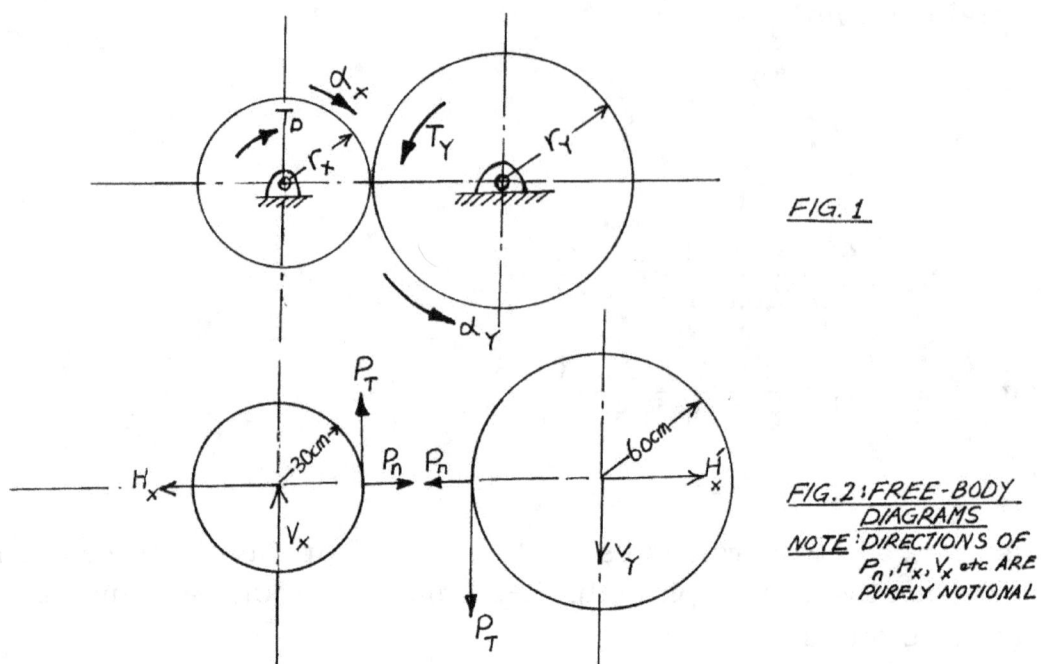

FIG. 1

FIG.2 : FREE-BODY DIAGRAMS
NOTE : DIRECTIONS OF P_n, H_x, V_x etc ARE PURELY NOTIONAL

Let T_D = Driver Torque

Equation of motion for 'X' is

$$\Sigma \text{Torque on 'X'} \quad \Sigma T_x = I_x \alpha_x \quad \dots \dots \dots \dots \dots \dots \dots \quad \text{(i)}$$

Where 'α'= angular acceleration and I_x = mass moment of inertia of Driver about axis of rotation

Now, $\qquad \sum T_x = T_D - P_T r_x \qquad \dots \dots \dots \dots \dots \dots \dots \quad \text{(ii)}$

where $\qquad r_x$ = radius of Driver Gear 'X'

$\therefore \qquad T_D - P_T r_X = I_X \alpha_x \qquad \dots \dots \dots \dots \dots \dots \dots \quad \text{(iii)}$

Similarly, equation of motion for 'Y' is

$$\sum T_Y = I_Y \alpha_Y \qquad \dots \dots \dots \dots \dots \dots \quad \text{(iv)}$$

where $\qquad I_Y$ = mass moment of inertia of driven gear 'Y'

$\qquad \alpha_y$ = angular acceleration of 'Y'; and

$\qquad T_y$ = torque on 'Y'

~ 444 ~

But $$\sum T_Y = P_T r_Y \qquad \dots\dots\dots\dots\dots\dots \text{(v)}$$

where r_y = radius of driven gear Y

$$\therefore \quad P_T\, r_y = I_Y\, \alpha_y \qquad \dots\dots\dots\dots\dots \text{(vi)}$$

From (vi) $$P_T = I_Y \cdot \frac{\alpha_Y}{r_Y}$$

Substituting this result in (iii) we get

$$T_D - I_Y \cdot \frac{\alpha_y}{r_y} \cdot r^x = I_x \alpha_x \qquad \dots\dots\dots\dots \text{(vii)}$$

Because, from geometry of the motion

$$r_x\, \alpha_x = r_Y\, \alpha_Y$$

we have $$\alpha_Y = \frac{r_x}{r_Y} \cdot \alpha_X$$

equation (vii) may be rewritten

$$T_D - I_Y \cdot \frac{r_x}{r_Y} \cdot \frac{r_X}{r_Y} \alpha_X = I_X \alpha_X$$

so that $$T_D = I_X\, \alpha_X + I_Y \left(\frac{r_X}{r_Y}\right)^2 \alpha_X$$

i.e. $$T_D = \left\{ I_X + I_Y \left(\frac{r_X}{r_Y}\right)^2 \right\} \alpha_x \qquad \dots\dots\dots\dots \text{(viii)}$$

Using the data given we may calculate I_x and I_y as follows:

The gears we are told may be treated as homogeneous cylinders, in which case

$$\therefore \quad I = \frac{1}{2} mr^2 \text{ in kg m}^2 \text{ units}$$

where   m = mass of gear in kg, and

        r = radius of gear

Accordingly

$$I_X = \frac{1}{2} 20 \cdot \left(\frac{30}{100}\right)^2$$

$$= 0.9 \text{ kg m}^2$$

Also

$$I_Y = \frac{1}{2} \cdot 50 \left(\frac{60}{100}\right)^2$$

$$= 9 \text{ kg m}^2.$$

$$\alpha_Y = 10 \text{ rad/sec}^2$$

$$\therefore \quad \alpha_X = \alpha_Y \cdot \frac{r_y}{r_X}$$

$$= 10 \cdot \frac{60}{30}$$

$$= 20 \text{ rad/sec2}$$

Substituting these values in (viii) produces

$$T_D = \left\{0.9 + 9 \cdot \left(\frac{30}{60}\right)^2\right\} 20$$

$$= (3.15)20$$

Ans: $\quad T_D = 63 N\, m.$

Q29. Determine the maximum shearing stress caused by a torque of 2500 Nm in the walls of a tube 3mm thick throughout, the cross-section of which as shown in Fig. 1 is made up of a 2 semi-elliptical ends welded to a rectangular section. Neglect any stress concentration.

Q30. A tube 2m in length and of square cross section but with walls of unequal thicknesses as shown below is under the influence of a torque of 2050 N m. Find the shear flow, maximum shearing stress and torsional strain energy in the tube. Assume that the ends of the tube are free to warp, and take C = 80 x 10⁹N/m².

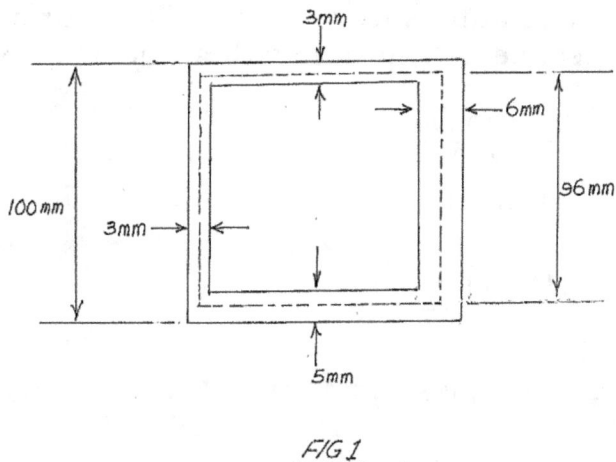

FIG 1

In Fig. 1, the centre-line of the cross section is shown dotted.

Now shear flow 'q' = $\dfrac{T}{2A_o}$

where A_o = area enclosed by the centre line of the cross

section

$$= \frac{96 \times 95.5}{10^6} m^2$$

$$= 9168 \times 10^{-6} m^2.$$

∴   Shear flow = $\dfrac{T}{2A_o}$

i.e. $q = \dfrac{2050 \times 10^6}{2 \times 9168}$

∴ $q = 111802$ Newton/metre

$\underline{q = 111.8 kN/m}$

Generally shear stress $(\tau_s) = \dfrac{Shear\ Flow}{Skin\ Thickness}$

i.e. $\tau_s = \dfrac{q}{t}$

Evidently, the maximum shear stress will occur in the skin with the smallest thickness, i.e. in the skin with a thickness of 3mm in this case.

$$\therefore \quad \tau_s = \frac{111802 N/m}{\left(\dfrac{3}{1000}\right)}$$

$$= \frac{111.8 \times 10^3 \times 10^3}{3}$$

i.e. $\quad \tau_s = 37.27 \times 10^6 \, N/m^2$

say 37.3 MN/m²

Using the expression developed previously for strain energy

i.e.

$$U = \frac{q^2}{2C} \int_o^L dx \sum \frac{\ell}{t}$$

We have $\quad \sum \dfrac{\ell}{t} = \dfrac{96}{6} + \dfrac{95.5}{5} + \dfrac{96}{6} + \dfrac{95.5}{5}$

$$= 70.2.$$

$$\therefore \quad U = \frac{(111.8 \times 10^3)^2}{2 \times 80 \times 10^9} \times 2 \times 70.2$$

$$= \frac{(111.8)^2 \times 10^6 \times 2 \times 70.2}{160 \times 10^9 10^3}$$

i.e. $\underline{U = 1096 Nm, \quad \text{say } 11N\,m.}$

Q31. The single-throw crank shown in Fig. 1 is supported by two bearings, the crank pin being 75mm in length. The centre of the crank pin is 300mm and 350mm from each of the bearings respectively. The maximum force on the crank pin is 50kilonewtons and occurs when the angle between it and the connecting rod of the piston is 90º. Energy is taken off at a point adjacent to bearing 'A'. Determine the bearing reactions and the twisting moment on the crank pin.

R_{AY}, R_{AX}, R_{AZ} are the three reactions at bearing A

Similarly, let R_{BY}, R_{BX} and R_{BZ} be the three reactions at bearing 'B'

Let T = torque available at a take-off point adjacent to bearing A

For equilibrium

$$\sum R_Y = 0 \quad ; \quad \sum R_X = 0 \quad ; \quad \text{and} \quad \sum R_Z = 0$$

also,

$$\sum M_Y = 0 \quad ; \quad \sum M_X = 0 \quad ; \quad \text{and} \quad \sum M_Z = 0$$

For the specific position of the crank shaft being considered, there are no components of force in the X-direction

$$\therefore \quad R_{AX} = 0 \quad ; \quad R_{BX} = 0$$

Similarly $\quad R_{AY} = 0 \quad ; \quad R_{BY} = 0$

Considering equilibrium of forces in the Z-direction

$$R_{AZ} + R_{BZ} = 50kN \qquad \dots\dots\dots\dots\dots\dots\dots\dots \text{(i)}$$

Taking $\quad \sum M_Y = 0$

$$+ R_{AZ}(650) - 50kN(350) = 0$$

i.e. $\quad R_{AZ} = 26.9kN$

$$\therefore \quad R_{AZ}(say) \quad 27kN$$

so that from (i)

$$R_{BZ} = 23kN$$

Fig.

Now, the torque transmitted through the crankshaft

$$= 50kN \times \frac{180}{1000}$$

$$= 9kN \qquad \text{metres (clockwise)}$$

Therefore if we start at bearing A and move towards a plane through 'M' just outside the web of the crank the torque on the shaft is 9 kNm. As 'M' is reached an anti-clockwise torque $= R_{AZ} = \left(\dfrac{180}{1000}\right)$ opposes the clockwise torque of 9kNm.

This means that the torque on the crank pin TCP is the nett sum of these two torques : 9kNm ↻ and $RAZ\left(\dfrac{180}{1000}\right) kNm$ ↺

$$= (9 - 4.9)kNm$$

i.e. $= 4.1$ kN m clockwise

This may be checked by starting at bearing 'B' with no twisting moment just before point 'N' on web, after which the twisting moment $=$ $R_{BZ}\left(\dfrac{180}{1000}\right)$

i.e. $\quad T_{CP} = 23 \times \dfrac{180}{1000} kNm$

$$= 4.1 \text{ kN m}$$

Collecting results, we have

$$R_{AX} = 0 \quad ; \quad R_{AY} = 0 \quad ; \quad R_{AZ} = 27kN$$

$$R_{BX} = 0 \quad ; \quad R_{BY} = 0 \quad ; \quad R_{BZ} = 23kN$$

Twisting moment on crankpin $= 4.1$kN m.

Q32. Determine the maximum shearing stress caused by a torque of 2500 N m in the walls of a tube 3mm thick throughout, the cross-section of which as shown in Fig. 1 is made up of 2 semi-elliptical ends welded to a rectangular section. Neglect any stress concentration due to welds.

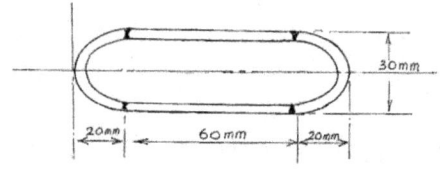

FIG. 1 (NOT DRAWN TO SCALE)

Ans: 87 MN/m²

CHAPTER 12

COMPOUND PLANAR STRESSES

In this chapter we shall consider a combination of stresses in plane sections. Such compound stresses arise in many practical engineering situations. For example, in thin-walled vessels under internal pressure a state of tri-axial stress is approximated to one of bi-axial stress. Another example is that of a shaft assumed directionally free, mounted on bearings, carrying a flywheel and transmitting power. Here there are shear stresses due to torque and to the weight of the shaft itself; and, also bending stresses. Other shafts performing similar duty, i.e. transmission of power, may also be subjected to an end thrust which introduces yet another direct stress. It is therefore appropriate to introduce the reader at an early stage to the subject of compound stress in plane sections in the hope that it would provide a soft landing on the runway to the study of stress analysis in three dimensions in advanced mechanics of materials.

In Chapter 5 the axial force 'F' caused a stress 'σ' equal to F/a, i.e. $\sigma = F/A$, perpendicular or normal to the cross-sectional area of the bar. But, what about stresses on oblique planes of the same bar? Let us examine a typical situation. To do this a plane at angle 'θ' to the cross-section of the bar is considered. This is shown in Fig. 1. The object is to find the value of the normal and shear stresses designated σ_n and τ_s respectively on this plane, the trace of which on the face of the bar is BC inclined to the axis of the loading as shown. In Fig. 1, the directions of σ_n, τ_s and the orientation of angle 'θ' with respect to the horizontal 'AB' were arbitrarily chosen.

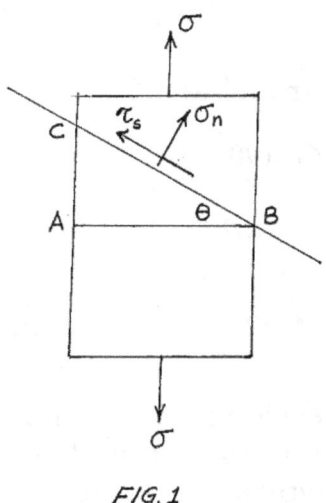

FIG. 1

Note carefully that σ, σ_n and τ_s are stresses. Therefore before any of the laws of statical equilibrium may be applied, these stresses must be converted to forces. To do this Popov's wedge analysis is applied. Accordingly, it is assumed that the cross-sectional area of the plane, the trace of which is BC is $1\,m^2$, that is, one square metre. Therefore, area of plane governed by $AB = 1 Cos\theta\ m^2$ and by AC, $1 Sin\theta\ m^2$. Assuming stresses are in, say, N/m^2 it follows that the force due to 'σ' on AB is $\sigma(1 Cos\theta)N$; to σ_n on BC, $\sigma_n(1)N$ and to τ_s, $\tau_s(1)N$. These forces are shown on the isolated element, Fig. 2. Note that stress in N/m^2 when multiplied by m² give force in newtons, N.

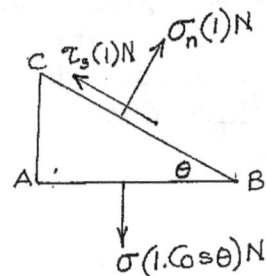

FIG. 2

Resolving forces normal to BC

$$\sigma_n(1) - \sigma(1 \ . \ Cos) \cdot Cos\theta = 0$$

from which

$$\sigma_n = \sigma\ Cos^2\theta \qquad \dots \dots \dots \dots \dots \dots \dots \dots \dots \text{(i)}$$

Similarly by resolving forces parallel to AC:

$$\tau_s(1) - \sigma \cdot (1 \cdot Cos\theta)\ Cos(90 - \theta) = 0$$

i.e. $\qquad \tau_s = \sigma \cdot Cos\theta\ Sin\theta$

$$\therefore \qquad \tau_s = \frac{\sigma}{2} Sin2\theta \qquad \dots \dots \dots \dots \dots \dots \dots \text{(ii)}$$

Thus we have obtained expressions for the normal and shear stresses on a different set of axes, i.e. on axes perpendicular to the oblique plane BC. Expressions (i) and (ii) are typical of what are called stress transformation equations.

We shall meet several more in the course of this chapter. The method of the Wedge analysis is an easy and rapid way of developing transformation equations in plane stress. To apply it, start with a sketch of the element with stresses acting as prescribed. Then indicate areas of planes on which stresses act in reference to the plane designated as the one with unit area; convert stresses to forces; and apply the laws of statical equilibrium. You do not have to memorise formulae. In this connection some readers may come across, in other texts, different expressions for σ_n and τ_s depending on how the element in Fig. 1 is oriented. Nothing is amiss. It all depends on how θ, σ_n and τ are set to act on your diagram. This means we ought to adopt a convention not only to defining 'θ' but one for stresses as well on elements. For generalized stress analysis considered later in this chapter I have introduced a convention as exemplified in Fig. 11. Do not be afraid. Now, what do equations (i) and (ii) reveal?

From (ii) it may be deduced that the maximum shearing stress $\tau_{s(max)}$ occurs on the plane defined by $Sin2\theta = 1$, i.e. where $\theta = 45°$; so that maximum shearing stress $= \frac{\sigma}{2}$; and on this plane $\sigma_n = \sigma\left(\frac{1}{\sqrt{2}}\right)^2 = \frac{\sigma}{2}$ also. This is a very important deduction. Accordingly the resultant stress 'σ_R' on a plane inclined at $45°$ to that carrying stress 'σ' is $\sigma_R = \sqrt{\left\{\left(\frac{\sigma}{2}\right)^2 + \left(\frac{\sigma}{2}\right)^2\right\}} = \frac{\sigma}{\sqrt{2}}$.

Illustrative Example 1

Find, working from first principles, the normal and shear stresses and their directions on the plane BC in the element shown in Fig. 3, the axial stress being 120 N/m², compressive.

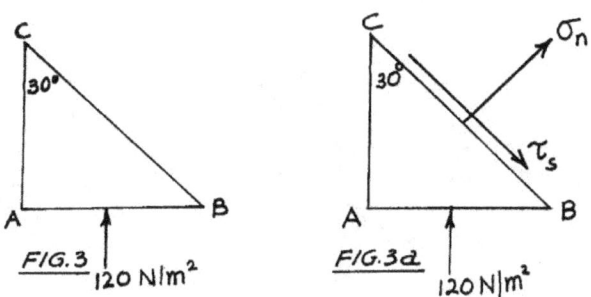

FIG.3   120 N/m²

FIG.3a   120 N/m²

The directions of σ_n and τ_s are assumed in Fig. 3b.

Resolving in direction of σ_n:

$$\sigma_n (1) + \sigma_{AB}.1.Cos\,(90°-30°)Cos\,(90°-30°) = 0$$

$$\therefore \qquad \sigma_n = -\sigma_{AB}Sin^2 60° = -120\left(\tfrac{1}{4}\right) = -30\ N/m^2$$

Resolving in direction of τ_s:

$$\tau_s(1) - \sigma_{AB}.1.Cos\,(60°)Cos\,30° = 0$$

$$\therefore \qquad \tau_s = +\sigma_{AB}\left(\tfrac{1}{2}\right)\left(\tfrac{\sqrt{3}}{2}\right) = 120\tfrac{\sqrt{3}}{4} = 52\ N/m^2$$

The results indicate that σ_n is in a direction opposite to that shown in Fig. 3. The vector $\tau_s = 52\ N/m^2$ is in the direction shown in the same diagram.

Now back to equation (ii) $\tau_s = \dfrac{\sigma}{2}\ Sin2\theta$. It may be deduced therefrom that at planes inclined at 45° to plane AB which carries the direct tensile stress, in this case the shear stress is one-half the direct stress, i.e. $\dfrac{\sigma}{2}$

It is interesting to see how normal and shear stresses are oriented in the bar for variation in 'θ'. In order to do so consider the element $BCDE$ in Fig. 4.

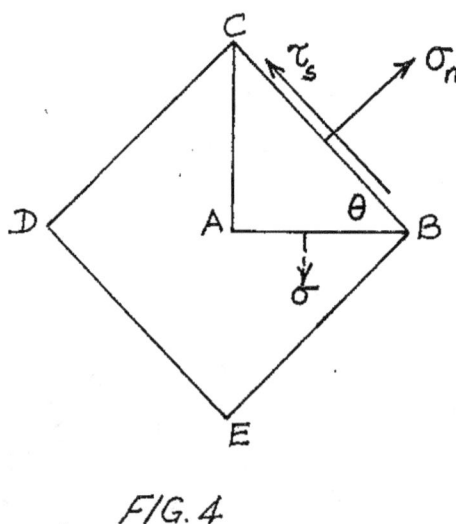

$$\underline{FIG.4}$$

To obtain σ_n and τ_s on plane BE, plane BC is rotated clockwise through 270°.

Accordingly,

$$\sigma_{n(270°+\theta)} = \sigma\,Cos^2\,(270+\theta)$$

i.e. $\quad \sigma_{n(270°+\theta)} = \sigma\,Sin^2\theta$

Also,

$$\tau_{s(270+\theta)} = \frac{\sigma}{2}\,Sin^2\,(270+\theta)$$

$$= \frac{\sigma}{2}\,Sin\,(540+2\theta)$$

$$= \frac{\sigma}{2}\,Sin\,(180+2\theta)$$

i.e. $\quad \tau_{s(270+\theta)} = -\frac{\sigma}{2}\,Sin2\theta$

For plane *DE* and *CD* additional rotation of 90° in each case is required. Thus for plane *DE*, angle 'θ' is replaced by $(360° + \theta)$ and for plane *CD* by $(450° + \theta)$. The values of σ_n and τ_s on the different sides of the whole element are shown in Fig. 5a; and for each of the four triangular components, in Fig. 5b.

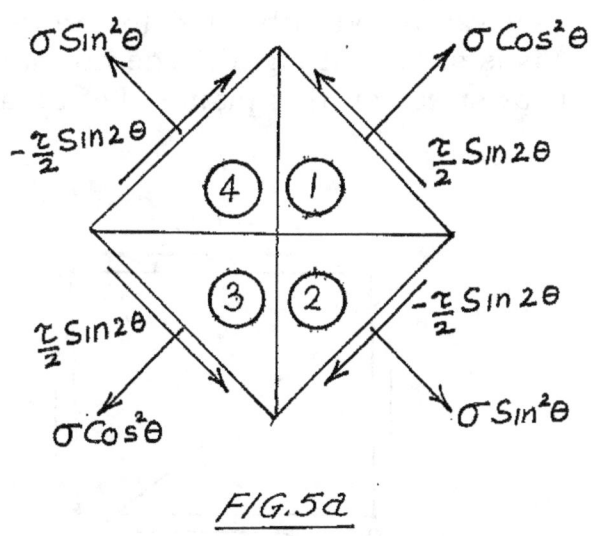

$$\underline{FIG.5a}$$

Thus in simple tension in plane-stress analysis different stress transformation equations may be derived by changing the orientation of the oblique planes in relation to the axes carrying stress 'σ'. For example if in Fig. 5a, 'θ' is angle *ACB* instead of angle *ABC*, then $\sigma_n = \sigma\,Cos^2(90-\theta) = \sigma\,Sin^2\,'\theta'$ and

~ 455 ~

$$\tau_s = \frac{\sigma}{2} Sin2(90-\theta) = \frac{\sigma}{2} Sin\ (180-2\theta) = \frac{\sigma}{2} Sin2\theta;\ \text{and the vectors for stresses on } BC$$

point in the same direction as in Fig. 5a.

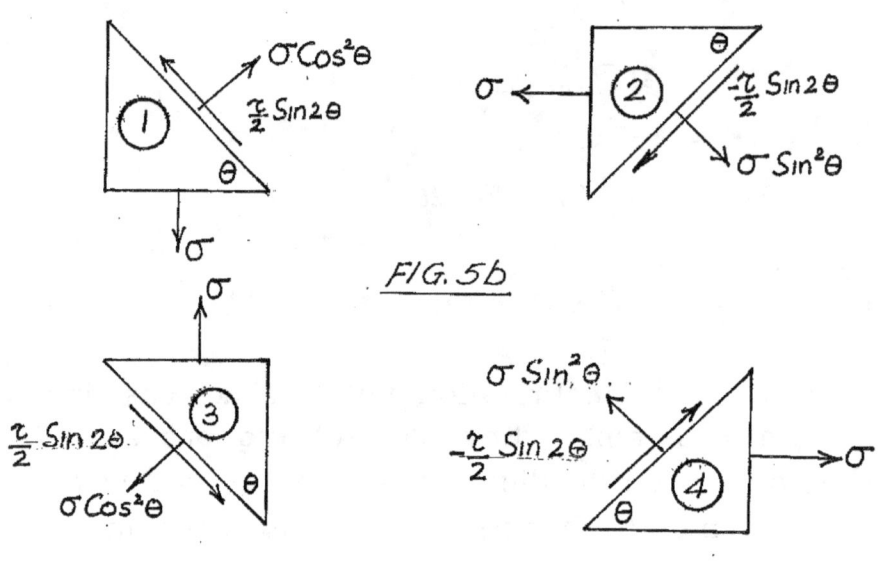

FIG. 5b

We consider next stresses on an oblique plane in an element subjected to pure shear as in the case of say a rod subjected to torque or twist. The element under shear stress q is is shown in Fig. 6. The normal stress on the plane BC is 'σ_n'. Is there a shear stress on this plane? Let us assume there is one and designate it τ_S.

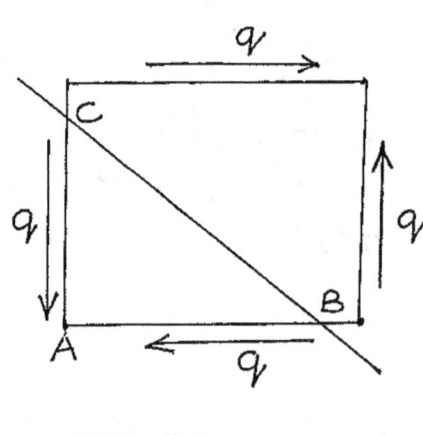

FIG. 6 a

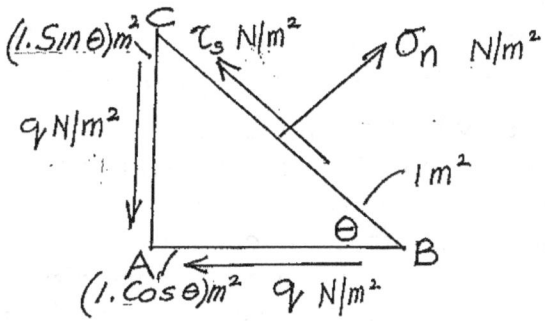

FIG. 6b

Proceeding in the usual manner with the resolution to forces: first normal to plane *BC* and second parallel to it. See Fig. 6b .

Accordingly

$$\sigma_n \cdot 1 - q \cdot Sin\theta\ Cos\theta - q \cdot Cos\theta\ Sin\theta = 0$$

$$\text{i.e.} \quad \sigma_n = 2q\ Sin\theta\ Cos\theta$$

$$\text{or} \quad \sigma_n = q\ Sin2\theta$$

and

$$\tau_s \cdot 1 + q\ Cos\theta\ Cos\theta - q\ Sin\theta\ Sin\theta = 0$$

When $\theta = 45°$, $\sigma_n = q$ and $\tau_s = 0$.

It is deduced therefrom that the normal stress on plane *BC* is a tensile stress of the same magnitude as the shear stress when $\theta = 45°$; and there is no shear stress on that plane. If instead of converging at '*A*', shear stresses '*q*' were diverging from each other, σ_n would turn out to be a compressive stress of the same magnitude '*q*' when $\theta = 45°$; Likewise $\tau_s = 0$ on the same plane.

PRINCIPAL STRESSES AND PRINCIPAL PLANES

In 2-dimensional stress analysis such as that which we shall now consider, it is of interest to determine the maximum and minimum stresses induced in a material as a result of the applied loads, acting normal to each other. It was seen for example with reference to equation (i) that σ_n was a maximum and equal σ, when $\theta° = 0$. That is on '*AB*' as in Fig. 1 and on planes parallel to it, i.e. on all sections normal to the axis of the bar. Also, when $\theta = 90°$,

~ 457 ~

$\sigma_n Cos^2 90^\circ = 0$. That is to say, on a plane at 90° to *AB* which is in fact the side of the bar, there is no stress. Thus the maximum and minimum direct stresses for our bar are σ and 0. When $\theta = 0$; $\tau_s = 0$ and also when $\theta = 90^\circ$; $\tau_s = 0$. Thus, a principal stress, whether maximum or minimum is associated with zero shearing stress. It can now be formally stated that in any system of plane stress there are two planes perpendicular to each other on each of which either a maximum or a minimum stress occurs and also on which no shearing stresses act. These special planes are called principal planes. Accordingly in uni-axial tension and uni-axial compression all sections normal and parallel to the axis of the bar being of constant cross-sectional area are principal planes. A somewhat generalized case is depicted in Fig. 7.

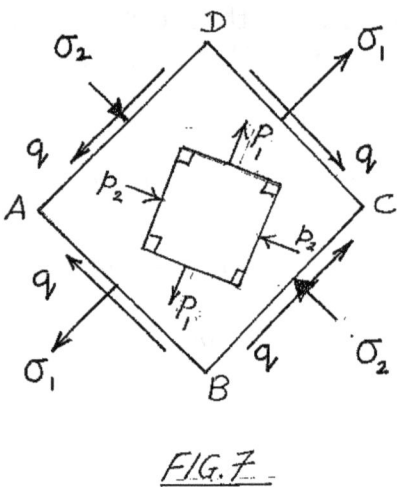

FIG. 7

The element is not only subjected to direct stresses σ_1, tensile; and σ_2, compressive, but also complementary shearing stresses 'q'. Within the element is shown another with principal stresses p_1 and p_2 acting on planes mutually perpendicular to each other; but there are no shearing stresses on these planes. These latter planes are principal planes.

This is a convenient juncture at which to introduce the topic of bi-axial stress.

BI-AXIAL STRESS

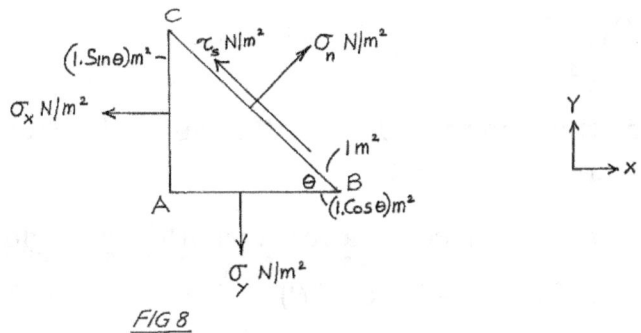

FIG 8

In Fig 8 the element is subjected to mutually perpendicular direct tensile stresses σ_Y and σ_X assumed to be expressed in N/m². The plane on BC is assumed to be 1m². Therefore areas on planes AB and AC are $(1 \cdot Cos\theta)m^2$ and $1 \cdot Sin\theta)m^2$ respectively. Note there are no shearing stress on either the planes of AB and AC. Thus these planes are principal planes. Likewise σ_x and σ_y are principal stresses.

Resolving forces perpendicular and parallel to BC we get, respectively:

$$\sigma_n(1) - \sigma_X(1Sin\theta)\ Sin\theta - \sigma_Y(1Cos\theta)\ Cos\theta = 0$$

i.e.

$$\sigma_{n(\theta)} = \sigma_Y Cos^2\theta + \sigma_X Sin^2\theta \quad \dots\dots\dots\dots\dots\dots \text{(iii)}$$

and

$$\tau_s(1) - \sigma_X \cdot (1Sin\theta)\ Cos\theta - \sigma_Y(1\ Cos\theta)\ Sin\theta = 0$$

i.e. $\qquad \tau_s = \dfrac{(\sigma_Y - \sigma_X)}{2}\ Sin2\theta \quad \dots\dots\dots\dots\dots \text{(iv)}$

Evidently, τ_s is at its maximum value when $Sin2\theta = 1$, i.e. when $2\theta = 90^\circ$ or $\theta = 45^\circ$.

Hence,

$$\tau_{s(max)} = \frac{\sigma_Y - \sigma_X}{2}$$

Also, when $\theta = 45^\circ$, equation (iii) becomes

$$\sigma_n = \sigma_Y Cos^2 45^\circ + \sigma_X Sin^2 45^\circ$$

~ 459 ~

$$= \sigma_Y \left(\frac{1}{\sqrt{2}}\right)^2 + \sigma_X \left(\frac{1}{\sqrt{2}}\right)^2$$

$$\therefore \qquad \sigma_{n(45^\circ)} = \frac{\sigma_Y + \sigma_X}{2}$$

The latter result is the magnitude of the normal stress on the plane of maximum shear, that is when $\theta = 45^\circ$.

Equation (iii) may be re-expressed in a form including angle 2θ by substituting $Sin^2\theta = \frac{1}{2}(1 - Cos2\theta)$ and $Cos^2\theta = \frac{1}{2}(1 + Cos2\theta)$

in which case

$$\sigma_n = \left(\frac{\sigma_Y + \sigma_X}{2}\right) + \frac{1}{2}(\sigma_Y - \sigma_X)\ Cos2\theta \qquad \dots\dots\dots\dots\dots\dots \text{(v)}$$

The resultant stress 'R_θ' on a plane BC due to $\sigma_{n(\theta)}$ and $\tau_{s(\theta)}$ is obtained from

$$\sigma_{R(\theta)} = \sqrt{\left(\sigma^2_{n(\theta)} + \tau^2_{s(\theta)}\right)}$$

$$= \sqrt{\left\{\left(\frac{\sigma_Y + \sigma_X}{2}\right) + \frac{1}{2}(\sigma_Y - \sigma_X)\ Cos2\theta\right\}^2 +}$$

$$\sqrt{\left\{\left(\frac{\sigma_Y - \sigma_X}{2}\right)Sin2\theta\right\}^2}$$

which after some minor mathematical weariness of the flesh works out to be :

$$\sigma_{R(\theta)} = \sqrt{\frac{1}{2}\left(\sigma^2_Y + \sigma^2_X\right) + \left(\frac{\sigma^2_Y - \sigma^2_X}{2}\right)Cos2\theta}$$

which may be simplified to

$$\sigma_{R(\theta)} = \sqrt{\left\{\sigma^2_Y\ Cos^2\theta + \sigma^2_X\ Sin^2\theta\right\}}$$

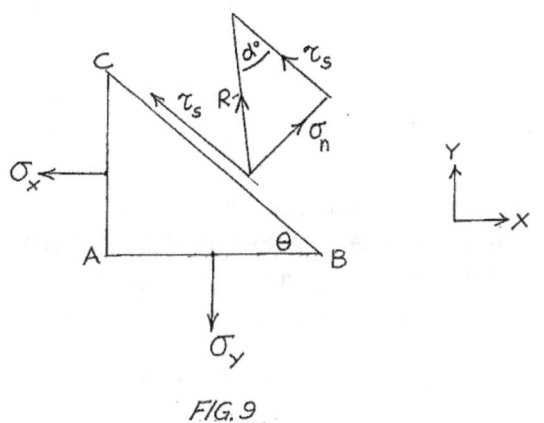

FIG. 9

In Fig. 9, angle 'α' is the inclination of the resultant $\sigma_{R(\theta)}$ to the plane. Accordingly

$$Tan\alpha = \sigma_{n(\theta)} / \tau_{s(\theta)}$$

$$= \frac{\sigma_Y \, Cos^2\theta + \sigma_X Sin^2\theta}{(\sigma_y - \sigma_X) \, Sin\theta \, Cos\theta}$$

$$= \frac{\sigma_Y \dfrac{Cos\theta}{Sin\theta} + \boldsymbol{\sigma}_x + \dfrac{Sin\theta}{Cos\theta}}{(\sigma_Y - \sigma_X)}$$

$$Tan\alpha = \frac{\sigma_Y Cot\theta + \sigma_X Tan\theta}{(\sigma_Y - \sigma_X)}$$

For minimum 'α' $\dfrac{d}{d\theta}(Tan\alpha) = 0$

$$\frac{d}{d\theta}(Tan\alpha) = \frac{-\sigma_Y Co\sec^2\theta + \sigma_X Sec^2\theta}{(\sigma_Y - \sigma_X)^2}$$

and for $\dfrac{d}{d\theta}(Tan\alpha) = 0$

$$\sigma_Y Co\sec^2\theta = \sigma_X Sec^2\theta$$

i.e. $$\frac{\sigma_Y}{Sin^2\theta} = \frac{\sigma_X}{Cos^2\theta}$$

from which $$Tan^2\theta = \frac{\sigma_Y}{\sigma_X}$$

or $\quad Tan\theta = \sqrt{\dfrac{\sigma_Y}{\sigma_X}}$

Illustrative Example 2

A prismatic ductile iron bar is subjected to principal stresses of 10MN/m² tensile and 25MN/m², compressive. Determine the normal and shear stresses on a plane inclined at 30° to the minimum principal stress. Calculate also the resultant stress on the same plane and its direction in relation to the same plane. See Fig. 10a.

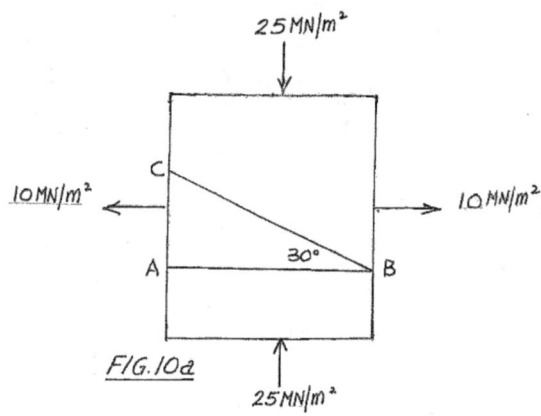

FIG. 10a

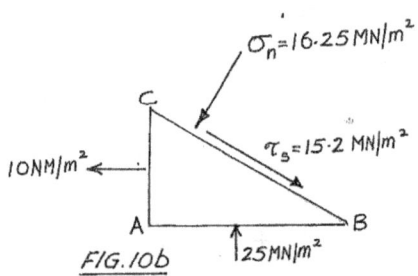

FIG. 10b

In a formal examination you may be asked to prove any formula you use. However since the configuration matches the one used in the foregoing analysis, I shall employ equation (iii). Here $\sigma_Y = -25MN/m^2$; $\sigma_X = 10MN/m^2$. But you can always work from past principles.

$$\therefore \quad \sigma_n = \sigma_Y \, Cos^2\theta + \sigma_X \, Sin^2\theta$$

$$= -25 \, Cos^2 30° + 10 \, Sin^2 30°$$

$$= -25 \left(\sqrt{3}/2\right)^2 + 10 \, (1/2)^2$$

$$= -25\frac{(3)}{4} + 2.5 = -18.75 + 2.5$$

$$\sigma_n = -16.25 MN/m^2$$

and

$$\tau_s = \left(\frac{\sigma_Y - \sigma_X}{2}\right) Sin2\theta$$

$$= \left(\frac{-25-10}{2}\right) Sin60^o = -17.5 \cdot \frac{\sqrt{3}}{2}$$

$$= -8.75\sqrt{3}MN/m = -15.2MN/m^2$$

Continuing $\quad R = \sqrt{\{(-16.25)^2 + (-8.75\sqrt{3})^2\}}$

$$= \sqrt{264 + 229.7} = \sqrt{493.7}$$

$$= 22.2MN/m^2$$

See Fig. 10c

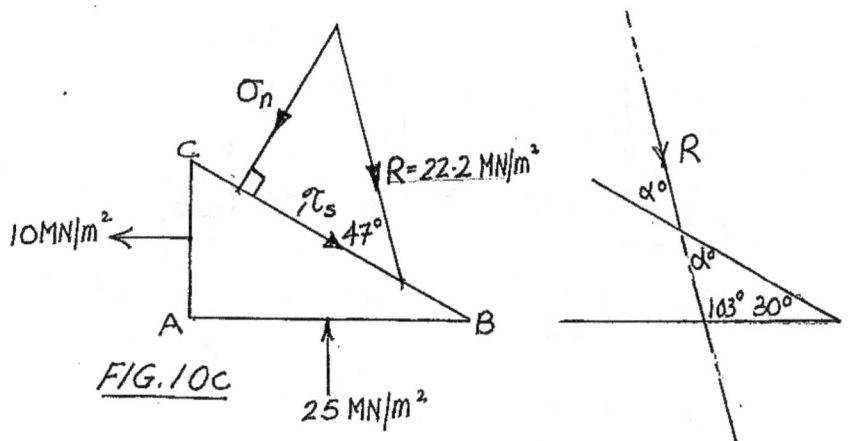

FIG.10c

$Tan\alpha = -16.25/-8.75\sqrt{3} = 16.25/15.2$

$$= 1.0691$$

$\therefore \qquad \alpha \approx 46.8^o$ to plane BC or 180º – (30º + 46.9º),

i.e. ≈ 103º to the horizontal. See Fig. 10c

PLANE STRESS

So far we have considered stresses within a material due to the application of direct stresses: firstly the case of uni-axial stress and latterly that of bi-axial stress. We note particularly that while shear stresses within the material were determined for these two loading configurations there were no such stresses on the planes on which the direct stresses, whether tensile or compressive were acting. Stresses σ in Fig. 1 and σ_X and σ_Y in Fig. 8 are, as was explained before principal stresses. The situation is somewhat different now in the sense that the objective is to determine principal stresses in a generalized stress configuration characterized by the presence of a shear stress on each of the planes on which the direct stresses are applied. This condition is that of plane stress, previously defined in Chapter 5. A commonplace example of such a configuration occurs in the case of, say, a closed circular cylindrical hollow tube under internal pressure which is subjected at the same time to an axial torque and load. Three stresses are involved here: an axial stress due to the load, a hoop stress due to the internal pressure and shearing stresses due to the torque. We shall deal with hoop stresses caused by internal pressure and shearing stresses due to axial torque later on in this text.

Consider the state of plane stress as represented by Fig. 11 and 12.

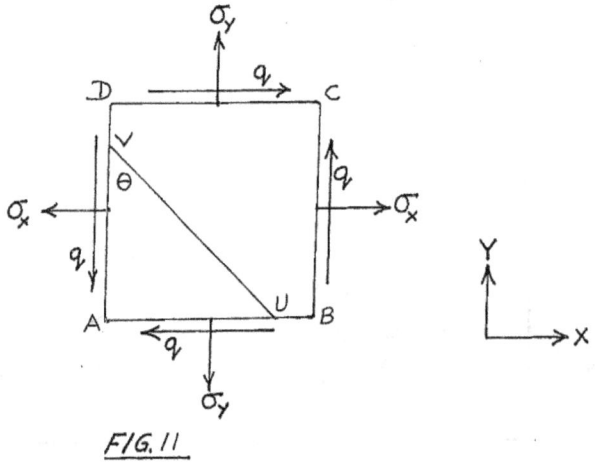

F/G. 11

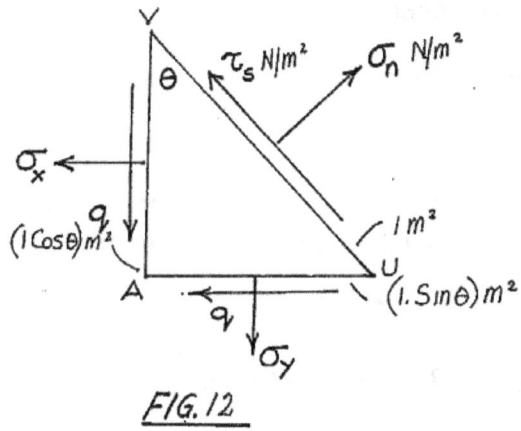

FIG. 12

As stated earlier, for the generalized analysis, we introduce the following convention. All the stresses acting in Fig. 11 are regarded as acting in a positive sense. In particular complementary shearing stress 'q' is positive when 'q', on the right-hand face of the element is acting upwards as shown in Fig. 11. Observe also that angle 'θ' is here regarded as positive when it is measured in an anti-clockwise direction from the vertical to the oblique plane. Evidently with 'q' acting as shown on face BC, all other 'q' in the diagram must be acting the directions indicated; these too act in a positive sense in relation to the entire diagram. Also, σ_n and τ_s : positive as indicated; and 'θ', positive anti-clockwise from the vertical.

To conduct the analysis, consider the element shown in Fig. 12: $+\sigma_X$, $+\sigma_Y$, $+\sigma_n + \tau_s$ and $+q$.

Employing Popov's Wedge Analysis as before and with the face $UV = 1 m^2$ and stresses in, say, MN/m^2, the equation for statical equilibrium normal to face UV may be expressed as:

$$\sigma_n(1) - \sigma_Y \ (1Sin\theta) \ Cos(90-\theta) - \sigma_X \ (1Cos\theta) \ Cos\theta$$

$$- q \ (1Cos\theta) \ Cos(90-\theta) - q \ (1Sin\theta) \ Cos\theta = 0$$

i.e. $\quad \sigma_n = \sigma_y \ Sin^2\theta + \sigma_X Cos^2\theta + 2qCos\theta \ Sin\theta \quad \dots\dots\dots\dots\dots$ (vi)

Making the substitutions:

$$Cos^2\theta = \left(\frac{1+Cos2\theta}{2}\right) \text{ and } Sin^2\theta = \left(\frac{1-Cos2\theta}{2}\right)$$

~ 465 ~

result (vi) may now be expressed as

$$\sigma_n = \left(\frac{\sigma_X + \sigma_Y}{2}\right) + \left(\frac{\sigma_X - \sigma_Y}{2}\right) Cos2\theta + q\ Sin2\theta \qquad \dots\dots\dots\dots \text{(vii)}$$

Continuing by resolving all forces due to stresses on the element in Fig. 12, but this time parallel to face *UV*, the following is obtained :

$$\tau_s(1) - \sigma_Y(1Sin\theta)\ Cos\theta - q(1 - Cos\theta)\ (Cos\theta) + q(1Sin\theta)\ Cos(90 - \theta)$$

$$+ \sigma_X(1Cos\theta)\ Cos(90 - \theta) = 0$$

i.e.

$$\tau_s = \sigma_Y Sin\theta\ Cos\theta - \sigma_X Cos\theta\ Sin\theta + qCos^2\theta - qSin^2\theta$$

$$= (\sigma_Y - \sigma_X)Sin\theta\ Cos\theta + q(Cos^2\theta - Sin^2\theta)$$

or

$$\tau_s = \left(\frac{\sigma_Y - \sigma_X}{2}\right) Sin2\theta + qCos2\theta \qquad \dots\dots\dots\dots\dots \text{(viii)}$$

Expressions (vii) and (viii) are the stress transformation equations for the generalized plane stress configuration represented by the elements in Figs 11 and 12. At this stage we do not know what the principal stresses acting in the element are. To find them we have to examine the results given by either equations (vi) or (vii). It is also necessary to determine the planes on which the principal stresses act. Let us start by choosing equation (vii) as the one to differentiate:

$$\frac{d}{d\theta}(\tau_n) = \frac{d}{d\theta}\left\{\frac{\sigma_X + \sigma_Y}{2} + \left(\frac{\sigma_X - \sigma_Y}{2}\right) Cos2\theta + qSin2\theta\right\}$$

$$= -(\sigma_X - \sigma_Y)\ Sin2\theta + 2q\ Cos2\theta$$

For $\quad \dfrac{d}{d\theta}(\tau_n) = 0$

$$(\sigma_X - \sigma_Y)\ Sin2\theta = 2q\ Cos2\theta$$

Hence

$$Tan2\theta = \frac{2q}{(\sigma_X - \sigma_Y)} = \frac{q}{\left(\dfrac{\sigma_X - \sigma_Y}{2}\right)} \qquad \dots\dots\dots\dots\dots \text{(ix)}$$

This value of 'θ' is a special angle in that it defines the planes on which the maximum or minimum normal stress acts. Let us designate it θ_1 as Popov does. So we write

$$Tan2\theta_1 = \frac{q}{\left(\dfrac{\sigma_X - \sigma_Y}{2}\right)} \qquad \dots \dots \dots \dots \dots \dots \dots \dots \dots \text{(x)}$$

i.e.
$$2\theta_1 = \tan^{-1}\frac{q}{\left(\dfrac{\sigma_X - \sigma_Y}{2}\right)} \qquad \dots \dots \dots \dots \dots \dots \dots \dots \text{(xi)}$$

Let us suppose evaluation of the right side of (x) gives a positive value and $Tan2\theta_1 = Tan\alpha$, then $2\theta_1 = \alpha$ or $\alpha + 180°$. Likewise if (x) gives a negative value then $2\theta_1 = (\alpha + 90°)$ or $(\alpha + 90°) + 180°$. For example if $Tan2\theta_1 = Tan60°$ then the 2 values of θ_1 are given by $2\theta_1 = 60°$ or $240°$, i.e. $\theta_1 = 30°$ or $120°$; if $Tan2\theta_1 = -\sqrt{3}$, then $2\theta_1 = 150°$ or $330°$ in which case $\theta_1 = 75°$ or $165°$. Thus principal planes are separated by $90°$.

Knowing the special value of θ, viz. θ_1 how do we find the values of $Sin2\theta_1$ and $Cos2\theta_1$ in order to make the appropriate substitution in the expression for σ_n in (vii) and thereby obtain the value of the principal stresses?

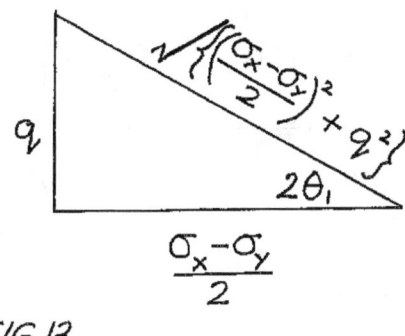

FIG. 13

To do this a hypothetical right-angle triangle is constructed as in Fig. 13 to fit the condition that $Tan2\theta_1$ is as determined in the foregoing.

Accordingly,

$$Sin2\theta_1 = \frac{q}{\pm\sqrt{\left\{\left(\dfrac{\sigma_X - \sigma_Y}{2}\right)^2 + q^2\right\}}}$$

and
$$Cos\,2\theta_1 \frac{(\sigma_X - \sigma_Y)/2}{\pm\sqrt{\left\{\left(\dfrac{\sigma_X - \sigma_Y}{2}\right)^2 + q^2\right\}}}$$

When these values are substituted in (vii) the following result is obtained:

$$\sigma_{n(max,min)} = \left(\frac{\sigma_X + \sigma_Y}{2}\right) + \left(\frac{\sigma_X - \sigma_Y}{2}\right)\cdot\frac{(\sigma_X - \sigma_Y)/2}{\pm\sqrt{\left\{\left(\dfrac{\sigma_X - \sigma_Y}{2}\right)^2 q^2\right\}}} +$$

$$\frac{q\cdot q}{\sqrt{\left\{\left(\dfrac{\sigma_X - \sigma_Y}{2}\right)^2 + q^2\right\}}}$$

$$= \left(\frac{\sigma_X + \sigma_Y}{2}\right) + \frac{(\sigma_X - \sigma_Y)^2}{4\left[\pm\sqrt{\left\{\left(\dfrac{\sigma_X - \sigma_Y}{2}\right)^2 q^2\right\}}\right]} +$$

$$\frac{q^2}{\pm\sqrt{\left\{\left(\dfrac{\sigma_X - \sigma_Y}{2}\right)^2 + q^2\right\}}}$$

$$= \frac{\sigma_X - \sigma_Y}{2} + \frac{\left[\left(\dfrac{\sigma_X - \sigma_Y}{2}\right)^2 + q^2\right]}{\pm\left[\sqrt{\left\{\left(\dfrac{\sigma_X - \sigma_Y}{2}\right)^2 + q^2\right\}}\right]}$$

from which

$$\sigma_{n(max,min)} = \frac{\sigma_X + \sigma_Y}{2} \pm \sqrt{\left(\frac{\sigma_X - \sigma_Y}{2}\right)^2 + q^2} \qquad \ldots\ldots\ldots\ldots\ldots \text{(xiii)}$$

Evidently
$$\sigma_{n(max)} = \left(\frac{\sigma_X + \sigma_Y}{2}\right) + \sqrt{\left\{\left(\frac{\sigma_X - \sigma_Y}{2}\right)^2 + q^2\right\}} \qquad \ldots\ldots\ldots\ldots \text{(xiv)}$$

and
$$\sigma_{n(min)} = \left(\frac{\sigma_X + \sigma_Y}{2}\right) - \sqrt{\left\{\left(\frac{\sigma_X - \sigma_Y}{2}\right)^2 + q^2\right\}} \qquad \ldots\ldots\ldots\ldots \text{(xv)}$$

It should be noted that from the expression for τ_s given by (viii), $\tau_s = 0$ when

$$qCos\theta = -\frac{(\sigma_Y - \sigma_X)}{2}Sin2\theta$$

i.e.

$$\left(\frac{\sigma_X - \sigma_Y}{2}\right)Sin\theta = qCos\theta$$

or

$$Tan2\theta = \frac{q}{\left(\dfrac{\sigma_X - \sigma_Y}{2}\right)}$$

which value of $Tan2\theta$ is the same value as $Tan2\theta_1$ which governs the orientation of the planes on which the principal stresses are located.

Evidently, the planes on which the principal normal stresses lie bear no shearing stress. Thus the declaration may now be made than $\sigma_{n(max)}$ and $\sigma_{n(min)}$ represented by equations (xiv) and (xv) are principal stresses. Note that the sum of the principal stresses is equal to the sum of the direct stresses, i.e.

$$\sigma_{n(max)} + \sigma_{n(min)} = \sigma_X + \sigma_Y \quad \cdots\cdots\cdots\cdots\cdots\cdots\cdots\cdots \text{(xvi)}$$

And as was demonstrated earlier, principal planes are perpendicular to each other.

Before attempting an example to illustrate application of the outcomes in this section, it is well to consider the case of principal shearing stresses. Just as there are principal normal direct tensile and compressive stresses there are principal shearing stresses. To obtain these we proceed in a manner similar to that followed in the case of maximum and minimum normal direct stresses. That is to say we differentiate equation (viii) with respect to 'θ' and equate to zero.

Accordingly,

$$\frac{d}{d\theta}(\tau_s) = \frac{d}{d\theta}\left\{\left(\frac{\sigma_Y - \sigma_X}{2}\right)Sin2\theta + qCos2\theta\right\}$$

$$= \left(\frac{\sigma_Y - \sigma_X}{2}\right)2Cos2\theta - 2q\,Sin2\theta$$

Equating this expression to zero results in

$$2qSin2\theta = (\sigma_Y - \sigma_X)Cos2\theta$$

from which

$$Tan2\theta = -\left(\frac{\sigma_X - \sigma_Y}{2q}\right) \qquad \dots\dots\dots\dots\dots\dots\dots \qquad \text{(xvii)}$$

Thus, it is on a plane governed by (xvii) that the shearing stress achieves its maximum or minimum value. Following the previous practice of drawing a right-angled triangle reflecting equation (xvii) as in Fig. 14, the values for $Sin2\theta_2$ and $Cos2\theta_2$ are:

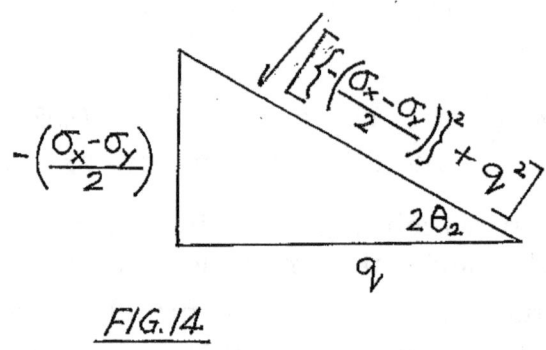

FIG. 14.

$$Sin2\theta_2 = \frac{-(\sigma_X - \sigma_Y)/2}{\sqrt{\{(\sigma_X - \sigma_Y)/2\}^2 + q^2}}$$

and
$$Cos2\theta_2 = \frac{q}{\sqrt{\{(\sigma_X - \sigma_Y)/2\}^2 + q^2}}$$

Notice that the special angle θ defining the principal shearing stress planes is designated θ_2.

Substituting these values of $Sin2\theta_2$ and $Cos2\theta_2$ in equation (viii) produces

$$\tau_{s(max,min)} = -\left(\frac{\sigma_X - \sigma_Y}{2}\right) - \left(\frac{\sigma_X - \sigma_Y}{2}\right) \cdot \frac{1}{\sqrt{\{(\sigma_X - \sigma_Y)/2\}^2 + q^2}}$$

$$+ \frac{q \cdot q}{\sqrt{\{\sigma_X - \sigma_Y)/2\}^2 + q^2}}$$

which reduces to

$$\tau_{s(max,min)} = \frac{\left(\dfrac{\sigma_X - \sigma_Y}{2}\right)^2 + q^2}{\sqrt{\left\{\left(\dfrac{\sigma_x - \sigma_Y}{2}\right)^2 + q^2\right\}}}$$

$$= \pm\sqrt{\left(\frac{\sigma_X - \sigma_Y}{2}\right)^2 + q^2} \qquad \dots\dots\dots\dots\dots \text{(xviii)}$$

Accordingly,

$$\tau_{s(max)} = +\sqrt{\left(\frac{\sigma_X - \sigma_Y}{2}\right)^2 + q^2} \qquad \dots\dots\dots\dots\dots \text{(xix)}$$

and

$$\tau_{s(min)} = -\sqrt{\left(\frac{\sigma_X - \sigma_Y}{2}\right)^2 + q^2} \qquad \dots\dots\dots\dots\dots \text{(xx)}$$

As stated earlier the planes on which these principal shearing stresses occur are governed by:

$$Tan2\theta_2 = -\frac{(\sigma_X - \sigma_Y)}{2q}$$

If as the case with $2\theta_1$, the right-hand side of this result is positive and $Tan2\theta_2 = say, \ Tan\delta$, then $2\theta_2 = \delta$ or $\delta + 180°$. Likewise, if the right-hand side of the equation is negative then $2\theta_2 = 180 - \delta$, i.e. $\theta_2 = 90 - \delta/2$. Thus the planes on which $\tau_{s(max)}$ and $\tau_{s(min)}$ occur are always 90° apart.

Note in particular also that $\tau_{s(max)}$ and $\tau_{s(min)}$ while of opposite sign are numerically equal. The difference in sign indicates that on the mutually perpendicular planes on which these maximum and minimum shear stresses act their vectors act either towards or away from the intersections of such planes as illustrated in Fig. 15.

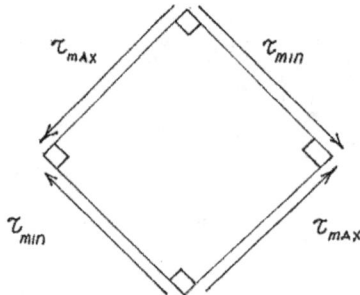

The question that now arises is: Are there any other stresses acting on the planes of principal shearing stresses? We know now that there are no other stresses, in particular any shearing stresses, on the planes on which principal direct stresses act. But what is the situation here? In order to answer the question we must now substitute the special values of θ given by:

$$2\theta_1 = Cos^{-1} \frac{\tau}{\pm\sqrt{\left(\dfrac{\sigma_X - \sigma_Y}{2}\right)^2 + \tau^2}} \quad \text{and}$$

$$2\theta_2 = Sin^{-1} \frac{-(\sigma_X - \sigma_Y)/2}{\pm\sqrt{\left(\dfrac{\sigma_X - \sigma_Y}{2}\right)^2 + \tau^2}}$$

in the general expression for σ_n labelled equation (vii). Call the value arising from such a substitution, say, σ_n'. Accordingly

$$\sigma_2' = \frac{(\sigma_X + \sigma_Y)}{2} + \frac{(\sigma_X - \sigma_Y)}{2} \cdot \frac{\tau}{\pm\sqrt{\left(\dfrac{\sigma_X - \sigma_Y}{2}\right)^2 + \tau_2}} + \left\{ \frac{-(\sigma_X - \sigma_Y)/2}{\pm\sqrt{\left(\dfrac{\sigma_X - \sigma_Y}{2}\right)^2 + \tau_2}} \right\}$$

The latter two terms cancel out and therefore

$$\sigma_n' = \frac{\sigma_X + \sigma_Y}{2} \qquad \dots\dots\dots\dots\dots\dots\dots\dots\dots\dots\dots \text{(xxi)}$$

and because by (xvi) $\sigma_{n(max)} + \sigma_{n(min)} = \sigma_X + \sigma_Y$

$$\sigma_n' = \frac{\sigma_{n(max)} + \sigma_{n(min)}}{2} \qquad \dots\dots\dots\dots\dots\dots\dots\dots\dots \text{(xxiii)}$$

Thus whereas on the planes of maximum and minimum direct stress, i.e. the planes of principal stress, there is no shearing stress acting, on the planes of principal shearing stress however, there are normal stresses of a constant magnitude acting, equal to $(\sigma_X + \sigma_Y)/2$.

Illustrative Example 3

An element of material is acted upon by direct stresses 10MN/m², tensile and 15MN/m², compressive on planes mutually perpendicular to each other. If the maximum principal stress in the material is limited to 18MN/m², then to what magnitude of complementary shearing stress may the material be subjected to on the planes bearing the direct stresses? Determine also the maximum shearing stress and the minimum principal stress. Also show the principal stresses on an oriented element.

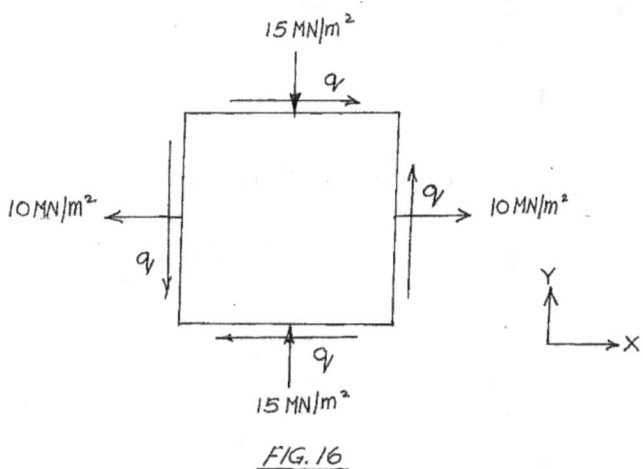

FIG. 16

A drawing of the element is shown in Fig. 16. It is assumed that shearing stress 'q' is positive in keeping with our convention. Because the relationship is of recent vintage here as one might say, equation (xiv) may be quoted. But I continue to reiterate, the student must always be prepared to work from first principles. In the problems at the end of this chapter you will be asked to solve a similar problem working from first principles.

From equation (xiv)

$$\sigma_{n(\text{max})} = \left(\frac{\sigma_X + \sigma_Y}{2}\right) + \sqrt{\left\{\left(\frac{\sigma_X - \sigma_Y}{2}\right)^2 + q^2\right\}}$$

Here $\sigma_X = 10MN/m^2$; $\sigma_Y = -15MN/m^2$; and we were given $\sigma_{n(\text{max})} = 18MN/m^2$

$$\therefore \quad 18 = \left(\frac{10 - 15}{2}\right) + \sqrt{\left\{\left(\frac{10 + 15}{2}\right)^2 + q^2\right\}}$$

i.e. $\qquad 20.5 = \sqrt{\{(12.5)^2 + q^2\}}$

or $\qquad 420.25 = 156.25 + q^2, \qquad$ i.e. $\quad q^2 = 264$

$\therefore \qquad\qquad\qquad q = +16.2 MN/m^2$

Therefore the assumed directions of the 'q' vector in the diagram are correct.

The maximum shearing stress is given by:

$$\tau_{s(max)} = \sqrt{\left(\frac{\sigma_X - \sigma_Y}{2}\right)^2 + q^2}$$

$$= \sqrt{\left\{\frac{10 - (-15)}{2}\right\}^2 + (16.2)^2} = \sqrt{420.25}$$

$$\therefore \tau_{s(max)} = 20.5 MN/m^2$$

Also, $\qquad \sigma_{n(max)} + \sigma_{n(min)} = \sigma_X + \sigma_Y = (10 - 15) MN/m^2$, i.e. $\quad 18 + \sigma_{min} = -5$

$$\therefore \quad \sigma_{n(min)} = -18 - 5 = -23 MN/m^2$$

i.e. $\quad$ 23MN/m², compressive.

Now, $\qquad Tan2\theta_1 = \dfrac{q}{(\sigma_X - \sigma_Y)/2}$

$$= \frac{16.2}{(10+15)/2} = 1.296$$

The two values of θ_1 are given by:

$\qquad 2\theta_1 = 52.3^o$ or $180^o + 52.3 = 232.3^o$, i.e. by $\theta_1 = 26.7^o$ and approximately 116.7º.

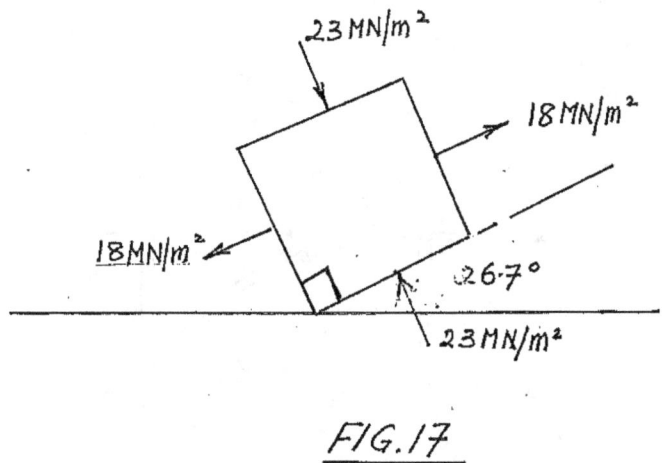

FIG.17

The orientation of $\sigma_{m(\text{max})}$ and $\sigma_{n(\text{min})}$ on a typical element is shown in Fig. 17.

MOHR'S CIRCLE OF STRESS

Mohr's circle of stress is a powerful geometrical method for solving problems in stress and strain analysis. Professor Mohr's perceptiveness in recognizing the circle as a means of doing so is to my mind quite extraordinary; a stroke of genius! His geometrical insights were unquestionably remarkable and would certainly have qualified him for red-carpet treatment at Plato's academy in Athens above which was written the now famous and immortal words "Let none ignorant of geometry enter here."

Mohr's circle is used in uni-axial, bi-axial and tri-axial stress and strain analysis. In this text, the method is exemplified in the case of generalized plane stress and strain of which the former shall now engage our attention.

Two typical elements are shown in Fig18 and I have separated the lower part and labelled it Fig. 19. Note that the big toe in Fig. 18 was excluded in the lower diagram, Fig. 19 on which our analysis is based. The toe is immaterial.

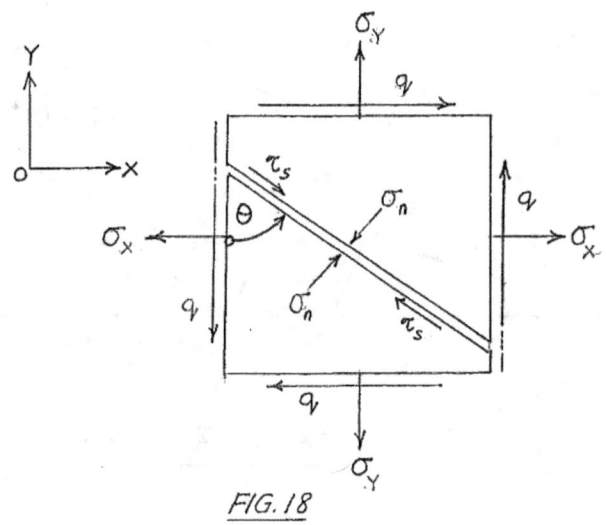

FIG. 18

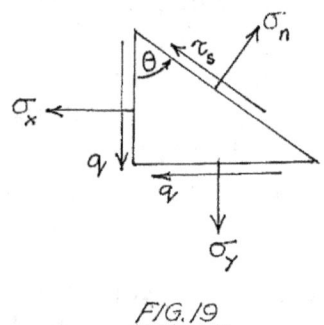

FIG. 19

Referring to the lower configuration it was shown earlier that:

$$\sigma_n = \frac{\sigma_X + \sigma_Y}{2} + \left(\frac{\sigma_X - \sigma_Y}{2}\right) Cos2\theta + qSin2\theta \qquad \dots \dots \dots \dots \text{(vii)}$$

and by rewriting (viii)

$$\tau_s = -\frac{(\sigma_X - \sigma_Y)}{2} Sin2\theta + qCos2\theta$$

Re-arranging (vii) and squaring both sides of the re-arranged equation

$$\left\{\sigma_n - \frac{(\sigma_X + \sigma_Y)}{2}\right\}^2 = \left(\frac{\sigma_X - \sigma_Y}{2}\right)^2 Cos^2 2\theta + q^2 Sin^2 2\theta +$$

$$(\sigma_X - \sigma_Y) \, qCos2\theta \, Sin2\theta \qquad \dots \dots \dots \dots \dots \dots \text{(xxiii)}$$

~ 476 ~

Now,
$$(\tau_s)^2 = \frac{(\sigma_X - \sigma_Y)^2}{2} Sin^2\theta + q^2 Cos^2\theta -$$

$$(\sigma_X - \sigma_Y) \, qSin2\theta \, Cos2\theta \qquad \dots\dots\dots\dots\dots \text{(xxiv)}$$

Adding results (xxiii) and (xxiv) yields in the following:

$$\left\{\sigma_n - \frac{(\sigma_X + \sigma_Y)}{2}\right\}^2 + \tau_s^2 = \left(\frac{\sigma_X - \sigma_Y}{2}\right)^2 + q^2$$

This latter expression may be simplified by writing it in the form :

$$(\sigma_n - a)^2 + \tau_s^2 = r^2 \qquad \dots\dots\dots\dots\dots\dots \text{(v)}$$

where
$$a = \left(\frac{\sigma_X + \sigma_Y}{2}\right) \text{ and } r = \sqrt{\left(\frac{\sigma_X - \sigma_Y}{2}\right)^2 + q^2}$$

The reader would recall that the equation of a circle of radius 'R' with its centre at the origin is expressed as $x^2 + y^2 = R^2$; and that the same circle with its centre removed to a point having coordinates *(a, b)* is given by $(x - a) + (y - b) = R.^2$ By a similar token the equation

$$(\sigma_n - a)^2 + \tau_{s\theta}^2 = r^2$$

may be used to construct a circle in an orthogonal coordinate system where, say, the X-axis could be used to represent direct stresses, 'σ' and the Y-axis, shear stress. Thus, the centre of the circle would be at point *(a, 0)*, i.e. at

$\{(\sigma_X + \sigma_Y)/2, \ 0\}$ and its radius of length $\sqrt{\left(\frac{\sigma_X - \sigma_Y}{2}\right)^2 + q^2}.$

Such a circle is shown as Fig 20.

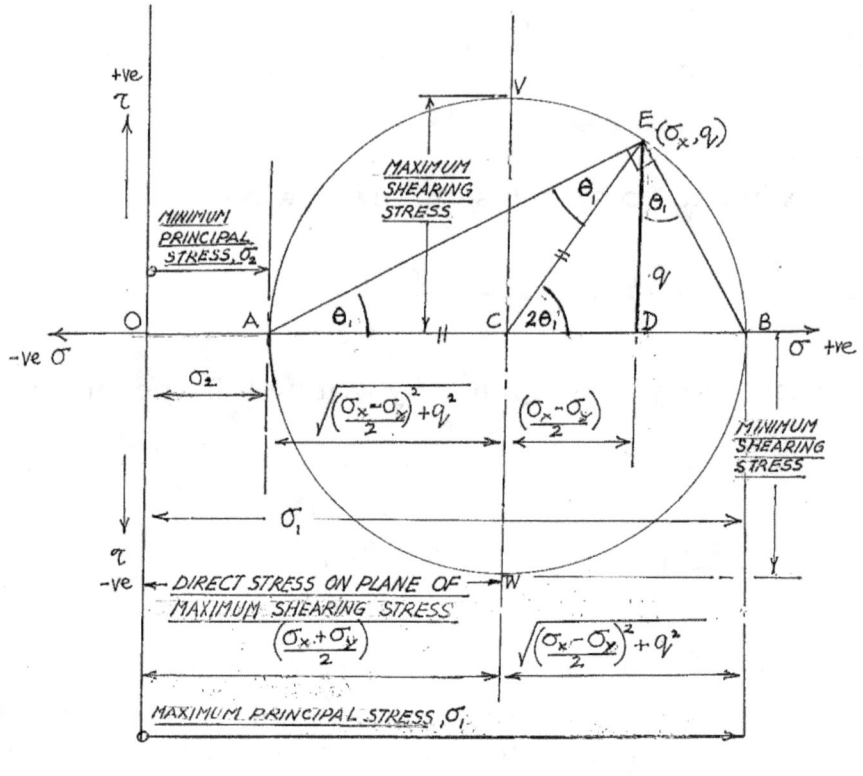

FIG.20

From previous working it is at once apparent that *OB* is the maximum principal stress, the coordinates of *B* being:

$$\left[\left\{\left(\frac{\sigma_X+\sigma_Y}{2}\right)+\sqrt{\left(\frac{\sigma_X+\sigma_Y}{2}\right)^2+q^2}\right\}, \; 0\right]$$

Similarly the minimum principal stress is *OA*, the coordinates of *A* being:

$$\left[\left\{\left(\frac{\sigma_X+\sigma_Y}{2}\right)+\sqrt{\left(\frac{\sigma_X-\sigma_Y}{2}\right)^2+q^2}\right\}, \; 0\right]$$

Noting that the distance $OD=\left(\dfrac{\sigma_X+\sigma_Y}{2}\right)+\left(\dfrac{\sigma_Y-\sigma_Y}{2}\right)=\sigma_X$ it follows that once the centre of the circle 'C' is located, the point with coordinates $(\sigma_x, \; q)$ is plotted. Hence 'E'. The circle with radius CE is thus drawn and the maximum and minimum shear stresses are respectively represented by *CV* and *CW*. Now let us determine orientation of the principal stresses. The angle *DCE* in Fig. 20 is 2θ, it having been shown that $Tan2\theta_1 = \dfrac{q}{(\sigma_X-\sigma_Y)/2.}$ By geometry the angles

CAB and *CEA* are each equal to θ_1 and angle *AED* being an angle in a semi-circle and therefore $= 90^\circ$. Accordingly principal stress σ_2 is in a direction perpendicular to the maximum principal stress σ_1.

Let us now put into practice the foregoing methodology and afterwards summarise the procedures in step-wise fashion. We begin here with two illustrative examples. Other problems are given at the end of the chapter.

Illustrative Example 4

The stresses acting on an infinitesimal element as shown in the accompanying diagram Fig. 21a are $\sigma_X = 120 KN/m^2$; $\sigma_Y = 60 KN/m^2$; $q = +50 KN/m^2$. Determine by means of Mohr's circle of stress: (i) the principal stresses, (ii) the principal shearing stress and its associated normal stress. Also show by means of neat diagrams all these stresses on properly oriented infinitesimal elements.

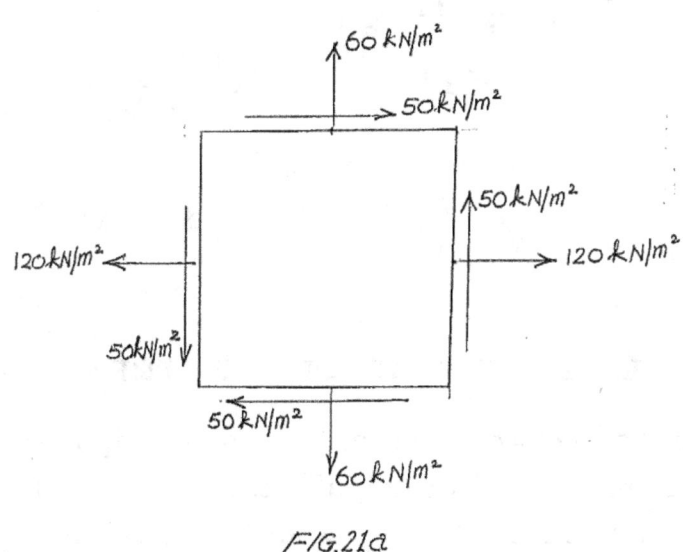

FIG.21a

The diagram is already drawn showing stresses acting ---- all positive in our convention--- and so it is not necessary to make a sketch of it which would normally be the case. The orthogonal coordinate stress axes are set up in the usual manner. Next σ_X and σ_Y are set out to scale on the direct stress axis.

The centre of Mohr's circle of stress is located at the point 'C' distant $(\sigma_X + \sigma_Y)/2$, i.e. at $(120 + 60)/2 = +90 MN/m^2$ from origin 'O' on this direct stress axis. Next, the point 'E' is plotted, its coordinates being (120, 50). Therefore, with 'C' as centre and *CE* as radius, Mohr's circle is drawn. See Fig. 21b.

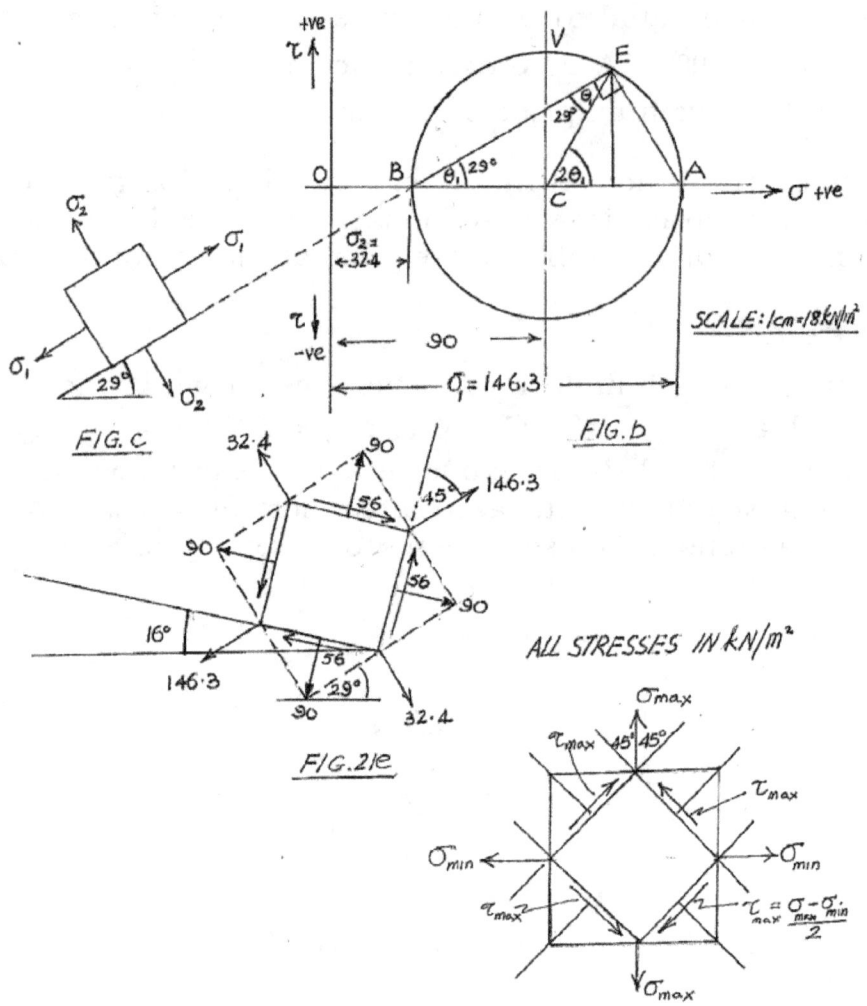

FIG.c

FIG.b

SCALE: 1cm=18kN/m²

FIG.21e

ALL STRESSES IN kN/m²

$\tau_{max} = \dfrac{\sigma_{max} - \sigma_{min}}{2}$

DETERMINATION FROM MOHR'S CIRCLE OF STRESS, FIG. 21B

(a)   By measurement $OA = \sigma_1$, the maximum principal stress $= 146.3 KN/m^2$; $OB = \sigma_2$, the minimum principal stress $= 32.4 KN/m^2$. Also, the maximum shear stress, $CV = \tau_{max} = 56 KN/m^2$, the minimum $\tau_{min} = -56 KN$; and the magnitude of the normal stress associated with both these values of shear stress being $OC = 90 KN/m^2$

(b)   Principal stress σ_1 acts in a direction perpendicular to the line AE; and σ acts in the direction perpendicular to line BE; these stresses act at $90°$ to each other, angle AEB being a right-angle. Thus, we can draw an element showing the disposition of the principal stresses. See Fig. 21c.

(c)   A computational check on the value of the maximum shear stress which was determined by direct measurement from Mohr's circle was made by employing a relationship worked out previously, viz. $\tau_{max} = (\sigma_1 - \sigma_2)/2$

~ 480 ~

which in this case is $\tau_{max} = (146.3 - 32.4)/2 = 56.9 KN/mm^2$; close enough to the value obtained by measurement from the diagram.

Also by measurement angle $ACE = 2\theta$ in Fig. 21b $\approx = 58°$. Accordingly angle $ABE = 29°$. This result was checked analytically using $Tan2\theta_1 = \tau / \dfrac{(\sigma_x - \sigma_y)}{2}$, $= 50/30 = 1.6667$, from which $2\theta_1 \approx 59.2°$ or $\theta_1 \approx 29°$. Check. An oriented element showing the disposition of the principal stresses is shown as Fig. 21c.

It was shown previously that maximums shear stress act on planes at $45°$ to the planes on which the principal stresses σ_1 and σ_2 act. See Fig. 21d.

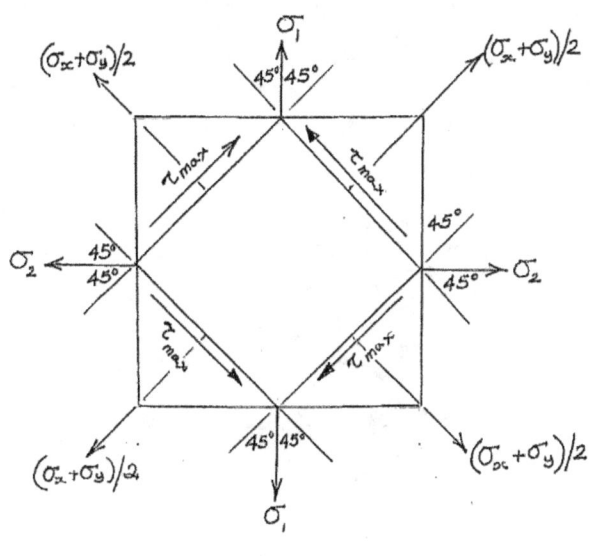

FIG. 21d

If therefore on the basis of Fig 21d, the element on which σ_1 and σ_2 are oriented at $\theta = 29°$ as shown in Fig. 21c, then the maximum stresses of $56KN/m^2$ act as shown in Fig. 21e, arrows pointing to the shear diagonal of the maximum principal stress σ_1 as shown in the diagram. The normal stress on each of the shear planes is $90KN/m^2$, tensile being $\left(\dfrac{\sigma_x + \sigma_y}{2} \right)$.

Illustrative Example 5

The stresses acting on the infinitesimal element shown in the accompanying diagram, Fig. 22a are: $\sigma_x = -3MN/m^2$; $\sigma_y = +7MN/m^2$; $\tau = +12MN/m^2$.

Determine by means of Mohr's circle of stress:

(i)   the principal stresses; (ii) the principal shearing stress and associated normal stress. Also show by means of neat diagrams all these stresses on properly oriented elements.

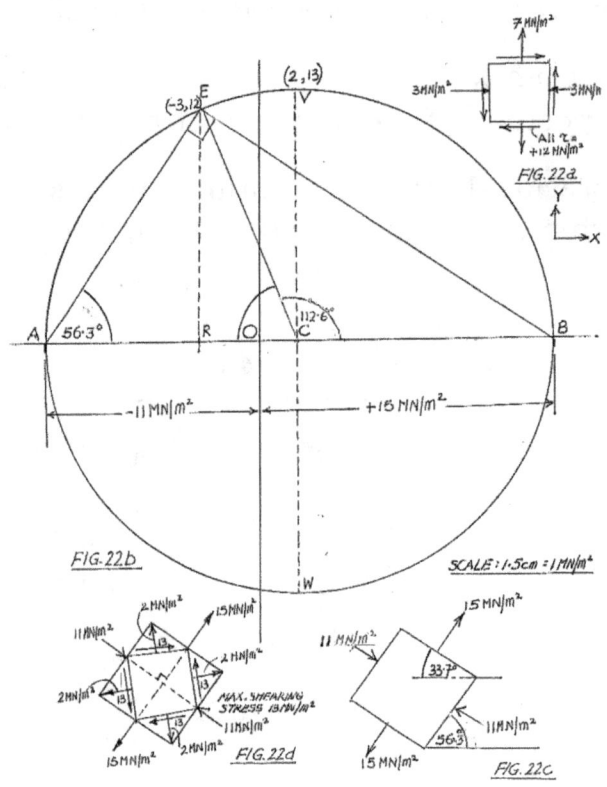

Referring to Fig. 22b, the centre of Mohr's circle is at $\left(\dfrac{-3+7}{2}\right)$, 0 i.e. (+2 0); hence centre C. Using the right-hand face of the element at Fig. 22a, point (-3, +12) is plotted. This fixes point E. Mohr's circle was accordingly drawn with CE as radius.

Determinations from Mohr's circle by direct measurement:

(a)   Maximum principal stress   = $OB = +15MN/m^2$;

Minimum principal stress   = $OA = -11MN/m^2$

Maximum shearing stress   = $CV = 13MN/m^2$

Minimum shearing stress   = $CV = -13MN/m^2$

Stress on plane with max shearing stress = $+2MN/m^2$,

(b) Angle BAE = 56.5°. Analytical check:
$Tan(180 - 2\theta_1) = 12(-3-7)/2 = -2.4$.

Therefore $2\theta_1 = 180° - 67.4 = 112.6°$ or $112.6° + 180°$, i.e. $\theta_1 = 56.3°$ or 146.3°.

Fig. 22c shows the planes of principal direct stress and Fig. 22d, the planes of principal shear stress.

GENERAL DIRECTIONS FOR SOLVING PROBLEMS BY MEANS OF MOHR'S CIRCLE OF STRESS (AFTER POPOV)

The basic steps are given hereunder, but first some advice. Do not be fazed by the problem. Be confident in your approach. Read the question carefully and "see" what is required.

(i)   If a diagram showing the force/stress configuration on an element is not provided, then sketch a suitable one and label it accordingly.

(ii) Recall your convention. The one followed here for stresses is: tensile, positive; compressive, negative. For angles: positive, anticlockwise from vertical to plane on which normal and shear stresses are required.

(iii) Set up an orthogonal coordinate system: origin at '0'. Positive: normal stresses to right of '0'; negative: normal stresses to left of '0' Positive: shearing stresses upwards of '0'; Negative: shearing stresses downwards of '0'.

(iv) Select a suitable scale so as to provide a diagram of reasonable size, i.e. one that fits into the space provided for the drawing.

(v)   Position the centre of Mohr's circle on the normal stress line at the point with coordinates with magnitude equivalent to $\left\{\left(\sigma_x + \sigma_y/2, \ 0\right)\right\}$, due regard being paid to the algebraic nature of normal stresses.

(vi) With reference to the right-hand face of the diagram at Step 1 above and following the convention for stress, plot the governing point i.e. the point having the coordinates (σ_x, q) on the diagram. To illustrate this, the following diagrams are examples of two possible situations, both for σ_x as a tensile stress. In Fig. 23a the shearing stress is positive, but negative in Fig. 23b.

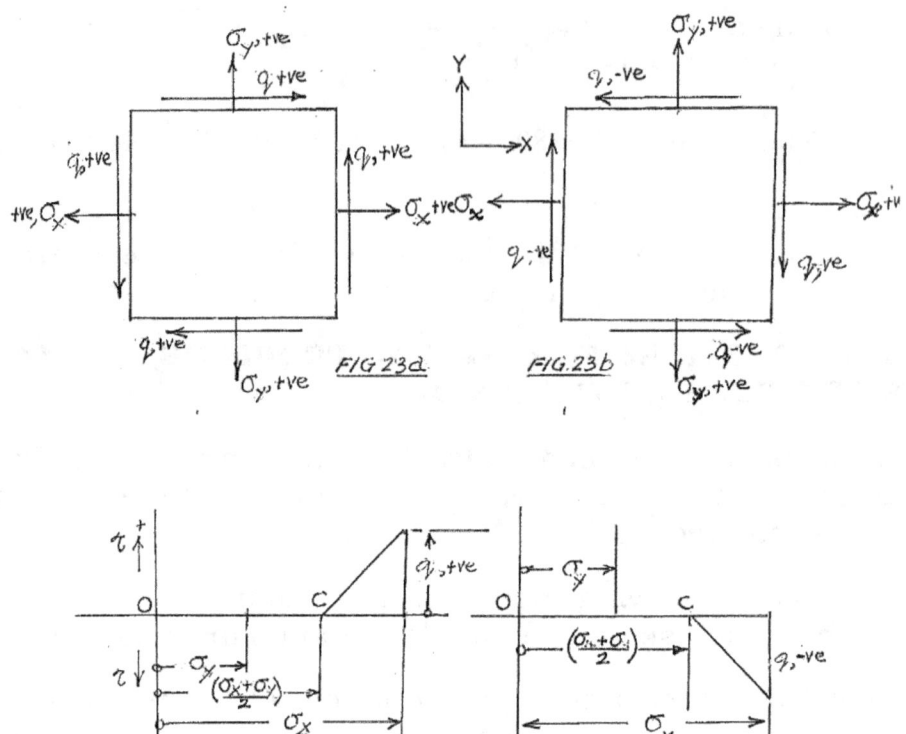

FIG. 23a   FIG.23b

FIG. 23c   FIG.23d

(vii)   Draw Mohr's circle with radius *CE*. The two intercepts measured from the origin *'O'* as a consequence of the circle cutting the normal stress axis give the magnitudes of the principal stresses. Remember for example that a principal stress of say $-5MN/m^2 > -10MN/m^2$.

(viii)   The upward vertical radius is equal in value by scale to the maximum shearing stress; the lower vertical the minimum. The distance between the origin *'O'* and the centre *'C'* of Mohr's circle is the normal stress on each of the surfaces bearing the maximum and minimum shearing stresses.

(ix)   The directions of the principal normal stresses σ_{max} and σ_{min} are found quite readily because they act perpendicular to the planes defined by the lines joining E to the points of intersection of Mohr's circle with the normal stress axis where   and   are defined as shown in Fig. 24a.

Accordingly, an element such as shown in Fig. 24b is drawn; oriented at to the horizontal. The planes on which the maximum shearing stresses act are at 45o to that on which the principal stresses act; and, all act in a direction towards the point where the maximum principal stress meets its plane. This is also shown in Fig. 24b. Observe also that the normal stress on each plane of maximum shearing stress is equal in magnitude to OC, being also equal

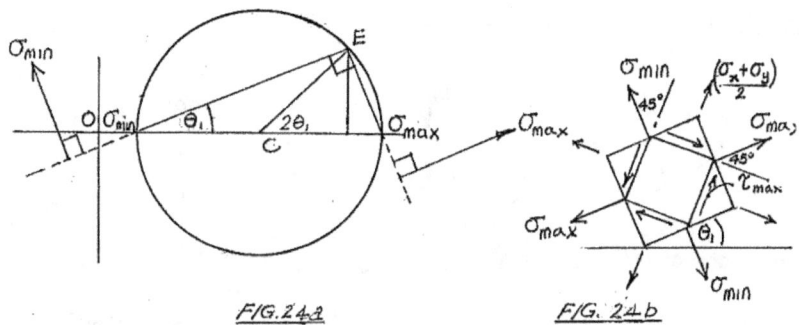

FIG. 24a   FIG. 24b

SOLVED AND OTHER PROBLEMS

Q1. A flat ductile iron bar of rectangular cross-section $100mm \times 25mm$ carries a tensile load of 100kN. Determine the direct and shear stresses on a plane inclined at 60° to the normal cross-section of the bar as shown.

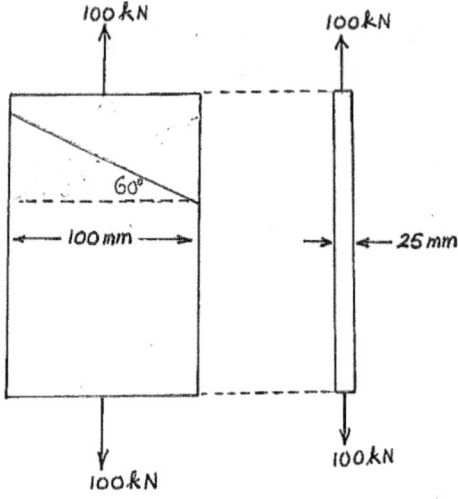

FIG. 1 (NOT DRAWN TO SCALE)

For the configuration shown it is known that 'σ_n' is given by:

$$\sigma_n = \sigma \, Cos^2\theta$$

Here

$$\sigma = \frac{100 \times 1000}{\left(\dfrac{100}{1000}\right) \times \dfrac{25}{1000}} \, N/m^2$$

$$= \frac{10^5 \times 10^6 \, m}{100 \times 25} = 40MN/m^2$$

$$\therefore \quad \sigma_n = 40Cos^2 60 = 40 \left(\frac{1}{4}\right)$$

i.e. $\sigma_n = 10MN/m^2$

Also, shearing stress τ_s on plane BC = $\dfrac{\sigma}{2} Sin2\theta$

$$\therefore \quad \tau_s = \frac{40}{2} Sin120^2 = \frac{40}{2} Sin60^o = \frac{40}{2} \frac{\sqrt{3}}{2}$$

$$= 10\sqrt{3}$$

$$\therefore \qquad \tau_s = 17.3 MN/m^2, \qquad \text{Answer.}$$

Q2. A ductile iron bar of cross-sectional area 750(mm)² bears a compressive stress of 100kN/m². Determine from first principles the magnitude of the normal and shear stress on a plane inclined at 30° to the vertical. What is the value of the maximum shear stress? See Fig. 1

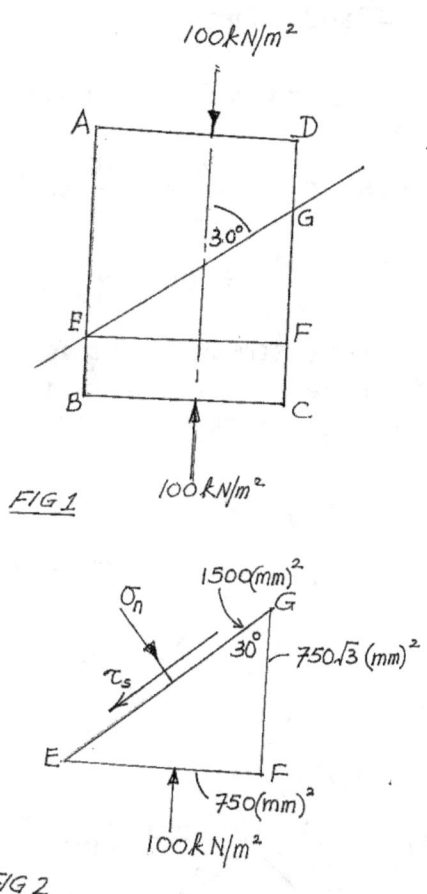

FIG 1

FIG 2

If 'EF' is the trace of the plane of cross-sectional area = 750(mm)² in Fig. 2 then the area of the plane on EG = 1500(mm)². Similarly plane GF = $750\sqrt{3}(mm)^2$. Resolving perpendicular to EG and working in kN/m²

$$\sigma_n \left(\frac{1500}{10^6} \right) - 100 \left(\frac{750}{10^6} \right) Cos60° = 0$$

$$\sigma_n = 100.750 \cdot \frac{1}{2}$$

~ 487 ~

or $\quad 3\sigma_n = 150 \quad$ or $\quad \sigma_n = 50 kN/m^2$

Resolving forces parallel to *EG*:

$$\tau_s \cdot \left(\frac{1500}{10^6}\right) - 100 \cdot \frac{750}{10^6} \cdot Cos30° = 0$$

i.e. $\quad 1500\tau_s = 100 \cdot 750 \cdot \dfrac{\sqrt{3}}{2}$

$$\tau_s = 25\sqrt{3} kN/m^2$$

or $\quad \tau_s = 43.3 kN/m^2, \quad$ Answer

Maximum shear stress, $\quad \tau_{s(max)} = \dfrac{\sigma}{2} = -\dfrac{100}{2} kN/m^2$

$$= 50 kN/m^2, \quad \text{Answer}$$

Q3. The direct and shearing stresses acting on an infinitesimal element *ABCD* are as shown in Fig. 1. Determine from first principles the normal and shear stresses on planes *QQ'* and *VV'* in the element.

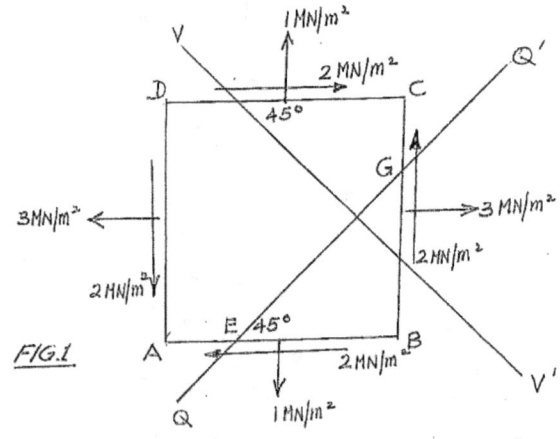

FIG.1

I shall follow Popov's "Wedge Analysis" to obtain a solution

The first step is to separate the wedges formed by the planes *QQ'* and *VV'* as shown in Fig. 2 and 3.

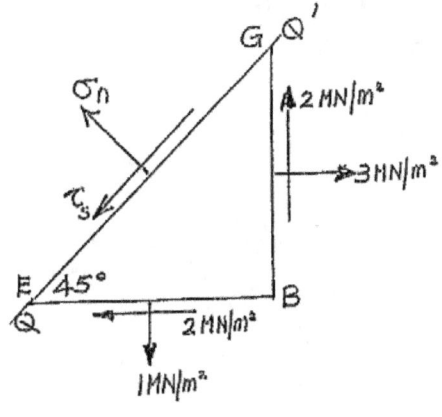

FIG.2

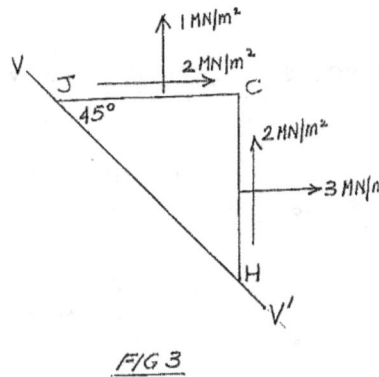

FIG 3

It is to be noted that the directions of the normal and shear stresses, respectively σ_n and τ_s on the planes, were arbitrarily assumed. If in the ensuring analysis they are evaluated as positive then the original assumption would be correct; if otherwise, then the true direction would be opposite to that assumed.

It is assumed also that the wedges are of unit thickness normal to the plane of this sheet.

Consider first wedge *EFB* and assume also that the area of the plane of which *EG* is a trace is *1m²*.

In order to emphasize the point about stresses being tensor quantities and in consequence the need to convert them to forces before applying the laws of static equilibrium, we shall first convert the stresses shown in Fig. 2 for this problem to forces. If area with side *EB = 1m²*, then areas within sides *FB* and *EF* are each $\frac{1}{\sqrt{2}} m^2$, being in each case *1Cos45°*.

On the basis therefore, the original stress may be converted to the forces shown on the element

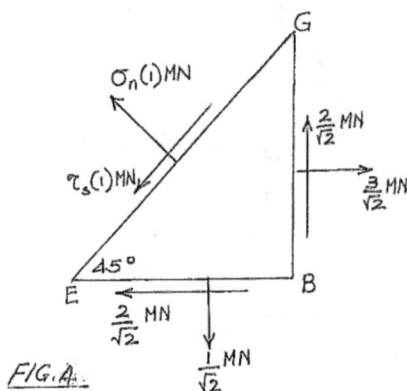

$FIG.4.$

Having converted stresses to forces it is now proper to apply the laws of statical equilibrium.

Resolving forces in directions perpendicular and parallel to EG, i.e. in the directions of σ_n and τ_s, respectively the following are obtained:

$$\sigma_n(1) + \frac{2}{\sqrt{2}} \; Cos45° - \frac{3}{\sqrt{2}} \; Sin45° + \frac{2}{\sqrt{2}} \; Cos45° - \frac{1}{\sqrt{2}} \; Sin45° = 0$$

$\therefore \qquad \sigma_n = \dfrac{3}{\sqrt{2}} \cdot \dfrac{1}{\sqrt{2}} + \dfrac{1}{\sqrt{2}} \cdot \dfrac{1}{\sqrt{2}} - \dfrac{2}{\sqrt{2}} \cdot \dfrac{1}{\sqrt{2}} - \dfrac{2}{\sqrt{2}} \cdot \dfrac{1}{\sqrt{2}}$

$$= \frac{3}{2} + \frac{1}{2} - 1 - 1 = 2 - 2$$

i.e. $\qquad \sigma_n = 0, \quad$ Answer

$$\tau_s(1) + \frac{2}{\sqrt{2}} \; Cos45° - \frac{3}{\sqrt{2}} \; Sin45° + \frac{2}{\sqrt{2}} \; Cos45° + \frac{1}{\sqrt{2}} \; Sin45° = 0$$

$\therefore \qquad \tau_s = \dfrac{2}{\sqrt{2}} \cdot \dfrac{1}{\sqrt{2}} + \dfrac{3}{\sqrt{2}} \cdot \dfrac{1}{\sqrt{2}} - \dfrac{2}{\sqrt{2}} \cdot \dfrac{1}{\sqrt{2}} - \dfrac{1}{\sqrt{2}} \cdot \dfrac{1}{\sqrt{2}}$

$$= 1 + \frac{3}{2} - 1 - 0.5 \; = \; 2.5 - 1.5 \; = \; 1$$

i.e. $\qquad \tau_s = 1MN/m^2, \quad$ Answer

Because the shear stress τ_s worked out to be positive it means that the direction assumed in Fig. 2 is correct.

Consider next Fig 3 for this problem in order to evaluate σ_n and τ_s for the configuration shown. As in the previous case it is taken that the plane upon which both σ_n and τ_s act has an area = $1m^2$. Consequently the areas on which the stresses on which HJ and CJ act = $\dfrac{1}{\sqrt{2}}m^2$. On this basis therefore the forces acting on the relevant planes are as shown in Fig. 5.

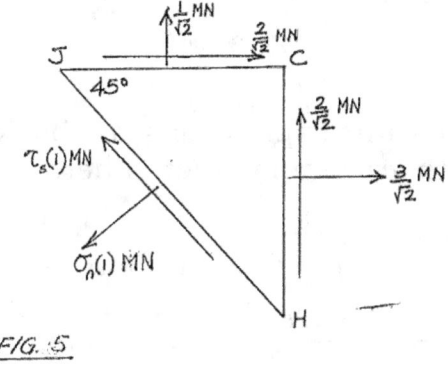

FIG. 5

Resolving forces in a direction perpendicular to HJ i.e. in the direction of σ_n :

$$\sigma_n(1) - \frac{3}{\sqrt{2}} \cdot \frac{1}{\sqrt{2}} - \frac{2}{\sqrt{2}} \cdot \frac{1}{\sqrt{2}} - \frac{1}{\sqrt{2}} \cdot \frac{1}{\sqrt{2}} - \frac{2}{\sqrt{2}} \cdot \frac{1}{\sqrt{2}} = 0$$

$\therefore$
$$\sigma_n = \frac{3}{2} + \frac{2}{2} + \frac{1}{2} + 1 = 4MN/m^2, \qquad \text{Answer}$$

Resolving forces in a direction parallel to HJ i.e. in the direction of τ_s :

$$\tau_s(1) - \frac{2}{\sqrt{2}} \cdot \frac{1}{\sqrt{2}} - \frac{1}{\sqrt{2}} \cdot \frac{1}{\sqrt{2}} - \frac{3}{\sqrt{2}} \cdot \frac{1}{\sqrt{2}} - \frac{2}{\sqrt{2}} \cdot \frac{1}{\sqrt{2}} = 0$$

$\therefore$
$$\tau = \frac{3}{2} + \frac{2}{2} - \frac{2}{2} - \frac{1}{2} = 2.5 - 1.5 = 1MN/m^2, \quad \text{Answer}$$

As before, because σ_n and τ_s are of positive value, it means the directions assigned to their vectors in Fig. 3 is correct.

Q4. The principal stresses at a point in a steel plate are tensile in kind being 8MN/m² and 5MN/m². Find from first principles expressions for the normal and shear stresses on a plane at 35° to the plane carrying the 8MN/m² stress. Determine also the maximum shear stress and the direct stress on the plane with the maximum shear stress.

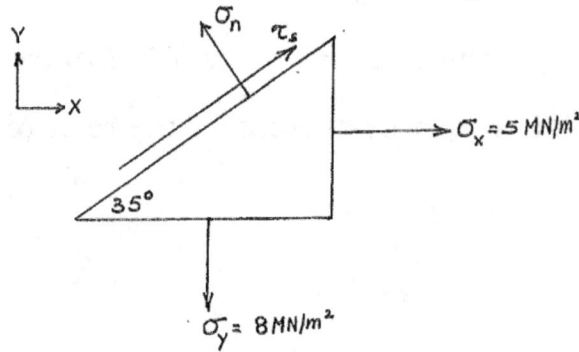

Developing the relevant expressions is bookwork and is left to the student. The results are simply quoted here:

$$\sigma_n = \sigma_x \; Sin^2\theta + \sigma_y \; Cos^2\theta \qquad \ldots\ldots\ldots\ldots\ldots \text{(i)}$$

and
$$\tau_s = (\sigma_x - \sigma_y)\frac{Sin2\theta}{2} \qquad \ldots\ldots\ldots\ldots\ldots\ldots \text{(ii)}$$

Let me stress the point here – no pun intended – that one should be able to work out the expressions relevant to any configuration, without necessarily relying on one's memory. Simply sketch the particular configuration and apply the laws of statics recalling that stresses must first be converted to forces before doing so.

Here, $\sigma_y = 8MN/m^2$; $\sigma_x = 5MN/m^2$ and $\theta = 35°$.

Therefore by (i)

$$\sigma_n = 5 \; Sin^2 35° + 8 \; Cos^3 35°$$

$$= 5(0.329) + 8(0.67)$$

$$= 1.645 + 5.36$$

i.e.
$$\sigma_n \approx 7 MN/m^2$$

Continuing

$$\tau_s = \left(\frac{8-5}{2}\right)Sin70°$$

$$= 1.5(0.9397)$$

$$\tau_s = 1.4 MN/m^2$$

From (ii), max shear stress occurs on plane where $Sin2\theta = 1$, i.e. where $\theta = 45$. Accordingly,

$$\tau_{s(max)} = \left(\frac{\sigma_y - \sigma_x}{2}\right)$$

$$\therefore \quad \tau_{s(max)} = \left(\frac{8-5}{2}\right) = 1.5 MN/m^2$$

To evaluate σ_n on plane with maximum shear stress, we substitute $\theta = 45°$ in (i)

$$\therefore \quad \sigma_n = 5 Sin^2\ 45° + 8 Cos^2\ 45°$$

$$= 5\left(\frac{1}{\sqrt{2}}\right)^2 + 8\left(\frac{1}{\sqrt{2}}\right)^2 = \frac{5}{2} + \frac{8}{4}$$

$$\sigma_n = 4.5 MN/m^2$$

Q5. A direct tensile stress 'σ_x' acts on an element subjected to pure shear 's'. Obtain expressions for the principal stresses and for the slopes of the principal planes.

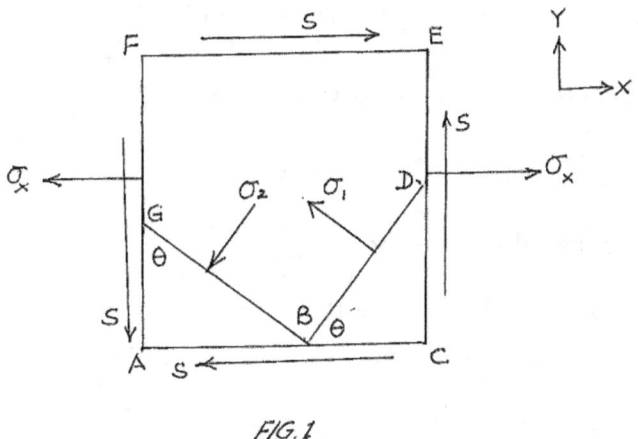

FIG. 1

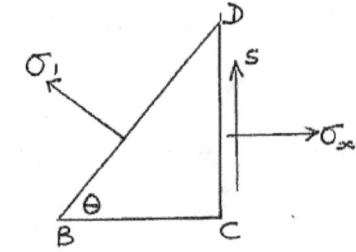

FIG. 2

Let the element be represented by the 2-dimensional block *ACEF*, subjected to direct tensile stress 'σ_s' and shear stress 's'. The traces of the principal planes are *BD* and *BG* and the principal stress on them, respectively σ_1 and σ_2. See Fig. 1

Consider element *BCD*: Fig. 2

Resolving forces horizontally, assuming unit thickness normal to this sheet.

$$\sigma_1 BD \, Sin\theta + s \cdot BC - \sigma_X \cdot CD = 0 \qquad \ldots\ldots\ldots\ldots\ldots \text{(i)}$$

i.e. $\qquad \sigma_1 BD \, Sin\theta = C_X CD - s \cdot BC \qquad \ldots\ldots\ldots\ldots\ldots \text{(ii)}$

But $CD = BD \, Sin\theta$ and $BC = Cos\theta$

Therefore (ii) becomes

$$\sigma_1 BD \, Sin\theta = \sigma_X BD \, Sin\theta - sBD \, Cos\theta$$

or $\qquad \sigma_1 \, Sin\theta = \sigma_X \, Sin\theta - s \, Cos\theta$

$\therefore \qquad \sigma_1 = \sigma_X - s \, Cot\theta \qquad \ldots\ldots\ldots\ldots\ldots\ldots\ldots \text{(iii)}$

When forces are resolved vertically,

$$\sigma_1 \, BD \, Cos\theta + s \cdot CD = 0$$

and because $CD = BD \, Sin\theta$

$$\sigma_1 \, BD \, Cos\theta + s \cdot BD \, Sin\theta = 0$$

$\therefore \qquad \sigma_1 \, Cos\theta = -s \, Sin\theta$

or $\quad Cot\theta = -\dfrac{s}{\sigma_1}$ $\qquad$ .... (iv)

Substituting this result in (iii)

$$\sigma_1 = \sigma_X - s\left(\dfrac{-s}{\sigma_1}\right)$$

$$\sigma_1 = \sigma_X + \dfrac{s^2}{\sigma_1} \quad \text{or} \quad \sigma_1^2 = \sigma_x\sigma_1 + s^2$$

i.e. $\quad \sigma_1^2 - \sigma_X\sigma_1 - s^2 = 0.$

Solving this quadratic produces

$$\sigma_1 = \dfrac{\sigma_X \pm \sqrt{\sigma_X^2 + 4s^2}}{2} \qquad .... \text{(iv)}$$

The two values of the principal stress are

$$\sigma_1 = \dfrac{\sigma_X + \sqrt{\left(\sigma_X^2 + 4s^2\right)}}{2}$$

and $\qquad \sigma_1 = \dfrac{\sigma_X - \sqrt{\left(\sigma_X^2 + 4s^2\right)}}{2}$

From (iii) and (iv)

$$\sigma_1 = \sigma_X - s\ Cot\theta \qquad .... \text{(iii)}$$

$$\sigma_1 = -\dfrac{s}{Cot\theta} = -sTan\theta \qquad .... \text{(iv)}$$

Subtracting (iv) from (iii)

$$0 = \sigma_X + s\ Tan\theta - s\ Cot\theta$$

i.e. $\quad \sigma_X = s\ Cot\theta - s\ Tan\theta$

$$\sigma_X = \dfrac{s\ Cos\theta}{Sin\theta} - \dfrac{s\ Sin\theta}{Cos\theta}$$

$$= \dfrac{s\ Cos^2\theta - s\ Sin^2\theta}{Sin\theta\ Cos\theta}$$

$$= \frac{s \; (Cos^2\theta - Sin^2\theta)}{2 \; Sin2\theta}$$

$$= \frac{-2s \; Cos2\theta}{Sin \; 2\theta}$$

$$\therefore \qquad \sigma_x = -2s \; Cot2\theta \qquad \dots\dots\dots\dots\dots\dots\dots \quad \text{(v)}$$

which may be re-expressed as

$$2\theta = Cot^{-1}\left(-\frac{\sigma_x}{2s}\right)$$

Note that if we substitute a positive value for s in (v) σ_X turns out to be negative for the configuration given in the diagram where θ is an acute angle. It simply means that the assumption of a downwards direction for shearing stress on the faces of the element was incorrect. With 's' downwards on face CD, a negative value for shearing stress would be entered in (v) and by so doing σ_X would emerge as a direct tensile stress.

Q6. In a planar stress configuration the maximum and minimum principal stresses in a steel plate are respectively 15MN/m² and 20MN/m². When a specimen of the material is tested in simple tension the elastic limit was determined to be 200MN/m². What is the factor of safety against overstrain assuming that it is caused by (i) maximum tensile stress and (ii) maximum shear stress?

Maximum direct tensile stress in plate is the maximum principal stress = 20M/m².

Answers:

Maximum stress at elastic limit = 200 MN/m².

$\therefore$ Factor of safety against failure by overstrain due to direct tensile stress = 200/20=10.

Maximum shear stress due to the principal stresses $=\left(\dfrac{20-15}{2}\right) = 2.5MN/m^2$

Shear stress in material at elastic limit $= 200/2 = 100MN/m^2$.

$\therefore$ Factor of safety against failure by overstrain due to shear stress $= 100/2.5 = 40$

Q7. The principal stresses acting at a point in a material are $\sigma_1 = 40MN/m^2$ and $\sigma_2 = 60MH/m^2$. Determine by working from first principles the normal and shear stresses on a plane inclined at *50°* to the plane on which the stress of 60MN/m² acts. What are the magnitude and direction of the resultant stress on the plane?

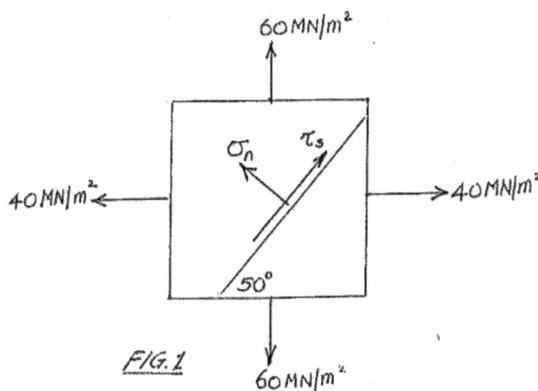

FIG.1

The stress configuration acting on an element of the material is shown in Fig 1. The direction of σ_n and τ_s were chosen arbitrarily.

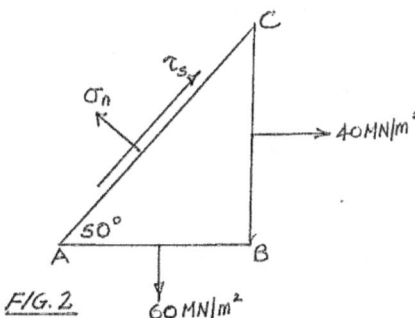

FIG.2

In Fig. 2, the element *ABC* is isolated. It is assumed to be of unit thickness.

Resolving forces in direction perpendicular to *AC*

$$\sigma_n(AC) - 40CB \; Cos40° - 60AB \; Cos50° = 0$$

i.e. $$\sigma_n = 40\frac{CB}{AC} \; Cos40° + 60\frac{AB}{AC} \; Cos50°$$

$$= 40 \; Sin50° Cos40° + 60 \; Cos50 \; Cos50°$$

$$= 40(0.7660) \; (0.7660) + 60(0.643) \; (0.643)$$

~ 497 ~

$$= 23.5 + 24.8$$

$$\therefore \qquad \sigma_n = +48.3 MN/m^2$$

Continuing by resolving forces parallel to AC

$$\tau_s \cdot AC - 60AB\ Cos40° + 40CB\ Cos50° = 0$$

i.e. $\qquad \tau_s = -40\dfrac{CB}{AC}\ Cos50° + 60\dfrac{AB}{AC}\ Cos40°$

$$= -40\ Sin50°\ Cos50° + 60\ Cos50°\ Cos40°$$

$$= -40(0.7660)\ (0.64270) - 60(0.64279)\ (0.7660)$$

$$= 29.5 - 19.7$$

$$\tau_s = +9.8 MN/m^2$$

Thus the assumed directions of σ_n and τ_s in Fig. 2 were correct.

The resultant 'R' of σ_n and τ_s is obtained from:

$$R = \sqrt{\sigma_n^2 + \tau_s^2} = \sqrt{(48.3)^2 + (9.8)^2}$$

i.e. $\quad R = 49.3 MN/m^2$

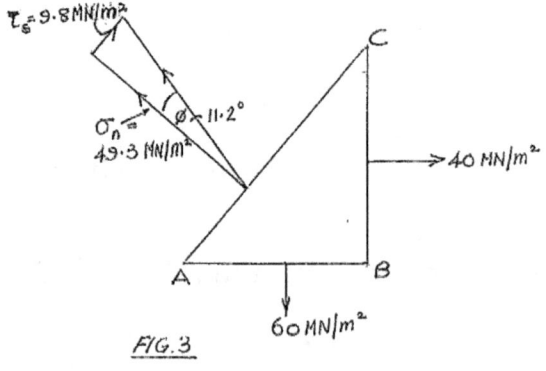

FIG.3

With reference to Fig. 3,

$$Tan\phi = \frac{9.8}{49.3} = 0.1988 \qquad \therefore \qquad \phi \approx 11.2°$$

Q8. Using Mohr's circle of stress confirm the results obtained for $\sigma_n = 48.3 MN/m^2$, $\tau_s = 9.8 MN/m^2$ and $R = 49.3 MN/m^2$ in Q.7. See Fig. 1

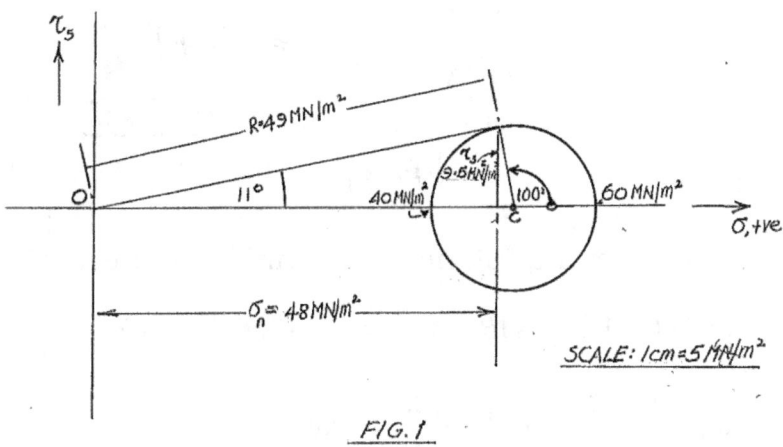

FIG. 1

Q9. By means of a well-illustrated diagram explain Mohr's circle of stress and prove how it represents the state of stress at a point in a material subjected to direct bi-axial stress and shearing stress.

At a point in a material the principal stresses are 2MN/m² and 12MN/m². Determine the normal and shearing stresses on a plane making an angle $\tan^{-1}\sqrt{3}$ with the plane on which the maximum principal is acting. Using Mohr's circle of stress to find these values and confirm your results analytically.

The first part of the question is bookwork. It is very strongly recommended, especially for those taking examinations in the subject that a thorough revision first be made of the theory of the method with particular emphasis on clearing up any points of difficulty.

Consult as many textbooks as necessary.

Make illustrative drawings and work through many examples.

Mohr's circle of stress for the data given in the question is represented by Fig. 1b. As concerns analytical confirmation, the element at Fig. 1a refers.

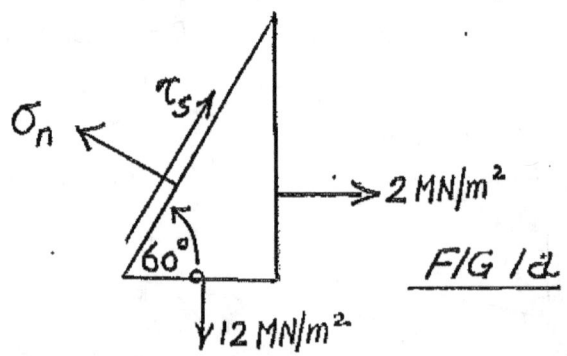

FIG 1d

Resolving forces normal to plane AC for unit thickness of element:

$$\sigma_n(AC \times 1) - 12(AB \times 1)\ Cos60^\circ - 2(BC \times 1)\ Cos30^\circ = 0$$

i.e. $\qquad \sigma_n = \dfrac{12AB}{AC}Cos60^\circ - \dfrac{2BC}{AC}Cos30^\circ$

$$= 12\ Cos60^2\ Cos60^\circ + 2\ Sin60^\circ Cos30^\circ$$

$$= 12\left(\dfrac{1}{4}\right) + 2\left(\dfrac{\sqrt{3}}{2}\right)\left(\dfrac{\sqrt{3}}{2}\right)$$

$$\sigma_n = 4.5MN/m^2$$

Resolving forces parallel to plane AC

$$\tau_s(AC \times 1) + 2(BC \times 1)\ Cos60^\circ - 12(AB \times 1)\ Cos30^\circ = 0$$

$\therefore \qquad \tau_s = \dfrac{12AB}{AC}Cos30^\circ - \dfrac{2BC}{AC}Cos60^\circ$

$$= 12\ Cos60^\circ Cos30^\circ - 2\ Sin60^\circ Cos60^\circ$$

$$= 12 \cdot \dfrac{1}{2} \cdot \dfrac{\sqrt{3}}{2} - 2\dfrac{\sqrt{3}}{2} \cdot \dfrac{1}{2}$$

$$= 5.2 - 0.87 = 4.13MN/m^2$$

$$\tau_s = 4.13MN/m^2$$

Thus the results obtained by Mohr's stress circle shown in Fig. 1b are acceptable; σ_n by calculation = 4.5MN/m² compares well with 4.4MN/m² obtained by measurement from diagram. Similarly shear stress of

4.13MN/m² by calculation compares well with 4.3MN/m² obtained by measurement from diagram.

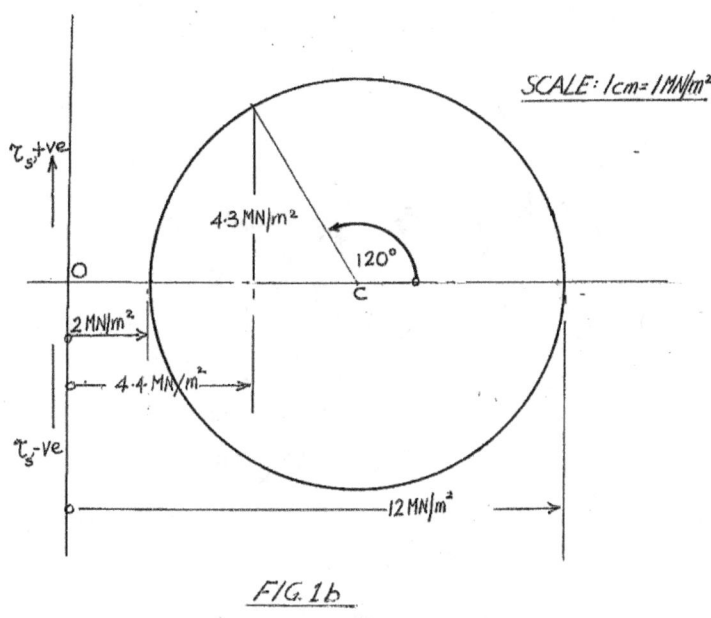

FIG. 1b

Q10. An element is subjected to pure shearing stress of 12MM/m² as shown in Fig. 1. Use Mohr's circle to obtain the values of the principal stresses.

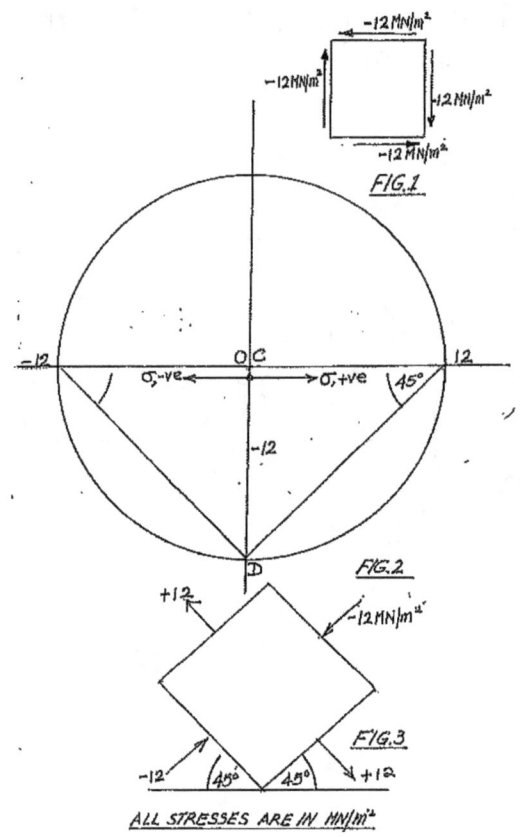

FIG.1

FIG.2

FIG.3

ALL STRESSES ARE IN MN/m²

Observe that in accordance with our convention, the shear stresses are of negative value. See Fig. 1. There being no direct stress on the element, the centre 'C' of Mohr's circle is at the origin 'O'. Therefore the point 'D' has coordinates (0, -12) and the circle is drawn with radius OD. The principal stresses are accordingly +12MN/m² and -12MN/m² as indicated on Fig. 2. An element showing the planes on which the principal stresses are acting is shown as Fig. 3. And as may be observed, there are no direct stresses on these planes.

Q11. Principal stresses of +50MH/m² and +30MN/m² act on an infinitesimal element in a 2-dimensional plane. Find the magnitude of the resultant stress on the plane on which it has its maximum obliquity. Refer to Fig. 1 and assume $\sigma_{n\theta} - \sigma_y Sin^2\theta + \sigma_x Cos^2\theta$ and $\tau_{s\theta} = \frac{1}{2}(\sigma_Y - \sigma_X)\ Sin2\theta; \quad \theta = 60°$

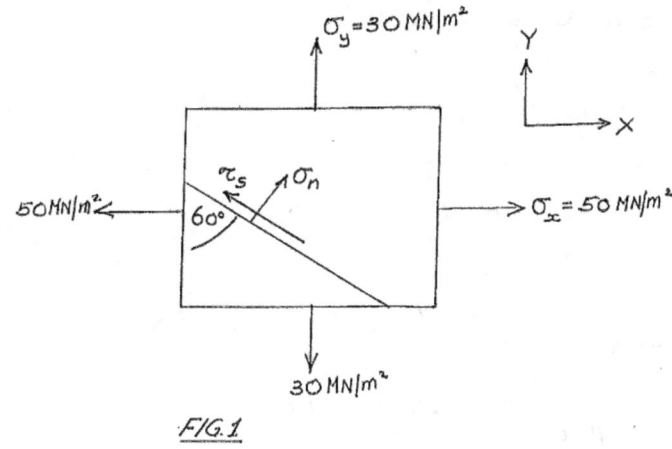

·FIG.1

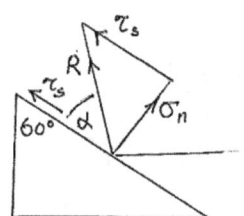

FIG. 2

With reference to Fig. 2, and with resultant stress designated σ_R

$$\sigma_R^2 = \sigma_{n\theta}^2 + \tau_{s\theta}^2$$

i.e. $\qquad \sigma_R = \sqrt{\left(\sigma_T Sin^2\theta + \sigma_X Cos^2\theta\right)^2 + \left\{\frac{1}{2}(\sigma_T - \sigma_X)\ Sin2\theta\right\}^2}$

Expressing $Sin2\theta$ as $2Sin\theta\ Cos\theta$

$$\sigma_R = \sqrt{\left(\sigma_Y Sin^2\theta + \sigma_X Cos^2\theta\right)^2 + \left(\sigma_Y - \sigma_X\right)^2 Sin\theta^2\ Cos^2\theta}$$

which with some mathematical weariness of the flesh boils don to:

i.e. $\quad \sigma_R = \sqrt{[\sigma_y^2 Sin^2\theta(Sin^2\theta + Cos^2\theta) + \sigma_x^2 Cos^2\theta(Cos^2\theta(Cos^2\theta + Sin^2\theta)]}$

$$\sigma_R = \sqrt{\{\sigma_Y^2 \ Sin^2\theta + \sigma_X^2 \ Cos^2\theta\}}$$

Referring to Fig. 2 and designating angle between the resultant σ_R and the plane as α, we may write

$$Tan\alpha = \frac{\sigma_{n\theta}}{\tau_{s\theta}}$$

Given $\sigma_n = \sigma_Y \ Sin^2\theta + \sigma_X \ Cos^2\theta$

and $\tau_{s\theta} = \frac{1}{2}(\sigma_Y - \sigma_X) \ Sin2\theta = (\sigma_y - \sigma_x) \ Sin\theta \ Cos\theta$

$$Tan\alpha = \frac{\sigma_Y \ Sin^2\theta + \sigma_X \ Cos^2\theta}{(\sigma_Y - \sigma_X) \ Sin\theta \ Cos\theta}$$

$$= \frac{\sigma_Y \ Tan\theta + \sigma_X \ Cot\theta}{(\sigma_Y - \sigma_X)}$$

From maximum obliquity of σ_R, angle 'α' must be a minimum.

$\therefore \quad \dfrac{d}{d\theta}(Tan\alpha)$ must be equated to zero and the value of 'θ' determined therefrom

Accordingly,

$$\frac{d}{d\theta}(Tan\alpha) = \frac{d}{d\theta}\left\{\frac{\sigma_Y Sin\theta}{(\sigma_Y - \sigma_X)} + \frac{\sigma_X Cot\theta}{(\sigma_Y - \sigma_X)}\right\}$$

$$= \frac{\sigma_Y}{(\sigma_Y - \sigma_X)} \ Sec^2\theta + \frac{\sigma_X}{(\sigma_X + \sigma_Y)}\left(-\frac{1}{Sin^2\theta}\right)$$

For $\quad \dfrac{d}{d\theta}(Tan\alpha) = 0$

$$\sigma_Y \ Sec^2\theta = \sigma_X \ \frac{1}{Sin^2\theta}$$

or $\quad \dfrac{\sigma_y}{Cos^2\alpha} = \dfrac{\sigma_x}{Sin^2\theta} \quad$ or $\quad \sigma_Y \ Sin^2\theta = \sigma_X \ Cos^2\theta$

i.e. $\qquad Tan^2 \alpha = \dfrac{\sigma_X}{\sigma_Y}$

or $\qquad Tan\theta = \sqrt{\dfrac{\sigma_X}{\sigma_Y}}$

Because $\qquad \sigma_R = \sqrt{\sigma_Y^2 \; Sin^2\theta + \sigma_X^2 \; Cos^2\theta}$

$$= \sqrt{\sigma_Y^2 \cdot \dfrac{\sigma_X}{\sigma_y} \; Cos^2\theta + \sigma_X^2 \; Cos^2\theta}$$

$$= \sqrt{\{\sigma_Y \sigma_X \; Cos^2\theta + \sigma_X^2 \; Cos^2\theta\}}$$

$$= Cos\theta \sqrt{\sigma_X^2 + \sigma_Y \sigma_X}$$

Substituting relevant data

$$\sigma_R = Cos60^\circ \sqrt{(50)^2 + (50)30}$$

$$= \dfrac{1}{2}\sqrt{4000}$$

$$= \dfrac{1}{2}(63.2)MN / m^2$$

or $\qquad \approx 31.6MN / m^2, \quad$ Answer

$$Tan\theta = \sqrt{\dfrac{50}{30}} = 1.2910$$

$\therefore \qquad \theta \cong 52.2^\circ$

For σ_R to have its maximum obliquity $\theta \approx 52.2^\circ$

Q12. Working from first principles, determine the normal and shear stresses on the plane $U - U'$ cutting the element shown in Fig. 1. Explain all the steps in your analysis.

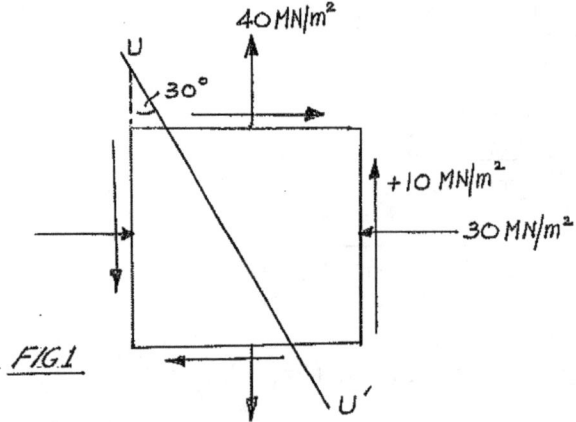

FIG.1

In solving the problem let us apply Popov's Wedge analysis.

First the element *ABC* is separated from the block. It is shown in Fig. 2 with all stresses including the unknowns σ_n and τ_s, the vectors representing the latter assumed to be acting in the directions shown.

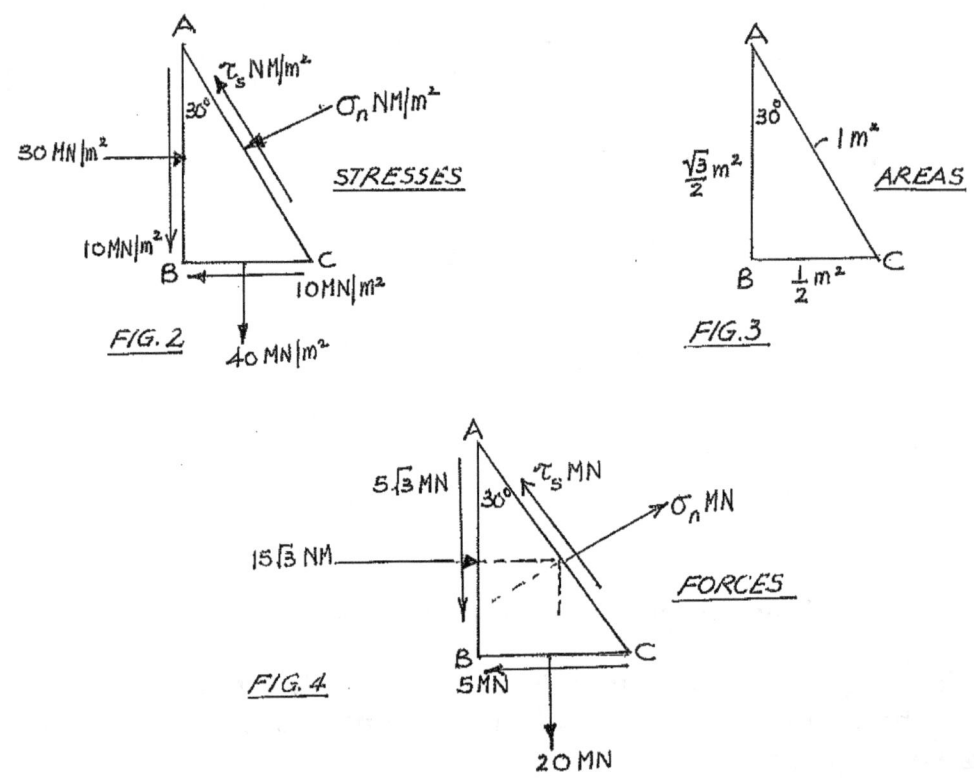

In Fig. 3 the area of the plane on which the required normal and shear stresses act is assumed to have an area = $1m^2$. Accordingly the areas on which *AB* and *BC* reside are respectively $1Cos30° = \dfrac{\sqrt{3}}{2}m^2$ and $\dfrac{1}{2}m^2$. These areas are marked on the diagram, Fig. 3. In Fig. 4, the stresses are

~ 506 ~

converted to forces by multiplying each by the area on and over which each one acts.

We come now to equilibrium considerations: Let us resolve forces along the lines of action of σ_n and τ_s in turn. Accordingly,

For $\quad \sum F_n = 0$, and with reference to Fig. 4:

$$\sigma_n + 20 \ Cos60° + 5 \ Cos30° - 15\sqrt{3} \ Cos30° + 5\sqrt{3} \ Cos60° = 0$$

i.e. $\quad \sigma_n + 20 \cdot \dfrac{1}{2} + 5\dfrac{\sqrt{3}}{2} - 15\sqrt{3} \cdot \dfrac{\sqrt{3}}{2} + 5\sqrt{3} \cdot \dfrac{1}{2} = 0$

$$\sigma_n + 10 + \frac{5\sqrt{3}}{2} - 15 \cdot \frac{3}{2} + \frac{5\sqrt{3}}{2} = 0$$

$$\sigma_n + 10 + 5\sqrt{3} - \frac{45}{2} = 0$$

$$\therefore \quad \sigma_n = \frac{45}{2} - 10 - 5\sqrt{3}$$

$$= 225.5 - 10 - 8.7$$

$$\sigma_n = 3.8 MN / m^2$$

Since this normal force is acting on 1m², the normal stress on the plane AC, coincident with plane $U - U' = 3.8 MN / m^2$.

For $\quad \sum F_s = 0$, and with reference to Fig. 4

$$\tau_s - 20 \ Cos30° + 5 \ Cos60° - 5\sqrt{3} \ Cos30° - 15\sqrt{3} \ Cos60° = 0$$

i.e. $\quad \tau_s - 20\dfrac{\sqrt{3}}{2} + 5 \cdot \dfrac{1}{2} - 5\sqrt{3} \cdot \dfrac{\sqrt{3}}{2} \cdot 15\sqrt{3} \cdot \dfrac{1}{2} = 0$

$$\tau_s = -10\sqrt{3} + 2.5 - \frac{15}{2} - \frac{15\sqrt{3}}{2} = 0$$

$$\tau_s - 10\sqrt{3} + 2.5 - 7.5 - 7.5\sqrt{3} = 0$$

$$\therefore \quad \tau_s = 10\sqrt{3} + 5 + 7.5\sqrt{3}$$

$$\tau_s = 30.3 + 5$$

from which shear stress $\tau_s = 35.3 MN/m^2$

Q13. In an element in a piece of stressed material there are direct stresses of 50MN/m², tensile and 30MN/m², compressive acting on two planes perpendicular to each other. Also acting on each of the planes is a shearing stress of 20MN/m². Working from first principles, determine the principal stresses in the material and the maximum shear stress. Refer to Fig. 1.

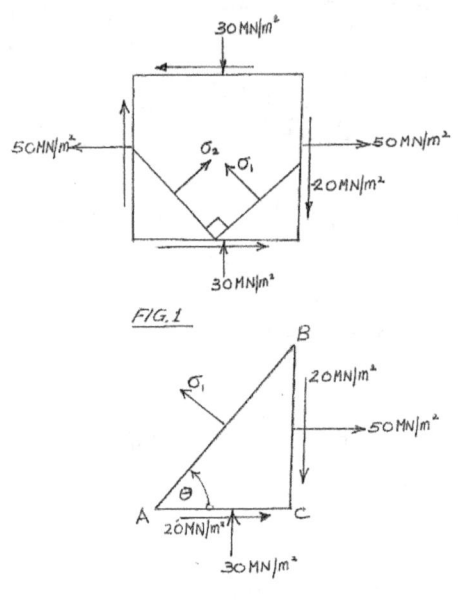

FIG. 1

FIG 2

Considering the horizontal equilibrium of the element ABC assumed to be of unit thickness and denoting σ_1 a principal stress.

Let s = shearing stress = 20MN/m²

$$\sigma_1 \cdot AB\ Cos(90-\theta) - s \cdot AC - 50 \cdot BC = 0 \qquad \ldots\ldots\ldots \text{(i)}$$

$\therefore \qquad \sigma_1 AB\ Sin\theta - s \cdot AB\ Cos\theta - 50AB\ Sin\theta = 0$

or $\qquad \sigma_1\ Sin\theta = 20\ Cos\theta + 50\ Sin\theta$

$\therefore \qquad \sigma_1 = 20\ Cot\theta + 50 \qquad \ldots\ldots\ldots\ldots\ldots\ldots\ldots\ldots \text{(ii)}$

Resolving vertically, remembering that stresses must first be converted to forces

$$\sigma_1 AB\ Cos\theta - s \cdot BC + 30 \cdot AC = 0 \qquad \ldots\ldots\ldots\ldots\ldots \text{(iii)}$$

or $\qquad \sigma_1 AB\ Cos = s \cdot AB\ Sin\theta + 30AB\ Cos\theta$

~ 508 ~

i.e. $\qquad \sigma_1 Cos\theta = 20 \; Sin\theta + 30 \; Cos\theta$

$\therefore \qquad \sigma_1 = 20Tan\theta + 30 \qquad \dots\dots\dots\dots\dots\dots\dots$ (iv)

Equating (ii) and (iv)

$$20 \; Cot\theta + 50 = 20 \; Tan\theta + 30$$

$$20 = 20 \; Tan\theta - 20 \; Cot\theta$$

$$\therefore \qquad 1 = Tan\theta - Cot\theta$$

$$= \frac{Sin\theta}{Cos\theta} - \frac{Cost}{Sin\theta}$$

$$= \frac{Sin^2\theta - Cos^2\theta}{Sin\theta \; Cos\theta}$$

$$= -\frac{2 \; Cos2\theta}{Sin2\theta}$$

or $\qquad 1 = -2 \; Cot\theta$

i.e. $\qquad Cot \; 2\theta = -\dfrac{1}{2}$

$$Tan2\theta = -2$$

$\therefore \qquad 2\theta = \approx -63.5^o \quad \text{or} \quad \theta = -31.8^o$

From (ii)

$$Cot\theta = \frac{\sigma_1 - 50}{20} \quad \text{or} \quad Tan\theta = \frac{20}{\sigma_1 - 50}$$

Putting this result in (iii) produces

$$\sigma_1 = 20 \left(\frac{20}{\sigma_1 - 50} \right) + 30$$

$$\sigma_1(\sigma_1 - 50) = 400 + 30(\sigma_1 - 50)$$

$$\sigma_1^2 - 50\sigma_1 = 400 + 30\sigma_1 - 1500$$

$\therefore \qquad \sigma_1^2 - 80\sigma_1 + 1100 = 0$

from which

$$\sigma_{1(max,min)} = \frac{80 \pm \sqrt{(6400-4400)}}{2}$$

$$= \frac{80 \pm \sqrt{2000}}{2}$$

$$= \frac{80 \times 44.7}{2}$$

$$\sigma_{1(max,min)} = 62.3 MN/m^2 \text{ and } 17.6 MN/m^2$$

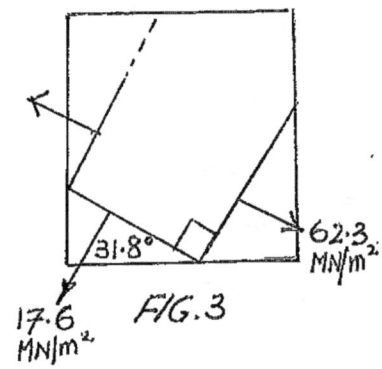

FIG.3

Maximum Shear Stress $= \dfrac{\sigma_1 + \sigma_2}{2}$

$$= \left(\frac{62.3 + 17.6}{2} \right) MN/m^2$$

$$\approx 40 MN/m^2$$

Q14. The stress condition existing at a point in a material is such that a shear stress 's' acts on a certain plane with a direct tensile stress 'σ_x'. Given that the shear stress has a magnitude of 12MN/m². Find the value of 'σ_x' and the magnitude of the principal stresses if 'σ_x' makes an angle of 30º with the major principal stress.

[You may use the result obtained in Q5]

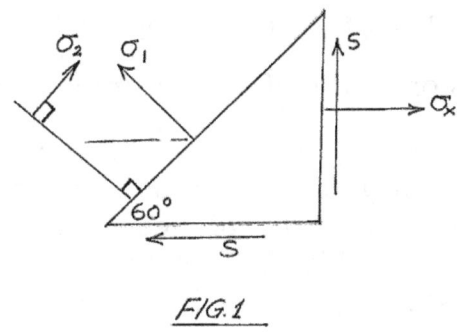

FIG.1

Answer: $\sigma_x = 13.9 MN/m^2$; $\sigma_1 = 20.8 MN/m^2$; $\sigma_2 = 6.9 MN/m^2$

Q15. The stresses acting on an element in a steel component are $\sigma_x = -3MN/m^2$; $\sigma_y = +1MN/m^2$ and $\tau = -2MN/m^2$. Draw Mohr's circle of stress to represent this stress condition and obtain the value of the maximum shearing stress and that of the normal stress on the plane of maximum shear stress. Show the results on a properly-oriented element.

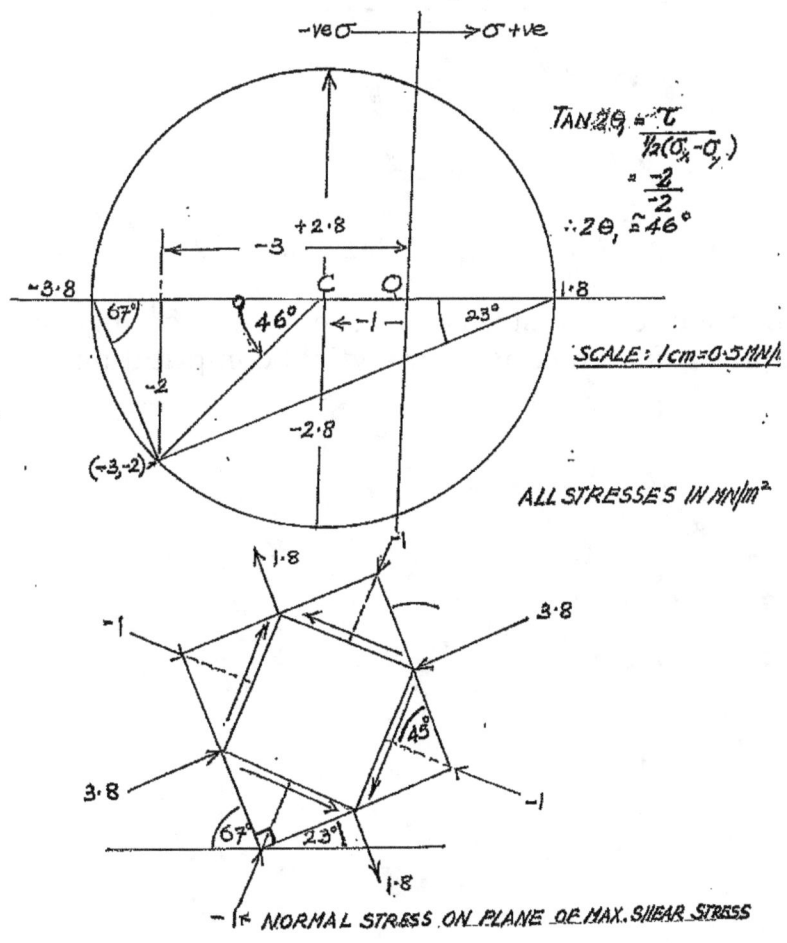

NORMAL STRESS ON PLANE OF MAX. SHEAR STRESS

Ans: Maximum shearing stress = 2.8MN/m²

Normal stress on plane of maximum shearing stress = -1MN/m²

Q16. In a specimen of ductile iron there are three stresses acting at a point: direct tensile and compressive stresses of magnitude 25MN/m² and 20MN/m² respectively at right angles to each other, and a shearing stress of +10MN/m². See Fig. 1. Determine the direction and magnitude of the principal stresses.

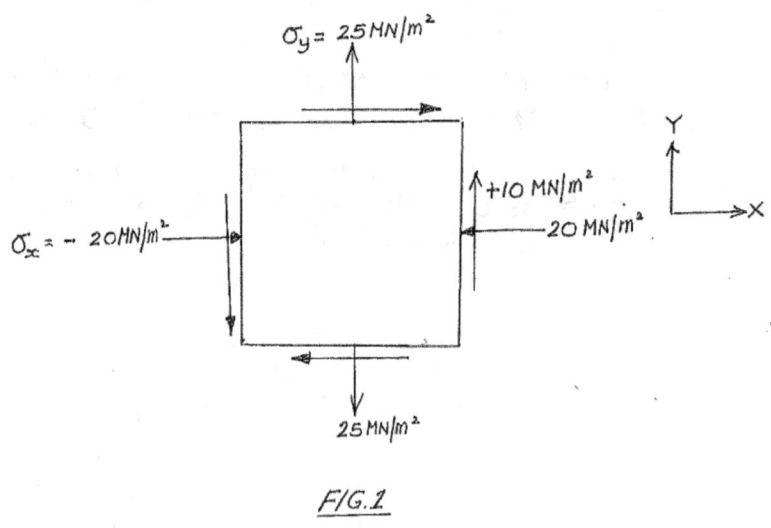

FIG.1

Ans: $\sigma_{max} = 12.8 MN/m^2$; $\sigma_{min} = 7.8 MN/m^2$; $\theta = 78°$ and $168°$.

Q17. Draw Mohr's circle of stress for the stress configuration in Q16 and check the results by means of analytical computation.

Q18. Fig. 1 represents an element in a beam with the stresses shown acting on it. Find the principal stresses using Mohr's circle of stress and show their sense on a properly-oriented diagram. Determine also the principal shearing stresses and the normal stress associated with them, and mark all these on the same diagram.

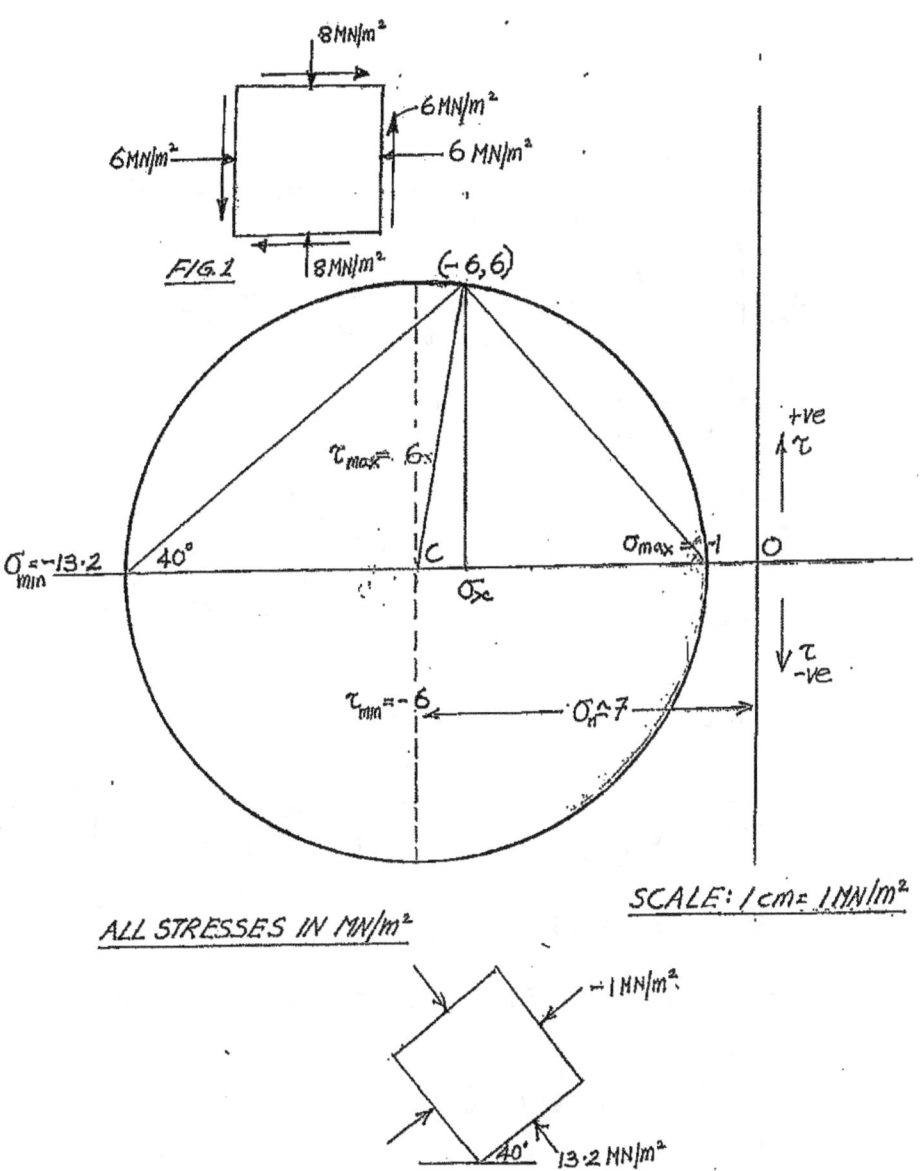

ALL STRESSES IN MN/m²

SCALE: 1 cm = 1 MN/m²

Q19. Use Mohr's circle of stress to determine the maximum values of direct and maximum shear stress in a shaft at which at a point there is a normal tensile stress of 90MN/m² and a shearing stress of 75MN/m². Check your result analytically.

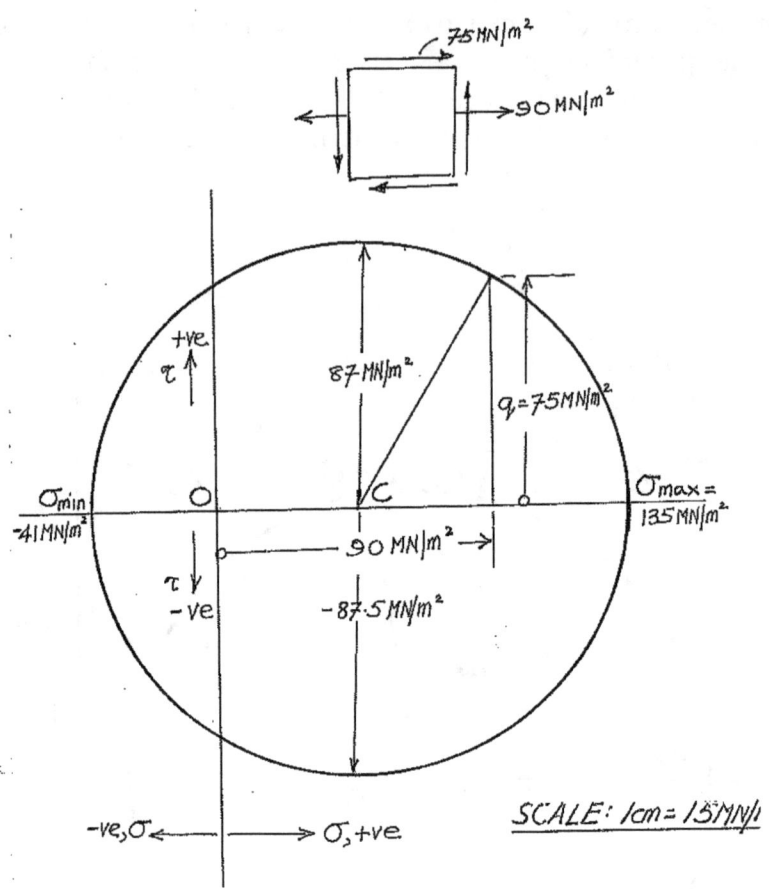

Analytical check =

$$\sigma_{min.max} = \frac{\sigma_X}{2} \pm \sqrt{\left(\frac{\sigma_X}{2}\right)^2 + q^2} = \frac{90}{2} \pm \sqrt{\left(\frac{90}{2}\right)^2 + (75)^2} = 45 \pm \sqrt{7650}$$

$$= 45 \pm 87.5 \ MNm^2$$

$$\therefore \qquad \sigma_{max} = 132.5 MN/m^2; \quad \sigma_{min} = -42.5 MN/m^2$$

$$\tau_{max} = \pm\sqrt{\left(\frac{\sigma_X}{2}\right)^2 + q^2} = \sqrt{7650} = \pm 87.5 MN/m^2$$

Q20. At a certain point in a component under load, the stress on these mutually perpendicular planes are: (i) a compressive stress of 30MN/m² and a shearing stress of 25MN/m²; (ii) a tensile stress of 40MN/m² and complementary shearing stress of 25MN/m²; and (iii) zero. Determine, working from first principles: (a) the principal stresses at the point and the orientation of the planes on which they act; and, (b) the orientation of the planes on which there is zero normal direct stress.

This is a typical examination question. As I have emphasized throughout the chapter, always be prepared to work from first principles.

The first thing to do is to make a sketch of the element showing the stresses acting at the point. There being zero stress on the third plane means that the problem is a planar one, in say, the X-Y plane, it is here assumed.

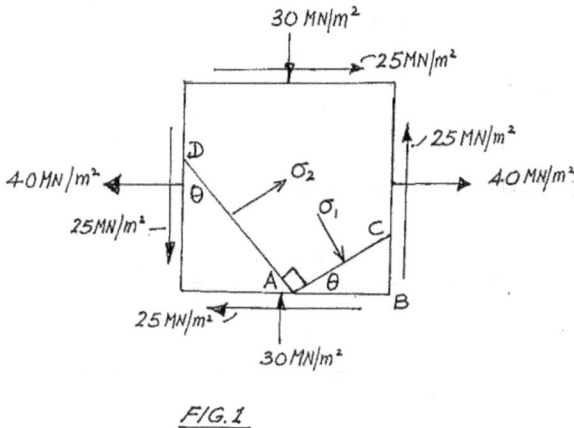

<u>FIG.1</u>

The stress configuration at the point is shown in Fig. 1. When any shear, say 'τ_s' on the plane AC = 0, only then can σ be a principal stress. When we find them they shall be designated σ_1 and σ_2; and their principal planes AC and AD respectively as shown in Fig. 1. Take element ABC and apply the laws of statics: $\sum F_X = 0$; $\sum F_Y = 0$. Refer to Fig. 2.

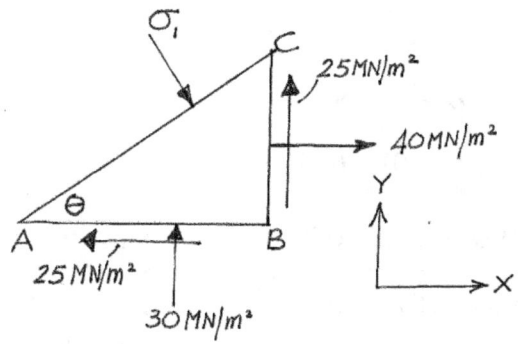

<u>FIG. 2</u>

For $\sum F_X = 0$

$$\sigma \; AC \; Sin\theta + 40 \cdot BC - 25 AB = 0 \qquad \dots\dots\dots\dots\dots\dots \quad (i)$$

i.e. $\qquad \sigma\ Sin\theta = \dfrac{25AB}{AC} - \dfrac{40BC}{AC}$

$$= 25\ Cos\theta - 40\ Sin\theta$$

or $\qquad \sigma = 25\ Cot\theta - 40$

$\therefore \qquad Cot\theta = \dfrac{\sigma + 40}{24}$ .... (ii)

For $\sum F_y = 0$

$\qquad \sigma\ AC\ Cos\theta - 30AB - 25BC = 0$

i.e. $\qquad \sigma\ Cos\theta = \dfrac{30AB}{AC} + \dfrac{25BC}{AC}$

$$= 30\ Cos\theta + 25\ Sin\theta$$

or $\qquad \sigma = 30 + 25\ Tan\theta$

$\therefore \qquad Tan\theta = \left(\dfrac{\sigma - 30}{25}\right)$ .... (iii)

or $\qquad Cot\theta = \dfrac{25}{\sigma - 30}$ .... (iv)

Equating (ii) and (iv)

$$\dfrac{\sigma + 40}{25} = \dfrac{25}{\sigma - 30}$$

i.e. $\qquad (\sigma + 40)(\sigma - 30) = 675$

or $\qquad \sigma^2 + 10\sigma - 1200 = 675$

$\therefore \qquad \sigma_2 + 10\sigma - 1875 = 0$ .... (v)

Solving (iv) $\qquad \sigma_{1,2} = \dfrac{-10 \pm \sqrt{\{(10^2 + 7500)\}}}{2}$

i.e. $\qquad \sigma_{1,2} = \dfrac{-10 \pm 87.17}{2} = -5 \pm 43.58$

so that $\qquad \sigma_1 = -48.58 MN/m^2, (compressive), \qquad$ say $\ -48.6 MN/m^2$

$$\sigma_2 = +38.58 MN/m^2, \text{ (tensile), say, } 38.6 MN/m^2$$

From (iii) $Tan\theta = (-48.6 - 30)/25$

$$= -3.1432$$

$\therefore \quad \theta_1 \equiv 180 - 72.3^\circ \equiv 107.7^\circ$

and for the other value of σ, viz. σ_2,

$$Tan\theta_2 = (38.6 - 30)/25 = \frac{8.6}{25}$$

$$= 0.3432$$

$\therefore \quad \theta \approx 18.9^\circ$

The values of θ_1 and θ_2 should differ from each other by 90°: 108.9° is as close to 107.7° as rounding errors would allow. Now to part (b) of the question.

If it is assumed that σ_n is the normal stress on plane AC, then resolving forces perpendicular to plane AC, using Popov's Wedge analysis:

$$\sigma_n - 30 \ Cos^2\theta - 50 \ Sin\theta \ Cos\theta + 40 \ Sin^2\theta = 0 \qquad \ldots \ldots \text{ (vi)}$$

When $\sigma_n = 0$, that is when the normal direct stress on $AC = 0$, equation (vi) reduces to

$$30 \ Cos^2\theta + 50 \ Sin\theta \ Cos\theta - 40 \ Sin^2\theta = 0 \qquad \ldots \ldots \ldots \text{ (vii)}$$

Dividing throughout by $Cos^2\theta$, (vii) reduces to

$$30 + 50 \ Tan\theta - 40 \ Tan^2\theta = 0$$

or $40 \ Tan^2\theta - 50 \ Tan\theta - 30 = 0$

$$\therefore \quad Tan\theta = \frac{50 \pm \sqrt{(50^\circ) + 4800}}{80}$$

$$= (50 \pm 85.44)/80$$

$$Tan\theta = 135.44/80; \quad -35.44/80$$

$$1.692; \quad -0.443$$

$$\therefore \quad \theta_1 \cong 59.4^o ; \quad 180^o - 23.9^o = 156.1^o$$

These are the planes on which normal direct stress is zero.

APPENDIX I

TABLES OF DIMENSIONS AND PROPERTIES OF SOME STRUCTURAL SECTIONS

APPENDIX 1 – TABLE 1

TABLE OF DIMENSIONS AND PROPERTIES
(EXTRACTED FROM BS EN 10219: 1997)

SQUARE HOLLOW SECTIONS

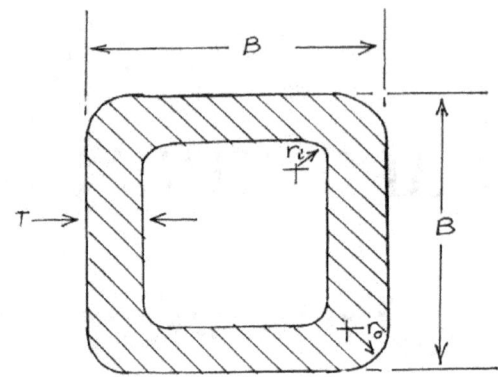

Size BxB	Thickness T	Corner Radii mm		Area of Section (cm)^2	Mass per Metre kg/m	I_{xx}, I_{yy}	Radius of Gyration	Section Modulus
		Ext r$_o$	Int r$_i$					
mm x mm						(cm)^4	cm	(cm)^3
20 x 20	2	4	2	1.34	1.05	0.692	0.72	0.692
25 x 25	2	4	2	1.74	1.36	1.48	0.924	1.19
30 x 30	2	4	2	2.14	1.68	2.72	1.1	2.1
30 x 30	3	6	3	3.01	2.36	3.5	1.08	2.34
40 x 40	2	4	2	2.94	2.31	6.94	1.54	3.47
40 x 40	2.5	5	2.5	3.59	2.82	8.22	1.51	4.11
40 x 40	4	8	4	5.35	4.2	11.1	1.44	5.54
50 x 50	2	4	2	3.74	2.93	14.1	1.95	5.66
50 x 50	3	6	3	5.41	4.25	19.5	1.9	7.79
50 x 50	4	8	4	6.95	5.45	23.7	1.85	9.49
60 x 60	2	4	2	4.54	3.56	25.1	2.35	8.38
60 x 60	3	6	3	6.61	5.19	35.1	2.31	11.7
70 x 70	4	8	4	10.1	7.97	72.1	2.67	20.6
80 x 80	8	20	12	20.8	16.4	168	2.84	42.1
90 x 90	6	12	6	19.2	15.1	220	3.39	49
100 x 100	12	36	24	36.1	28.3	408	3.36	81.6
120 x 120	8	20	12	33.6	26.4	677	4.49	113
140 x 140	12.5	37.5	25	57	44.8	1425	5	204
150 X 150	16	48	32	74.8	58.7	2009	5.18	268
180 X 180	5	10	5	34.4	27	1737	7.11	193

APPENDIX 1 – TABLE 2

TABLE OF DIMENSIONS AND PROPERTIES
(EXTRACTED FROM BS EN 10056: 1999)

ASB (ASYMMETRIC) BEAMS

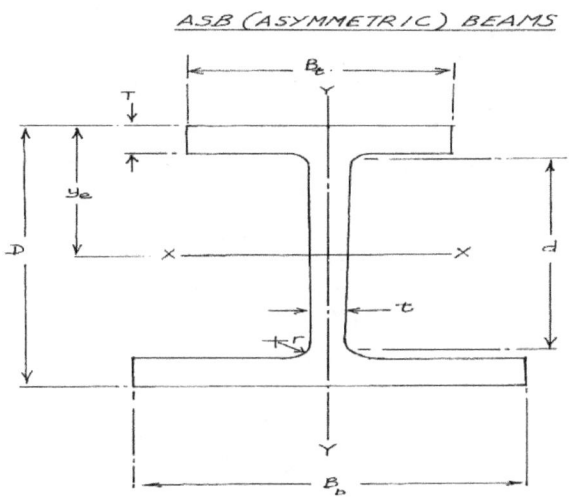

Designation Serial Size	Mass per metre kg/m	Depth of Section d mm	Width of top flange B_t mm	Width Of bottom flange B_b mm	Thickness of Web t mm	Thickness of Flange T mm	Root Radius r mm	Depth Between Fillets D mm	I_{xx} (cm)^4	I_{yy} (cm)^4	k_{xx}	k_{yy}	Elastic Neutral Axis Position, Y_e cm
300 ASB (FE) 249	249.2	342	203	313	40	40	27	208	52920	13190	12.9	6.45	19.2
300 ASB 196	195.5	342	183	293	20	40	27	208	45870	10460	13.6	6.48	19.8
300 ASB (FE) 185	184.6	320	195	305	32	29	27	208	35660	8752	12.3	6.1	18
300 ASB 155	155.4	326	179	289	16	32	27	208	34510	7989	13.2	6.35	18.9
300 ASB (FE) 153	152.8	310	190	300	27	24	27	208	28400	6840	12.1	5.93	17.4
280 ASB (FE) 136	136.4	288	190	300	25	22	24	196	22220	6256	11.3	6	16.3
280 ASB 124	123.0	296	178	288	13	26	24	196	23450	6410	12.2	6.37	17.2
280 ASB 105	104.7	288	176	286	11	22	24	196	19250	5298	12	6.3	16.8
280 ASB (FE) 100	100.3	276	184	294	19	16	24	196	15510	4245	11	5.76	15.6
280 ASB 74	73.6	272	175	285	10	14	24	196	12190	3334	11.4	5.96	15.7

APPENDIX 1 – TABLE 3

TABLE OF DIMENSIONS AND PROPERTIES
(EXTRACTED FROM BS EN 10056: 1999)

STRUCTURAL TEES SPLIT FROM UNIVERSAL BEAMS

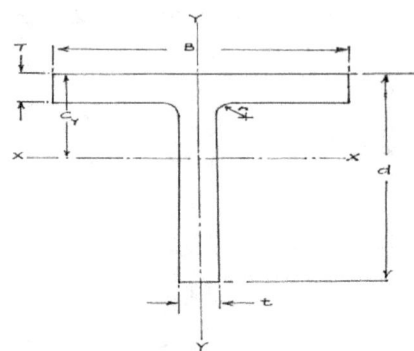

Serial Size	Mass per metre	Width of Section mm	Depth of Section mm	Thickness		Root Radius mm	Position of Centre of Mass From x-axis	Second Moment of Area (cm)^4	
				Web t	Flange T			I_{xx}	I_{yy}
B x d x kg/m	kg/m	B	d			r	C_y		
254 x 343 x 63	62.6	253	338.9	11.70	16.2	15.2	8.85	8980	2190
305 x 305 x 75	74.6	304.8	306.1	11.8	19.7	16.5	6.45	7410	4650
229 x 305 x 70	69.9	230.2	308.5	13.1	22.1	12.7	7.61	7740	2250
229 x 305 x 63	62.5	229	306	11.9	19.6	12.7	7.54	6900	1970
229 x 305 x 57	56.5	228.2	303.7	11.1	17.3	12.7	7.58	6270	1720
229 x 305 x 51	50.6	227.6	301.2	10.5	14.8	12.7	7.78	5690	1460
210 x 267 x 61	61	211.9	272.2	12.7	21.3	12.7	6.66	5160	1690
210 x 267 x 55	54.5	210.8	269.7	11.6	18.8	12.7	6.61	4600	1470
210 x 267 x 51	50.5	210	268.3	10.8	17.4	12.7	6.53	4250	1350
210 x 267 x 46	46	209.3	266.5	10.1	15.6	12.7	6.55	3880	1190
210 x 267 x 41	41.1	208.8	264.1	9.6	13.2	12.7	6.75	3530	1000
191 x 229 x 49	49.1	192.8	233.5	11.4	19.6	10.2	5.53	2970	1170
191 x 229 x 45	44.6	191.9	231.6	10.5	17.7	10.2	5.47	2680	1040
191 x 229 x 41	41	191.3	229.9	9.9	16	10.2	5.47	2470	935
191 x 229 x 37	37.1	190.4	228.4	9	14.5	10.2	5.38	2220	836
191 x 229 x 34	33.5	189.9	226.6	8.5	12.7	10.2	5.46	2030	726
152 x 229 x 41	41	155.3	232.8	10.5	18.9	10.2	5.96	2600	592
152 x 229 x 37	37.1	154.4	230.9	9.6	17	10.2	5.88	2330	523
152 x 229 x 34	33.6	153.8	228.9	9	15	10.2	5.91	2120	456
152 x 229 x 30	29.9	152.9	227.2	8.1	13.3	10.2	5.84	1880	397
152 x 229 x 26	26.1	152.4	224.8	7.6	10.9	10.2	6.04	1670	322
127 x 152 x 24	24	125.3	155.4	9	14	8.90	3.94	662	231
127 x 152 x 19	18.5	123.4	152.1	7.1	10.7	8.9	3.78	501	168
102 x 152 x 17	16.4	102.4	156.3	6.6	10.8	7.6	4.14	487	97.1
102 x 152 x 13	12.4	101.6	152.5	5.8	7	7.6	4.43	377	61.5
102 x 127 x 14	14.1	102.2	130.1	6.3	10	7.6	3.24	277	89.3
102 x 127 x 13	12.6	101.9	128.5	6	8.4	7.6	3.32	250	74.3
102 x 127 x 11	11	101.6	126.9	5.7	6.8	7.6	3.45	223	59.7
133 x 102 x 15	15	133.9	103.3	6.4	9.6	7.6	2.11	154	192
133 x 102 x 13	12.5	133.2	101.5	5.7	7.8	7.6	2.1	131	154

APPENDIX 1 – TABLE 4

TABLE OF DIMENSIONS AND PROPERTIES
(EXTRACTED FROM BS EN 10219: 1997)

RECTANGULAR HOLLOW SECTION

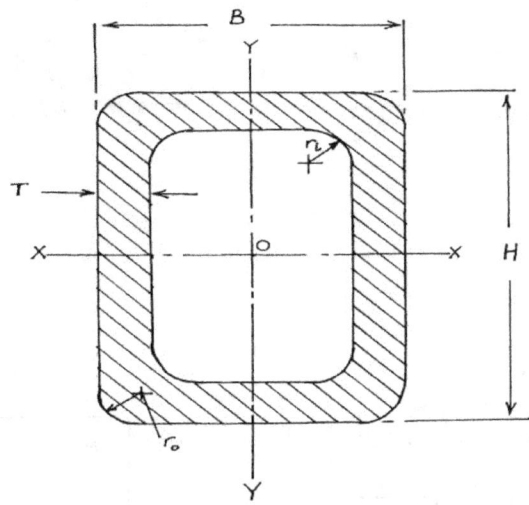

Size H x B mm x mm	Th'k's T mm	Corner Radius mm		Mass/m kg/m	Cross Sectional Area (cm)^2	Second Moment of Area (cm)^4		Radius of Gyration cm		Section Modulus (Elastic) (cm)^3	
		Ext r_o	Int r_i			I_{xx}	I_{yy}	k_{xx}	k_{yy}	Axis X-X	Axis Y-Y
40 x 20	2	4	2	1.68	2.14	4.05	1.34	1.38	0.793	2.02	1.34
40 x 20	3	6	3	2.36	3.01	5.21	1.68	1.32	0.748	2.6	1.68
50 x 25	2	4	2	2.15	2.74	8.38	2.81	1.75	1.01	3.35	2.25
50 x 25	3	6	3	3.07	3.91	11.2	3.67	1.69	0.969	4.47	2.93
50 x 30	2.5	5	2.5	2.82	3.59	11.3	5.05	1.77	1.19	4.52	3.37
50 x 30	3	6	3	3.3	4.21	12.8	5.7	1.75	1.16	5.13	3.8
60 x 30	4	8	4	4.2	5.35	15.3	6.69	1.69	1.12	6.1	4.46
60 x 40	2	4	2	2.93	3.74	18.4	9.83	2.22	1.62	6.14	4.92
60 x 40	2.5	5	2.5	3.6	4.59	22.1	11.7	2.19	1.6	7.36	5.87
60 x 40	5	10	5	6.56	8.36	35.3	18.4	2.06	1.48	11.8	9.21
70 x 50	2	4	2	3.56	4.54	31.5	18.8	2.63	2.03	8.99	7.5
70 x 50	4	8	4	6.71	8.55	54.7	32.2	2.53	1.94	15.6	12.9
70 x 50	6	10	5	8.13	10.4	63.5	37.2	2.48	1.9	15.6	12.9
80 x 40	2	4	2	3.56	4.54	37.4	12.7	2.87	1.67	9.34	6.36
80 x 40	3	6	3	5.19	6.61	52.3	17.6	2.81	1.63	13.1	8.78
80 x 60	4	8	4	7.79	10.1	87.9	56.1	2.94	2.35	22	18.7
90 x 50	2	4	2	4.19	5.34	57.9	23.4	3.29	2.09	12.9	9.35
90 x 50	4	8	4	7.97	10.1	103	40.7	3.18	2	22.8	16.3
10 x 100 x 40	3	6	3	6.13	7.81	92.3	21.7	3.44	1.67	18.5	10.8
10 x 100 x 50	6	12	6	12.3	15.6	179	58.7	3.38	1.94	35.8	23.5
100 x 60	6	12	6	13.2	16.8	205	91.2	3.49	2.33	41.1	30.4
100 x 80	4	8	4	10.5	13.3	189	134	3.77	3.17	37.9	33.5
120 x 60	2.5	5	2.5	6.74	8.59	161	55.2	4.33	2.53	26.9	18.4
120 x 80	6.3	15.75	9.45	17.5	22.2	408	217	4.28	3.12	68.1	54.3
140 x 80	4	8	4	13	16.5	430	180	5.1	3.3	61.4	45.1
15 x 150 x 100	4	8	4	14.9	18.9	595	319	5.6	4.1	79.3	63.7
150 x 100	8	20	12	27.7	35.2	1008	536	5.35	3.9	134	107
160 x 80	4	8	4	14.2	18.1	598	204	5.74	3.35	74.7	50.9
160 x 80	5	10	5	17.5	22.4	722	244	5.68	3.3	90.2	61

APPENDIX 1 – TABLE 5

TABLE OF DIMENSIONS AND PROPERTIES
(EXTRACTED FROM BS EN 10056: 1999)

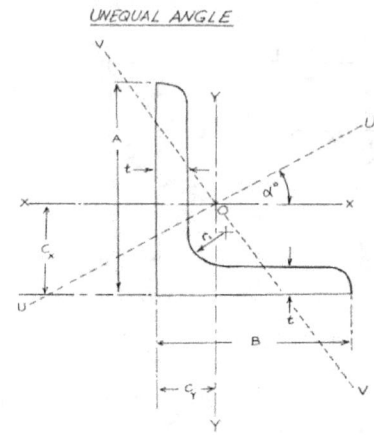

UNEQUAL ANGLE

Designation A x B x t	Mass per Metre	Root Radius r_1	Toe Radius r_2	Area of Section	Distance of Centre of Mass		I_{xx}	I_{yy}	I_{uu}	I_{vv}	k_{xx}	k_{yy}	k_{uu}	k_{vv}	Angle between Axes X-X and U-U tan α°
					c_x	c_y									
mm x mm x mm	kg/m	mm	mm	(cm)²	mm	mm	(cm)⁴	(cm)⁴	(cm)⁴	(cm)⁴	cm	cm			
200 x 150 x 18′	47.1	15	7.5	60	6.33	3.85	2376	1146	2920	623	6.29	4.37	6.97	3.22	0.549
200 x 150 x 15′	39.6	15	7.5	50.5	6.21	3.73	2023	979	2480	526	6.33	4.4	7	3.23	0.551
200 x 150 x 12′	32.0	15	7.5	40.8	6.08	3.61	1653	803	2030	430	6.36	4.44	7.04	3.25	0.552
200 x 100 x 15′	33.7	15	7.5	43	7.16	2.22	1759	299	1860	193	6.4	2.64	6.59	2.12	0.260
200 x 100 x 12′	27.3	15	7.5	34.8	7.03	2.10	1441	247	1530	159	6.43	2.67	6.63	2.14	0.262
200 x 100 x 10′	23.0	15	7.5	29.2	6.93	2.01	1219	210	1290	135	6.46	2.68	6.65	2.15	0.263
150 x 90 x 15′	26.6	12	6	33.9	5.21	2.23	761	205	841	126	4.74	2.46	4.98	1.93	0.354
150 x 90 x 12′	21.6	12	6	27.5	5.08	2.12	627	171	694	104	4.78	2.49	5.02	1.94	0.358
150 x 90 x 10′	18.2	12	6	23.2	5	2.04	533	146	591	88.3	4.8	2.51	5.05	1.95	0.360
150 x 75x 15′	24.8	12	6	31.7	5.52	1.81	713	119	753	78.6	4.75	1.94	4.88	1.58	0.253
150 x 75 x 12′	20.2	12	6	25.7	5.4	1.69	589	99.6	623	64.7	4.78	1.97	4.92	1.59	0.258
150 x 75 x 10′	17	12	6	21.7	5.31	1.61	501	85.4	531	55.1	4.81	1.99	4.95	1.6	0.261
125 x 75 x 12′	17.8	11	5.5	22.7	4.31	1.84	354	95.5	391	58.5	3.95	2.05	4.15	1.61	0.354
125 x 75 x 10′	15	11	5.5	19.1	4.23	1.76	302	82.1	334	49.9	3.97	2.07	4.18	1.61	0.357
125 x 75 x 8′	12.2	11	5.5	15.5	4.14	1.68	247	67.6	274	40.9	4	2.09	4.21	1.63	0.360
100 x 75 x 12′	15.4	10	5	19.7	3.27	2.03	189	90.2	230	49.5	3.1	2.14	3.42	1.59	0.540
100 x 75 x 10′	13	10	5	16.6	3.19	1.95	162	77.6	197	42.2	3.12	2.16	3.45	1.59	0.544
100 x 75 x 8′	10.6	10	5	13.5	3.1	1.87	133	64.1	162	34.6	3.14	2.18	3.47	1.6	0.547
100 x 65 x 10′	12.3	10	5	15.6	3.36	1.63	154	51	175	30.1	3.14	1.81	3.35	1.39	0.410
100 x 65 x 8′	9.9	10	5	12.7	3.27	1.55	127	42.2	144	24.8	3.16	1.83	3.37	1.4	0.413
100 x 65 x 7′	8.8	10	5	11.2	3.23	1.51	113	37.6	128	22	3.17	1.83	3.39	1.4	0.415

APPENDIX 1 – TABLE 6

TABLE OF DIMENSIONS AND PROPERTIES
(EXTRACTED FROM BS EN 10219: 1997)

CHANNELS

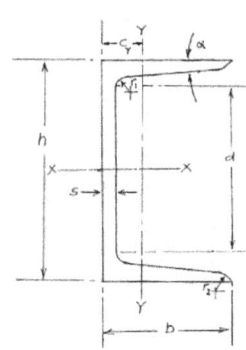

Designation h x b x kg/m	Mass/m kg/m	Depth of Section h mm	Width of Section b mm	Thickness		Radii		Flange Taper x Degrees	Depth Between Fillets d mm	Cross-Sectional Area (cm)²	Second Moment of Area (cm)⁴		Radius of Gyration (cm)		Section Modulus (Elastic) (cm)³		Elastic Neutral Axis C_y cm
				Web s mm	Flange t mm	Root r_1 mm	Toe r_2 mm				I_{xx}	I_{yy}	Axis X-X	Axis Y-Y	Axis X-X	Axis Y-Y	
432 x 102 x 65	65.5	431.8	101.6	12.2	16.8	15.2	4.8	5	362.5	83.4	21373	627	16	2.74	990	80	2.31
381 x 102 x 55	55.01	381	101.6	10.4	16.3	15.2	4.8	5	312.6	70.1	14869	579	14.6	2.87	781	75.7	2.52
305 x 102 x 46	46.21	304.8	101.6	10.2	14.8	15.2	4.8	5	239.3	58.9	8208	499	11.8	2.91	539	66.5	2.65
305 x 89 x 42	41.81	304.8	88.9	10.2	13.7	13.7	3.2	5	245.4	53.3	7078	326	11.5	2.48	464	48.6	2.18
254 x 89 x 36	35.66	254	88.9	9.1	13.6	13.7	3.2	5	194.7	45.4	4445	302	9.89	2.58	350	46.7	2.42
254 x 76 x 28	28.18	254	76.2	8.1	10.9	12.2	3.2	5	203.9	35.9	3355	162	9.67	2.12	264	28.1	1.85
229 x 89 x 33	32.68	228.6	88.9	8.6	13.3	13.7	3.2	5	169.9	41.6	3383	285	9.01	2.61	296	44.8	2.53
229 x 76 x 26	26.08	228.6	76.2	7.6	11.2	12.2	3.2	5	177.8	33.2	2615	159	8.87	2.19	229	28.4	2
203 x 89 x 30	29.77	203.2	88.9	8.1	12.9	13.7	3.2	5	145.2	37.9	2492	265	8.11	2.64	245	42.4	2.65
203 x 76 x 24	23.85	203.2	76.2	7.1	11.2	12.2	3.2	5	152.4	30.4	1955	152	8.02	2.24	192	27.7	2.14
178 x 89 x 27	26.79	177.8	88.9	7.6	12.3	13.7	3.2	5	121	34.1	1753	241	7.17	2.66	197	39.3	2.76
178 x 26 x 21	20.84	177.8	76.2	6.6	10.3	12.2	3.2	5	128.8	26.6	1338	134	7.1	2.25	151	24.8	2.2
152 x 89 x 24	23.87	152.4	88.9	7.1	11.6	13.7	3.2	5	96.9	30.4	1168	216	6.2	2.66	153	35.8	2.87
152 x 76 x 18	17.91	152.4	76.2	6.4	9	12.2	2.4	5	105.9	22.8	852	114	6.11	2.23	112	21	2.21
127 x 64 x 15	14.92	127	63.5	6.4	9.2	10.7	2.4	5	84	19	482	67.2	5.04	1.88	76	15.2	1.94
102 x 51 10	10.4	101.6	50.8	6.1	7.6	9.1	2.4	5	65.8	13.3	207	29.1	3.95	1.48	40.8	8.14	1.51
76 x 38 x 7	6.71	76.2	38.1	5.1	6.8	7.6	2.4	5	45.8	8.56	74.3	10.7	2.95	1.12	19.5	4.09	1.19

APPENDIX 1 – TABLE 7

TABLE OF DIMENSIONS AND PROPERTIES
(EXTRACTED FROM BS EN 10056: 1999)

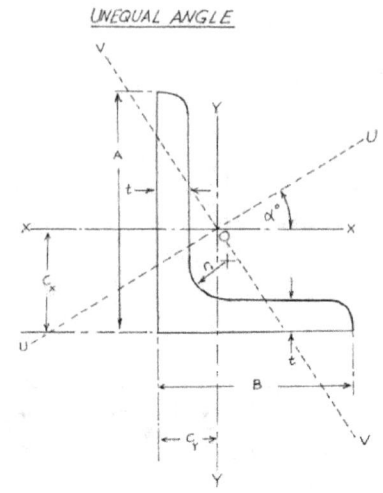

UNEQUAL ANGLE

Designation A x B x t	Mass per Metre	Root Radius r_1	Toe Radius r_2	Area of Section	Distance of Centre of Mass		I_{xx}	I_{yy}	I_{uu}	I_{vv}	k_{xx}	k_{yy}	k_{uu}	k_{vv}	Angle between Axes X-X and U-U
					c_x	c_y									
mm x mm x mm	kg/m	mm	mm	(cm)²	mm	mm	(cm)⁴	(cm)⁴	(cm)⁴	(cm)⁴	cm	cm			tan α°
200 x 150 x 18'	47.1	15	7.5	60	6.33	3.85	2376	1146	2920	623	6.29	4.37	6.97	3.22	0.549
200 x 150 x 15'	39.6	15	7.5	50.5	6.21	3.73	2023	979	2480	526	6.33	4.4	7	3.23	0.551
200 x 150 x 12'	32.0	15	7.5	40.8	6.08	3.61	1653	803	2030	430	6.36	4.44	7.04	3.25	0.552
200 x 100 x 15'	33.7	15	7.5	43	7.16	2.22	1759	299	1860	193	6.4	2.64	6.59	2.12	0.260
200 x 100 x 12'	27.3	15	7.5	34.8	7.03	2.10	1441	247	1530	159	6.43	2.67	6.63	2.14	0.262
200 x 100 x 10'	23.0	15	7.5	29.2	6.93	2.01	1219	210	1290	135	6.46	2.68	6.65	2.15	0.263
150 x 90 x 15'	26.6	12	6	33.9	5.21	2.23	761	205	841	126	4.74	2.46	4.98	1.93	0.354
150 x 90 x 12'	21.6	12	6	27.5	5.08	2.12	627	171	694	104	4.78	2.49	5.02	1.94	0.358
150 x 90 x 10'	18.2	12	6	23.2	5	2.04	533	146	591	88.3	4.8	2.51	5.05	1.95	0.360
150 x 75 x 15'	24.8	12	6	31.7	5.52	1.81	713	119	753	78.6	4.75	1.94	4.88	1.58	0.253
150 x 75 x 12'	20.2	12	6	25.7	5.4	1.69	589	99.6	623	64.7	4.78	1.97	4.92	1.59	0.258
150 x 75 x 10'	17	12	6	21.7	5.31	1.61	501	85.4	531	55.1	4.81	1.99	4.95	1.6	0.261
125 x 75 x 12'	17.8	11	5.5	22.7	4.31	1.84	354	95.5	391	58.5	3.95	2.05	4.15	1.61	0.354
125 x 75 x 10'	15	11	5.5	19.1	4.23	1.76	302	82.1	334	49.9	3.97	2.07	4.18	1.61	0.357
125 x 75 x 8'	12.2	11	5.5	15.5	4.14	1.68	247	67.6	274	40.9	4	2.09	4.21	1.63	0.360
100 x 75 x 12'	15.4	10	5	19.7	3.27	2.03	189	90.2	230	49.5	3.1	2.14	3.42	1.59	0.540
100 x 75 x 10'	13	10	5	16.6	3.19	1.95	162	77.6	197	42.2	3.12	2.16	3.45	1.59	0.544
100 x 75 x 8'	10.6	10	5	13.5	3.1	1.87	133	64.1	162	34.6	3.14	2.18	3.47	1.6	0.547
100 x 65 x 10'	12.3	10	5	15.6	3.36	1.63	154	51	175	30.1	3.14	1.81	3.35	1.39	0.410
100 x 65 x 8'	9.9	10	5	12.7	3.27	1.55	127	42.2	144	24.8	3.16	1.83	3.37	1.4	0.413
100 x 65 x 7'	8.8	10	5	11.2	3.23	1.51	113	37.6	128	22	3.17	1.83	3.39	1.4	0.415

APPENDIX 1 – TABLE 8

TABLE OF DIMENSIONS AND PROPERTIES
(EXTRACTED FROM BS EN 10056: 1999)

EQUAL ANGLES

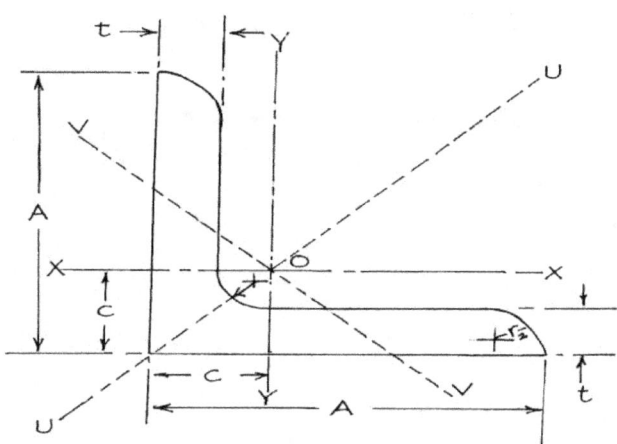

Designation Serial Size A x A x t	Mass per metre	Root Radius r_i	Toe Radius r_2	Area of Section	Distance of Centre of Mass c	I_{xx}	I_{vy}	I_{uu}	I_{yv}	k_{xx}, k_{yy}	k_{uu}	k_{vv}
mm x mm x mm	kg/m	mm	mm	(cm)^2	cm					cm		
200 x 200 x 24	71.1	18	9	90.6	5.84	3331	3331	5280	1380	6.06	7.64	3.9
200 x 200 x 20	59.9	18	9	76.3	5.68	2851	2851	4530	1170	6.11	7.70	3.92
200 x 200 x 18	54.2	18	9	69.1	5.6	2600	2600	4150	1050	6.13	7.75	3.9
200 x 200 x 16	48.5	18	9	61.8	5.52	2342	2342	3720	960	6.16	7.76	3.94
150 x 150 x 18	40.1	16	8	51	4.37	1050	1050	1680	440	4.54	5.73	2.92
150 x 150 x 15	33.8	16	8	43	4.25	898	898	1430	370	4.57	5.76	2.93
150 x 150 x 12	27.3	16	8	34.8	4.12	737	737	1170	303	4.6	5.8	2.95
150 x 150 x 10	23	16	8	29.3	4.03	624	624	990	258	4.62	5.82	2.97
120 x 120 x 15	26.6	13	6.5	33.9	3.51	445	445	710	186	3.62	4.57	2.34
120 x 120 x 12	21.6	13	6.5	27.5	3.4	368	368	584	152	3.65	4.6	2.35
120 x 120 x 10	18.2	13	6.5	23.2	3.31	313	313	497	129	3.67	4.63	2.36
120 x 120 x 8	14.7	13	6.5	18.7	3.23	256	256	411	107	3.69	4.67	2.38
100 x 100 x 15	21.9	12	6	27.9	3.02	249	249	395	105	2.98	3.76	1.94
100 x 100 x 12	17.8	12	6	22.7	2.9	207	207	328	85.7	3.02	3.8	1.94
100 x 100 x 10	15	12	6	19.2	2.82	177	177	280	73	3.04	3.83	1.95
100 x 100 x 8	12.2	12	6	15.5	2.74	145	145	230	59.9	3.06	3.85	1.96
90 x 90 x 12	15.9	11	5.5	20.3	2.66	148	148	235	62	2.7	3.4	1.75
90 x 90 x 10	13.4	11	5.5	17.1	2.58	127	127	201	52.6	2.72	3.42	1.75
90 x 90 x 8	10.9	11	5.5	13.9	2.5	104	104	166	43.1	2074	3.45	1.76
90 x 90 x 7	9.6	11	5.5	12.2	2.45	92.6	92.6	147	38.3	2075	3.46	1.77

This page is intentionally left blank.

APPENDIX II

TENSILE FRACTURE TOUGHNESS

APPENDIX II

TENSILE AND FRACTURE TOUGHNESS PROPERTIES OF SOME MATERIALS AT ROOM TEMPERATURE

Material	Yield Stress MN/m^2 or MPa	Tensile or Ultimate Strength MN/m^2 or MPa	Elongation at Fracture based on Original Length %	% Reduction in Area %	Fracture Toughness (MN/m^2)√m or MPa√m or MN/m$^{3/2}$
AISI 1144 Steel	540	840	5	7	66
AISI 4130 Steel	1090	1150	14	49	110
Titanium Alloy	925	1000	16	34	66
Aluminium Alloy (7075)	415	485	13	N/A	24
ASTM A517-F Steel	760	830	20	66	187

APPENDIX III

BIBLIOGRAPHY

BIBLIOGRAPHY

Askeland, D. *The Science of Engineering Materials*. 2nd SI Edition. Chapman and Hall, 1990

Atrops, J. L. *Strength Properties of Trinidadian Timbers*, Faculty Publication No. 12/1970. The University of the West Indies, St. Augustine, 1970

Bailey, A. R. *The Structure and Strength of Metals*. England: Metallurgical Services Laboratories Limited, 1967

Beaumont, R. A. *Mechanical Testing of Metallic Materials*. 3rd Edition. London: Sir Isaac Pitman & Sons Limited

Benham, P. P. and Crawford, R. J. *Mechanics of Engineering Materials*. 1st Published 1987. Reprinted 1988. London: English Language Book Society (ELBS) Longmans

Brand, T. and Sherlock A. *Matrices: Pure and Applied*. London: Edward Arnold, 1979

Bright, J. *Freemasonry: Some Deeper Considerations*. England: Toye, Kenning & Spencer (Butterworth) Limited, Gartree Press Limited, 1992

Brooks, W. H. *Strength and Elasticity of Materials and Theory of Structures*. Vols. 1-4. London: Macdonald & Company (Publishers) Limited, 1956

Bruce, R. G.; Dalton, D. K.; Kibbe, R. R; Neely, J. E., *Modern Materials and Manufacturing Processes*. Third Edition, Pearson Prentice Hall, 2004, Pp. 468

Case, J. *Strength of Materials*. 3rd Edition, Reprinted 1957. London: Edward Arnold Publishers Limited

Dowling, Norman E. *Mechanical Behaviour of Materials*. 3rd Edition. New Jersey: Pearson Prentice Hall, 2007

Edser, Edwin. *Heat for Advanced Students*. Revised Edition. Bligh, N. M. Macmillan & Company Limited, 1950 Pp. 487

Fawcett, J. N and Burdess, J. S. *Basic Mechanics with Engineering Applications*. 2nd Edition, London: Edward Arnold, 1993

Fenner, R. T. *Mechanics of Solids*. 1st Edition. London: Blackwell Scientific Publications, 1989

Fisher Cassie, W. *Structural Analysis*. 2nd Edition, Sixth Impression. London: Longmans, Green and Company, 1956

Ford, H and Alexander, J. M. *Advanced Mechanics of Materials*, 1st Edition. Longmans, 1963

Geary, A., Lowry, H. V., Hayden, H. A. *Advanced Mathematics for Technical Students, Part I,* 1st Edition, Reprinted 1954. London: Longmans, Green and Company, 1954

Gere, J. M. and Timoshenko, S. *Mechanics of Materials*. 3rd Edition. London: Chapman & Hall, 1993

Hammond, Rolt. *Engineering Structural Failures*. 1st Edition. London: Odhams Press Limited, 1956

Hearn, E. J. *Mechanics of Materials*. Vols. 1 and 2, 2nd Edition. Oxford: Pergamon Press, 1993

Jagger, J. G. *A Textbook of Mechanics*. 1st Edition. London: Blackie & Son Limited, 1952

Krishnamachari, *Applied Stress Analysis of Plastics: A Mechanical Engineering Approach.* Van Nostrand Reinhold, 1993

Levinson, J. J. *Introduction to Mechanics*. 1st Edition. Prentice-Hall Incorporated, 1961

Mc Evily, A. J. *Metal Failures: Mechanisms, Analysis, Prevention*. John Wiley & Sons, Inc. 2002, Pp. 324

Parry, His Honour Edward Abbott. *The Seven Lamps of Advocacy*. London W.C.2.: T. Fisher Unwin, Adelphi Terrace, 1923

Petroski, H. Design Paradigms: *Case Histories of Error and Judgment in Engineering*. 1st Edition. Cambridge University Press, 1994

Pippard, A. J. S and Baker, J. F. *The Analysis of Engineering Structures*. 3rd Edition. London: Edward Arnold (Publishers) Limited, 1957

Popov, E. P. *Mechanics of Materials*. 1st Edition. London: Macdonald & Company Limited, 1952

Rees, D. W. A. *Basic Solid Mechanics*. 1st Edition. Macmillan Press, 1997

Rice, Harold. S and Knight, Raymond M. *Technical Calculus and Analysis*. Mc Graw-Hill Book Company Incorporated, 1959

Roark, R. J. *Formulas For Stress and Strain*. 3rd Edition. Mc Graw-Hill Book Company Incorporated

Rollason, E. C. *Metallurgy for Engineers*. 1st Edition. London: Edwards Arnold (Publishing) Limited, 1952

Shin-ichi Nishida, *Failure Analysis in Engineering Applications*. Oxford OX2 8DP: Butterworth-Heinemann Limited, Linacre House, Jordan Hill, 1992

Stroud, K. A. *Engineering Mathematics*. 4th Edition. Hampshire and London: Macmillan Press Limited, 1996

Stroud, K. A. *Further Engineering Mathematics*. 3rd Edition. Hampshire and London: Macmillan Press Limited, 1996

Timoshenko, S. *Strength of Materials Parts I and II*. 3rd Edition. London: D. Van Nostrand, 1957

Timoshenko, S. and Goodier, J. N. *Theory of Elasticity*. 2nd Edition. Mc Graw-Hill Book Company Incorporated, 1951

Timoshenko, S. and Mac Cullough, G. H. *Elements of Strength of Materials*. 3rd Edition. D. Van Nostrand Company Incorporated, 1956

Warnock, F. V. *Strength of Materials*. 7th Edition. Sir Isaac Pitman and Sons Limited

Wolfe, Tom. *The Bonfire of the Vanities*. Bantam Books, 1988

INDEX

This page is intentionally left blank.

www.ingramcontent.com/pod-product-compliance
Lightning Source LLC
Chambersburg PA
CBHW081102170526
45165CB00008B/2294